Essays and Papers in the History of Modern Science

ESSAYS AND PAPERS
IN THE HISTORY
OF MODERN SCIENCE

❂

Henry Guerlac

THE JOHNS HOPKINS UNIVERSITY PRESS
Baltimore and London

Manufactured in the United States of America

The Johns Hopkins University Press, Baltimore, Maryland 21218
The Johns Hopkins Press Ltd., London

Library of Congress Cataloging in Publication Data
Guerlac, Henry.
Essays and papers in the history of modern science.
   Bibliography
   Includes index.
   1. Science—History—Collected works. 2. Chemistry,
Physical and theoretical—History—Collected works.
3. Science—France—History—Collected works.
I. Title.
Q125.G887   509   77–4555
ISBN 0–8018–1914–8

FOR RITA

Contents

Foreword

To set a preface before the achievements of a man whose enduring scholarship here assembled speaks so clearly for itself is perhaps to ask too much of a former student and a lifelong friend. The privilege of presentation, however, immediately evokes memories of the history of science seminar at Cornell and the great teacher whose pedagogy is not directly visible in the papers brought together in this volume. There we, his fortunate students, were introduced to a subject much more recondite than we knew. For, assuming that the solid content of some science was safely within our grasp, Henry Guerlac put to the test not only our meager command of letters and languages as tools of the trade but also our capacity to grow under criticism, given and received. Most of all, the evolving model of his own accomplishments was there to be seen, responding to the same conditions of learning and, without fanfare or pose, demonstrating the appropriate relationship between research and graduate education.

The *apologia pro vita sua* necessarily suggested by the assembly and republishing of papers produced during more than thirty years of scholarly life is sufficiently set out by Professor Guerlac's own introduction. The essays, which sweep broadly across the whole of the history of science at first, then concentrate carefully on the two central areas of interest that characterize Guerlac's work in depth—Lavoisier and the chemical revolution and Newtonian science —then broaden again to deal with science in the cultural context that is also part of his personal heritage, reflect the logic of retrospective organization, not the life lived. For as the publication dates indicate, these several themes have always been present, woven into the seamless fabric of an ongoing intellectual life. Perhaps we may be permitted to draw, from a source to which Guerlac is often drawn, not only elements of an ideology that earned his allegiance but also, from a text he cited frequently, details of a description that depicts much of his own work.

In Thomas Sprat's *History of the Royal Society*—an essay that stands as no history but a spirited defense of the scientific way not without parallel to Jefferson's apotheosis of political liberty in the *Declaration*—the purpose and practice of the new experimental science is set forth. Although speaking formally of the members of the Royal Society, but actually for all in the name of science, Sprat says that it is their task

> to put a Mark on the Errors which have been strengthened by long Prescription; to restore the Truths that have lain neglected; to push on those, which are already known, to more various Uses; and to make the way more passable, to what remains unrevealed. . . .

And to accomplish this, they have endeavor'd to separate the Knowledge of Nature, from the Colours of Rhetorick, the Devices of Fancy, or the delightful deceit of Fables. They have labor'd to inlarge it, from being confined to the Custody of a few, or from Servitude to private Interests. They have striven to preserve it from being over-press'd by a confus'd Heap of vain and useless Particulars; or from being straightened and bound too much up by general Doctrines.

They have tried to put it into a Condition of perpetual increasing, by setting an inviolable Correspondence between the Hand and the Brain. They have studied to make it not only an Enterprise of one Season, or of some lucky Opportunity; but a Business of Time: a steady, a lasting, a popular, an uninterrupted work.

Those who will read the papers now conveniently brought together in this single volume—whether for the first time or in renewal of good things past—will surely see how well Sprat's *apologia* for the life of science applies to Henry Guerlac's life of scholarship. In his prolonged study of Lavoisier and the chemical revolution he has "put a Mark on the Errors which have been strengthened by long Prescription," and in making "the way more passable" for others, encouraged the continuous correction of fact and the refinement of interpretation that is the essence of the learned tradition. Relatively few historical subjects are more difficult to delineate than the movement of ideas, especially when the likely filiations among them are neither direct nor acknowledged and are often disguised by argument over priority or "the delightful deceit of Fables." The close construction and careful reasoning of "The Continental Reputation of Stephen Hales" or "Joseph Priestley's First Papers on Gases and their Reception in France" are masterpieces of the genre, as is "The Origin of Lavoisier's Work on Combustion." In the latter paper, for example, Professor Guerlac works out the order in which Lavoisier performed his experiments and demonstrates the convergence of influences upon him from Stephen Hales and and Guyton de Morveau, thus establishing substantive knowledge where earlier only the "colours of Rhetorick" had prevailed. If at times, as in "Lavoisier and his Biographers," the critic appears paramount, the balancing elements are restored with the historiographical paper "Some Historical Assumptions of the History of Science," or history put to serve political judgment in "Science and French National Strength."

When Henry Guerlac began teaching the history of science after the Second World War, only a small coterie of scholars was sufficiently capable and competent to undertake significant research and teach in the subject. Since then, the prophecy in the Book of Daniel that "many shall run to and fro, and Knowledge shall be increased" has come to pass. The history of science has made its way into the university curriculum and spawned companion disciplines along the way. That national success story rests upon a handful of scholars, among whom Henry Guerlac is a central figure. In his paper "Science and the Historian," Guerlac spoke of science as the characteristic Natural Philosophy of the liberal, competitive era of early modern history; . . . [that] our ideals of reasonableness, freedom, tolerance, respect for the individual, even our principles of representative government and democracy . . . grew up step by step with the rise of modern science. . . . In our highly technologized society, science too has become technologized, and has fallen out of touch with its own great tradition, with its essentially humanistic past. . . . science is called upon to play its essential part in a vastly more complex society of giant corporations, great labor unions, top-heavy bureaucracies, and heavily armed Leviathan states. It may have to change, because the world has changed. Science, as men once knew it—a leisurely, free and essentially humane investigation into nature—may prove to have been an admirable, but ephemeral, episode of early Modern History.

I have quoted this passage at some length because it reveals an aspect of Henry Guerlac not usually to be seen, but always there—a concern for the contribution the history of science can make to an understanding of the human condition. In that passage, as elsewhere, he reveals in himself that quality he so admired in Carl Becker when he described his capacity to enlighten "the most thorny subject . . . by the passing insight and the ironic *aperçu.*"

HARRY WOOLF

The Institute for Advanced Study
Princeton, New Jersey

Acknowledgments

Because it is not general practice to include personal acknowledgments in published research articles (except to give credit for scholarly assistance on specific matters), the appearance of this collection of papers provides a welcome opportunity to express my gratitude for the help I have received over the years from so many different quarters.

Needless to say, I owe a long-standing debt to the men and women who attended my graduate seminar at Cornell, endured my criticism, argued strenuously for their point of view, and did not hesitate to challenge me on the spot and correct me later in print.

Many persons have aided me in typing the early manuscripts; others at various times served as my research assistants. Among those who provided much-needed help were the secretaries of the Department of History, Mary Menke and Anita Reed, whose years of service at Cornell almost exactly parallel my own. Others who served as personal secretaries or assistants were Vivian Sessions, Marion Heron, Dorothy Nelkin, Anne-Marie Garcia.

Several of the papers dealing with Lavoisier were written when I was the beneficiary of a grant from the National Science Foundation, during which period I received much help from Dr. Carl Perrin, who was completing his doctorate in chemistry, and from my then graduate students, J. B. Gough and Leslie Burlingame. For typing papers written while I served after 1970 as Director of Cornell's Society for the Humanities, I cannot sufficiently express my thanks to members of my able staff for their dedication and forbearance: Pamela Armstrong, Louise Noble, Betty Tamminen, and Olga Vrana.

I wish also to record my debt to the staffs of the Bibliothèque Nationale, the Archives Nationales, the library of the Muséum National d'Histoire Naturelle, and the Institut de France, and notably to Mme. Pierre Gauja of the Archives of the Académie des Sciences in Paris, who has generously assisted so many historians of science. For access to certain Lavoisier papers I am indebted to the librarian of the Bibliothèque Municipale de Clermont-Ferrand; to the late M. René Fric, editor of the Lavoisier Correspondence; and to Count Guy de Chabrol.

For assisting me in work carried on in Great Britain, I wish to thank the officials of the Library of the University of Edinburgh, the British Museum, and the Ashmolean Museum at Oxford, and at Cambridge the librarians of the Fitzwilliam Museum, of Kings College, of Trinity College, and of the Cambridge University Library. In this country I owe especial gratitude to the tireless, and always cooperative, personnel of the Cornell University Library.

In the preparation of the present manuscript I wish particularly to thank Martha Linke for carrying out the formidable task of standardizing the notes of articles published in a dozen different books or journals, all with different house styles, and for correcting misprints that escaped my eye. Betty Slater shared the onerous task of retyping the notes, and Antonia Rachiele whipped the typescript into presentable shape. I greatly appreciated the invitation of the Johns Hopkins University Press to submit this manuscript and the kind assistance of its editor-in-chief, Michael Aronson, and of his staff, especially senior editor Barbara Kraft.

Finally, I am grateful to the Hull Memorial Fund of Cornell University for helping substantially to defray the production costs of this volume.

Introduction

The papers gathered here span more than a quarter century. They are of diverse kinds, some that may have some general appeal for historians of science, whatever their specialty, and may even interest the historian of ideas; and others which are more detailed and directed towards the historian of chemistry and of the physical sciences. My excuse for bringing them together here is the diversity of journals, *Festschriften*, published symposia, and the like in which they first appeared, as well as the repeated urging of my students and others that these papers be made more readily accessible. I have included essays (some of them originally public lectures) and research articles in the fields that interested me specially or into which I was momentarily lured, but no book reviews, even those that seemed to be of some intrinsic interest, or which became the vehicle for suggestions I could not resist throwing out.

A further circumstance may justify the republication of the papers: they span a personal career and a period of time during which, notably in the United States, the history of science emerged from what was virtually Dr. George Sarton's personal domain into an accepted academic discipline widely, if perhaps gingerly, accepted in our universities. The earliest papers in this collection (not, to be sure, quite the earliest of all my efforts) were published soon after my graduate years at Harvard and during a brief but happy tenure at the University of Wisconsin, Madison, where I served as chairman—and solitary member—of the first Department of the History of Science in an American university. This department, given distinguished forward impetus by my friend and successor, Professor Marshall Clagett, has remained (I am pleased to record) even after his departure an active center of the history of science. The latest articles, including several lengthy contributions to the *Dictionary of Scientific Biography*, have been written since my retirement from active teaching at Cornell, whose faculty I joined at the close of World War II, and where, among stimulating colleagues and students and with an outstanding research library, most of this work was done.

The war, needless to say, interrupted the meager flow of my scholarly efforts. As I was the historian of the MIT Radiation Laboratory, these years resulted in a monumental, if generally unreadable (and fortunately unpublished), hectographed work entitled *Radar in World War Two*, not a veritable history but a source that should be of use to anyone attempting to write the history of the unique way science was used by the Allies in that conflict. My great regret was that during those years I was obliged to decline an invitation to contribute to the collection of essays in honor of George Sarton.

At Cornell during the late forties and early fifties I concentrated my first efforts on developing a year-long undergraduate survey course, "Science in Western Civilization," and the beginnings of a graduate program. With the help of a grant from the Rockefeller Foundation, I published a syllabus (which has served as a kind of teacher's outline for the subject, since it proved too detailed for student use) and a two-volume mimeographed set of source readings. Except for short book reviews and two articles derived from my radar work, none of which seemed appropriate for this collection, I published little. This seemed odd to my co-workers in view of my all too obvious visibility and active participation in the affairs of the History of Science Society and the professional emergence of the first Ph.D.s from the Cornell program. Once a solicitous young friend even suggested (he was undergoing analysis at the time, and partial to psychopathological jargon) that I had some psychological block preventing me from writing. In any event, the situation soon changed as I had planned, or at least hoped, that it would. The turning point was an invitation to spend two years at the Institute for Advanced Study at Princeton, which was followed, some years later, by a Senior Research Grant from The National Science Foundation.

Apart from those who *do* have an aversion to seeing themselves permanently enshrined in print, and prefer simply to read and teach, scholars, I imagine, can be roughly divided into two sorts: book men and article men. I have to confess to having been an article man, a choice that is, of course, partly a matter of temperament and prior training but fully as much, I suspect, determined by the character and maturity of the scholar's particular discipline. Young fields like the history of art or musicology, or the history of science, put a premium upon the detailed research article to correct accepted but unexamined facts and interpretations and open new paths. Though there have been brilliant exceptions, notably biographies or syntheses of recent research, general books in rapidly growing fields have a short life expectancy: they feed upon the prevailing mythology and are soon outdated. Who today in the history of science seriously consults (let alone reads for relaxation) the *History of Science* (1929) of W. C. Dampier Dampier Whetham of Upwater Lodge, Cambridge, albeit evidently written under the benign sign of Aquarius? Even that remarkable *tour de force* Sir Herbert Butterfield's *Origins of Modern Science* (London, 1949) was out of date not long after its appearance, although put together by a skilled craftsman and versatile historian with a sound instinct for the reliable sources even in a field that up to that time had been strange to him.

At the start of my career I soon came to suspect that the corner of the field I proposed to cultivate needed unromantic spadework. Yet, I must confess, this was in tune with my temperament, disposition, and prior training; like others in different sorts of endeavor I work most happily in small form. So it was fortunate, I suppose, that the scholarly world of today can tolerate only a limited number of Toynbees. But my years as a young science student, at Cornell, the Woods Hole Marine Biological Laboratory, and Harvard, probably exerted the determining influence.

Scientists then and now, by and large, generally eschew the book as a vehicle for publishing fresh research. Partly, I am certain, this results from the intense competition among those working on the outer fringes of our knowledge of the natural world; but partly, also, because creative scientists are by nature problem solvers: they deal with discrete observations or experiments or mathematical theories that are intended to answer bothersome questions, resolve apparent contradictions, and put our picture of a small corner of nature into some kind of order. Somewhere I have advanced the notion that a willingness to concentrate on the narrow but significant problem, and to forgo the comfort of great systems of the universe, was one of the keys to the so-called scientific revolution of the seventeenth century. Yet at first, before the advent of the independent scientific journal, books were a natural medium for presenting discoveries and theories. One has, I believe, to return in spirit at least to the nineteenth century to point to books in which major discoveries or theories were first presented: Darwin's *Origin of Species* (1859), James Clerk Maxwell's *Treatise on Electricity and Magnetism* (1873), and Lord Rayleigh's *Theory of Sound* (1877–78) may perhaps serve as examples.

Even as a graduate student, I found that much that had been written about the period of the history of science which particularly interested me appeared shallow, repetitive, and filled with a host of unanswered questions. Historical investigation as a form of detective work had for me, then as now, an irresistible appeal—another expression, I suppose, of my earlier training. And on those few occasions when a really ambitious project was contemplated, the reliable foundation material often seemed not merely unconvincing but sometimes filled with utterly improbable statements accepted as indisputable fact. In scholarship, as in science, a useful trait is a degree of skepticism toward what J. K. Galbraith has called the "conventional wisdom." In both, success often depends upon a ready disposition to be unconvinced, as well as the ability to ask the right question, especially when it appears to be an obvious, simple-minded question that ought to have been asked and answered long ago.

Let me give two examples where the obvious questions had not been asked, and where asking them led to worthwhile investigations. First, anyone who had glanced through the several volumes of either of the eighteenth-century editions of the writings of Robert Boyle should have been struck by the range of apparently unrelated experimental inquiries he published and should have asked, as nobody to my knowledge did, what common thread could possibly relate them. What, in other words, justified Boyle's reputation as the outstanding English scientist of his age? I urged this question upon my first Ph.D. candidate to see if she could find an answer; the result of her own independent efforts was one of those rare doctoral dissertations that have become classic. A similar question was to ask why nobody had read or at least studied with care the manuscript lectures of Lavoisier's teacher, G.–F. Rouelle. Having acquired on microfilm the so-called Diderot copy of these lectures from the city library of Bordeaux, I urged this problem upon another of my able graduate students, with very fine results.

Other examples of asking seemingly obvious questions led to a number of papers included in this volume. For example, a contemporary of Boyle's, the physician John Mayow, propounded a theory that there was a substance in the air (a nitrous substance, he believed) whose presence is essential if combustible materials are to burn and if animals and man are to breathe, and which is responsible for turning dull-colored venous blood into the bright scarlet of arterial blood. All scholars have recognized the superficial similarity of Mayow's "nitre in the air" to oxygen gas, which was not discovered, or its role in chemistry determined, until more than a century later. The obvious question was, "Why did not Mayow's hypothesis bear fruit?" In seeking an answer to this question (which in fact is rather complex) the theory was found not to be so original after all, in fact was traceable back to an early seventeenth-century work, with quite obvious Paracelsian antecedents.

Similar questions led me, and eventually my students, collaborators, and critics, to a radical reassessment of the origin and early stages of the chemical revolution associated with the name of Lavoisier. It was obvious first of all to ask why such a revolution would have occurred in France, not in England or Scotland, when the stimulus for Lavoisier's recasting of chemistry was generally supposed to derive from the work of the British pneumatic chemists Joseph Black, Joseph Priestley, and Henry Cavendish. This inquiry led me to pinpoint the British influence as chiefly, and at first exclusively, that of the Rev. Stephen Hales, whose *Vegetable Staticks* (1727), ostensibly a work on plants, contained an important chapter dealing with chemistry. Exploring the same question further resulted in the paper entitled "Some French Antecedents of the Chemical Revolution," an exercise, as it turned out, in studying the peculiarity of France's industrial revolution, notably the *métiers de luxe*—the textile and dye industry, the manufacture of Gobelins and Beauvais tapestries, ceramic factories like Sèvres—and a governmental concern to develop the mining and metallurgical industries, all of which transformed the chemist's profession by freeing him from his attachment to pharmacy and medicine.

Research is, of course, an ongoing, never-ending process. One valuable feature of the research article which purports to present a novel discovery or a new interpretation is that, to my mind at least, it is never conclusive, but should serve as a stimulus to further research by others—by one's students and of course by one's colleagues and critics. Ideally, such a paper would define its purpose (the puzzle to be solved or the question to be explored), indicate the prior studies that formed the point of departure, present the case fairly, marshal the available evidence, and offer at least a conclusion of sorts, while recognizing implicitly its tentative nature, with no claim to have reached the truth. It is my firm conviction that such articles, especially in a rapidly growing field, should be, as it were, *open-ended* and recognized by both author and reader as a mere stage in the great process of discovery, a point along a curve. This, by the way, would seem to be a powerful argument against the artificial separation of teaching from scholarship.

Obviously I hope that my papers will be read in that spirit and with a view to the state of our knowledge (or ignorance) at the time they were first published. Some evidence of how some of them led on one to another should be apparent in this book. But it is also pleasant to know that certain of the papers have directly produced investigations by others, pushing the inquiry into directions I had neither the inclination nor the occasion (or sometimes the imagination) to pursue.

I have left the early papers virtually unaltered, with no attempt to bring them in line with later knowledge, confining myself to minor corrections and the elimination of manifest misprints. Nevertheless, I have felt free to make more substantial changes in the recent articles.

I
ASPECTS OF THE HISTORY OF SCIENCE

1
Humanism in Science

❀

There has never been a period of modern history when what we commonly call science—meaning, of course, the natural sciences—has not struck many persons as the antithesis, perhaps even the enemy, of what we loosely lump together as the humanities. Two popular stereotypes about science have contributed to the prevalence of this attitude. The first, which we can call the Gradgrind theory, always assumes that the scientist deals only with facts, usually qualified for some reason as "stubborn" or "brute" facts, which it is his sordid custom to transform into columns of figures or express in unaesthetic graphs and charts. According to this theory, science is hopelessly materialistic and deals mainly with dull, prosaic matters, a failing which is excusable when the studies prove useful to us but which seems absurd when we cannot detect any practical purpose beyond the innocent amusement of the investigator. The second view is not much more favorable, though it is more sophisticated, for it recognizes that science deals in reality more with ideas about things than with things themselves. This view complains that although scientific inquiry is a search for truth, it is often too abstract and too mathematical to be really in touch with human life and man's individual and social problems. Here again science and the humanities seem to have lost all contact: for science is concerned with the barren wastes of inanimate nature, if not its mathematical shadowland, while the humanities deal with the rich and varied spectrum of human experience.

Of course these are both caricatures, which is a very good reason why they are likely to have both a wide currency and a considerable life expectancy. The two opinions, it should be noted, have one thing in common, though they approach the problem from opposite sides: both assume that either science is indifferent to man's peculiarly human problems, or it is concerned only with providing inordinately, with a never-ceasing flow of goods, for his animal needs and socially created desires.

Let us ask how far these criticisms are justified and inquire to what degree science—especially in the seventeenth and early eighteenth centuries, the age when it first became aware of its separate identity and was struggling to formu-

From Julian Harris, ed., *The Humanities: An Appraisal* (Madison, Wis.: University of Wisconsin Press, 1950). Copyright 1950 by The Regents of the University of Wisconsin.

late its ideals and its goals—did in fact tend to exclude or indefinitely postpone the study of man. Did its methods or attitudes inevitably lead to ignoring human values, and could it have been at the same time grossly materialistic, unduly remote from reality, and wholly indifferent to man's fate?

That scientific discovery is a complex human activity was already apparent by the end of the seventeenth century, when the two tendencies I have mentioned had clearly revealed themselves and were already being ridiculed. Thomas Sprat, the historian of the Royal Society, lamented in 1667 that the experimenters of the Society, called in the language of the day the *virtuosi*, were criticized by some for aiming too high and by others for aiming too low. The adversaries and worldly critics of what was styled the New Experimental Learning found its interests to be at once too vulgar and too abstract and lampooned it unmercifully on both counts. In his satirical comedy *The Virtuoso*, Shadwell poked coarse fun at the investigators, to whom nothing was too revolting to be painstakingly studied.[1] The protagonist, Sir Nicholas Gimcrack, "the finest speculative gentleman in the whole world," is disrespectfully described by one of his nieces, Clarinda, as "A sot, that has spent 2000 pounds in Microscopes, to find out the nature of Eels in Vinegar, Mites in Cheese, and the blue of Plums, which he has subtilly found out to be living creatures." Dean Swift in his description of the Academy of Lagado did not overlook this weakness, but his sharpest barbs were reserved for the opposite tendency when he conjures up the absent-minded, impractical, and easily deceived mathematicians of Gulliver's floating island of Laputa, "in the common actions and behavior of life . . . a clumsy, awkward, unhandy people."

There is some justice in these criticisms, even though they are criticisms of extremes, for they point to tendencies in science to which some men, some times, some scientific fashions have shown great partiality. It must be confessed that both Shadwell and Swift struck home with deadly accuracy, and the often uncritical, acquisitive, bumbling Baconian curiosity of the early Royal Society is well taken off in Sir Nicholas Gimcrack. Yet the other tendency was equally evident, for at the same time—with all its talk of observation and experiment— the ideal of science in the seventeenth century (only think of the names of Descartes, John Wallis, Newton, and Huygens) was mathematical, Archimedean.

An emphasis on mathematical formalism, by a paradoxical and unholy alliance with the hypothesis of materialism, was daily bringing intellectual successes that staggered the imagination. These great triumphs effectively banished spiritual forces and final causes from the domain of physics and from the serious consideration of scientists. To the distress of the conservative, the occult qualities, the substantial forms, and the other semantic traps of the scholastic imagination were relegated to the boneyard. With the air thus cleared, the mechanical philosophy was applied with heady optimism on every hand; the physical world seemed to be a complex engine, a great clock of innumerable moving material parts; much that was thought to be part and parcel of this world proved on closer examination to be only our own subjective evaluation of it. By denying the objective reality of everything in nature except matter in motion, and excluding everything that could not be spatially or temporally measured, the new physics of Galileo, Descartes, Boyle, and Newton reduced the scientific importance of the so-called "secondary qualities" (color, sound, odor, the sensations of heat and cold)—all that gave color, meaning, and richness to ordinary life.

Prematurely and at greater peril these mechanical ideas were introduced into the realm of biology: a school of physicians called the iatrophysicists tried to explain all bodily functions and disarrangements on purely mechanical terms. Giovanni Borelli, a disciple of Galileo, carried out a classic analysis of muscular and skeletal motion in the human body in terms of simple machines and explained locomotion, respiration, and digestion as purely mechanical processes.

A more extreme member of this school was Giorgio Baglivi (1668–1705), a pupil of the great anatomist Malpighi, who conceived the bodily machine as itself an assemblage of smaller machines and compared the teeth to scissors, the stomach to a flask, the heart and blood vessels—as Harvey had done—to a system of water works, the thorax to a pair of bellows. The more complicated bodily processes were rashly explained, by men like Baglivi, in terms of the fine structure of fibers, glands, ducts, and secretions which the earliest microscopic examinations had made visible. Physiological theories of this sort had an immense vogue and a very considerable influence in the early eighteenth century.

These were indeed crude attempts, which lent support to the opinion that the mechanical philosophy was crude and naïve. But to the physicist, be it said, there was nothing "crass" about matter. Its behavior might be followed, but its nature and essence were forever hidden. Still more remarkable, it proved quite unnecessary—and perhaps neither desirable nor possible—to describe nature only in its tangled and varying complexity. Galileo, the most amazing genius of this surprising century, had shown how fruitful it could be to account for the behavior of the real world in terms of ideal mechanisms, of material systems that existed in full perfection only in his own mind: a world of perfect spheres, frictionless surfaces, empty space, and unceasing motions. It was this, by the way, that made possible the fullest extension of the concept of *scientific law*, in many ways the greatest achievement of the first scientific century. But men often forgot to caution others, or did not themselves fully realize, that these simple mathematical rules that nature apparently followed, and that exhibited the mathematical elegance with which God seemed to have planned the physical universe, were in fact the absolute laws of only an imaginary, ideal, nonexistent Platonic world of the Galilean imagination.

Here we have it! On the one hand the casual botanizing, the passionate curiosity, the moth-eaten specimens, the fetal monsters, and all the crowded dusty disorder of the virtuoso's cabinet; and on the other, a mathematical fairyland, floating above the earth like Gulliver's island. If either or both together exhaust the content of science, then surely humanism, by any definition, suggests a richer diet, and something nobler and closer to man's real interests. But you sense, I am sure, that these are the same caricatures with which I began, portraits of two quite familiar but opposing tendencies in science. With no intention of pushing the comparison too far, my caricatures, I think, correspond to the division of scientific temperaments into the classic and romantic types long ago proposed by the chemist Wilhelm Ostwald. With good effect, William Morton Wheeler, the grand old Harvard entomologist, while admitting that most scientists of his acquaintance were a mixture of these two tendencies, applied this classification in order to contrast the laboratory biologist with the more desultory naturalist:

> The naturalist is mentally oriented toward and controlled by objective, concrete reality, and probably because his senses, especially those of sight and touch, are highly developed, is powerfully affected by the esthetic appeal of natural objects. He is little interested in and may even be quite blind to abstract or theoretical considerations. . . . When philosophically inclined, he is apt to be a tough-minded Aristotelian. . . .
>
> The biologist *sensu stricto*, on the other hand, is oriented toward and dominated by ideas, and rather terrified by or oppressed by the intricate hurly-burly of concrete, sensuous reality and its multiform and multicolored individual manifestations. He often belongs to the motor rather than to the visual type and obtains his esthetic satisfaction from all kinds of analytical procedures and the cold desiccated beauty of logical and mathematical demonstration. . . . He is a denizen of the laboratory. His besetting sin is oversimplification and the tendency to undue isolation of the organisms he studies from their natural environment. As a philosopher he is apt to be a tender-minded Platonist. . . .[2]

Perhaps it is significant that the terms for these extreme tendencies in science have been borrowed from what I take to be extreme tendencies in literature, music, and art, which is quite proper, since the taxonomy of scientific history is far less rich than the historical vocabulary of the humanities. I hope I shall not seem impertinent—and I am certainly in no danger of being original— if I suggest that the taxonomy of literary history—like all taxonomy, for that matter—is based on the easily characterized extremes, and that the words "classic" and "romantic" no more describe suitably the greatest figures of literature and art and music than they do those of science, however well they may serve to lump together the lesser men. The classic and the romantic are extreme embodiments of tendencies found in harmonious equilibrium, though in varying proportions, it is true, in different men and different times, in the masters of any art and period. The classical mentality, we say, loves reason and order, values restraint, and deals with the formal and readily communicable norms of beauty. The romantic prefers nature, variety, the rich expression of an emotional world that is often deeply personal. But how satisfactory is this division even to professors of literature? What masterpiece is not to a large extent describable in both sets of terms—in terms of reason and nature and of restraint with vitality? The giants defy classification in such simple terms. Perhaps, after all, as good a modern definition as any of that protean word *humanism* is to use it to denote the mean between the classic and romantic extremes, between formal frigidity and emotional rapture, between reason and nature. At all events the term is so useful that I shall usurp it to explain why the greatest names of the history of science seem to me to fail miserably of inclusion in Ostwald's classification.

I am well aware that humanism is a dangerous term of many meanings. To begin, I shall not use it in its original sense, that of a man devoted to seeking out and restoring to general esteem the humane letters of antiquity, though it is in this limited sense that we apply the term to a host of Renaissance classical scholars from Petrarch to Scaliger.

But out of this meaning of the word humanist there inevitably unfolds another, which is often seized upon to express one aspect of the elusive and complex Renaissance spirit. The humanist was a man who, having studied the writings of antiquity, became imbued with the life-giving spirit which is so often found in them and turned to them for encouragement and inspiration in his struggle to give the Western world a fresh start and a new basis of sound learning. Humanism thus became identified with a protest against the decadent ecclesiasticism, the arid and disputatious learning—the wreck of the great medieval system of instruction—by which the humanists found themselves surrounded.

But this humanist spirit is by no means confined to the men of the Renaissance, though it flourished mightily in those great centuries. A feeling for the excitement and the promise of life; a sense of proportion; a dislike of cant, hypocrisy, and pedantry; a profound sense of life and a desire to partake of it totally—these are familiar aspects of the humanist spirit of the Renaissance. But something similar belongs in some degree to every age and is found in the best of human nature at all times. Modern scholarship has not failed to detect it in medieval man himself. The possession of this indefinable spirit is an indispensable condition for those great moments of intellectual achievement which we recognize as the noblest product of the human mind.

William James, in his famous lecture *On the One and the Many*, has, I think, pointed to the fundamental trait that sets apart the man whom we may identify, in whatever branch of human culture we find him, as the true humanist. He finds among philosophers tendencies that correspond to our loose but convenient categories of classic and romantic. Somewhat resembling the classic is the man obsessed with the Parmenidean quest for the *one*, the monistic urge to establish the unity of knowledge, at whatever cost to life and sanity. By contrast there is the pluralist, "your 'scholarly' mind, of encyclopedic, philological type," James calls him, "your man essentially of learning." Once again, these are extremes, and neither one satisfied the Jamesian need to explore the richness and the complexity of things. "What our intellect really aims at," he says, and he might well have specified his own, "is neither variety nor unity taken singly, but *totality*. . . . Acquaintance with reality's diversities is as important as understanding their connexion."

Totality: now here is the word which seems to express the humanist aspiration. James gives us a somewhat apologetic simile, which fits our scientific problem admirably:

> We are like fishes swimming in the sea of sense, bounded above by the superior element, but unable to breathe it pure or penetrate it. We get our oxygen from it, however; we touch it incessantly, now in this part, now in that, and everytime we touch it we turn back into the water with our course redetermined and re-energized. The abstract ideas of which the air consists are indispensable to life, but irrespirable by themselves.[3]

How well this fits our scientists! Abstract investigations are necessary for re-energizing scientific life, but they cannot make up all of scientific experience, so science must dip back into the stream of life after each stratospheric ascension. There are men, we have seen, who by natural propensity spend an undue amount of time swimming in the aeriform fluid of abstraction and who not rarely float away wholly out of sight. These are the mathematical philosophers of Laputa. But there are others who never leave the lower depths of the aquarium.

The greatest men of science seem to have possessed this Jamesian sense of totality. They have kept in close touch with nature, while deriving sustenance and direction from their speculative flights. What greater humanist in this very important sense can one possibly imagine than Galileo, the principal author of that idealized mechanical universe of which I have just spoken, and yet the most resourceful, gifted, and tireless of experimenters. Although he possessed a mind of great abstract power, he did more than any other man to extend the range of the human senses, since in the main the refracting telescope, the compound microscope, the pendulum clock, the thermometer, and the barometer can be traced to him. This sense of totality, this harmonious cultivation of both reason and nature, is a mark of the greatest scientific minds of all ages. We have it in Lavoisier, and we can find it also in Newton, in Helmholtz, and in Darwin, no less than in Galileo. Classics and romantics indeed!

Why, then, not call them humanists, since we find in them that rare concentration of power and balance of forces it has been our purpose to describe? If my analysis is even partially correct, this explains, and I think justifies, the inclusion among the classics of the world's literature certain, though perhaps not many, of the milestones of science. This practice at the University of Chicago and St. John's (which, by the way, was long ago advocated by Matthew Arnold) has called down upon Mr. Hutchins and Mr. Stringfellow Barr some severe criticism, and while I should not like to be called upon to defend the selection of, say, Ptolemy's *Almagest* or the *Conics* of Apollonius of Perga for the Great Books courses, the principle is sound. On what basis should we choose scientific books as representing a portion of the "best which has been thought and said in the world"? Only, it seems to me, when they show to a high degree this humanistic quality of dealing in some fashion with the world as a whole.

Let us now take Mr. James's advice and descend a bit from the level of abstraction to which he lured us, attempting to trace more prosily the influence of humanism upon the early growth of science.

Renaissance humanism, in both the broad and the restricted sense that I have used, had an intimate historical connection with the emergence of modern science in the period from the fifteenth to the eighteenth centuries. Concerning the influence upon science of the humanist movement, in our narrow definition, there can be little doubt. The scientific curiosity of Europe was immensely stimulated when scholarly scientists like Regiomontanus, Linacre, and Tartaglia and literary humanists with scientific interests like Giorgi and Commandino edited or translated the great classics of Hellenistic and Greco-Roman science which the Middle Ages had known imperfectly or not at all. An acquaintance with the original works of Ptolemy and Galen, of Vitruvius and Celsus, of Archimedes, Diophantus, and Hero Alexandrinus spread abroad in well-edited and beautifully printed editions, lifted European science to a new level of technical accomplishment in astronomy, medicine, mathematics, and physics. The direct influence of this movement upon the work of Copernicus, Harvey, Galileo, and others is not difficult to discern.

But even humanism in the broader and more important sense, so far from diverting attention from scientific interests as many scholars would have us believe, contributed, I think, to that more militant mood of scientific emancipation we associate with the generation of Gilbert, Galileo, and Francis Bacon. For it was from the earlier humanists that the scientific rebels borrowed the mordant phrases, already somewhat shopworn, with which they attacked the dim perspectives, the logic-chopping, the gigantic circularities of the scholastic mind. There is a Protestant spirit in science; and perhaps humanism paved the way for the scientific revolt in much the same way as it cleared the ground for Calvin and Luther and the religious Reformation.

The constant refrain of the early humanists had been for a restoration of honest learning. From the time of Vives, that extraordinary and little-known forerunner of Francis Bacon, a succession of able men tried to diagnose the causes of what they termed the "corruption of the arts" and concentrated their energies on the grandiose project of reforming all learning. In this program the reform of natural science was to play a prominent part. Vives in 1531 advocated a return to the original inquiring spirit of Aristotle, the adoption of the experimental method, and—to reform both medical training and the study of natural history—a return to nature and observation. The humanists were also by the same token pioneers in exposing the futility of the pseudosciences. Pico wrote the classic refutation of astrology, while Vives, in treating of the corruption and decline of natural knowledge, emphasized how each pseudoscience had contaminated its corresponding valid science.

Out of the early, unsystematic criticism emerged the great revolutionary nature philosophies of Telesio, Bruno, Campanella, and Francis Bacon, all hoping to construct a new philosophy upon a reformed science of nature. The humanist inspiration is quite evident, and it was the aim of all these systems to draw as fully as possible upon the new vistas which the expansion of Europe had revealed. Bacon, to cite the best-known example, expected his New Instauration to draw upon an exhaustive inventory of the resources of the world: experiments, observations, and histories without end; the reports of the travelers; the slowly accumulated secrets of the arts and crafts; and even the muddied waters of the pseudosciences, certain of which he rightly suspected of concealing useful truths.

Scientific progress was never wholly dependent upon these grandiose plans for a revival of learning, as detractors of Francis Bacon are always quick to point out. But it is an inescapable historical fact that by linking themselves consciously with this movement—whose leaders were not scientists, but philosophers and agitators tinged with the great secular movement of humanism—scientists less bold and original than Gilbert or Galileo were able to attain a sense of their collective identity, their common purpose, and the revolutionary significance of their activities for the modern world. It is for this reason that Francis Bacon became the household deity of the Royal Society of London and the Royal Academy of Sciences in Paris. Under his banner the movement for a reform of philosophy through natural science became instead an antimetaphysical movement confined, as far as the new Academies were concerned, largely to the promotion of experimental science.

That the New Learning was antagonistic to the study of man, or even indifferent to it, would never have been admitted by any philosopher of the scientific way of life from Francis Bacon to Hume. But by the close of the seventeenth century the Baconian New Learning had already become fully equated, to a degree Bacon might not have relished, with the Experimental or Mechanical Philosophy, and one may well ask whether attention was not then so completely focused upon external nature that the ancient problems of man and of society were in danger of utter neglect. The enemies of the *virtuosi* were convinced that this was the case. In Shadwell's play, which I should like to quote once more, another disrespectful and exasperated niece describes her virtuoso uncle as "one, who has broken his Brains about the nature of Maggots, who has studied these twenty years to find out the several sorts of Spiders, and never cares for understanding mankind."

Have we more reliable evidence that this is a fair picture of the scientist's indifference to human problems? Is there anything to make us believe that the man of science of the late seventeenth century cared a fig for these weighty problems or felt that as scientists they were in any way responsible for their solution? Alas, the first evidence that comes to hand, the testimony of John Wallis and Robert Hooke as to the range of activity of the Royal Society, would seem to indicate not. "The business and design of the Royal Society," wrote Hooke, "is to improve the knowledge of natural things, and all useful Arts, Manufactures, Mechanik practices, Engynes and Inventions by Experiments—(not meddling with Divinity, Metaphysics, Moralls, Politicks, Grammar, Rhetorick, or Logick.)"[4]

Yet in 1667 in his official defense of the Royal Society, the Reverend Thomas Sprat put matters more cautiously. After expounding the Baconian subdivision of all knowledge into the study of God, of Man, and of Nature, he did not hesitate to point out deferentially that the first was excluded from the work of the Society and that the study of external nature, as the proper field for the new fashion of experiment, claimed the principal allegiance of the investigators:

In Men, may be consider'd the *Faculties* and Operations of their *Souls,* the *Constitution of their Bodies,* and the *Works of their Hands.* Of these, the *first* they omit; both because the Knowledge and Direction of them have been before undertaken by some *Arts* on which they have no mind to intrench, as the *Politicks, Morality,* and *Oratory*; and also because the *Reason,* the *Understanding,* the *Tempers,* the *Will,* the *Passions* of Men, are so hard to be reduc'd to any certain Observation of the *Senses,* and afford so much room to the *Observers* to falsify or counterfeit; that if such Discourses should be once entertain'd, they would be in Danger of falling into *talking,* instead of *working. . . .*

Such subjects, he remarks, have therefore been kept out, but unlike Hooke, he suggests that this is only a temporary, tactical move. When the members of the Society

shall have made more Progress in *material* Things, they will be in a Condition of pronouncing more boldly on them [these subjects] too. For though Man's *Soul* and *Body* are not only one *natural Engine* (as some have thought) of whose Motions of all Sorts, there may be as certain an Account given, as of those of a Watch or Clock; yet by long studying of the *Spirits,* of the *Blood,* of the *Nourishment,* of the Parts, of the *Diseases,* of the *Advantages,* of the Accidents which belong to *human Bodies* (all which will come within their Province) there may, without Question, be very near Guesses made, even at the more *exalted* and *immediate* Actions of the Soul. . . .[5]

In this paper I have been arguing that the scientist and the humanist have more in common than is usually supposed. I have tried to show how humanism affected the growth of science, and have even suggested that Renaissance humanism may have contributed to the great scientific rebellion of the Age of Galileo. It remains now to ask if the scientific movement did exert an appreciable influence upon the study of man and human society, and whether within a reasonable time someone attempted to fulfill Sprat's prophecy that physiological studies might contribute to a better understanding of man's nature.

The first serious attempt to make a scientific study of human society is due to a man intimately associated with the scientific and historical currents of his time, and it is based on a theory of human nature which the discussions of the humanists and the rise of modern science—in particular the famous Quarrel of the Ancients and Moderns—had brought to the fore: the theory of the essential invariance of human nature through time and space, and the belief—which Fontenelle put into the mouth of Socrates in his *Dialogues des Morts*—that man was a part of the "general order of nature."

Montesquieu's *Esprit des Lois,* published just two hundred years ago, is, I believe, the first book of any consequence on political and social questions to be influenced to any degree by the scientific movement. Although Machiavelli's *The Prince* has been loosely referred to as a "scientific" manual of practical politics, and Thomas Hobbes hoped *The Leviathan* would make him the Copernicus or the Harvey of moral philosophy, Montesquieu's *Spirit of the Laws* is yet the first book to be genuinely saturated by the spirit, the example, the point of view, the method, and even the findings of natural science. Auguste Comte, for better or for worse the accredited father of sociology, has pointed out that Montesquieu was one of the first to approach the whole study of human institutions from the point of view of modern science and to develop the idea that men were as subject to invariable laws as the rest of animate and inanimate nature.

The *Esprit des Lois* opens with Montesquieu's classic definition of Law as expressing in the broadest sense "the necessary relations that arise from the nature of things." God created and conserves the world and all its creatures, including man himself, according to fixed laws which resemble, and in fact include, the mathematical uniformities discovered by the scientist. Man in society is subject to these laws too, and man-made legal systems have unconsciously reflected them. Montesquieu's development of his theory is designed to escape the deterministic consequences which the broadly phrased statement of it would seem to imply. The words "natural law" in reality embrace two sorts of things. First there are the rigid, inflexible, inescapable *physical* laws of the universe which govern man as part of material nature. Then there are the natural *moral* laws which exist, as it were potentially, and govern man as a rational creature through his right reason. They are normative principles detectable in the rules of justice and equity. These laws are invariable like those of the scientist, but since man is a spontaneous being subject to error they are not invariably obeyed as the scientific laws must be.

It is, of course, the main purpose of the *Esprit des Lois* to study the systems of positive law that the peoples of the world have set up for themselves. Montesquieu is not concerned, like Grotius or Pufendorf or Vattel, to expound the normative principles and say what should be the practices common to all men. He is concerned with the deviations from these norms observable in different parts of the world and in different ages. What causes these differences? If human nature and human reason depend only upon the moral natural law and human reason, the positive laws should be everywhere the same. But in every nation the systems of positive law are applied by fallible human beings living under somewhat different conditions. The variations and particularities are introduced by the action upon man of the physical environment and the deterministic laws of the physical world. The legal systems and institutions of the world thus result from the simultaneous operation of the two sets of natural laws.

The problem is posed as a problem in biology might be, and is tackled by a method similar to the comparative method which a biologist might use in studying the homologous structures of related animal groups in order to discover the special environmental adaptation in each case. Montesquieu is interested in the great patterns detectable in human society, but he also wishes to describe the variations and particularities. He sets up a whole program for the comparative sociological study of the peoples of the world far beyond the capacities of a single man, even of his great genius and diligence, to carry out. Human laws should be interpreted:

> relative to the physical characteristics of the country, and to whether its climate is frigid, torrid or temperate, to the quality of its soil, to its location, to its size, to the mode of life of its people (whether laborers, hunters, or shepherds). They should be considered in terms of the amount of liberty the constitution can allow, to the religion of the inhabitants, their inclinations, wealth, commerce, customs and manner.[6]

In the *Esprit des Lois* Montesquieu takes up these problems one after the other. Several famous and much criticized sections deal with the influence of climate. It is here that he undertakes the sort of application of current physiological ideas which Sprat had adumbrated, and by the same token commits the ordinary error of many who have tried to borrow from science; namely, that of taking as literally and permanently true the scientific theories and hypotheses of one's own time; whereas about the only thing that the social sciences can safely take from the biological sciences is the knowledge that the biological sciences are already much more complicated than the physical sciences and their theories correspondingly more tentative.

To understand the influences that led Montesquieu to adopt the now long-extinct fiber theory of the iatrophysicists requires a word about his scientific background.[7] His education, at the hands of the Oratorians of Juilly, was mainly classical, but early in his brief legal career he devoted himself passionately to the study of science. Though seemingly ignorant, in the main, of Newtonian science, he was an avowed Cartesian both in physics and biology. He was a lifelong member and supporter of the Academy of Bordeaux, before which as a young man he read a number of scientific papers: on the cause of the echo, on the possible function of the mysterious suprarenal glands, and on the "gravity of bodies." He projected a great Natural History of the Earth to be based on physical, i.e., presumably Cartesian, principles. His published correspondence reveals a wide scientific acquaintance, and his *Pensées*—voluminous notebooks of jottings on all subjects, unpublished until 1899—show an incredibly diverse scientific curiosity.[8] He has notes on the absorption of the thymus; on the coronary and fetal circulations; on the significance of fossils; on soil exhaustion; on the mechanics of bird flight; on the teeth of mammals and the correlation of their morphology with their feeding habits. He seems even to have had a clearly formulated theory of the "multiplication" of species, that is, of the evolution of the present multiplicity of forms from a few ancestral types; he even believed that the extinction of species was somehow involved in the process, a view only accepted much later. He was also interested in microscopy and in such medical problems as inoculation for smallpox, theories of contagion, the course of epidemics, and above all the relation of climate to disease.

That climate could determine the traits and institutions of mankind was a very old theory. It had been advanced in antiquity in the Hippocratic treatise *Airs, Waters and Places,* in which there was a revived interest shortly before Montesquieu's time.[9] It had been used by Jean Bodin in the sixteenth century, and more recently had been advanced by people as different as the traveler Chardin, the Abbé du Bos, and the scientific publicist Fontenelle. But Montesquieu offers in addition a physiological theory—a mechanical, iatrophysical theory—to explain just how climate can bring about changes in the human constitution and hence account for the inclinations, laws, customs, and manners of different peoples. This explanation is based on the presumed action of air upon the fibers of which the body was believed to be composed. The fibers of the body respond to the temperature, humidity, and impurities of the air—cold, dry air shortening and compressing the fibers, with moist, warm air having the opposite effect. The motor response of voluntary muscles, the pumping action of the heart, the efficiency of venous return are all influenced by the tone of the fibers. Because of this, people in cold climates are men of action, courage, and creativeness, lovers of liberty and national independence, given to military prowess but also enjoying a high incidence of suicide. The people of warm climates are correspondingly slothful and conservative; among them the institutions of slavery and monasticism flourish, and they tend to produce philosophic and religious systems which favor contemplation and inaction. A similar action of the air is invoked to account for the difference in temperament of peoples. Sensory nerve trunks are composed of bundles of nerves which Montesquieu believed to be spread out, like the frayed end of a thread, at the skin or sensory organ. Heat expands the epidermal area and splays out the nerve endings, producing finer sensitivity and discrimination. Cold, on the contrary, brings them together as the epidermal layer contracts; this tends to deaden sensation. It follows that denizens of hot climates must be more sensual and greater voluptuaries than persons condemned to live in cold climates. It is in this connection that Montesquieu cites in the *Esprit des Lois* his own microscopic observations of a sheep's tongue, where he saw the organs of taste seem to disappear and shrink together under the action of cold.[10]

A French scholar, Abbé Joseph Dedieu, long ago identified the principal source from which Montesquieu drew his fiber theory of the influence of climate.[11] This is a little book by John Arbuthnot—physician and member of the Royal Society of London, wit and satirist, close friend of Gay and Swift, and inventor of the British national symbol, John Bull—entitled *An Essay Concerning the Effects of Air on Human Bodies.* It was first published in 1733 and became available in a French translation in 1742.

It is a rare and interesting little octavo.[12] In showing how the air can exert its effect on the fibers and fluids of the body, Dr. Arbuthnot gives a masterful abridgment of all that had been discovered about the properties of air up to his own time. The ideas are based upon the discoveries of Robert Boyle, Robert Hooke, Francis Hauksbee, Stephen Hales, and Edmond Halley—all members of the English school that had centered its attention on the study of air—and upon the great chemical textbook of the Dutch physician Hermann Boerhaave, the first work dealing mainly with what today we would call physical chemistry. Through this little book an extraordinary amount of recent scientific information was thereby made compactly available to Montesquieu. Arbuthnot, moreover, combined the anatomist's fiber theory with a physicochemical theory of the action of air upon these fibers and performed experiments to confirm his theory. "I have found," he wrote, "that the single fibers, both of vegetables and animals, are lengthened by water or by moist air; a fiddle string moisten'd with water will sink a note in a little time."

Arbuthnot presents in a nutshell, and if possible more bluntly than Montesquieu, the fiber theory of the action of climate, even going so far as to discuss its effects on human societies:

> Governments [Arbuthnot wrote] stamp the Manners, but cannot change the Genius and Temper of the Inhabitants; and as far as they are unrestrain'd by Laws, their Passions, and consequently their National Virtues and Vices will bear some Conformity with the Temperature of the Air.[13]

At one point Arbuthnot illustrates his argument with an example Montesquieu may not have appreciated. He attributes the characteristic frivolity of the French to the persistence of climatic factors. He cites the Emperor Julian as saying that Lutetia had "more Comedians, Dancers and Fidlers than there were Citizens," a picture he felt to be still valid for the eighteenth century, "and I believe," he continued, "if a Race of Laplanders were transplanted thither, in a few Years they would be found in the condition describ'd by the Emperor Julian."[14]

Perhaps you remember that Voltaire disposed of Montesquieu's idea that courage and military prowess depend only upon temperature by pointing out that Laps and Samoiedes were not remarkable for their courage, while in eighty years the Arabs of Mohammed conquered a territory larger than the whole Roman Empire. I wonder what he would have said about Arbuthnot's climate theory of language:

> I will venture to add another Observation, which, tho' it may seem a little too much refin'd [we would say far-fetched], is not improbable: That the Air has an Influence in forming the Languages of Mankind: The serrated close way of Speaking of Northern Nations, may be owing to their Reluctance to open their Mouth wide in cold Air, which must make their Language abound in Consonants; whereas from a contrary Cause, the Inhabitants of warmer Climates opening their Mouths, must form a softer Language, abounding in Vowels.
>
> Another Observation is, that People in windy Countries naturally speak loud, to make themselves be heard in the open Air.[15]

I have perhaps dwelt too long upon Arbuthnot's theory and the use that Montesquieu made of it, but I would not want to leave you with the feeling that it is the best that Montesquieu took from the scientific influences that surrounded him and that helped shape his intellect, for I believe it rather to have been the worst. "All the sciences are good," Montesquieu wrote in a sentence that reveals much, "and mutually assist one another."[16] A great motto, it is true, if one is suitably wary of his borrowings.

Let me then conclude on the note with which I began this discussion of Montesquieu. It was his determination to study man as a part of physical as well as moral nature, to view him as responding to invariable physical laws, that constitutes Montesquieu's great achievement. He asserted the axioms and laid down the principles that should lead to a science, or more accurately a natural history, of man. He made a bold and historic attempt to show by writing a sociology of law how this sort of thing should be done. We must forgive him his climates and his fiber theory. He borrowed them as freely, and trusted them as guilelessly, as many modern scholars borrow and trust the equally tentative guesses of modern psychology. It is the spirit and method of Montesquieu, I feel, by which we should judge him.

But what of his method? There are those who believe with Albert Sorel that Montesquieu is mainly a *génie généralisateur*, more Euclidean and Cartesian than scientific in any modern sense, that his book—despite the significant lack of classic order in its composition—is a good example of the dogmatic, a priori reasoning of a sheltered scholar, a typical product of the mathematical rationalism of the Age of Reason. This I feel confuses the method of exposition with the method of discovery. The *Esprit des Lois* must be read in the light of the *Pensées*, which are the chips and shavings from Montesquieu's workshop. Here he appears more Baconian than Cartesian, professes his distaste for the mathematical mind, proclaims himself as "not fond of opinions" and an enemy of rationalist systems. But did he have an insight into the so-called inductive aspects of natural science? I believe he did. The *Esprit des Lois* itself was not spun simply out of his inner consciousness, though there is a deceptive sentence in the book which has led some persons to think this was the case. Reference to his classic blunder about the British Constitution can blind us to the enormous reading, the extensive travel, the tireless energy, the active correspondence which went into the *Esprit des Lois*.

But as proof of his grasp of the scientific approach to difficult problems let me borrow from the *Pensées* his contribution to a burning medical question of the day, the study of the epidemiology of the plague. This is what he says:

> It seems to me that in Europe we are not in a position to make proper observations on the plague. This disease which is transplanted here does not display its natural characteristics. It varies a great deal with the differences in climate, without taking into account the fact that since it is not always present, and there being intervals of whole centuries between its appearances, we cannot

make continual observations. Moreover, the observers are so distraught with fear that they are not able to make any at all.

But accurate, enlightened and well-paid observers should be sent into the region where this disease is epidemic and appears every year, as in Egypt and several places in Asia.

We should ascertain what are the causes of it, what seasons are favorable or unfavorable, and observe the winds, the rains, the nature of the climate; what ages and temperaments are the most vulnerable; what remedies, preventives and variations of the disease there are. We should have observations from many different times and places, and use any information that certain countries might be able to give us.[17]

This proposal—with its echoes of Hippocrates—can only command the respect of anyone of scientific training and experience. How can the same man have written in the *Esprit des Lois*: "When I discovered my principles, everything came to me. . . . I laid down the principles and I saw the special cases fit in by themselves." Was Montesquieu a man of the classic type, or did the Baconian "romantic" disposition predominate? Who can tell? But perhaps the answer to the enigma is that like Galileo, the founder of modern physics, Montesquieu, the father of the social sciences, had something of the spirit William James attempted to describe, a spirit too big and complex to be readily pigeonholed, with a love for both abstract constructions and the tangled wilderness of the real world: in a word, a scientific humanist.

Notes

1. Claude Lloyd, "Shadwell and the Virtuosi," *PMLA* 44 (1929), 472–94.
2. Quoted in Charles P. Curtis, Jr. and Ferris Greenslet, *The Practical Cogitator or the Thinker's Anthology* (Boston: Houghton Mifflin Co., 1945), 207–8. The original appeared in William Morton Wheeler's "What Is Natural History?" in *Bulletin of the Boston Society of Natural History* 59 (April 1931), 9–10.
3. William James, *Pragmatism,* ed. Ralph Barton Perry (New York: Longmans Green, 1943), 128.
4. Quoted in Martha Ornstein, *The Role of Scientific Societies in the Seventeenth Century* (Chicago: University of Chicago Press, 1928), 108–9.
5. Thomas Sprat, *The History of the Royal Society of London, For the Improving of Natural Knowledge,* 3d ed. corrected (London, 1722), 82–83.
6. *Esprit des Lois,* bk. 1, chap. 3.
7. The scientific interests of Montesquieu are mentioned by his principal biographers but have never been given adequate treatment, especially in view of the materials made available by the publication of the papers from La Brède and the *Correspondance* brought out by Gebelin and Morize in 1914.
8. *Pensées et Fragments inédits de Montesquieu,* published by Baron Gaston de Montesquieu, 2 vols. (Bordeaux, 1899).
9. Thomas Sydenham (1624–89) and Hermann Boerhaave (1668–1738) were leaders in this Hippocratic revival. The revival is reflected in the space given by Daniel Leclerc to Hippocrates in his *History of Medicine* (1699). In 1734 Francis Clifton brought out an English translation of *Airs, Waters and Places.*

10. *Esprit des Lois*, bk. 14, chap. 2.
11. Joseph Dedieu, *Montesquieu et la tradition politique anglaise en France—les sources anglaises de "L'Esprit des Lois"* (Paris, 1909). See especially chapter 7, "Les rapports des lois avec le climat."
12. John Arbuthnot, *An Essay Concerning the Effects of Air on Human Bodies* (London, 1733).
13. *Ibid.*, 150.
14. *Ibid.*, 149–50.
15. *Ibid.*, 153–54.
16. Montesquieu, *Pensées*, I, 460.
17. *Ibid.*, I, 487–88.

2
Teaching and Research in the History of Science

❁

I am somewhat intimidated at finding myself, so the program says, representing the humanities in our discussion "Research and the Curriculum." The humanities constitute a vast area, and with most of these important fields I have no familiarity whatsoever.

You will forgive me, then, if I confine myself to the field of historical study—for unlike some of my colleagues I consider history not a social science but a humanistic discipline—and you must allow me to talk to you about my own subject, the history of science. It will illustrate, I believe, the kind of work done in humanistic research and the relation that exists between research and teaching in these subjects.

Since the history of science is one of the newest branches of historical study to win a place in the curriculum of our colleges and universities, and since therefore many of you may have had little occasion to see it at work, I should perhaps begin with a few general remarks about the field.

As an organized scholarly discipline it is not much more than fifty years old. Its recognition as a valid object of university instruction is more recent still. Unless I am greatly mistaken, the earliest established undergraduate course in the subject was one launched at Harvard not by an historian but by a distinguished physiologist, Lawrence J. Henderson, soon after the end of the First World War. During the 1920s and 1930s there were other enterprising individuals who taught the history of science, some very successfully, in their universities. Such a course was given at Minnesota for many years; there was a famous one given at Columbia by Frederick Barry, a chemist; and the one offered until her retirement by Miss Dorothy Stimson, an historian, at Goucher College; and there must have been others of which I am ignorant. But until 1941 there was only one man in America—one of the few persons anywhere—whom we should qualify as a full-time, professional specialist in the history of science: Professor George Sarton, the distinguished Belgian scholar who came to Harvard about 1920. Sarton did more than any other man to give the subject professional standing and to provide some of the basic research implements. He founded and edited the most distinguished journal in the field, *Isis*, which I am glad to say is still a going con-

From *The North Central Association Quarterly*, vol. 33 (1959), 242–47. Published by the North Central Association of Colleges and Secondary Schools.

cern; he instituted an annual Critical Bibliography of publications in the history of science; he wrote widely and persuasively; and he offered courses on both the undergraduate and graduate levels. Harvard conferred the first Ph.D. in the subject in the 1930s. In 1941, the University of Wisconsin established the first separate Department of the History of Science. Reactivated after a hiatus caused by the Second World War, by the energy and learning of my confrère Marshall Clagett, it has become, with the strong support of the Administration at Madison, a great center of research and teaching.

The last twelve years have seen a most rapid, indeed an almost explosive, development of the subject. At least a dozen distinguished universities have men on their faculties whose prime concern is teaching and research in the history of science. And in three or four of these institutions there is an active program of graduate training leading to the Ph.D. Since 1949 more than a baker's dozen of first-rate young people have received the Ph.D. from such institutions as Cornell, Harvard, and Wisconsin. Young though it is, the subject now has all the accouterments of the older scholarly fields: it has its organized society, with more than 1,500 members, its specialized journals, its periodic international congresses, and even its honorific international academy. The philanthropic foundations have discovered this growing field and have been generous with fellowships and research grants. Even the National Science Foundation, which is primarily concerned to support the physical and biological sciences, has awarded fellowships to deserving historians of science and to graduate students. Just this past year the National Academy of Sciences–National Research Council has established a U. S. National Committee for the History and Philosophy of Science, to facilitate, as similar committees do in the other scientific fields, the international aspect of our work.

At this point you may well ask what all the fuss is about and what precisely it is that historians of science do. I think most persons in this field would agree that by *science* they mean chiefly the physical and natural sciences, and that they pay little attention to the social or behavioral sciences, so-called, though including them should be, and probably will be, the next step.

Nor are we centrally concerned to any extent with the history of technology and invention; like the history of medicine, this is thought of as a cognate, but largely independent, discipline. We think of science, furthermore, in three related ways: as *a body of knowledge* about the natural world; as the *method of inquiry* by which this body of knowledge has been built up; and as the *human or social or professional activity* through which this knowledge has been acquired, disseminated, and steadily enriched and refined.

It is the central business of the historian of science to reconstruct the story of the acquisition of this knowledge and the refinement of its method or methods, and—perhaps above all—to study science as a human activity and learn how it arose, how it developed and expanded, and how it has influenced or been influenced by man's material, intellectual, and even spiritual aspirations. It is a humanistic study, as George Sarton liked to remind us, of the *only really cumula-*

tive and progressive intellectual activity of man. But in all other respects the historian of science does not differ in his approach from the scholars who study the histories of other human endeavors for the light they throw on man, his failures and his accomplishments: the history of law or philosophy or art or literature or religion, for example.

Before discussing the research being carried on in the history of science, and the vast and complicated areas awaiting exploration, I should perhaps seek to justify—if only briefly and inadequately—the place of the history of science in the university curriculum. Some of you may have an uneasy feeling that perhaps this subject is an academic curiosity or a luxury that can serve the needs of only a few students. This is not the case. All of you will certainly accept the proposition that no young person should emerge into our scientific and technical twentieth century without an appreciation of what science is about, what it is that scientists really do, how they think and work, and how science came to exert this overwhelming influence on our modern life and thought. But what you may well ask is whether the history of science can do this job, or at least contribute to it in any substantial fashion.

Your institutions may be more fortunate than my own in the quality of the science courses taken by the nonscientific undergraduates, the so-called terminal courses in college science. But I suspect there is a rather general pattern suggesting that all is not too rosy, which is why there is so much experimentation with general science courses, general education courses in science, and so on. I myself happen to believe that the solution lies in a root-and-branch recasting of the old introductory courses, so that they are not simply watered down versions of the standard preprofessional courses. I believe that the best way for a student to learn about science is to study science—a particular science. But the goal should be not to teach techniques or the latest so-called facts, but through the particular science to convey something like a genuine understanding of the scientific enterprise as a whole, its spirit, its methods, its manner of thinking and working. Some of the really fundamental problems ought to be raised and debated, and for this some exposure to the history and the philosophy of science ought to be included, as well as some discussion of the human and social impact of science.

For various reasons, I prefer this solution to one which would make exclusive use of a history of science course to present the case for science to the nonscientist. But in certain circumstances it has been amply demonstrated that a good, and not too technical, course in the history of science can do a very effective job in conveying the scientific attitude, making clear the self-imposed limitations of science, and presenting science for what at its best it has always been: a rigorously self-disciplined, essentially modest, and open-ended business of ceaseless inquiry into nature, with its own aesthetic and intellectual rewards; in short, one should emphasize the humanistic side of science. But this can best be done, I am convinced, *after* the student has taken the kind of rejuvenated course in a special science which I have described. But here is the difficulty: we simply do not have, certainly not in sufficient numbers, men or women who have received—in addition to their specialized preparation—the kind of training in the history of science and the philosophy of science which is needed to give the requisite depth and breadth to these terminal science courses. Too many, skilled though they may be, are products of our intensely vocationalized undergraduate and graduate training in science. They cannot see the forest for the trees, and their own particular kind of tree at that. This leads me to my next point: the importance of some exposure to the history of science on the part of *science students*, both undergraduate and graduate.

It is appalling to me to contemplate the kind of highly trained but half-educated young persons we commonly turn out of our science programs. Many, to be sure, take up humble technical positions in our complex modern industry, where they can do much good and little intellectual damage. But many of them, let us not forget, remain in universities and colleges as teachers, often perforce as teachers on the elementary level. They are in the main wholly unequipped to present science to the nonscientist in the manner I have described. In recent years I have taught quite a number of such young men and women after they have progressed quite far in their science programs. It is astonishing how little understanding of science such young people possess, what misapprehensions they harbor, how naïve they are about the logical and conceptual methods they learn by rote, and, above all, how ignorant they are about the great scientific tradition that should illuminate and inspire them. They are, I might add, totally inarticulate and inept when called upon to defend their subject, science, from the increasingly vocal, and often very adroit, critics and antagonists of science. For none of these critics can be convinced or persuaded by arguments which defend science only on the grounds that it amuses its devotees, gives them well-paid security (at least in industry), and sheds material benefits upon the world. The utilitarian argument for science, I need hardly add, has become ambiguous and double-edged. And it has never been a very good argument.

So one strong and compelling reason for teaching the history of science on both the undergraduate and graduate level is to make sure that the elementary teachers of science in our schools and colleges will have not merely a technical proficiency in their subjects but some genuine understanding of science in its human and humanistic dimension.

I have always believed that one of the chief educational experiences provided by the humanistic studies comes from confronting the student with examples of great and exceptional human accomplishment: in the political, the intellectual, and the moral spheres. This comprehension of greatness sets goals, gives standards of excellence, and provides some relief from the depressing spectacle of the current Human Comedy. I do not need to insist that a view of human greatness which leaves out—or turns into a mere wax-work symbol—a Galileo, a Newton, a Darwin, or an Einstein gives a quite impoverished view of human culture and the potentialities of the human mind and spirit.

But there is a more practical reason for trying to give some appreciation of greatness in his own field to the young science student. So vast is the number of science students in our graduate schools—so many, even, who expect to do creative work in the sciences—that it is a rare and privileged young person who can work intimately in the laboratory in close association with a really first-class investigator, let alone a really great and authentic genius. Yet, more often than not, this sort of close apprenticeship was the way men learned their scientific craft in the simpler and less complex age of a century ago. By working closely with some great scientific teacher, a Gauss, a Liebig, or a Claude Bernard, men learned the tactics and strategy of investigation at first hand and picked up also, more often than not, some insight into the history and the philosophy of science. It was thus that they acquired not only higher skills but also an awareness and appreciation of the great tradition which they could and did pass on to others in the same fashion. It was a genuine laying on of hands.

Today, graduate instruction in science is big business, mass production. Though there are certainly many more able investigators at work today than at any time in the past, they are so grievously outnumbered by the crowd of students that the old apprentice system (from which our system of graduate study is patently derived) has become in this sense unworkable.

I believe that the history of science can supply, at least at second hand and vicariously, the missing ingredient. A critical reading of Galileo or Newton or Claude Bernard can bring the student into close rapport with a really first-rate mind at work, struggling with confusion and difficulty, and eventually, out of his own inner resources, reaching a viable solution. I see no reason why science students should not benefit from the close study of these men, carried on in small groups or seminars, just as the student of literature studies the writings of Shakespeare, of Milton, or of Dostoievsky.

It should not need repeating, though I'm afraid it often does, that really creative teaching in any field, especially on the graduate level, is inseparably connected with constant study and research. I need not insist that this is especially true in a field as new and unexplored, and as wide in its ramifications, as the history of science. Though much valuable research has been done, our ignorance— if you will allow the term—is almost encyclopedic. Much that passes for knowledge of the history of science in the introductory chapters of science books, in encyclopedia articles, and in the popular press is shallow, uninspired, and often downright misleading. But, by the same token, there are few historical fields of research that offer more rewards for painstaking, first-hand work. There is scarcely a single major figure, to say nothing of lesser ones, in the history of science of whom it can be said we have adequate information and deep understanding, and whose study is not full of daily surprises. You can understand the situation, perhaps, if I remind you that for many of these great figures there is still not a biography worthy of the name. For some crucial figures, like Newton for example, there is no modern critical edition of the collected works, and no publication of the collected correspondence. One big task of the future is to prepare just such essential working material.

Today, research in this area is being carried on in most of the countries of the world, not solely, by any means, by professional historians of science, but by men and women of diverse background and varied linguistic and technical knowledge. There are historically minded scientists, mathematicians, and medical men; there are professors of philosophy or literature or history; there are trained classicists and medievalists, as well as experts in the Far Eastern and Middle Eastern cultures and languages. A high degree of specialization is of course the rule, and even the small number of professional historians of science, those who work in science museums or who teach the subject in our colleges and universities, are necessarily specialized when it comes to their own research.

The central business of the historian of science, as I have said, is to reconstruct the actual sequence of events in the growth of scientific thought in a particular branch, or related branches, of the physical and natural sciences. In this "mechanical" reconstruction we can distinguish at least two approaches: (1) the careful study of the internal technical factors in the rise of a science, the use made of antecedent knowledge, and the patterns of transmission, stimulus, and influence which led to certain important theories or discoveries; and (2) the social and cultural influence exerted upon the scientists of a particular period which affected the choice of problems and the manner of their solution. A good example is the recent book of Alexandre Koyré, *From the Closed World to the Infinite Universe*, which throws into strong relief the philosophic and religious influences upon the thought of Isaac Newton. Other studies of equal significance have sought to investigate the influence which technology and social conditions in war and peace have often exerted upon the growth of science.

These are studies of science itself, and of the forces that determined its matter, its form and its methods. But there are two other wide areas where much work remains to be done, though promising beginnings have been made. The first is the study of the cultural effects of science, of what the success and example of the science of any period has meant for philosophy, religious belief, or social thought or—as in the famous studies by Professor Marjorie Nicolson—for the literary imagination. And secondly there is an immense problem, where the historian of science must join hands with the social historian and the historians of medicine and technology. This is the vast assignment of studying the material effects of the application of scientific discoveries for useful purposes, a matter particularly crucial in the nineteenth and twentieth centuries.

From the point of view of chronology, research in the history of science covers an immense span of time. Yet is it fair to say that up to now our efforts have been concentrated largely on the early periods—i.e., on the growth of science before about 1800?

Greek science, where much of our scientific knowledge began, was perhaps the first area to receive the close attention of serious scholars. Towards the end of the nineteenth century, distinguished classicists began to give Greek science a closer scrutiny, and specialists like Paul Tannery in France and Sir Thomas Heath in England did a prodigious job in reconstructing the exact science, the mathematics and astronomy, of the Greek world.

Basic though this was, it produced nothing so astonishing as the discoveries made by scholars like Otto Neugebauer who have worked with Babylonian and Assyrian materials. From the clay tablets of these peoples came evidence of their extraordinary skill in mathematics and practical astronomy, well before the Greeks. Neugebauer and his disciples are at present moving into papyrology in an effort to disentangle the Middle Eastern from the purely Classical elements in Hellenistic science. Already they have profoundly modified our opinions.

The scientific accomplishments of the Middle Ages were long ignored except by an occasional student of early alchemy. With their positivistic prejudices, men in the nineteenth century hardly admitted that much that was useful for science could have been accomplished in this theologically dominated age. Today we know better, and have come to realize that important medieval pioneers in physics—in statics, dynamics, and cosmology especially—paved the way for Copernicus and Galileo. At present the study of medieval science is one of the most active and productive fields of historical investigation.

The great Scientific Revolution of Galileo's day has been re-examined in the light of these new discoveries, and we have come to realize how little we actually know about it, and how difficult it is to grasp how it came about. It is, and will long remain, our central problem; just as, for example, the Reformation is the chief concern of historians of Protestantism.

The tendency of much recent work on the great Scientific Revolution has been to emphasize, indeed I think to overemphasize, the extrascientific and philosophic influences on these early scientists. We hear a great deal, for example, about the Platonism of Galileo and the influence upon him of his medieval precursors, and much about the religious and mystical side of Newton's thought. I sense a turning of the tide, with more attention being paid to the central scientific factors, and to a re-examination of the role of technology and the contribution of the Renaissance artisan and craftsman, about whose influence we used to hear so much.

The study of scientific progress in Europe and America during the eighteenth century is being actively pursued by many investigators. We have recently seen appear the first exhaustive study of Benjamin Franklin as a scientist, and are beginning to understand something about the foundation of modern chemistry by Lavoisier and his precursors, a subject largely surrounded hitherto by ignorance and mythology.

A number of scholars have turned their attention also to studying the influence of science on the political and social reformers of the French Enlightenment and on our own Founding Fathers—for example, Thomas Jefferson. And we are beginning to investigate the links between eighteenth-century science and the early Industrial Revolution, though we have scarcely begun to find answers to the most important questions that arise in this connection.

Devoted and sustained research must be continued in all the areas and periods I have mentioned. But the real work that lies before us is to study the science of the nineteenth century—and eventually of the early twentieth century—with the same care that has been devoted to earlier periods.

I can report only the beginnings of such work. A good deal has been written on obviously important figures like Charles Darwin and Louis Pasteur, but the history of nineteenth-century biology has still to be written. This is also true for the physical sciences. There are few if any definitive studies of the leading figures, and needless to say we still await the scholar who will be equipped to dive headlong into the study of the great Second Revolution in physics, which began in the 1890s and transformed not only physics but our whole image of science and the grounds on which it rests. There is an immense amount still to be done.

I hope I have been able to convey something of the accomplishments that can already be credited to investigators in this exciting field and, even more, something of the sense of adventure and limitless opportunity that lie ahead of any researcher in this new and still largely unexplored area of historical investigation. But I shall be satisfied if I have convinced you that the man who tries to teach the subject, even on the elementary level, without carrying on his own research and keeping abreast of what his colleagues are doing, will be a confused and pathetic figure. In the new field of the history of science, research is the staff of life and the condition of creditable accomplishment as a teacher.

[This paper was presented in 1958. A year later, when it was published, there appeared the first volume of *The Correspondence of Isaac Newton*, which by now, in 1976, has reached its sixth volume. Newton scholarship is proceeding at a rapid pace. And it should be added that since this paper was written, a new generation of scholars has begun to explore the complexities of nineteenth- and early-twentieth-century science.]

3

Some Historical Assumptions of the History of Science

❖

History, of course, has something to do with the past. As we commonly use the word—neglecting such "historical" subjects as paleontology or evolutionary biology—we mean human history, the past of mankind, especially since the mastery of the art of writing.

But the word has a built-in ambiguity. It can refer to what Charles Beard called "history as past actuality," that is, the sum and total of everything men must have done in the past. Or it can mean written history: what men have thought or written about the human past.[1] The distinction is of more theoretical than practical significance. The past itself—history as past actuality—has fled forever; we cannot experience it directly; only our thoughts about it have any real existence for us. We can only construct an image of the past, or rather of selected parts and instants of it, by an act of mind, just as the scientist constructs a simplified model of the physical reality into which he cannot penetrate. We give the past, as Cassirer says, "a new, ideal existence" in our minds. Properly carried out, this ideal reconstruction is inferential in character—a "connaissance par traces," as F. Simiand[2] called it—based upon our study of the documents, monuments, and artifacts which have defied the erosion of time. It is the result of probable inference guided by our imagination, but not by an unfettered imagination. Our method is the method of Zadig.

History, then, as I shall use the term, is a more or less disciplined inquiry into the past of mankind. Nevertheless we would not recapture, even if we could, everything we imagine to have happened in history as past actuality. Written history can only be highly selective, at best a mere sketch or outline of past reality. But what dictates our selection? An answer to this question could tell us a good deal about the objectives of the historian, his opportunities as well as the restrictions imposed upon him.

Chance, to be sure, has played a major role in what we are permitted to know about the past; for though man is a record-making and record-preserving animal, he has set down only an insignificant fraction of his doings. And most of what he has recorded has, providentially, been destroyed and is forever lost. I say "providentially" because there is much about man's activities—one is tempted to

From A. C. Crombie, ed., *Scientific Change* (London: Heinemann; New York: Basic Books, 1963). © Heinemann Educational Books Ltd., 1963.

say, by far the greatest part—which is so trivial and commonplace, so infinitely repeated without significant variation, that we would never think to dwell upon it.

But is there—has there always been—a criterion for what is not trivial and commonplace, for what has "historical significance"? To a greater extent than we often realize, what we can know about the past is what our ancestors—the participants in events or those who came soon after—determined that we should know. They placed in the intentional record—in annals, memoirs, and commemorative inscriptions—those men and events which appeared to them as exceptional, striking, and wholly outside the ordinary dull routine of private existence. In the main, they singled out for preservation in the collective memory those events which they saw to have markedly affected the way of life, the thoughts and actions, of the larger social groups and political entities: a tribe, a city-state, a nation, or an empire. So it is that the main scaffolding and framework of our view of history consists of those deeds, thoughts, and productions which others besides ourselves deemed worthy of preservation because of their effect upon man in society. So it is, too, that early historians dealt for the most part with notable human actors and the part they played in political decision, war, and dynastic struggle. It was upon such events—in the monarchies, oligarchies, and feudal aristocracies of earlier time—that the destinies of men most obviously seemed to depend.

The Broadening of History

From the beginning, then, written history focused on public matters. Individual actions were of interest only if they demonstrably affected the lives and thoughts of the social collectivities. And this, I feel, is still the characteristic of history, setting it apart from biography, romance, and antiquarian curiosity. What has changed (for the writing of history has demonstrably changed in the past two centuries) has been chiefly the recognition of what varied and often subtle sorts of human action do in fact influence the fabric and the destinies of human groups.

What is most often stressed about the flowering of historical study in the nineteenth century—a new sympathy with neglected periods of history, like the Middle Ages, and the effort to make history "scientific" by the systematic use of official and administrative documents—is perhaps not the most revealing. Fully as interesting is that the enterprise of writing history was broadened and deepened. When the historian relaxed his dependence upon what I called the intentional record—the annals, memoirs, and contemporary narratives—and turned instead to the unwitting testimony of official records and private correspondence, he not only demonstrated the bias and unreliability of his traditional sources but also caught a glimpse of a new and more subtle approach to the past. History was seen to consist not merely in the deeds of a few dominant personalities but in the largely impersonal unfolding of a nation's political and social institutions, in the cumulative action of many forgotten and often anonymous persons, and in the condition and aspirations of different social classes. The great European Revolution, which had cast up so many documents from their hiding places in looted chancelleries, confiscated estates, and suppressed institutions, was itself a vivid demonstration of the complex forces producing historical change. Paradoxically, this greatest of all historical discontinuities—where venerable institutions were toppled and pent-up energies burst forth to take command of men and events—made it evident that the historian could no longer neglect the hidden agencies, the gradual as well as the spectacular changes, which affect a society in the flow of time. A heightened sense of nationhood which the Revolution brought in its train, and the new democratic aspirations of which it was the expression, brought home the fact that history must be the history of peoples, not merely of their leaders. The slow, often imperceptible, alteration in men's lives; the change in their material condition, and in their ideas about God, Nature, Man, and the State: these were seen to be hidden stuff of history. This uniformitarianism, and this tendency to see history as social history, is already evident in the writings of Sismondi, Augustin Thierry, and Michelet, with their picture of life in medieval town and country, and their attention to the role of the communes and to the rise of the Third Estate.[3] Something of the same sort is found in the Whig historians of Britain—Hallam, Macaulay, and Grote—for whom history is the story of political liberty. Against this stream Carlyle set himself with his well-known, and quite reactionary, theory of the Hero in History. But Carlyle's was a hopeless stand. Opposing him were arrayed not only a new breed of constitutional and institutional historians, but the social philosophers, whose theories were destined to leave their mark on the work of the professional historian. Under the spell of men like Comte and Spencer, Hippolyte Taine and Thomas Henry Buckle made the first conscious attempts to write history in a societal, if not quite a sociological, fashion, and to give a proper place to social factors and to the role of ideas and ideologies. Buckle's admirer Lecky broke new ground with his classic study of the ideas of Western man which could be linked to the "practical, active and social" aspects of history.[4] And after Karl Marx, no historian, least of all men living in an age of rapid industrial and social change, could wisely neglect the role that class interest, economic institutions, and the changing modes of production play in shaping the human condition.

This expansion of the historian's canvas had clear implications. The most obvious was the need for specialization, for a detailed inquiry into those aspects of human history which had been largely neglected: for example, the history of commerce and exploration, of social customs and manners, of agriculture and village life, the rise of an urban middle class, and the condition of the laborer.

How, then, were these multiform aspects to be related one to another? Was it possible to delineate, in something approaching its living complexity, the temper, tone, and cultural pattern of earlier societies? Here, I think, the work of the pioneer sociologists, and eventually the rise of ethnology, may have exerted a subtle influence. At all events after mid-century we encounter some bold excursions into historical writing that employed a new form and displayed a different texture.

With a few prophetic exceptions like Voltaire's *Siècle de Louis XIV*, historical writing had been exclusively narrative, consisting of little but a running account of events and human actions. But the societal aspect of history demanded that the historian not only narrate the events transpiring in an early society but also display the pattern and structure of that society, the organic cohesiveness of its institutions, and the interdependence of its cultural elements. Only in this way could the events themselves be seen in their proper light, and their true character appreciated. Moreover, if the distant and highest goal of the historian was to study the development of societies in space and time—as Pirenne in our day has defined the purpose of history, and as Toynbee has sought to realize it on a panoramic scale—then the motion picture must perhaps be arrested, and its separate frames studied in detail. The solution was to make a cut through temporality, and write what we may call "horizontal" history, offering a kind of instantaneous photograph or portrait where the characteristic and stable elements of a society are singled out and described in their interrelationships. We find some early examples of this kind of writing, foreshadowed in Voltaire and in parts of Gibbon, brought to the fore in the famous third chapter of Macaulay's *History of England*, and in Taine's volumes on the *Ancien Régime*. But entire works, now classic, were written in this new form: Fustel de Coulange's *La Cité Antique*, for one, and a generation later Samuel Dill's *Roman Society from Nero to Marcus Aurelius*.

Specialization on the one hand, and attempts at cultural synthesis on the other: these have been the two polarities of historical writing during the last century and a half, two consequences of the expanded scope of historical inquiry. Both were—and are—essential. They should of course be related in an obvious way, with special studies providing the detailed knowledge and some of the insights necessary for intelligent and meaningful synthesis. Regrettably this is not always the case. Specialties, as we all know, acquire a life of their own, a jealous independence, a private jargon, and an esoteric concern with the smallest technical detail. This has been true of those historical subdisciplines which split off from history itself, and which—like the examples I have given—had as their original justification a deeper understanding of the processes underlying historical change. It is even more true of those quasi-historical fields, like our own, which had, as I shall now try to show, a different origin, and which dealt with human activities less obviously connected with those sociopolitical concerns that remain central to the work of the historian.

It is well to remind ourselves that the flowering of historical writing was an expression of a deeper and all-pervasive realignment of thought and method in the study of man and nature we see emerging at the close of the eighteenth century and the beginning of the nineteenth. Historical explanation as a mode of understanding through retracing a development (real or imagined); history, in the loose sense of genetic treatment of any subject; these pervaded philosophy and social theory from the time of Turgot through Hegel and Marx, Comte and Spencer. With Thomas Wright, William Herschel, and Laplace it entered into sidereal astronomy and cosmology. It supplied the geologists with their great explanatory principle. And it was applied—by embryologists, naturalists, and anatomists—to elucidate the form and structure, the relationships and adaptations, of living creatures.

In what we now call the social sciences, the sciences of man, the same shift of attitude and method is to be noted. In the eighteenth-century Age of Reason it was customary to treat, in a theoretical and normative way, such problems as the nature of law or economic behavior or the character of political institutions, by testing their conformity to some ideal plan, or by judging them as approximating to, or falling away from, the dictates of the Moral Law or the Laws of Nature. This tendency was far from abandoned in the nineteenth century.[5] But "historical schools" arose in the study of law—with Savigny and Henry Sumner Maine, for example—and in economics and in political theory.[6] The purpose was to illuminate the present by turning to the past, noting the conditions which had given rise to particular forms and practices, and the changes these underwent. While such studies were clearly of value to the historians, with their heightened curiosity and expanded outlook, it was not always for or by the historians that they were written, but often by legal scholars for lawyers, or by economists for their confrères and for statesmen.

The same "historical" tendency expressed itself in the study of man's artistic and cultural accomplishments: his language and literature, his art and his music, his philosophic doctrines, his practical inventions and his scientific knowledge. Here, too, the genetic approach supplied a new kind of understanding in areas which had been the special preserve of the theoretical and normative modes. Again, this was the work not of historians but of specialists in the particular subjects.

It is important to emphasize, without depreciating their value, that these historical specialties, these genetic and developmental studies of special areas, lay outside the domain of history proper. The books of the pioneers—of Winckelmann, of Dr. Burney, of Tiraboschi and Victor Cousin—were intended to deepen the understanding of the student of art, music, literature, or philosophy. With few exceptions—and they came much later, like Taine's *History of English Literature*—such narratives bore the hallmark of their origin. They were often dry, sometimes technical, usually innocent of historical reverberations. The historical environment, the *moment* and the *milieu*, were studiously ignored.[7] It is hardly surprising that historians were slow to make use of them; slow, despite the example of Burckhardt, to consider the part played by art and letters in the social complex; slower still to appreciate the pioneer works in the history of science, which also were specialized works written for the specialist, when they were not popular introductions to the subject.

Some Remarks on the History of Science

We still consult today, and with profit, histories of science—or rather, of the several sciences—written in the late eighteenth and early nineteenth centuries in the spirit, and with the limited objective, to which I have just referred.[8] It is obvious that Montucla's *Histoire des mathématiques*, and Delambre's classic volumes on the history of astronomy—like the books of Joseph Priestley—were intended for the scientific reader. So also were the books published at Göttingen, between 1790 and 1810, which seem to have been introductions to the several scientific fields taught there. Among these works are Gmelin's *Geschichte der Chemie*, Kästner's *Geschichte der Mathematik*, and Johann Karl Fischer's *Geschichte der Physik*. We know, too, that Johann Beckmann's pioneer history of inventions was based on the lectures he gave at Göttingen to his students of political economy.[9] Each subject, so it was felt, should be introduced by an historical conspectus.

This remained, I think it is safe to say, roughly the character of much writing in the history of science during the remainder of the century. With a wealth of factual detail, the several branches of science were traced out in their development by scientists and mathematicians—men like Sprengel and Sachs in botany, Thomson and Kopp in chemistry, Daremberg in medicine, Moritz Cantor in mathematics, and Poggendorff in physics. These contributions were notable, but the unity of science, to say nothing of the cultural environment in which it arose, was hardly suggested. This was also true of the ambitious work that has been described as the first modern history of science, William Whewell's *History of the Inductive Sciences*, which is largely an assemblage of separate histories of the several sciences.[10] The unity Whewell was capable of giving the subject was obscured by his decision to treat "facts" and discoveries in this book, and to deal separately with "ideas" in his historically oriented *Philosophy of the Inductive Sciences*.[11] His aim, at all events, was to throw light on the sciences through their history, and to provide "a basis for the philosophy of science."[12]

To all this work the general historian of the nineteenth century remained stolidly indifferent. Even Burckhardt, whose net was cast so wide, is a case in point. He virtually ignored the scientific side of the *Quattrocento*, said little or nothing of Leonardo da Vinci as a man of science, and cited only once, I think, and slurringly, Libri's *Sciences mathématiques en Italie*. When Buckle, under positivist inspiration, used some of the works on the history of science we have mentioned when he wrote his *History of Civilization in England*, and attempted to fit the history of science into modern cultural history, the majority of historians received his effort with almost total incomprehension.[13]

It is August Comte—for all the narrow rigidity of his thought—who first offered a new and more meaningful conception of the history of science. He believed, as his famous doctrine of the three historical stages amply testifies, that the true history of mankind is the history of the human mind. He believed, too, in the underlying unity, or unifiability, of the sciences, and in the possibility of what he was the first to call the *histoire générale des sciences*. As a perusal of his rather turgid writings reveals, he had a broad, if superficial, knowledge of the development of the sciences. And though he wrote no history, or history of science, he had a truly historical sense and came up with some perceptive insights. He saw the first anticipations of modern science among the Greeks of the Hellenistic period, though he had great respect for Hippocrates and Aristotle. With little knowledge of or sympathy for the Middle Ages, he saw science developing from a "révolution générale et continue" dating from the end of the sixteenth century, although prepared by workers of previous centuries, especially the Arabs. So far as I can discover, Comte was the first to conceive of, and to baptize, the Scientific Revolution. Like Whewell, he saw this revolution largely in terms of great men. It was set on its course, he wrote, by the new impetus simultaneously given to the human mind by the *conceptions* of Descartes, the *precepts* of Francis Bacon, and the *discoveries* of Galileo.[14] The sciences did not, however, emerge all at once. Indeed Comte's famous hierarchy of the sciences reflects his theory of their successive appearance: first mathematics, then astronomy and physics, and finally—in the eighteenth century—chemistry with Lavoisier and his paladins, and physiology (a term he used where we should use the word "biology") with Haller, Spallanzani, Lamarck, and Bichat.

Yet for all his emphasis upon outstanding creative personalities, Comte was—needless to say—not blind to social forces affecting the growth of science. Scientific progress, he argued, has never been due solely to man's rational faculties. It has been "heureusement accélérée par une stimulation étrangère et permanente," chiefly the "impulsion énergique qui résulte des besoins de l'application." While lending cautious assent to the view that each of the sciences had arisen from a corresponding art or craft, he gave this familiar speculation a special twist. In its infancy every science was nurtured by its close relation to some art or craft, which supplied it with positive data and impelled speculation into a real and accessible domain. Later, however, the rapidity of progress in any science depended upon its emancipation from its related craft.[15] Thus chemistry emerged as a science only when, in Comte's own day, it acquired a largely independent and speculative character. Writing in 1838, he believed that physiology had reached a point where it must cast off the bonds linking it to its associated art, medicine, if it was to become a science.

Comte was an articulate promoter of the history of science as a learned discipline. He urged upon Guizot, then Minister of Public Education, the establishment of an academic chair devoted to the "histoire générale des sciences." And he influenced some work of real distinction in the history of science, for example Littré's translation of the Hippocratic Corpus—the chief work of ancient science recommended in Comte's *Bibliothèque positiviste*—and the study of Aristotle by Comte's English disciple, George Henry Lewes.[16] Buckle, another disciple as we have seen, wrote the first work of general history in which the history of science found a place—a brilliant and neglected pioneer effort.[17]

To see science as an historical phenomenon, responsive to, and influencing, the course of social change, one must view it, so far as the complexity and diversity of science permits, in unitary terms. It was through the logic and method of science, as well as its spirit and common purpose—in a word, through the philosophy of science broadly conceived—that a certain unification appeared possible. Men like Whewell, Mill, John Herschel, and Jevons—according to their various preconceptions—sought to analyze the intellectual machinery which had brought science to the stable plateau of accomplishment they felt it had attained by the mid-nineteenth century. We are all familiar with their optimistic belief that they could set forth this method with convincing finality. But they could hardly foresee the profound changes, the deep insecurity, which settled over science—especially physics—near the turn of the twentieth century, and which impelled philosophers and scientists to a pregnant re-examination of the foundations of science. This effort to hammer out a new philosophy of science—by Mach, Poincaré, and others—gave a powerful incentive to a more detailed, and an intellectually more sophisticated, inquiry into the past of science. This activity supplied new depth to the work of Paul Tannery, whom we all look to as the true founder of the modern history-of-science movement, yet whose debt to Comte is well attested.

It was his reading of the *Philosophie positive*, Tannery tells us, which first suggested to him what the history of science might become. Like Comte, he saw in the progress of science a key to modern history; and he discerned, more clearly than Comte, that a general history of science must be, first and foremost, a history of scientific *thought*, of scientific *ideas*, and not merely a chronology of great men, or, as in the books of Siegmund Guenther, a genealogical compendium of discoveries in the several sciences.[18] But, unlike the philosophers among his associates—André Lalande, Emile Meyerson, and the rest—Tannery was sensitive to the *social* role of science. He urged attention to the history of technology in any course of lectures in the history of science. And he criticized Comte for ignoring the history of medicine, one of the earliest, and for long one of the most significant, points of contact between the man of science and the society in which he lived.[19] Considerations such as these contributed to his belief that the history of science belonged with history, not with philosophy. In 1900 he organized, as a section of a Paris congress of historians, the first international gathering of men working in the history of science; and between that year and 1905 he contributed articles to the newly established *Revue de synthèse historique*. Tannery's vision of an *histoire générale des sciences* was a giant step towards giving our subject meaning for the historian. How his heirs attempted to bring this vision to fulfillment, and how successful they were, is another question.

The History of Science Today

The striking feature of work in history of science during the last fifty and more years has been its diversity. Much history of the older sort continues to be written, and of course should be written; for only by detailed study can the tangled skein of discoveries and influences be unraveled. But I think it safe to say that studies of this sort, under the influence of men like Tannery and George Sarton, have been carried out with a keener historical sense, a greater awareness of the complexity of scientific progress, and a real sensitivity to earlier modes of thought and to the context of contemporary ideas than was the case before. We have had, it is true, our share of encyclopedists and cataloguers, of happy dilettantes and popularizers, and of scientific antiquarians in passionate quest of the smallest technical detail. But if I were to single out the most notable achievement, it is the leadership of those men who have taught us how to focus upon the evolution of key scientific ideas and concepts. In this, Pierre Duhem is the acknowledged teacher of us all, whether medievalist or not. In this he has been followed by a number of our keenest minds—men like Burtt and Koyré—who have gone further and demonstrated the close relationship, during the formative era of modern science, between science and philosophy, and science and religious thought. Their contribution has been immense, and our debt to them is great indeed.

This enlarged and deepened conception of the history of science had brought us closer to the philosophers of science. Indeed it is men of philosophic training who have been chiefly responsible for, and philosophers who have been most responsive to, the notable progress that had been made. From the standpoint of the historian we may not have done so well. The newer history of science, with its strong flavor of idealism and superrationalism, its often exclusive preoccupation with the genesis and development of key concepts, has about it the aura of a new specialism, a kind of metahistory of science.

We run the risk, in consequence, of seeing our subject spawn new subspecialties, the history of technology affording a good example. The encyclopedic range of George Sarton, some of us can recall, did not willingly extend into this prickly subject; he had little sympathy for it and deemed it unworthy of his attention. He once replied by postcard to a distinguished medieval historian who had the misfortune to forward a reprint treating some aspects of medieval technology and bearing the title: "In Praise of Medieval Tinkers." Dr. Sarton's acknowledgment was brief. It read: "Dear Profesor X: Thank you for your paper. I am interested in medieval thinkers, not medieval tinkers. Yours very truly, George Sarton." His attitude, I suspect, is quite widely shared. So much so—to refer to a current symptom—that the historians of technology in the United States have banded together and founded their own society and their own journal, just as the historians of medicine had done earlier. I think that Tannery would not have welcomed these developments.

At all events, the relation of the history of science to the growth of technology, and the debt that early science may have owed to the practical arts and crafts, are questions with a diminished appeal for many leading historians of science today. It is quite clear to me that the causes of this indifference are complex and deep-rooted. There is the matter of changing fad and fashion, but there is another tendency that, as an historian, I sincerely deplore. To discuss social influences, especially economic and technological factors, in the growth of science—or so it appears to many historians of science in Western Europe and America—is to assume a political and ideological posture. A short while ago I published a modest paper which sought to relate the rise of chemistry in eighteenth-century France to some aspects of French industrial progress. The somewhat teasing comment of one of my good friends and admired colleagues was that it was interesting, "mais un peu Marxiste." I found that remark quite revealing, for I doubt that a general historian would have made it; certainly no good Marxist would have discovered much to applaud in my approach to the subject. In large measure, I share the impatience with much that English and Continental Marxists have written. I find a good deal of it naïve, rhetorical, un-historical, and crude. But I do want to be free to use, when I see fit, the insights these books and articles can provide me, and to be allowed to evaluate according to my lights the facts on which they lay so much stress. Take the case of Francis Bacon. A recent and often penetrating book on the rise of science, written in the idealist vein by an American scholar, takes a haughty line towards him, disposing of him with sneers. Yet scientist or not, profound philosopher or not, the his-torical influence of Bacon, in his day and long after, was immense. To cast him out—to view him disparagingly from the heights of the new idealism—is, very simply, to distort the history of science, to view it in only one of its aspects. Perhaps the reason for doing so is that Bacon is a special pet of Marxist his-torians. I am glad to say that such inhibitions are less evident among historians, the majority of whom (though they are not always the most unprejudiced of men) long ago made a certain accommodation with Marx. If we nurture such timidity, and even covertly feel that there exist forbidden subjects, and safe con-ventional attitudes towards certain great figures and events, we can hardly lay claim to the objectivity that should be our pride.

Conclusion

In this paper I have taken the point of view that historical writing, by its very nature, should be synthetic as well as analytic; that it increasingly involves the study of cultural patterns of past societies; and that its goal—admittedly a distant and perhaps unattainable goal—is the study of cultural and societal change. I have tried to suggest that one does not write history, or even necessarily con-tribute to its understanding, merely by studying in a quasi-historical or genetic way some selected and isolated aspect of human endeavor, however absorbing and valuable for itself. The history of science, like the history of theology, can be written—has been written—for its own sake. But from the standpoint of the historian, this is to write private, insulated history, and to see in science a mere cultural embellishment, as remote from crucial human problems as the history of gardening, dress, or cricket, subjects that nevertheless can be treated in what, to the historian, is a meaningful fashion. All that is required is that, in the specialization which is inevitable in all historical fields, the larger aims of history be kept in mind, and the special subject approached so that its place in the greater synthesis can be at least dimly seen. It is the specialist, no less than the generalist (the historian), who must be sensitive to the points of attachment to the larger cultural fabric.

How is this to be done? Obviously, a broad knowledge of the period about which the specialist is writing is a prerequisite. Beyond this, our problem is no different from that facing any historian of ideas, for I would hasten to agree that the history of science is primarily (but not exclusively) the history of *thought* about nature. But what is the place of ideas in history? Nobody, it seems to me, has given a satisfactory answer. Nor do I pretend to do so. But I can only insist upon one truth, or what I take to be a truth: that it is fallacious to make an arbitrary separation between ideas and experience, between thought and action, and to treat ideas as if they had a totally independent life of their own, divorced from material reality. Very possibly, indeed I suspect inevitably, the most abstract ideas—about God and Man and Nature—reflect or express, no less than art and music, the characteristic "forms" of a given culture, and are shaped by them. Furthermore, abstract ideas of the more exalted kind have a habit of being transformed, indeed often debased, into ideas of cruder coinage and therefore of wider circulation and influence. It is these which men most readily translate into action or, sometimes, by a kind of historical "Third Law," into justifications for inaction. I suspect it is these kinds of ideas that the historian, in contrast to the philosopher, finds most fruitful for his purposes. To study such operative ideas, it has been argued, is what we often mean, or perhaps should mean, when we speak of "intellectual history."

The intimate connection between thought and action is a notable characteristic of science, the secret indeed—in the case of modern science—of its forward thrust and its immense power to affect our lives. I find the methodological distinction between "pure" and "applied" science, between science as thought and science as action upon material things, to be fundamentally deceptive. It forces a choice between viewing science as a kind of philosophy or as mere technical advance. Science is neither of these things, though it partakes of both. As a collective human endeavor—which is one of the ways the historian would like to see it studied, in terms, for example, of its recruitment, support, and institutions—it has called upon the talents of many kinds of men, the lone thinker, the gifted experimentalist, the active physician, the skilled draftsman, the instrument maker—men of peace and men of war.

No human intellectual activity less deserves to be studied in arbitrary isolation. None, unless it be religion, has played a greater social and cultural role. Yet it is these points of contact with the general flow of history that the historian can justly accuse many of us of neglecting. When we overlook the social connections of scientific thought and action—and I am prepared to argue that no aspect of our subject is more in need of sophisticated treatment—we are not only raising a barrier between the historian's work and ours but we are also in danger of distorting—and this is far more grave in its consequences—our image of science itself.

Notes

1. Charles A. Beard, "Written History as an Act of Faith," *American Historical Review* 29 (1934), 219–29.
2. Cited by Marc Bloch, *Métier d'historien* (Paris: Armand Colin, 1952), 21.
3. One of Michelet's most important innovations was the attention he paid to geographical factors influencing the history of the French people. See the "Tableau de France" which opens the second volume of his *Histoire de France*.
4. See Lecky's *History of the Rise and Influence of Rationalism in Europe* (1865) and his *History of European Morals from Augustus to Charlemagne* (1869). The phrase appears in a letter to Charles Hartpole Bowen apropos of Sir Leslie Stephen's *English Thought in the Eighteenth Century* (1876). Here Lecky wrote: "I hope we two may rather help than injure each other, he being concerned with the intellectual and speculative side, I with the practical, active, and social." See his wife's *Memoir of the Right Hon. William Edward Hartpole Lecky* (London, 1909), 133.
5. The classical economists—James Mill, Nassau Senior, and the rest—are well-known examples. In law and jurisprudence, extreme cases are the "analytical jurists" Jeremy Bentham and John Austin.
6. Maine's criticism of Austin is enlightening on this altered approach. See his *Lectures on the Early History of Institutions* (New York, 1878), esp. Lectures 12 and 13.
7. A neglected early exception, and one of great interest, is Mme. de Staël's *La littérature considérée dans ses rapports avec les institutions sociales* (Paris, 1800).
8. For what follows, I have made use of my paper in *IXᵉ Congrès International des sciences historiques*, I, Rapports (Paris: Armand Colin, 1950), 182–211.
9. An amusing account of Beckmann's lectures on political economy, and their lack of relevance to the needs of future Prussian financiers, may be found in Karl Bruhns, ed., *Alexander von Humboldt*, 3 vols. (Leipzig, 1872), I, 51.
10. George Sarton, *A Guide to the History of Science* (Waltham, Mass.: Chronica Botanica Co., 1952), 15, 49, 86.
11. Or, as he put it himself, his *Philosophy of the Inductive Sciences* "contains the history of the Sciences so far as it depends on *Ideas*; the present work contains the history so far as it depends upon *Observation*." *History of the Inductive Sciences*, 3 vols. (London, 1837), I, 51.
12. Whewell also had a practical purpose:
      The present generation finds itself the heir to a vast patrimony of science. . . . The eminence on which we stand may enable us to see the land of promise, as well as the wilderness through which we have passed. The examination of the steps by which our ancestors acquired our intellectual estate . . . may teach us how to improve and increase our store . . . and afford us some indication of the most promising mode of directing our future efforts to add to its extent and completeness. To deduce such lessons from the past history of human knowledge, was the intention which originally gave rise to the present work. [*Ibid.*, 41–42]
13. See, for example, Bishop Stubbs, who remarked, "I don't believe in the philosophy of history, so I don't believe in Buckle." Cited by G. P. Gooch, *History and Historians in the Nineteenth Century* (London: Longmans, Green, 1935), 345–46.
14. Auguste Comte, *Cours de philosophie positive*, 6 vols. (Paris, 1830–42), IV, 217.
15. *Ibid.*, 278–79.
16. G. H. Lewes, *Aristotle: A Chapter from the History of Science* (London, 1864). This work is by no means wholly favorable to its subject.

17. Henry Thomas Buckle, *History of Civilization in England*, 2 vols. (London, 1857–61 and later editions). The title, as has frequently been pointed out, was singularly inappropriate. Conceived as the introduction to a 15-volume history that was never written, the work treats not only England, but deals at length with Scotland and France, and more briefly with Spain.

18. For Tannery's conception of the history of science, see his *Mémoires scientifiques*, X, especially the "Programme pour un cours d'histoire des sciences" (pp. 1–9); "De l'histoire générale des sciences" (pp. 163–82); and "Auguste Comte et l'histoire des sciences" (pp. 196–218).

19. Tannery objected to Comte's excessive preoccupation with the *sciences théoriques et abstraites*. At one point he wrote:

> "En principe, si l'histoire générale des sciences doit retracer tout mouvement intellectuel scientifique, si elle doit s'attacher à mettre en lumière les diverses influences qui déterminent le progrès, elle ne peut évidemment écarter ni les sciences concrètes ni les sciences appliquées. . . . Les techniques les plus diverses, et non seulement les techniques proprement scientifiques peuvent obtenir, dans l'histoire générale des sciences, une mention que leur refuserait l'histoire spéciale." *Mémoires scientifiques*, X, 220–21.

[The circumstances under which papers are written can be extremely odd. This one deserves a special word of clarification. It was read at a meeting held at Oxford University in July 1961, and was commented upon with his customary wit by my admired friend Alexandre Koyré. Koyré and I had been asked to present two concluding papers dealing in general terms with the history of science, he from the standpoint of the philosopher, I as one trained (at least in part) as a general historian. Shortly before the conference Koyré informed the organizer, A. C. Crombie, that he would be unable to carry out his assignment but would comment upon mine. This I did not know when I drafted the paper; remembering my charge, I took the occasion to urge a greater concern for external factors —socioeconomic factors, developments in technology, etc.—in treating the history of science expecting, when I wrote the piece, that Koyré would doubtless present the case for the essentially internalist (or idealist) view of science as a largely self-propagating activity, influenced—if at all—largely by kindred intellectual currents in philosophy and religion. To some listeners, my paper appeared to be a criticism of Koyré, and there was much murmuring by his disciples. This was not my intent, which was merely to point out a neglected approach and one, indeed, not especially exemplified in my own work. It was quite wrong to take my words as expressing my mature philosophy of the history of science.

To compound the confusion, the representatives from the Marxist countries found it to their liking (perhaps a closer reading would have tempered their enthusiasm), and the editor of the principal Polish journal asked if he could publish the paper, with a reply or refutation by one of Koyré's admirers. Dr. Crombie quite understandably declined to release my paper until the volume of contributions to the conference had been published. If the refutation had actually appeared, it would have mystified any reader: a refutation of a paper still not printed, and which had been conceived as a companion piece for a paper never written in the first place.]

4
Science and the Historian

❁

At the risk of being charged with coming here tonight to advertise the subject of the history of science, I would like to begin with a rhetorical question, and then try to suggest some answers in the course of the next hour. The question is this: of what conceivable use to the student of history, for a deeper understanding of history, is some familiarity with the emergence and development of science?

It may seem that to ask the question is to answer it, but as I intend the question, this is not so. Obviously every intelligent person would profit by learning how it came about that we live in an age where all aspects of our lives have felt the impact of, and been radically altered by, the discoveries of scientists.

But this is a lecture honoring a devoted and distinguished teacher of history, and it is the profound difference between history, as an intellectual discipline, and science as a quite different intellectual discipline, that should first concern us. For they are different. As Collingwood has said, it is a mere play on words to insist that the historian is a kind of scientist, merely because he employs quite rigorous and quite demanding techniques of investigation.

History is the study of man, of man in society: of his actions, of his political and social objectives, of the institutions in which these objectives have been embodied. It is written—sometimes clumsily, sometimes with brilliance—in the language of common discourse.

Science, on the other hand, is the impersonal and highly technical study of the world external to man and of man himself as a physical object, as the highest of the animals in the evolutionary scale. Increasingly, in the outer reaches of physics and cosmology, science has ceased to be (if it ever was) what Huxley called "systematized common sense," for it deals with ideas and concepts that transcend our ordinary experience. The language of science—even when it is not the symbolism of mathematics—is a strange language full of specialized terms: even common words like *force, power, action,* or *field* have highly restricted and uncommon meanings. Goethe might have said of all scientists what he said of mathematicians: "Mathematicians," he wrote, "are like Frenchmen:

This was delivered as the Grace A. Cockcroft Memorial Lecture, Skidmore College, November 30, 1966. Copyright © 1967 by Skidmore College.

whatever you say to them they translate into their own language, and forthwith it is something entirely different."

Above all, the scientist is a generalizer; he schematizes and abstracts, thereby eliminating much that the ordinary man or woman would like to consider. He is less interested in specific things than in the general statements he can make about them. Special circumstances—the unique events—vanish: the individual object is absorbed into the genus. In so doing, the scientist simplifies as far as possible: what he cannot simplify he cannot use and tends to ignore and discard; oversimplification, when he speaks of matters outside his special realm—politics and social problems, for example—is often his besetting sin.

The historian, though he too looks for generalizations, and speaks of "trends" and "forces" and "influences," is always concerned with the unique event, even when such events are of vast extent and complexity, like the collapse of Rome, the French Revolution, or the rise of Naziism. He may, and often must, center his attention on individual personalities. He cannot turn into "Everyman" figures like Caesar Augustus, Robespierre, Lenin, or Hitler, nor reduce them to mere products of social forces.

More requisite still is the historian's sensitivity to, and respect for, the complexity of historical events. The better historian he is, the less he is disposed to oversimplify the "tangled bank" of human affairs. He knows, like the modern poet, that the

World is crazier and more of it than we think
Incorrigibly plural.

Perhaps I have exaggerated the contrast between science and history, but I do not think so. If I am right, then we are obliged to ask what service can the history of science possibly render the general historian? Is not science a kind of epiphenomenon, above and apart from the flow of history, a closed, private world, having its own internal laws of growth, largely independent of the course of social history, and only influencing society, as it has in recent centuries, when its discoveries have been turned into useful, practical, and sometimes frightening applications? Is the history of science, then, any more interesting for the general historian than, say, the history of ballet or the history of chess playing?

Until the last few decades this, I believe, has been the attitude of many, if not most, historians. The great pioneering books in the history of science, published in the nineteenth century, were written for scientists: histories of astronomy for astronomers, histories of chemistry for chemists and so on. Philosophers, like Auguste Comte, William Whewell, and John Stuart Mill, made good use of these works, but the historians passed them by. When Henry Thomas Buckle, under positivist inspiration, made effective use of these studies in his *History of Civilization in England*, and saw the development of science as a key to modern cultural history, to a philosophy of history, the great majority of historians received his work with total incomprehension. It was Bishop Stubbs, the great constitutionalist, who commented: "I don't believe in the philosophy of history, so I don't believe in Buckle."

Such an attitude was still prominent in the first decades of this century. Take, for example, the older *Cambridge Modern History*, whose twelve volumes, appearing between 1903 and 1910, began with a volume on the Renaissance, and ended with "The Latest Age" (i.e., the period that saw the work of Faraday, Maxwell, Darwin, and the early achievements of Mme. Curie, Einstein and Ernest Rutherford). In this massive work only 130 pages out-of nearly 10,000— that is, about 1 percent—deal with science. And at the end of Chapter 23 of the last volume, entitled "The Scientific Age," we find the interesting comment: "For this chapter it seemed beyond its purpose to supply a bibliography"!

The situation has markedly altered in the last half-century, yet we still have some distance to go. As Ernest Renan once remarked, "it requires time to command the attention of men." That more time is still needed is shown by the example of Arnold Toynbee—that most sweeping and general of general historians. In his vast panorama of civilizations one interesting name is missing from the index. It is the name of Sir Isaac Newton.

We all know, of course, that historians are no longer writing only about wars and revolutions. Man's creative accomplishments—in literature, music, and the arts—are recognized as the stuff of history, no less than treaties, constitutions, and the diaries of statesmen. The fields of the history of art and the history of music, for example, have emerged as respectable, and respected, university subjects, with all the accouterments of professional societies, specialized journals, annual get-togethers, international congresses, and so on. The same is true of the history of science. Yet it must be confessed that in all these fields, including my own, there is a temptation to write only for one's professional colleagues; still, much has filtered out to enrich and deepen our understanding of cultural history, indeed of history *tout court*. The history of art has been particularly fruitful in showing us that changing tastes, and something called "style," reflect and manifest those general attitudes and tendencies we refer to, for lack of a better phrase, as the "spirit of an age."

The special domains of history upon which the history of science has already exerted a marked influence are the history of literature and that broader discipline—if indeed it is a discipline—we call the history of ideas, or intellectual history. Marjorie Nicolson and her school have shown us how richly the science of an earlier day illuminates the literary productions of the seventeenth and eighteenth centuries in England: how imperfectly Milton, John Donne, Dean Swift—even Shelley and Wordsworth—can be understood without some intimate knowledge of the scientific ideas with which they were familiar. None, today, would write about the French *philosophes* of the eighteenth century, about the Enlightenment itself, without appreciating, to some extent at least, the veneration in which this age held the discoveries of Hermann Boerhaave, Isaac Newton, and Benjamin Franklin. And who would try to understand the thinkers of the late nineteenth century without a grasp of the great Darwinian controversy? "Darwin," Gertrude Stein reminded us, "was all over my youth."

All this is, of course, quite familiar and quite obvious, or ought to be. So I wish to talk about history and history of science in a larger perspective. The chief points I wish to make are quite different. I would like to suggest, first of all, that some knowledge of early science, and of how modern science arose, can markedly change our perspective towards the flow of history in general. Second, that there have been pronounced changes in what men believed science to be—what should be its goals, how it should be carried on—that may, without undue distortion, be called changes in scientific "style." These I believe are remarkably analogous to—influencing and being influenced by—attitudes and values in other domains of the cultures that produced them. I shall conclude by suggesting that, while science since about 1600 has shown a steadily progressing and "cumulative" character—the character that George Sarton used to say set it off from all other human intellectual endeavors—even this "modern" science had been marked by significant changes of "style." Though it has built upon the achievements of the past, science is different now—in spirit and essence—from what it was, say, two hundred years ago. It is likely, if we read the signs correctly, to become different in the future.

My first point is not original with me. It was put forward some years ago, not by an historian of science, nor by a scientist, but by one of the most brilliant and versatile of modern historians, Professor Herbert Butterfield of Cambridge.

Professor Butterfield—asking why the history of science was not taught in his university (the university of Francis Bacon, Isaac Newton, J. J. Thomson, and Ernest Rutherford)—delivered a series of remarkable lectures devoted, in large measure, to what he called the *scientific revolution* of the seventeenth century. What struck him most forcibly in the course of his reading was the radical new perspective from which he found himself viewing the course of history. And he wrote some passages that created a considerable furor among historians; he chided his fellow historians for neglecting in their writings the story of the growth of science; he even urged them to revise their customary divisions into historical periods in order to give the emergence of modern science the place he felt it deserved. And he boldly asserted that the scientific revolution "outshines everything since the rise of Christianity and reduces the Renaissance and Reformation to the rank of mere episodes, mere internal displacements, within the system of Medieval Christendom." And he added that "it looms so large as the real origin both of the modern world and of the modern mentality that our customary periodization of European history has become an anachronism and an encumbrance."

I should like to expand, or perhaps twist, Professor Butterfield's thesis in another direction, and give some examples to show how a concern for the scientific accomplishments of early periods of history can enrich our understanding of these cultures and, sometimes, lead us to modify our over-all appraisal of them.

Take fifth-century Athens, the men of the Golden Age of Greece, as an obvious first example. We are always reminded of their stress upon the ideal, the canons of balance, logic, and form expressed in their art, their literature, their political thought. But who can grasp this better than one who appreciates the rigorous clarity of Greek geometry, surely one of their greatest intellectual accomplishments, or one who remembers how closely Plato's interest in mathematics led him to, or gave support to, his famous doctrine of the Eternal Ideas. It is a vivid detail, also, that the earliest Greek work in optics grew out of the study of Greek scene painters, like Agatharcus, who produced sets for the plays of Aeschylus.

The great expansion of the human horizon that came from the conquests of Alexander the Great, left its mark, if not in the *History of Animals* of Aristotle, surely on the botanical work of his successor as head of the Lyceum, Theophrastus of Eresus, who describes for the first time exotic plants from India and the East.

The ensuing Hellenistic Age looks different to us, too; takes on a special significance. The intellectual tone of this new cosmopolitan age of embattled succession states, of orientalized monarchs, can no longer be summed up in artistic imitativeness, or in a pedantic concern with the literary past, when we realize that—during the early centuries, at least—it was the great culminating age of Greek scientific achievement: of Eratosthenes, who first measured the circumference of the earth; of Aristarchus, who anticipated the heliostatic universe of Copernicus; of Euclid; and of Archimedes, the greatest mathematician, and probably the greatest scientist, to appear for two thousand years—and not merely, as one of my students once wrote on an examination, the inventor of the bathtub.

Roman culture has a different aspect, too. Rome, indeed, appears strikingly diminished, a derivative and—from this point of view—even an anti-intellectual culture, wholly dependent, for the best of its thought, upon Greek inspiration. Against the Roman's practical skills, his capacity as builder, lawmaker, and ruler of men; and against his contribution to world literature and the formation of the Western languages, we ought to weigh his backwardness in science and philosophy. Rome produced no great natural philosopher, no scientific figure of even the second rank, no man who could master, let alone enrich, the sophisticated and highly mathematical writings of the Hellenistic scientist. The Romans' partiality was for the easier pseudosciences (astrological writings, in fact, abound in Roman times) and their mathematics scarcely rose above the level of Neo-Pythagorean numerology.

Even more surprising is the view we acquire of the Middle Ages, when we look at these centuries from the standpoint of scientific achievement. Feudal Europe, the Christian Latin West, seems backward, even primitive and barbaric, when compared with the two great cultures of the Byzantine Empire and the vast Arab world. Byzantine history no longer seems only a long succession of palace revolutions and court scandals, nor can Byzantinum be praised only for its contributions to Christian art. Byzantine scholars were the chief preservers of classical Greek literature and learning, especially science and mathematics. Our finest texts of Aristotle and Euclid are ultimately derived from Byzantium, indeed our only texts of much that Archimedes wrote. These works they not only preserved but mastered, and accompanied them with astute commentaries of great value to the historian.

The Arab world, extending from India to Spain, has attracted the attention of scholars partly because of the high level of scientific creativity at a time when Byzantium was standing still, and the Western Latin countries were culturally deprived and scientifically ignorant. Within a century of their spectacular emergence on the historical scene, the Arabs not only sought out and assimilated a rich harvest of Greek philosophy, mathematics, and science but also added to it and enriched it as in the following: in mathematics they adopted and spread abroad the numerals we now use, and moved on from practical arithmetic to algebra; in astronomy, chemistry, and medicine, they also outshone not only the impoverished West, but even ancient Rome herself. If the Arab scientific achievements reflected their general cultural level, it is no wonder that when medieval Rome was a rabbit warren of decaying ruins, and London and Paris were mere overgrown villages with muddy, unpaved streets, Baghdad and Cordova were rich and prosperous cities, with impressive buildings, a high-pressure water supply, public libraries, and busy universities. It was, too, as every historian knows, chiefly from the intellectual riches of the Arab world that Western Europe in the twelfth and thirteenth centuries drew sustenance for a slow and painful renewal and upward progress.

These are, for the most part, my own impressions. But there is one area in which the findings of the history of science have left a marked impress upon recent historical scholarship. This is the study of the later Middle Ages, and the disputed question of the uniqueness and originality of the Renaissance. Was the Renaissance a sudden, unprepared, discontinuous event—a revival of letters which marked a sharp break with the cloistered, priest-ridden, static world of the Middle Ages? This was the view put forward by historians like Voltaire in the eighteenth century and hardly altered by men of the nineteenth like Burckhardt and John Addington Symonds.

In the first decade of this century, an eminent French scientist turned historian, Pierre Duhem, wrote a series of articles, followed by a series of monumental volumes, in which he disclosed a high level of scientific activity in the schools of Paris and Oxford in the fourteenth century, which nobody had hitherto suspected. In the writings of two French scholastics, Jean Buridan and Nicole Oresme—at that time only names to the medieval historian—Duhem discovered striking anticipations of certain ideas in mathematics, physics, and astronomy (the theory of a moving earth, a physical principle like that of inertia, something that resembled the graphical representation of mathematical functions), ideas up to that moment linked with the names of Copernicus, Galileo, and Descartes. And Duhem was able to establish that a continuing tradition, derived from this early work, persisted down to the time of Galileo. This line of investigation has since been carried on by the eminent Austrian scholar Anneliese Maier, by the American philosopher J. H. Randall, Jr., and by the historian of medieval science Marshall Clagett and his students. Not only were the later Middle Ages shown to be more intellectually alive and scientifically creative—albeit by a radical minority of Aristotelian scholars—than had been believed, but the sharp division between the Middle Ages and the Renaissance was seen to be, if not a pure invention of the historians, at least greatly exaggerated. In a related domain, the history of technology, other scholars, among them Bertrand Gille in France and Lynn White, Jr., in this country, are persuading us that the technical inventiveness of medieval man lay back of that Renaissance exploitation of the machine that inspired the notebooks of Leonardo da Vinci and such amply illustrated books as Georgius Agricola's *De re metallica*, with its vivid portrayal of an early machine age.

So far, I have used the term *science* very loosely. Now that I have come to my second point, which has to do with the different "styles" of science discernible at various periods of time, I ought to clarify my use of the word.

We all of us can agree to use the word *science* to mean the study of the natural world. In this loose sense—of a rational, ordered effort to study nature in her own terms, not in terms of illustrative myths or religious tradition—the Greek philosophers were the inventors of science. Yet what we mean more commonly and more narrowly is what I shall call "modern science": that very peculiar, immensely successful way of studying nature that Western man hit upon, and in a measure codified, only about three and a half centuries ago.

Between this modern science—product of Butterfield's "scientific revolution of the seventeenth century"—and the way the Greek or medieval scholar studied nature, there is a world of difference. The divergence reflects, I believe, not only differences in sophistication and in knowledge, but differences of objective, of attitude, that reflect the values of the particular age. In other words, a difference in style.

Neither the Greeks nor medieval man conceived of a single, more or less unified discipline for seeking to comprehend, let alone control, nature. Aristotle, for example, saw man's knowledge to be of two dissimilar kinds, the concerns of different kinds or classes of people. There was *techne*—the practical, knowledge of the artisan, the navigator, the architect, and the physician—and there was *episteme*, or theoretical knowledge, a word the Latins translated as *scientia*, science. It meant, not just the ordered study of nature but any well-reasoned, well-ordered, well-disciplined system of knowledge whatsoever. Theology, in the Middle Ages, as we all know, was called the "Queen of the Sciences."

Much that we often speak of as Greek science was not *episteme* or *scientia* at all. When we examine it closely we find that what we should describe as scientific inquiry is broken into several distinct disciplines, carried on by different professions. Astronomy, optics, and mechanics were the domain of the mathematician, and like medicine, were often thought of as *technai*, carried on by specialists. Natural history, and surely botany, were ancillary aids to medicine or agriculture.

Episteme, on the other hand, was the preserve of the philosopher, whose fundamental science of nature was *physice*, the word from which we derive "physics," but which we ought to translate as *natural philosophy*. Much of what we would call physics (the optics of Euclid, the mechanical investigations of Archimedes) lay outside *physice*, indeed by Aristotle mathematics was peremptorily excluded from it.

In Aristotle's great system of knowledge his "physics" was a kind of foundation stone, and it cannot be viewed apart from his other writings: on rhetoric, politics, ethics, and aesthetics. Much of this physics strikes us as a tortuous verbal exercise: it was in fact a sort of grammar of nature, in which grammatical and subjective categories were projected upon the world around him. Furthermore, Aristotle employed, in dealing with the physical world, what has been called the "animate model." Nature, for him, was almost a living thing: the inanimate flows into the animate; the differences between them are essentially differences in degree. The universe, like man and the societies he forms, has goals, is purposive.

All this, if we know anything about Greek culture, was to be expected. After Socrates, at all events, the primary preoccupation of the Greek thinker was with man, his communication with his fellow citizen in the *polis*, and with those arts, including the art of citizenship, which make for the happiness of man. Aristotle's works on physics and cosmology prepared man for citizenship in the greater *polis*, showed man a meaningful, purposeful cosmos in which he could be at home.

These tendencies were, if anything, enhanced when the medieval scholar fell heir to the massive corpus of the Aristotelian writings and set to work to make this philosophy the foundation of a Christian life.

This natural philosophy, whether of Classical Antiquity or the Middle Ages, had a design far different—and more elevated, perhaps—than the study of nature for her own sake. It was part of the greater enterprise of understanding *both* man and nature, or perhaps I should say of seeing man *in* nature. For the medieval scholar, the natural order is the great theater in which is enacted the recurring drama of human life and death, and the greater drama of salvation. For this reason, mainly, natural philosophy had its important place in the system of scholastic thought. To the scholastics, as to Aristotle, the cosmos was eminently hospitable: a closed, tiny universe of nesting spheres, embracing and focused upon a fixed and central earth. It was not inert and lifeless, but richly contrived and animated with meaning and purpose, indeed with something very like man's own aspirations and values. Upon this animate model, and the physics and the cosmology of Aristotle, Saint Thomas Aquinas could erect a psychology, an ethics, a political theory, and a Christian theology. It was through this ordered cosmos that Beatrice, in Dante's great poem, led the poet upward through the spheres to the vision of God. Beatrice was not being a pedant or blue-stocking when, before the journey to Paradise, she tutored the poet rather unmercifully in physics and astronomy. She was merely being practical, for these sciences were keys to Heaven.

It is this subordinate, propaedeutic role, no less than the dogmatic assurance with which it was taught, that gives medieval natural philosophy its closed and fixed character, its status as doctrine to be elaborated, clarified in scholastic debate, but—since so much depended on it—to be staunchly defended. Scientific knowledge, we may conclude, was less valuable for itself than as part of a great cathedral of knowledge, natural and revealed: it was the rational keystone of a beautifully articulated view of man, his universe, and his destiny on earth and in Heaven.

Science as we have come to know it—a free, autonomous, and open-ended inquiry into nature—was possible only when this bond of subservience to higher ends was severed. This was achieved, not without struggle, by men who lived and worked from the middle of the sixteenth century to the end of the seventeenth. We now call this age, after Butterfield and others, the Age of the Scientific Revolution, or (like Whitehead) the Century of Genius. And in so doing, we have recognized and labeled an historic event no less momentous in its consequences than the rise of Christianity.

How this New Science arose, what was its special appropriateness to the age that spawned it, how its "style" reflected the tendencies of that age: these are complex and difficult historical questions. We sometimes read that this New Science, with its startling discoveries, its revolutionary method of inquiry, was the principal agent, the efficient cause, of those great transformations that prepared and furnished the modern mind: the collapse of medieval unity, the new spirit in religion, the substitution of new critical philosophies to replace the Aristotelian and scholastic synthesis. I find this completely naïve and unsatisfactory. And quite obviously it leaves unanswered, and perhaps unanswerable, the question of where the New Science itself came from, how it arose, and why it was so swiftly accepted.

I prefer to look at the problem the other way around, and to ask whether science might be better seen as a product, a necessary result, of all those historic forces that marked a break with the medieval past: the voyages of discovery, the demands of a new and more representational form of art, the practical needs of warring national states, and perhaps most important, all the new intellectual options made available during the fifteenth and sixteenth centuries by the rediscovery of such ancient philosophic doctrines as those of the Greek skeptics, the Stoic philosophers, the ancient atomists, and—last but not least—Plato and the Neo-Platonists. The advent of these rivals did more, perhaps, than science, and earlier than science could have done it, to weaken the allegiance to Aristotle. Most significant of all, I firmly believe, was the startling revelation during the Renaissance of those Hellenistic achievements in mathematics, medicine, and astronomy which the Middle Ages had either not known or had neglected: Ptolemy and Galen, uncontaminated by transmission through the Arabic; Apollonius of Perga; and lastly, Archimedes, whose advent was like the discovery of a new world.

What I am suggesting is that the New Science was simply the natural philosophy appropriate to a new, fact-minded age of competitive nation-states, of burgeoning capitalism, of religious fragmentation, and of worldly, rather than other-worldly, considerations.

The New Science was, as Bacon put it, "useful knowledge," knowledge "for the relief of man's estate," knowledge which, though it had to be acquired piecemeal, would surely lead to still more knowledge. A *scientia activa*, rather than a *scientia contemplativa*, it would be the work of many hands. Galileo wrote "Let us confess quite truly 'Those truths which we know are very few in comparison with those we do not know.'" And his contemporaries were wont to quote the lines from the Book of Daniel: "Many shall run to and fro, and Knowledge shall be increased."

The peculiar feature of the New Science was that it worked, that Nature herself responded to it. Certain questions were no longer asked; instead new questions were asked of Nature to which she was forced to reply. Thorstein Veblen, the philosopher-economist who wrote so wisely about technology and modern industrial society, once commented upon the mysterious fit between the new ideas about Nature and Nature's responses. How, he asked, did men of the seventeenth century learn to ask the appropriate questions? Veblen's answer, though it is oversimplified, is still worth pondering; even as developed by others it is only a part of the story. But it is something like this. Instead of assuming that the natural world should be a sort of projection of the human personality, with purposes and desires like his, instead of applying the "animate model" to inanimate nature, the men of the seventeenth century came to conceive of the world as ultimately composed of particles of matter in motion, and of nothing else. The causes that operate in Nature—even in what is remote and hidden from view—must be purely mechanical: they must be like the pushes and pulls at work in the water mills, the pumps, tilt hammers, and clockwork machinery of late Renaissance industry. Not, as with Aristotle, like the drives and compulsions of human beings. The uniform behavior that was to be described, the laws of nature, should be less like the normative injunctions of a lawgiver than like the practical rules of the artisan. Drawn from that aspect of Nature which men, out of their accumulated experience in the world of industry and the crafts, had come to know best, these principles, applied elsewhere in the order of Nature, did indeed fit and work.

I need not add that the price of this pragmatic success was a steep one. This mechanization of the world picture—to use the title of a recent and splendid book—meant eliminating from among the questions that men asked some that seemed most important of all. To plan a valid experiment, to apply mathematics to Nature's countinghouse, meant assuming that the significant realities which concerned the scientist were only those things which, in principle or in fact, can be measured and expressed in mathematical symbols.

The New Science of Galileo, which Newton forged into a great new System of the World, deprived man in great measure of his illusory bond with nature. The underlying realities, or primary qualities (as Locke was to call them), must be embodied in particles or larger masses of matter, at rest or in motion in empty Euclidean space. All else—everything that gives richness, texture, and value to our everyday world of experience—the colors, sounds, odors which had real substantial existence in Aristotle's universe, lie only within us. They were seen to be, as we now say, purely subjective: interior and private responses to the stimulus of moving particles of inert matter. It is perhaps not surprising, as we are beginning to learn from a study of Newton's unpublished manuscripts, that to Newton himself such a world was one in which he could not really believe; or rather, as he confided in his private musings (and when Newton mused, he always scribbled), such an image of the world was far from exhausting reality.

I have said that every age has its own kind of image of the world, its own natural philosophy, mirroring the deepest aspirations and affinities of the age. And I have suggested that one way to look at science, at early modern science, at science perhaps down to our own time, is to view it as the natural philosophy of the new age that dawned with the Renaissance. If so, it must express the same values, the same "modern" character, that we encounter in political thought, in philosophy, in art, and in religion. It is a point worth remembering that the New Science flourished with special vigor in such active, urban, middle-class cultures as we find in England and in Holland and that it prospered particularly, when one looks at Catholic Italy, in the liberal, forward-looking manufacturing and trading centers: in Venice and in Florence.

Many scholars have argued, and not without some persuasiveness, that the Protestant ethic—and in England the Puritan and latitudinarian strain—was particularly responsive to science: that Protestants and men of Puritan persuasion contributed to the New Science to an extent out of proportion to their numbers. Whether this is so or not, I shall not argue. But it is a simple fact of chronology—and hardly a coincidence—that the first enunciations of the doctrine of religious independence and toleration, the first strong symptoms of political liberalism, arose *pari passu* with the rise of the New Science. When, towards the end of the seventeenth century, the pioneers of the New Science attempted to explain and justify their work and the manner of their doing it—when, in other words, they first articulated what we would call their "ideology"—the image they projected echoed their deepest social values. The crucial document is that remarkable book, Thomas Sprat's *History of the Royal Society,* a defense of that newly formed scientific Society and of the New Experimental Philosophy.

Two sorts of men, Sprat told his readers, had criticized the newly founded scientific Society: some "over-zealous Divines" who considered natural philosophy a carnal knowledge "and a too much minding wordly things" and some men of business who saw in it "an idle matter of Fancy, and as that which disables us from taking right measures in human affairs." Sprat, a liberal clergyman and Fellow of the Society, based his defense primarily upon showing what manner of men make up the Society, how they conduct their work, and what principles and values impel them. Their purpose, he writes:

> . . . is . . . to make faithful Records of all the Works of Nature or Art, which can come within their Reach; so that the present Age, and Posterity, may be able to put a Mark on the Errors, which have been strengthened by long Prescription; to restore the Truths, that have lain neglected; to push on those, which are already known, to more various Uses; and to make the way more passable, to what remains unrevealed. This is the Compass of their Design.
>
> And to accomplish this, they have endeavour'd to separate the Knowledge of Nature, from the Colours of Rhetorick, the Devices of Fancy, or the delightful deceit of Fables. They have labor'd to inlarge it, from being confined to the Custody of a few, or from Servitude to private Interests. They have striven to preserve it from being over-press'd by a confus'd Heap of vain and useless Particulars; or from being straightened and bound too much up by general Doctrines.
>
> They have tried to put it into a Condition of perpetual increasing, by settling an inviolable Correspondence between the Hand and the Brain. They have studied, to make it not only an Enterprise of one Season, or of some lucky Opportunity; but a Business of Time: a steady, a lasting, a popular, an uninterrupted work.

Into their Society, Sprat goes on, they have freely admitted men "of different religions, countries, and professions of life. . . . For they openly profess not to lay the foundation of an English, Scotch, Irish, Popish, or Protestant philosophy, but a philosophy of mankind." If the majority of the members are "Gentlemen, free and unconfin'd," this very fact ensures a free and equal exchange, and forbids paltry considerations of profit. Yet the doors are open to merchants and shopkeepers, for among the Fellows, Sprat reminds us, is John Graunt, a humble artisan we remember as a founder of the field of vital statistics.

The meetings of the Society are pervaded, Sprat insisted, by a spirit of democratic equality and mutual toleration; visitors are impressed by the gravity, plainness, and calmness of their debates. They "enjoy the benefits of a mix'd Assembly," Sprat declares, "which are Largeness of Observation, and Diversity of Judgments, without the Mischiefs that usually accompany it: such as Confusion, Unsteadiness and the little Animosities of divided Parties."

For Thomas Sprat, then, the strongest argument for the New Experimental Learning, and the body which promoted it, was to be found in the unique professional values which its pursuit imposed: dedication to truth and evidence; freedom of expression and inquiry; tolerance and equality in discussion. These were the values which, in another sphere, Sprat's fellow member, the philosopher and political theorist John Locke, would seek to define as the most precious values of political life.

These were the values that, I believe, generally prevailed in the scientific community during much of the eighteenth century and the nineteenth century as well. Yet gradually, as society changed, the character of scientific inquiry changed with it. The men of the French Enlightenment came to stress the general intellectual, especially the creative social, import of the spirit and method of science. For them it appeared, in the words of Adam Smith, as "the great antidote to the poison of enthusiasm and superstition." And they firmly believed that the critical, factual method of science—what Voltaire, following Newton, loved to call the "method of analysis"—could be applied to an examination and improvement of social and political life. Scholars have noted, for example, that after the middle of the eighteenth century, more and more emphasis is placed on the *application* of the findings of science, on its practical social utility. With the flowering of the Industrial Revolution, and surely by the end of the nineteenth century, an almost exclusive weight came to be placed upon the utilitarian value of science. Scientific progress more and more seemed to be justified by the fact that it worked so well, by the innumerable practical applications that flowed from its discoveries: the electric telegraph, gas lighting, the preservation of foods, the trans-Atlantic cable, the striking improvements in medical practice, to mention only the earliest. Meanwhile science itself became increasingly complex and specialized, more and more a congeries of esoteric arts, a bundle of techniques. Physical science, in particular, grew still more abstract and mathematical, still farther removed from the comprehension of the average educated man. Yet the more incomprehensible and abstract it became, the more powerful it seemed to be. It was no longer a natural philosophy in the older sense: a view of the world that man could understand and appreciate.

I have suggested that modern science, as it flourished, expanded, and deepened from the late sixteenth century until the end of the nineteenth century, can best be described as the characteristic natural philosophy of the liberal, competitive era of early modern history; and that it was no historical coincidence that our deepest political beliefs—our ideals of reasonableness, freedom, tolerance, respect for the individual, even our principles of representative government and democracy—grew up step by step with the rise of modern science. Their histories are parallel and intertwined. If there was a basic harmony between our older liberal, democratic tradition and the professional values and attitudes of early science, it was because both expressed in their own fashion the underlying aspirations and goals of early modern man.

Yet, as I pointed out, the character of scientific endeavor has changed, and is changing, in step with the great mutations of modern society. It has become increasingly difficult to view the enterprise as a whole, and today we hear less than we might wish of the purely intellectual and humane values of science, and more about the professional opportunities it affords or the useful results of its application to practical problems. In our highly technologized society, science too has become technologized, and has fallen out of touch with its own great tradition, with its essentially humanistic past. Perhaps already a new kind of science is emerging in which those old values must recede into the background as echoes of an earlier age. But I am not proposing that we lament a simpler and more idealistic past. Science is called upon to play its essential part in a vastly more complex society of giant corporations, great labor unions, top-heavy bureaucracies, and heavily armed Leviathan states. It may have to change, because the world has changed. Science, as men once knew it—a leisurely, free, and essentially humane investigation into nature—may prove to have been an admirable, but ephemeral, episode of early modern history.

5

A Backward View

❖

Historians of science are wont to insist, somewhat defensively, that as an approved academic subject their discipline is new and, in various quarters, suspect: a parvenu among the fields of history; less sophisticated and presumptuous than the philosophy of science; less timely than the sociology of contemporary science; and, of course, less glamorous than science itself. Yet it is not so new, after all, for it penetrated academe more than half a century ago at the University of London and at Harvard. Still earlier, from 1892 to 1903, Pierre Laffitte, a disciple of Auguste Comte, taught the history of science in Paris "with dignity but without distinction," as George Sarton once wrote. Indeed, if we may speak of *histories of the sciences*, rather than the *history of science*, the subject is nearly as old as the study of nature itself. And this is perhaps the way to look at the matter, at least at first, for in our language the word *science*—implying a fundamental unity of outlook and general method in the sciences of nature—was a neologism of the early nineteenth century, echoed reluctantly, if at all, in other languages.

Aristotle's Lyceum produced the first histories of the sciences; aside from much scattered information in Aristotle's own writings,[1] there was the *History of Geometry* of Eudemus of Rhodes, and Theophrastus's *History of Philosophy*, an account of Greek thought from Thales to Plato. Both of these works, products of the Lyceum, are lost, but some of the substance is preserved by the doxographic tradition. Bits of historical lore, derived from sources long since lost, are found in Cicero's *De natura deorum*, in the *De re architectura* of Vitruvius, in the elder Pliny's *Historia naturalis*, and in Diogenes Laërtius's (third century A.D.) *Lives of Eminent Philosophers*, a work we are always told to distrust, yet which everyone seems to cite nonetheless. It is from Plutarch, in his life of Marcellus, that we glean our most intimate glimpse of Archimedes, that greatest figure of ancient exact science. Yet we can point to no genuine history of a science written in the centuries of Roman domination. Later, the Semitic Middle Ages did better than the Latin West, which had minimal understanding of the past and, until the later centuries, little scientific curiosity, or at least little to show for it. Yet from the

This is a conflate text combining a paper read in Paris in 1950: "Rapport de M. H. Guerlac [sur l'histoire des sciences]," *IXᵉ Congrès International des Sciences Historiques*, 28 août–3 septembre, I. *Rapports* (Paris: Armand Colin, 1950), 182–211, with a paper in the *Times Literary Supplement*, no. 3764 (Friday, April 26, 1974), pp. 449–50.

East came that remarkable bibliographical conspectus, the Kitab-al-Fihrist, which is the source of much that we know about early Islamic science—for example, alchemy.

As for the Renaissance centuries, despite their love affair with antiquity, and the recovery, editing, and translating of the surviving scientific works of Greece and Rome,[2] they produced little beside that odd compendium, at once sketchy and wide-ranging, of social, intellectual, and technological "inventions," the *De rerum inventoribus* of Polydore Vergil (1499).[3]

The seventeenth century—not surprisingly, since it saw so many scientific advances that one speaks of a "scientific revolution" and the birth of modern science—showed a keen interest in the origin and growth of the sciences. It yielded, to cite some worthy examples, histories of chemistry by Olaus Borrichius in Denmark; John Wallis's history of algebra (1685),[4] which has been called the first serious attempt in England to write the history of mathematics; and the earliest single volume devoted largely to the histories of the sciences: William Wotton's *Reflections upon Ancient and Modern Learning* (1694).[5]

Written by a young English divine, famous in his day for his precocity and learning,[6] Wotton's book was a pioneer attempt to evaluate the scientific advances of his century and to compare its achievements with those of antiquity. It was a tract for the times, a product of the famous *Quarrel of the Ancients and Moderns* (Swift's "Battle of the Books") and specifically a reply to Sir William Temple's essay *Of Ancient and Modern Learning* (1690). Temple lauded the ancients at the expense of the moderns and did not disguise his distaste for the New Learning and its institutional embodiment, the Royal Society of London. Wotton's *Reflections*, on the other hand, devotes its thirty chapters to the principal fields of human knowledge, with special attention to the sciences of nature.[7] The chapters, to be sure, are of uneven quality, but it was in the recent triumph of modern science that Wotton found the most effective reply to those who asserted the unquestioned superiority of ancient Greco-Roman culture. He surveys the whole range of the natural sciences but treats more intensively and with most success the recent progress in natural history and what we would call medical biology. The most remarkable chapter is the one in which, treating William Harvey's discovery of the systemic circulation of the blood, this young English scholar called attention to the partial anticipations in the writings of the Spanish physician Michael Servetus and of the Italians Realdus Columbus and Cesalpinus. The tone of Wotton's book is moderate, and he treats fairly and objectively the major scientific accomplishments of the ancient world. But he makes it clear—in pages like those dealing with the modern developments in anatomy, and in the chapter entitled "On the Several Instruments Invented by the Moderns, which have Helped advance Learning"—how superior in many respects the modern world is to antiquity.

The best fruit of the next century's great interest in the history of science, taken to be the most vivid exemplification of the great principle of universal progress, is not to be found in the sweeping inaccuracies of Condorcet's *Esquisse,* nor in the pages of the *Encyclopédie*, but in the practice of writing histories of the several sciences, as much for the benefit of the specialist as for the contemporary reader of broad culture, which is characteristic of the best works of the Enlightenment. Even today one cannot ignore Montucla's *Histoire des mathématiques* (2 vols., 1758), which treats not only pure mathematics but also the exact sciences of mechanics, optics, physical astronomy, acoustics, etc.; Joseph Priestley's *History and Present State of Electricity* (1767; second edition 1769) or his equally valuable *History of Vision, Light and Colours* (1772). And still indispensable to the scholar are the volumes on the history of astronomy by J.-B. Delambre (1749–1822), published shortly before the death of this distinguished scientist.[8]

At the University of Göttingen, late in the eighteenth century, there emerged the practice, if not an actual policy, of accompanying instruction in each science by lectures on its history, or on what we should call its "literature." This brought forth several useful (and heavily bibliographical) reference works, among them J. F. Gmelin's history of chemistry (1797–99), A. G. Kästner's history of mathematics (1796–1800), and J. K. Fischer's *Geschichte der Physik* (1801–8), which deserve mention alongside a more original work, Johann Beckmann's *Beyträge zur Geschichte der Erfindungen*, published in five volumes between 1786 and 1805. This pioneer historian of technology included much historical material in his lectures in *Oeconomy*, to the annoyance of the more practically minded students. It was with reference to Beckmann's course that one critic complained in 1805 that instead of learning economics and the skills needed to administer a modern state, one heard only about agricultural products and mechanical inventions.

The nineteenth century was, as we all know, enamored of history and of the genetic, as opposed to the purely theoretical, approach to problems. One would expect, accordingly, that it would have been quick to surpass the sort of achievements we have just described. Rather surprisingly it did not. The book often singled out as the first modern history of science is the *History of the Inductive Sciences*, by that formidable, contentious Master of Trinity, William Whewell. Published in 1837, the year Victoria came to the throne, Whewell's book, written to provide an historical basis for a philosophy of the sciences along Baconian lines, remained the favorite and most quoted history of science until well after Victoria's reign. Yet by modern standards it is an inferior work, an assemblage of histories of the separate sciences, some—like "thermotics" and "atmology"—christened with characteristically Whewellian names and classified according to a highly artificial scheme. The book is a mere calendar of scientific facts and their discoverers; as Whewell warned in his preface, *ideas* were reserved for a companion work, his *Philosophy of the Inductive Sciences*. Based wholly on often unreliable secondary sources, his *History* was out of date even as it was composed.[9]

During the rest of the nineteenth century progress was distinctly uneven, at least until the last two decades. We encounter a number of scientific antiquarians, whose love of rare books and manuscripts led some of them into grave temptation, like the notorious and light-fingered thief, Guillaume Libri, whose *Histoire des sciences mathématiques en Italie* (1834–41) is something of a landmark, and James Orchard Halliwell, founder of the Historical Society of Science, which died aborning (1841) and had on its council enthusiastic amateurs like the Rev. Baden Powell, Oxford's Savilian Professor of Geometry, and the mathematician Augustus de Morgan, author of that mine of scientificohistorical anecdotes, the *Budget of Paradoxes*, as well as of sober contributions to the bibliography of mathematics.

Charles Daremberg should be mentioned as a distinguished pioneer in the history of medicine, who wrote with breadth and insight studies that were portents of better things to come. But the typical products of the time are the often monotonous, factual accounts of the special sciences, among which should be mentioned Heinrich Haeser's *Lehrbuch der Geschichte der Medizin und der epidemischen Krankheiten* (3 vols., 1845), Hermann Kopp's classic *Geschichte der Chemie* (1843–47), August Heller's *Geschichte der Physik* (1882–84), J. V. Carus's *Geschichte der Zoologie* (1872), and Julius von Sachs's *Geschichte der Botanik* (1872), some of which one can still consult with profit. Yet in Germany the general reader was likely to draw his stock of information from the voluminous writings of J. H. M. Poppe, whose works on the history of technology and the exact sciences left their impress on Karl Marx; or, in France, from the encyclopedic, tireless, and often inaccurate pen of Dr. Ferdinand Hoefer (1811–78), whose compendious histories, not only of his own science of chemistry but also of physics, astronomy, zoology, botany, mineralogy, geology, and the mathematical disciplines were widely popular.

These men of the nineteenth century were, for the most part, annalists of science, not really historians; though they often boasted a vast time scale ("depuis les temps les plus reculés jusqu'à nos jours"), they did not relate their chronicles to the sister sciences, or to the stream of cultural history. They neither painted broadly nor penetrated deeply; they deployed the discoveries in the several sciences in monotonous single file, unleavened by historical imagination or philosophic insight. Nevertheless they fashioned some of the raw materials for a richer and more critical approach to the subject. Specialized studies devoted to a single period or phase of the history of science were uncommon, although scattered articles of importance on specific problems can be found in scientific and learned publications, and in ephemeral journals, like the *Janus* (1846–48 and 1851–53) of Henschel and Bretschneider in the history of medicine, or the *Bulletino di bibliografia e storia delle scienze matematiche*, edited between 1868 and 1887 by Prince Baldassare Boncompagni. The *Bulletino* was sufficiently rich in valuable source material to have been reprinted in our own time.

Although the early and middle decades of the nineteenth century assembled much useful material, especially for the history of early modern science, intensive critical work was rarely carried out for any single period, or any broad aspect, of the history of science; and whole areas, like Classical Antiquity and the Latin Middle Ages, were only superficially treated, or were used to provide a contrasting background against which to display the triumphant entry of modern science. Islamic science, like the science of the preclassical Middle East, was virtually a closed book. The scientific complacency of the nineteenth century, its faith in the absolute perfection of scientific laws, was reflected in contempt for the errors and wild speculation of earlier times; and this in turn solidified the sense of superiority and the simplistic trust in the philosophical impregnability of their world view. In characterizing the Middle Ages as a "stationary period," marked by "indistinctness of ideas," "dogmatism," "mysticism," and the "commentatorial spirit," William Whewell shows a prejudice that parallels that of Auguste Comte who, though a force to reckon with in the history of science movement, did not display the requisite historical detachment and empathy needed to understand the predominantly "theological" and "metaphysical" periods of human thought.[10] This defect is even more grave than the one we have already mentioned: thinking of the history of science as only the sum of the several sciences, each in turn pictured, in deference to the rapidly growing practice of scientific specialization, as an essentially self-contained and independent discipline.

On the other hand, in the last third of the century we note a steady rise in scholarly and professional standards, exemplified by the studies of alchemy by Marcellin Berthelot and E. O. von Lippmann, and the brilliant investigations of Greek exact science by J. L. Heiberg, Sir Thomas Heath, and Paul Tannery. The history of mathematics received detailed and superior treatment at the hands of Moritz Cantor in Germany and H. G. Zeuthen in Denmark.

An important development in these years was the publication, under governmental auspices, of the collected works of the great creative figures of early modern science. I have in mind the publication in the years 1864–93 of Lavoisier's *Oeuvres*, begun under the editorship of J. B. Dumas and continued by Lavoisier's biographer, Edouard Grimaux. There followed (1888–1950) the Dutch National Edition of the works and correspondence of Christiaan Huygens and Antonio Favaro's great edition of the works of Galileo.

The philosophic impulse was in many respects central to the emergence of the history of science movement, for the appreciation of the methodological unity of science was a clear prerequisite. In the early years of the twentieth century, the men who tried to transform the history of science into a reputable scholarly discipline sought to remedy its defects and make it both broader and deeper. It was clearly necessary to invigorate the study anew and to reconstitute it radically in accordance with the accepted standards of philological and historical investigation. But did this collection of separate specialties have enough in common to enable us to speak of *a* history of science? We have seen that Comtian positivism was not conducive to a sympathetic investigation of the less obviously productive periods of scientific development; yet it is to Comte himself and his *Cours de philosophie positive* that we owe the unified approach and the point of view towards the sciences that made the new movement possible. Two key ideas—in fact the two root ideas of Comte's great scheme—pointed towards the creation of a history of science. The first was the notion of the unity, or unifiability, of the sciences: that it is possible to view all the fields of science in terms of a common method, a common spirit, and to a degree a common origin; the second, that the true history of mankind is the history of the human intellect. Since in his well-known "law of the three stages" the scientific or positive phase of the human intellect seemed to Comte the highest and most significant level of man's development, the history of science took on commanding interest. His classification of the sciences rests in fact upon an elaborate genetic theory of their sequential emergence; to Comte the unity of science was best illustrated (inductively) through the historical development of the sciences seen as related to each other like the twigs and branches of a great tree. In 1852 Comte actually applied to Guizot for the creation of a chair devoted to the *Histoire générale des sciences*.

Paul Tannery, and after him George Sarton, have readily admitted the influence of Comte upon the history-of-science movement and upon themselves; Tannery has even confessed that a reading of the *Philosophie positive* first aroused his sense of the importance of the history of science. But it should not be overlooked that a critical philosophy of science, inspecting seriously for the first time the assumptions of nineteenth-century scientific thought, and puzzling over the growing contradictions within physics, came into being at roughly the same time. The writings of Boutroux, Poincaré, and Ernst Mach deeply stirred the intellect of Tannery's generation and confirmed the need of an historical approach and the possibility of treating the sciences from a fundamental and unified point of view. This new philosophy of science was itself deeply influenced by the Comtian historical approach. At the hands of Pierre Duhem and Ernst Mach the historical method became a powerful instrument of philosophic criticism just as their writings proved of profound importance for the emergence of the history of science. The lines between the two movements were never sharply drawn. The group which in France assembled around Tannery in his efforts to organize the history of science included historically minded philosophers like André Lalande, Emile Meyerson, Gaston Milhaud, Léon Brunschvigg, and Abel Rey. These men took an active part in the history-of-science movement, writing in some instances more like historians than philosophers.

The central figure in the history-of-science movement—that autodidact of genius, as Wilamowitz called him—was an engineer employed by the French government tobacco monopoly, Paul Tannery (1843–1904). His claim to this central position has a threefold justification: he made contributions of fundamental importance to the whole range of scientific history, in particular lending depth and understanding to our knowledge of Greek mathematics; he defined the field of the history of science in a manner that gave it scope and significance; and finally he took the first steps to organize it.

As the many volumes of his *Mémoires scientifiques* testify, Tannery's great contribution was in the form of technical articles, scattered over nearly thirty years, through a variety of different publications. In 1887 he brought together related articles to make up his first two books, *Pour l'histoire de la science hellène* and *La Géométrie grecque*. In the former he treats the Pre-Socratic thinkers not from the standpoint of philosophy alone, but as *physiologoi* or as philosophers of mathematics. Here his greatest contribution was to demonstrate that the paradoxes of Zeno the Eleatic were not idle sophisms but arguments of the very greatest importance for the foundations of mathematics. In the *Géométrie grecque* he made a very careful study of the doxographic tradition in order to evaluate as a source the commentary of Proclus on Euclid. Having satisfied himself as to the general value of Proclus's testimony, Tannery then used the Commentary to reconstitute the history of Greek mathematics before Euclid, showing how, book by book, the *Elements* could be understood as the rigorous formulation of the work of earlier hands, and the epitome of a mathematical tradition dating back to Pythagoras. Most of this edifice of historical reconstruction still stands, corrected here and there by the hands of men who owed their inspiration to the tradition of Tannery, by Zeuthen, Heiberg, Heath, Enriques, Loria, and others. In the *Astronomie ancienne* (1893) Tannery performed a similar feat of historical analysis by showing the extent to which the great achievement of Hipparchus—upon which Ptolemy relied so heavily—stems in great part from an astronomical tradition running back to Eudoxus and the Platonic Academy.

Tannery's competence was not confined to antiquity. He wrote on Byzantine science, the Latin Middle Ages, and aspects of seventeenth-century mathematics. He prepared the critical Teubner edition of Diophantus (Leipzig, 1893–95, 2 vols.); he was entrusted by the Ministère de l'Instruction Publique with the preparation of a definitive edition of the writings of Fermat (1891–96); and with Charles Adam he undertook the great edition of the works of Descartes that he did not live to see completed. Tannery never realized his ambition of writing a general work in this history of science, but his ideas on the development of science, his notions of what the *histoire générale des sciences* should be, can be found in certain of his less technical and specialized writings: his numerous contributions to the *Grande Encyclopédie*; the chapters he wrote on science since the thirteenth century for the *Histoire générale* of Lavisse and Rambaud; and his articles on the histories of the exact sciences, published in the newly created *Revue de synthèse historique* between 1900 and 1905.

Under Tannery's leadership there was held in 1900 in Paris the first international gathering of historians of science, organized as a section of the *Congrès international d'histoire comparée*, held in connection with the *Exposition Universelle* of that year. As President of the Section, Tannery assumed the work of organization and edited a small volume of memoirs read at the Congress. The contributors include a stellar array of those working in the field: André Lalande, the philosopher of science; Gaston Milhaud, author of excellent studies on Greek science; the Danish historian of mathematics J. L. Heiberg; Siegmund Günther, a leading German authority; Antonio Favaro, the great Galileo scholar; Moritz Cantor, author and editor of the monumental *Vorlesungen über Geschichte der Mathematik*; Gustav Eneström, editor of the *Bibliotheca mathematica*, and a number of others, including Tannery himself.

Throughout most of his career Paul Tannery was not a teacher but a technical civil servant whose prodigious scholarly accomplishment was the work of leisure hours. Only at the end of his career did teaching the history of science appear to be a possibility. In 1892 there was created at the *Collège de France* for the aged Pierre Laffitte, a positivist of the strict observance and the recognized heir of Auguste Comte, a chair of the history of science. Here in the spirit of the master, and within his limitations, Laffitte until his death in 1903 sought to present the history of civilization in the light of the progress of scientific knowledge. After Laffitte's death, as the candidate put forward both by the *Collège de France* and the *Académie des Sciences*, Tannery presented his credentials for the post. His appointment seemed a foregone conclusion, for there was nobody in France with equal qualifications. He set to work to plan a great course in the general history of science—which he expected to publish as the crowning achievement of his career—and completed an inaugural lecture in which he developed his view of it as a synthesis of the special histories of science along Comtian lines. Astonishment, incredulity, and a deep sense of outrage greeted the announcement late in 1903 that the Minister of Public Instruction had disregarded the weight of expert opinion and made what was widely considered to be a political, or at least an ideological, appointment. Paul Tannery did not long survive this shattering blow to his plans and to the future of the history of science. He died within the year.

Tannery's younger contemporary, Pierre Duhem, was over forty when he published the first of his historical writings, yet these were soon carried out with the same breadth and imagination we encounter in the case of Tannery. Scientist first of all—his chief contributions were in the new field of chemical thermodynamics—and only later a philosopher and historian of science, Duhem spent his whole career in provincial universities. Combative personal qualities, together with his political and religious views—he was a devout Roman Catholic in a France dominated by anticlericals—explain his failure to gain a chair in Paris: the goal, then as now, of most French scholars. His first historical works dealt with the evolution of scientific problems and concepts, rather than with the history of specific disciplines. In 1902 he published his *Le mixte et la combinaison chimique, Essai sur l'évolution d'une idée* and his *L'Evolution de la mécanique*. These works immediately preceded his major philosophical work, *La théorie physique, son objet et sa structure* (1906). Clearly, like other contemporaries—Ernst Mach, André Lalande, and Emile Meyerson—he found in the history of science support for his philosophy of science.

Between 1906 and 1913 he published three volumes that were to shift his emphasis and alter profoundly our ideas of medieval thought, his *Etudes sur Léonard de Vinci*. This study of Leonardo led him to a radical reassessment of the supposedly antiscientific Middle Ages and brought to light such generally neglected figures as Albert of Saxony, Jordanus Nemorarius, Jean Buridan, and Nicole Oresme. The work of these fourteenth-century figures, Duhem argued, not only foreshadowed the achievements of Leonardo and Galileo but anticipated as well the invention of Cartesian coordinates and the cosmology of Copernicus.

These clues were elaborated and placed in historical perspective in his monumental history of cosmology, Le Système du monde de Platon à Copernic (1913–59). If his conclusions now seem somewhat overblown, Duhem nevertheless deserves full credit for rehabilitating the later Middle Ages and picturing them as a time of scientific achievement. He was partly responsible, too, for the faddish assertion of some medievalists that there was really no novelty in the Renaissance, merely an elaboration of medieval themes. Even George Sarton, at one time, described the Renaissance as an antiscientific period, or an age of scientific sterility, sandwiched between the creative fourteenth century and the seventeenth century, the age of the so-called scientific revolution.[11]

If Tannery was the hero of Sarton (1884–1956), Sarton in turn, at least for a time, was the unchallenged guru of my own generation of historians of science. A Belgian who emigrated to the United States in 1915, and who after a few years took up residence at Harvard, he was powerfully influenced by the ideas of Auguste Comte and the example of Paul Tannery; like Tannery he conceived of his subject as "l'histoire générale des sciences," as the "history of science," not the "history of the sciences." In fact he broadened his definition of science to a degree that might have startled Tannery.

Hindsight tells us that Sarton will have left his mark primarily as one who supplied the scholarly equipment for the field and whose achievements were first and foremost those of a bibliographer and propagandist for the history of science. In 1913, he founded Isis, the first successful journal devoted to the subject and still, perhaps, the most influential and comprehensive. He published various prolegomena to the field as well as a bibliographical guide; and his lifework was the massive and indispensable Introduction to the History of Science (1927–48), which carried the story from antiquity through the fourteenth century. The last volume, published in two parts, is a monstrous thing, filling nearly 2,000 pages.

Sarton was at his best in writing persuasive lectures and delightful biographical essays about early scientists; he was less effective when he was defending the subject from imaginary (or real) antagonists. Perhaps his major intellectual contribution was his stress upon medieval Arab culture, and his insistence upon our scientific debt to it. This he set forth strongly in one of his Colver Lectures at Brown University, which were published as The History of Science and the New Humanism (1937), and in a lecture delivered at the Library of Congress in 1959. Although Sarton's Introduction deals exhaustively with the science of the Semitic Middle Ages, what he wrote was entirely based upon secondary sources. Although he read Arabic to some extent, he did no original monographic research nor did he edit a single Islamic scientific text.

Sarton had two weaknesses: he was not particularly concerned in a technical sense with the history of scientific ideas—personalities (and bibliography) interested him more; and he was genuinely opposed to the history of technology, as somehow, he believed, unworthy and unimportant for the growth of science. During the 1930s writers of a sociological or Marxist leaning began the attempt to integrate the history of science into the economic and social history of the early modern period. Several were popularizers or journalists, and the impact upon the main activity of historians of science was, with a single exception, minimal. This exception was Robert Merton's classic monograph, *Science, Technology and Society in Seventeenth Century England*, which was published in *Osiris* (another of Sarton's journals, designed for papers too long for *Isis* or, one suspects, otherwise unsuitable). Merton's paper has had a lasting, if for a time underground, influence, largely because the new trend was towards what has been called the "internalist" or "idealist" approach to the history of science.

A major stimulus to considering the history of science as the history of scientific *thought* came from the later writings of Alexandre Koyré. Born in Russia, where he received his early education, Koyré completed his studies in Paris under, among others, the philosophers André Lalande and Léon Brunschvigg. As a creative scholar, he centered his work for a time on the history of philosophy, notably the thought of Plato, Anselm, and Descartes. Perhaps the earliest work that attracted the attention of the learned world was his doctoral thesis, a book on the seventeenth-century mystic Jacob Boehme. Except for an early study of Zeno's paradoxes (1922), it was the philosophy of religious thinkers that first completely engaged Koyré's attention. His conversion to the history of science took place about 1932 when, as *Directeur d'études* at the Ecole Pratique des Hautes Etudes in Paris, he gave a course of lectures on science and faith in the sixteenth century.

Just what combination of influences led him back to what he later called his "first love," the history of science, is not easy to assess. But in personal conversation with this writer he once remarked that his reading of E. A. Burtt's remarkable book, the *Metaphysical Foundations of Modern Physical Science* (1925), played an essential role. And this is easy to understand since Burtt's book, dealing with the religious implications of scientific thought from Copernicus through Newton, presented, almost for the first time, a picture of early modern science as inseparable from the philosophical and religious currents of the sixteenth and seventeenth centuries. As Burtt abandoned the history of science for the rest of his career to teach the philosophy of religion at Cornell University, so Koyré moved in the opposite direction. In 1934, then a visiting professor at the University of Cairo, he published a translation of Book One of the *De Revolutionibus* of Copernicus, his first real contribution to the history of science. His next work, and the book which brought him fame among historians of science, his *Etudes Galiléenes*, appeared in Paris in 1940 just after the fall of France. Koyré saw in Galileo not the first master of experimental physics, as he was pictured by most nineteenth-century writers, but primarily a thinker deeply influenced first by those medieval writers rediscovered by Duhem and then by his admiration for the Archimedean texts, and finally reaching by his own intellectual genius the law of free fall and the principle of inertia.

In all he wrote, and he wrote admirably and with wit, Koyré remained close to the texts, illuminating his arguments with extended quotations. This approach he followed in his later years in a series of extremely important papers on Newton and what he called the "Newtonian synthesis." A Platonist and an idealist in his philosophical posture, he tended to underestimate the role of experiment, so exaggerated by the earlier historians of "inductive" science. Science to him was a sort of epiphenomenon, an independent activity moved forward by its inner resources, its built-in challenges and unsolved problems, and—except in the case of religion and philosophy—virtually isolated from the stresses and demands of the world. While for contemporary science he would admit the reciprocal inter-action of material culture and scientific thought, he doubted that this was a significant factor before our own time. He would probably not approve, or would view with characteristic irony, the popular direction the history of science seems to be taking: inquiries into, and a deep concern for, the relations of science, technology, and society.[12]

Notes

1. Harold Cherniss, *Aristotle's Criticism of Presocratic Philosophy* (Baltimore: The Johns Hopkins Press, 1935), and Werner Jaeger, *Aristotle: Fundamentals of the History of his Development*, trans. Richard Robinson, 2d ed. (Oxford: the Clarendon Press, 1948). It is in Aristotle that we find our earliest summary of the views of Pythagoras and his immediate followers, as well as a famous critique of the atomic theory of Leucippus and Democritus.

2. Useful in this connection is George Sarton, *The Appreciation of Ancient and Medieval Science During the Renaissance (1450–1600)* (Philadelphia: University of Pennsylvania Press, 1955). Sarton had earlier ignored this aspect of the scientific interests of the Renaissance, which he treated as an antiscientific period, a dip in the curve of progress between the fourteenth century and the innovative seventeenth. See his "Science in the Renaissance," in James Westfall Thompson et al., *The Civilization of the Renaissance* (Chicago: University of Chicago Press, 1929; reprinted New York: Ungar Publishing Co., 1959). Before this work was reprinted (Sarton died in 1954), he had retreated from this position.

3. A student of the mathematician Guido Ubaldo del Monte and of Federico Commandino, Bernardino Baldi (1553–1617), a specialist in mechanics, wrote lives of famous mathematicians, but they were not published, even in part, until long after his death. His *Cronica de' matematici*, an abridgment of his *Vite de' matematici*, was published in Urbino in 1770. Parts of the *Vite* were printed in Prince Boncompagni's *Bulletino* between 1872 and 1877.

4. See the historical preface of Wallis's *Treatise of Algebra, Both Historical and Practical* (London, 1685). The Latin edition of this work (1693) contains a treatise on the "Quadrature of Curves," the earliest printed account (and in Newton's own words) of the doctrine of fluxions.

5. A second edition (London, 1697) had appended to it Richard Bentley's famous exposure of the *Epistles of Phalaris*, a fraudulent work much admired by Sir William Temple and other defenders of the Ancients.

6. For Wotton's precocity see John Evelyn's *Diary* under the date of July 6, 1679. Wotton's father celebrated his child's linguistic ability in a pamphlet entitled *On ye Education of Children*, addressed to King Charles II.

7. A good account of Wotton's book is by A. R. Hall, "William Wotton and the History of Science," *Archives Internationales d'Histoire des Sciences* 28 (1949), 1047–62.

8. Delambre's *Histoire de l'astronomie ancienne*, 2 vols. (Paris, 1817), has been reprinted by the Johnson Reprint Corp. with an introduction by Otto Neugebauer;

I. B. Cohen has supplied the introduction for the reprint of Delambre's *Histoire de l'astronomie moderne*, 2 vols. (Paris, 1821), as Harry Woolf has for the *Histoire de l'astronomie au dix-huitième siècle.* The six reprinted volumes were published between 1965 and 1969.

9. On Whewell see George Sarton, *Horus: A Guide to the History of Science* (Waltham, Massachusetts: Chronica Botanica, 1952), 49–50.

10. A good illustration is the case of Comte's English disciple, G. H. Lewes, whose *Aristotle: A Chapter from the History of Science* (1864) is a scathing and unbalanced attack on Aristotle's worth as a scientific figure. A similar "positivism" pervades, and to some extent also perverts, that extremely learned but contentious work, Andrew D. White's *History of the Warfare of Science with Theology* (1896), which is perhaps the first important book on the history of science to be published in the United States. An exception to this attitude towards antiquity among positivists is the work of Littré, whose great edition of Hippocrates is one of the monuments of nineteenth-century scholarship in the history of science.

11. See n. 2.

12. See Koyré's amusing reply to my paper read at the Oxford symposium on the History of Science in 1961: "Commentary by Alexandre Koyré," in A. C. Crombie, ed., *Scientific Change* (New York: Basic Books, 1963), 847–57. For an eloquent account of Koyré and his work—and there is none better—see the sketch by Charles C. Gillispie in the *Dictionary of Scientific Biography*, vol. 7 (1973), 482–90.

II
NEWTONIAN SCIENCE

6
Newton's Changing Reputation in the Eighteenth Century

❂

Few pieces of historical writing by American historians of the older generation have left a more profound mark upon our own than Carl Becker's *The Heavenly City of the Eighteenth-Century Philosophers*. And few works that have won this position of influence seem to demand for their intelligent comprehension as much understanding of the author's temperament and individuality as does this series of lectures. I knew Professor Becker far less well than many who are contributing to this symposium; I was never his student, to say nothing of being his colleague. Yet having grown up in Ithaca when Becker was in his prime, I met him on numerous occasions, heard him talk, and got some feeling for the temper of his mind. It is difficult for me to exclude these impressions from what I think about *The Heavenly City*, so perhaps I should begin by recalling them.

I think I met Mr. Becker first in my humble capacity as youthful cupbearer to small gatherings of historians that my father, who taught French at Cornell in Becker's time, used to assemble in his library. This group often included George Lincoln Burr, Preserved Smith, our neighbor Charles Hull, and, of course, Carl Becker. As my reward for buying and distributing the Swiss cheese and the near beer—for those were prohibition days—I was allowed to sit in a corner and listen to the talk, with firm instructions from my father to watch for emptied glasses and to maintain a proper silence. Much of the conversation turned on history, and much of it—needless to say—was beyond me. But the talk, unlike the beer, was intoxicating, even when I had difficulty following it.

What I chiefly remember of those evenings with Carl Becker—aside from the fact that he was the least loquacious of the group—was his impatience with sham and pomposity and his demonstration that even the most important matters could be discussed without pontification and the most thorny subject illumined by the passing insight and the ironic *aperçu*.

This is why I find my present assignment difficult and even slightly embarrassing. From what little I know of Becker I suspect he would be quietly amused to hear his *Heavenly City*, in which these characteristics are so manifest, discussed with great solemnity as if it were a pretentious monograph on the great age he relished and admired so much.

From Raymond O. Rockwood, ed., *Carl Becker's Heavenly City Revisited* (Ithaca, N.Y.: Cornell University Press, 1958). © 1958 by Cornell University Press.

The Heavenly City, I hardly need remind you, is not a monograph or a synthetic study but a minor masterpiece of that precious but obsolescent form of writing, the historical essay. Whatever else we may think of it, few would deny that it is a work of literary art—a proof that historians can write well and need not, indeed should not, write like sociologists. To dissect a minor work of art—above all a work of deft irony and marked elusiveness—with the blunt tool of scientific criticism often seems, even when it is necessary, to be a somewhat barbarous task. And it becomes also a bit absurd when one has the uncomfortable feeling that perhaps the author took the work less seriously than does the critic. So it is fair to ask, at the start, how seriously Becker took the conclusions of *The Heavenly City*, for the book is surely something of a *jeu d'esprit*. But even if Becker did not take his conclusions literally, in one sense at least I am convinced he took the book as a whole quite seriously. As a distinguished teacher—indeed, for graduate students, one of the great teachers—Becker was well aware how hard it is to make people *think*, especially those historians and teachers of history who view their task solely in terms of bibliographies and footnotes. I suspect the chief purpose of *The Heavenly City* was to make us think: to move us to a new vantage point from which to contemplate some of the familiar facts of eighteenth-century intellectual history and to shed a beam of paradox over well-traveled ground.

Such a radical change of position is often what is needed; and this makes me think of the M.I.T. professor who has found that many mechanical puzzles prove insoluble when examined in the ordinary way but that the solutions—or at least the elements of a problem—reveal themselves at once if the student will only have the kindness to stand on his head. I think that Becker has, in a manner of speaking, made us stand on our heads to contemplate the Age of Reason, and in doing so he has shown us aspects we would otherwise not perceive. Yet Becker would have been the last to recommend this posture as a permanent one for *Homo sapiens*, and I think he would have been unhappy to learn that many of his readers have remained ever since in the position in which he placed them. Sad to relate, this is precisely what has happened. A generation of history students has been brought up to take *The Heavenly City* more literally than I believe Becker intended. Its paradoxes have been hardened into doctrine by a succession of textbook writers.

There is some point, then, in examining, as we are doing here, what Becker said, the manner of his saying it, and the conclusions that have been widely drawn from this stimulating book. I shall leave to Peter Gay, who has done the job expertly, as you shall see, the task of discussing the book as a whole. I shall limit myself to one aspect only, an important aspect, to be sure, and one that falls properly within the province of the historian of science. This is Becker's interpretation of the influence on eighteenth-century thought of what he calls—as many in that century also did—the "Newtonian philosophy."

Becker carefully placed this term in quotation marks, and he was much too wise to identify this "Newtonian philosophy" either with Newton's own discoveries and opinions or with the totality of eighteenth-century scientific thought. He is careful to speak of Newton as a convenient *symbol* of the influence of science on the world view of the Age of Reason, much as Darwin was a symbol in the great debates about the role of science that raged through the middle and late nineteenth century.[1] And he surely did not believe that the science of either period could be adequately summed up in the work and influence of a single man, however great. But our textbook writers have been unmindful of Becker's caveat; they have drawn from *The Heavenly City* certain persistent notions about the influence of Newton in the eighteenth century and the debt to him of the social philosophers of the Enlightenment.

What Becker says of Newton and the "Newtonian philosophy" is familiar to all of you, and I do not propose to cite him at length. Instead, I should like to set down in the form of six propositions certain ideas which I believe are widely and erroneously accepted and which appear to me to sum up the prevailing impressions concerning the place of science in the Enlightenment. Here they are:

1. That the science which interested the men of the eighteenth century was "Newtonian science," which is to say chiefly physical astronomy and cosmology.

2. That the substance of this scientific outlook was derived from the discoveries and opinions set down in Newton's *Principia* of 1687.

3. That this Newtonian physics introduced men for the first time to the notion of a mechanistically determined universe of external nature and that Newton's discoveries lay behind the materialistic picture of a physical universe made up only of massy particles in motion, operating like a vast piece of clockwork.

4. That the lessons of this physics largely explain eighteenth-century man's abiding faith in the regularity of nature and the inexorableness and simplicity of the "Laws of Nature and of Nature's God."

5. That the chief writers of the Enlightenment—*philosophes* and *économistes* alike—had their world view shaped by this Newtonian philosophy; that these men were all Newtonians, in the sense that they accepted, in simplified and diluted form, the cosmological and physical assumptions of Newton's *Principia*.

6. That the *philosophes*, like their loyal readers, cared little for what was taking place in science but much for the use they could make in their liberal propaganda of this "Newtonian philosophy"; and that their knowledge of science was scanty and came largely through such Newtonian popularizers as Pemberton, Maclaurin, Algarotti, and Voltaire.

You will recognize that these propositions are deliberately oversimplified, and you will probably see that while Becker may have said some of these things, and seems to imply others, he certainly did not say them all or mean them all. Yet, without undue exaggeration, these are some of the conclusions commonly drawn from *The Heavenly City*, and we have heard them all in one form or other.

I shall return in a moment to these propositions, only stopping here to remark that they present a wholly misleading picture of the role of science in eighteenth-century thought, as any reading of Cassirer, Lovejoy, and Chinard will convince you. Today we know far more than Becker could have known, with the resources at his disposal, about this important problem. We know more about Newton himself and his accomplishments,[2] and the reputation he enjoyed in the Age of Reason. From Margaret Libby, Ira Wade, and others we have learned to take Voltaire's excursion into science more seriously than Becker did.[3] New and important studies have appeared on most of the great figures of the Enlightenment, and the authors have been at pains to examine the scientific interests of their subjects.[4] Diderot has been the object of intensive research, and so has the persistent influence of Cartesian thought in the course of the century.[5] And I should add that the history of eighteenth-century science is receiving increased attention from specialists in this field.[6]

If it is not fair to judge Becker harshly from our improved position, it is nevertheless our business to ask how much use he made of materials available when he set to work to prepare these now-famous Storrs Lectures. What had he read? What could he have read?

First of all, let us remember that what Becker has to say of Newton and eighteenth-century science in *The Heavenly City* was taken, with no perceptible alteration, from the more extended treatment he had included ten years before in his *The Declaration of Independence*, where he gives one of the first accounts in English of the spread and influence of Newtonian thought in France.[7] He seems to have sought out little of significance in the intervening period. He did not tackle, or see the importance of, Ernst Cassirer's *Erkenntnis Problem*, the second volume of which would have proved illuminating. Except for the Newtonian popularizers, whom he cites with delight, he seems to have taken his general picture of eighteenth-century science largely from William Whewell's antiquated *History of the Inductive Sciences* or from the pedestrian *History of Science* of Dampier-Whetham. Although, between the two drafts, Becker encountered Whitehead's *Science and the Modern World* (1925), a book which left a strong imprint on *The Heavenly City*, he found little to cause him to alter his earlier opinions.

Yet at the time he wrote, there was already a substantial secondary literature of which he seems to have been oblivious or about which he was indifferent. The latter, I think, is probably the case, for I am not persuaded that Becker had the slightest interest in the harsh and sometimes grubby materials of the history of science. Two important books appeared, one in the same year as *The Heavenly City*, the other a year later, which show that the facts were not totally inaccessible, and also that they led to conclusions quite at variance with his own. The first was Cassirer's *Philosophie der Aufklärung*. Little read by historians in this country until the author arrived in person just before World War II as a refugee from Nazi Germany, it is only now reaching a wide audience as a result of the English translation brought out in 1951. It is a penetrating book, by a historically minded philosopher of great genius who was steeped in the primary materials and the secondary literature.

A quite different sort of book was published in 1934 by Becker's colleague in the Cornell Department of History, Preserved Smith. This was the second volume of Smith's *A History of Modern Culture*. It did not, I understand, elicit much enthusiasm from the reviewers; yet its merits lie precisely in those chapters on eighteenth-century science which the reviewers were least likely to appreciate. Historians of science, and I am among them, have often praised it. Its documentation is formidable, and Smith's conclusions, insofar as he troubled to draw conclusions, are largely sound. If he did not penetrate as profoundly as did Cassirer into the thought of the age, his account supplements the *Philosophie der Aufklärung* in several important respects: it conveys the richness and variety of science in the eighteenth century; shows that this century was by no means held in bondage solely by the memory of Isaac Newton;[8] and, together with the work of Daniel Mornet, convinces us that the general appreciation of science was a social and intellectual phenomenon in the eighteenth century without which the thought of the Enlightenment can scarcely be understood.[9]

Let us now consider the six propositions in the light of what is now becoming known about our problem:

1. The science which interested the men of the eighteenth century was not merely "Newtonian science," nor was it confined to physics and cosmology. Other branches of physics—like the discoveries in electricity—captured the public imagination, especially during the second half of the century. Natural history—stimulated by the work of men like Réaumur, Buffon, Bonnet, and Linnaeus—was devotedly studied and became a source of pleasure and fruitful speculation to many a painstaking collector, both noble and *roturier*.[10] The mysteries of reproduction—as elucidated in part by the work of Bonnet on parthenogenesis and Trembley's study of the hydra[11]—were heatedly debated in the salons as well as the laboratory. Chemistry, mineralogy, and broader geological problems not only yielded some of the century's most important scientific accomplishments, but these advances were discussed in the periodical press, and crowded public lectures were devoted to them. In one year, for example, a popular topic of conversation was the new noble metal, platinum; a year later it was the discovery that the diamond—thought to be not only the hardest, but the most resistant, of natural substances—could be utterly destroyed by fire. The public sessions of the Academy of Sciences in Paris, held once in the spring and once in the fall, were social events of great popularity. These examples, and a casual perusal of the gazettes or of the *Encyclopédie*, should convince the most skeptical that we are dealing with a sincere, if sometimes amateurish, interest in all manner of timely scientific questions.[12]

2. Insofar as there existed throughout the century such a commonly held "world view" as Becker describes under the term "Newtonian philosophy"—and a close reading of the texts does not quite bear this out—it would seem to owe as much to Descartes, and to other influences from the previous century, as it does to Newton.

3. It was certainly not Newtonian physics that first introduced thoughtful men to the notion of a mechanistically determined physical universe. This, again, was largely the work of Descartes, who, more than any other philosopher or scientist, had made systematic use of this basic article of scientific faith. Still less can we lay at Newton's door the crude materialism we find in some of the later writers of the eighteenth century. This materialism drew heavily upon Descartes —even when its tenets were expressed in Newtonian language—but it also derived to a surprising extent (as I have tried to show elsewhere) from classical sources, so that we may with some justice speak of a Neo-Epicurean revival among the thinkers of the second half of the eighteenth century.[13] This is an important matter if we wish to be clear about the religious implications of scientific progress in the eighteenth century. It is not from scientists (certainly not from Newton) but from philosophers that the atheistic materialism of d'Holbach and Diderot can be legitimately and logically derived.

Newton's physics—like that of Galileo before him—was far less materialistic and concrete than is commonly supposed and than Becker believed. Becker's strongly empiricist interpretation of the work of Galileo and Newton would hardly satisfy a modern historian of science. To describe these men as making "relatively little use" of deductive logic and as enthroning instead a technique of observation and measurement and a reliance on brute fact[14] is to distort our knowledge of the founders of modern exact science. Newton's physics was highly abstract, a true mathematical physics, dealing with idealized bodies moving in a wholly idealized time and space. Indeed, for his cautious use of the word "attraction" to describe the mysterious force exerted between all bodies in the universe, Newton was accused of departing from the solid ground of Cartesian materialism and of introducing ideas that were neither clear nor precise. He was even charged with introducing mysterious entities in a manner shockingly reminiscent of the Scholastics.

4. Becker was well aware, of course, that the concepts of nature and natural law have a long history and that what we encounter in the eighteenth century is a change in meaning of these venerable terms. For what he calls the ideal image of nature ("too ghostly ever to be mistaken for nature herself") which had prevailed in earlier centuries, there is substituted, according to Becker, "a more substantial image." The ideal image of nature was abandoned as a result of the recent scientific discoveries: "natural law, instead of being a construction of deductive logic, is the observed harmonious behavior of material objects."[15]

But it was certainly not to Newtonian physics, or even to Cartesian "physics" before it, that eighteenth-century man owed his faith in the simple regularity of nature. This Becker surely knew. Such a conviction had been basic to Greek science; Aristotle's great scheme of things was reared upon it; and, as Lynn Thorndike has recently stressed, the pseudoscience of astrology may have played a powerful role, from the Stoic philosophers onward, in the spread of this idea.[16] Belief in scientific law—even if the term itself was not popular until after its use by Descartes—is at least as ancient as the related tradition of a *moral* natural law, although it was from the moral sphere that the term "scientific law," a conscious metaphor, was itself manifestly borrowed in the sixteenth and seventeenth centuries.[17]

What was new, as a result of the Scientific Revolution, was that these uniformities were seen to be *mathematical* in character. And if this is true, as I believe most historians of science would now agree,[18] then what took place is precisely the reverse of what Becker believed to have happened. An ideal image of nature is substituted for the crude scheme of ordered common sense which was the system of the Schools. It is manifest that the scientists of the new age, like all scientists before or since, were concerned that the principles they hit upon should be in accord with the observed behavior of nature. But now, indeed, measurement and the quantitative view of nature supplied a tool for demonstrating this accord. Yet there is a more important difference: precisely because their mathematical image of nature was an ideal one, these scientists recognized that the conformity of theory to reality had to be approximate and limiting. No projectile actually follows the ideal parabolic path that Galileo *deduced* from his laws of moving bodies, but under favorable circumstances it can approach it. That scientific truth is not absolute but probabilistic and asymptotic; that it begins in experience and ends there, too, after passing through the mysterious alembic of mathematical deduction, Newton expressed clearly in his *Opticks*. And many men in the eighteenth century, Voltaire among them, perceived in this the great novelty of the Newtonian philosophy and saw in the "method of analysis" a powerful critical instrument to be used against the system builders and the blind adherents of tradition, who thought of truth as unique, absolute, and perfectly attainable.

5. In their scientific activities, the *philosophes* did not confine themselves by any means to a superficial study of the Newtonian popularizers. Their interest—like the public interest—in natural science was wide-ranging, and they were certainly not solely concerned with science because it might supply arguments for their social, political, and religious preconceptions. They were proud of science's steady advance; it testified to the power and possibilities of the rational human mind untrammeled in its work. They saw in the *method* of science—and it is here that Newton exerted his greatest influence—an analytical instrument that could be extended into other intellectual spheres, to extinguish superstition and emancipate the human mind from traditional error.[19] They were aware also of the immense practical possibilities of improving man's welfare through the useful applications of scientific discovery, a proposition to which the Encyclopedists were especially dedicated.[20] Above all, perhaps, they saw in the great forward strides taken by science the strongest argument for their belief in a law of human progress.[21]

Though most of the *philosophes* can lay little claim to high scientific distinction, it is wrong to think them content to read a handful of popular books. Fontenelle, Montesquieu, Voltaire, Diderot, d'Holbach, Turgot, Thomas Jefferson, Richard Price—to mention some outstanding men of letters—had a surprising sophistication and breadth of knowledge about the science of their day. Fontenelle was, of course, for many years the official expositor of the work of his colleagues in the Academy of Sciences. Voltaire, though unable to follow Newton through the maze of his mathematics, had nevertheless a keen appreciation of Newton's basic methodology and was a surprisingly accurate judge of the epistemological position taken by Newton. Diderot studied chemistry and natural history; d'Holbach, a disciple of Lavoisier's teacher Rouelle, wrote some of the best chemical and metallurgical articles for the *Encyclopédie*; Turgot, a close friend of the outstanding scientists, drew the best of them into the service of government when he was *Intendant* and *Contrôleur des Finances*; and in his correspondence with Condorcet we find him writing astutely on some current scientific controversies.[22]

And perhaps it is well to be reminded that the line between *philosophe* and scientist is often impossible to draw. There were scientists whose writings on general "philosophic" questions, whose contributions to the *Encyclopédie*, or whose activities in public life and in the service of liberal causes ought to allow us to classify them as *philosophes*: d'Alembert, Maupertuis, Joseph Priestley, Benjamin Franklin, Condorcet, and Lavoisier are familiar examples.

6. Lastly—and this is the point that has given the title to my paper—it is of some interest to realize that not all the social philosophers of the Enlightenment were Newtonians; that is, when they invoked for ideological purpose the findings of physics, it was not always the physics of Newton. While the thinkers were generally more or less loyally Newtonian in Britain and the American colonies,[23] the situation in France was more complicated. Even in professional scientific circles, the reputation of Newton, or let us say his influence, suffered some surprising changes in the course of the century.

To begin with, we should remember the unchallenged influence that Descartes exerted over physical science in France during the first third of the century. This influence was remarkably persistent. Fontenelle, who remained until his death in 1757 a staunch Cartesian in celestial physics, is not the only outstanding case. It is less fully appreciated that Montesquieu also remained loyal to Descartes, that he had little knowledge and only limited appreciation of the accomplishments of Isaac Newton, and that he did not accept Newton's System of the World.[24]

The great wave of Newtonian enthusiasm in France coincided with the battle against the dominant Cartesians: the high point of this enthusiasm was the period from about 1732 to about 1755, when it was stimulated by the campaign on behalf of Newton by scientists like Maupertuis and Clairaut and by the success of popularizers like Algarotti and Voltaire. This is the period from which Becker takes most of his evidence and which he would make typical of the entire century. Yet if a tradition of Newtonians of strict observance can be traced to the end of the century—from d'Alembert and Condillac to Condorcet and Lavoisier—and remained particularly strong among the professional scientists, there is nonetheless a pronounced change of mood after mid-century.

This shift is marked by a growing suspicion among scientists that too much emphasis had been placed upon mathematical ingenuity and not enough on experimentation and observation.[25] Even mathematicians such as d'Alembert begin to express their doubts. This antimathematical tendency, accompanied by pleas for more attention to natural history and the experimental sciences, finds a strong champion in Diderot and the Encyclopedists, but it is also reflected in a shift of focus within the Academy of Sciences itself[26] and in scientific activity generally. More attention is henceforth paid to chemistry, geology, and electrical experiments. The discoveries of Benjamin Franklin seemed to many to epitomize the new movement and justify the new emphasis. Just as Newton was the guiding star of the earlier decades, Franklin becomes in many quarters the new hero. He comes close, in the second half of the century in France, to replacing Newton as the symbol of science and its importance for mankind.[27]

The drift away from Newton is clearly evident in the writings of Diderot and of d'Holbach. The "scientific materialism" of d'Holbach's *Système de la nature* owes little to Newton, toward whom he is by no means very friendly. Instead, his materialism owes much to the writings of the English deist John Toland, and it shows a revived interest in the materialist philosophers of classical antiquity. These same influences, plus a tincture of Neo-Stoicism, are found in the scientific speculations of Diderot. Newton's thought seemed, to these men, too mathematical, too abstruse, and too clearly tied to the deism of the older generation.

And lastly, in the closing years of the century, a vociferous attack is leveled against Newton from still another quarter. It comes from those who believed that Newton had been excessively deified, that he had become the symbol of official academic science, an authority as unquestioned as Descartes had been earlier and Aristotle before that. This overt anti-Newtonianism won little support among the qualified scientists, but it was nonetheless a force to be reckoned with. It began with Rousseau and was carried on as a sort of minor guerrilla warfare by some of Rousseau's admirers, chief among them the future revolutionaries Marat and Brissot. In France this movement found its most systematic expression in Bernardin de Saint-Pierre's *Etudes de la nature*; and it is to this tradition that Goethe's scientific and anti-Newtonian writings may owe a substantial debt.

I have tried, in the preceding pages, to focus on one matter that Carl Becker dealt with in *The Heavenly City* and to summarize the results of some recent work by historians of science and historians of eighteenth-century thought.

My argument has been directed less against certain misleading notions in Carl Becker's delightful and stimulating classic than against those who, I feel, have misread his intentions and have mistaken a stop sign for a declaration of martial law. If I am right that Professor Becker hoped to make us think, I can only conclude that he would be depressed by the result. The charm, lucidity, and convenient brevity of *The Heavenly City* have worked their spell. And a book he may well have intended to stimulate and provoke readers of some sophistication, already familiar with the eighteenth century, has become, for all too many inexperienced students, the only book they open on the subject.

But *The Heavenly City* has suffered a more serious fate. When Carl Becker talked playfully and irreverently about a century whose great accomplishments he valued and whose faith he shared more than he would openly concede, he could hardly have imagined the use that would be made of his remarks by the depreciators of much that he esteemed. Even Becker could not predict the trick that a drifting climate of opinion would play on him. In 1931 it seemed good sport to expose the humbug and the stereotypes of the superrationalist historian —was he perhaps thinking of his old master, James Harvey Robinson?—and almost a public service to deflate the supposed pretensions of science and its idolaters. I doubt that he would feel quite the same about it today. At least I am sure he would not have relished the fate that has made his book ready ammunition for the foes of reason and the antagonists of science.

Notes

1. "The 'Newtonian philosophy' was, accordingly, as familiar to common men in the middle eighteenth century as the 'Darwinian philosophy' is in our day. . . . No need to open the *Principia* to find out what the Newtonian philosophy was—much better not, in fact. Leave that to the popularizers, who could find in the *Principia* more philosophy than common men could, very often more, I must say, than Newton himself did." *The Heavenly City of the Eighteenth-Century Philosophers* (New Haven: Yale University Press, 1932), 60–61.

2. See, for example, Florian Cajori's usefully annotated English version of the *Principia* (Berkeley: University of California Press, 1934) and, appearing in the same year, Louis Trenchard More's *Isaac Newton* (New York: Scribner's, 1934), a detailed biography. On the technical side the reader should consult the numerous contributions of Alexandre Koyré. Recent specialized studies have stressed the important influence of Newtonian thought, especially the *Opticks*, on chemistry and other branches of experimental science. See I. B. Cohen, *Franklin and Newton* (Philadelphia: American Philosophical Society, 1956), and my "Continental Reputation of Stephen Hales," *Archives Internationales d'Histoire des Sciences* 4 (April 1951), 393–404.

3. Margaret S. Libby, *The Attitude of Voltaire to Magic and the Sciences* (New York: Columbia University Press, 1935); Ira O. Wade, *Voltaire and Madame du Châtelet: An Essay on the Intellectual Activity at Cirey* (Princeton: Princeton University Press, 1941); the same author's *Voltaire's Micromégas: A Study in the Fusion of Science, Myth and Art* (Princeton: Princeton University Press, 1950), and his *Studies on Voltaire* (Princeton: Princeton University Press, 1947), especially chap. 2, sec. 5.

4. For example, the recent studies of Montesquieu by Pierre Barrière (1946) and Sergio Cotta (1953); Pierre Naville's *Paul Thiry d'Holbach et la philosophie scientifique au XVIIIe siècle* (Paris: Gallimard, 1943); Daniel J. Boorstin's *The Lost World of Thomas Jefferson* (New York: H. Holt, 1948); Douglas Dakin's *Turgot* (London: Methuen, 1939); Carl Van Doren's well-known *Benjamin Franklin* (New York: Editions Transatlantique, 1938); and Paul Dimoff, *La vie et l'oeuvre d'André Chenier*, 2 vols. (Paris: E. Droz, 1936).

5. Of importance are the papers of Herbert Dieckmann and the well-documented study by Aram Vartanian, *Diderot and Descartes: A Study of Scientific Naturalism in the Enlightenment* (Princeton: Princeton University Press, 1953). See also Paul Mouy, *Le Développement de la physique cartésienne* (Paris: J. Vrin, 1934); Robert Lenoble, *Mersenne, ou la naissance du mécanisme* (Paris: J. Vrin, 1943); and Raymond Bayer, ed., *Etudes cartésiennes* (Paris: Hermann, 1937), a collection of valuable papers presented to the 9ᵉ Congrès International de Philosophie.

6. See the important work of the late Hélène Metzger, especially her *Newton, Stahl, Boerhaave et la doctrine chimique* (Paris: F. Alcan, 1930); A. Wolf's encyclopedic *A History of Science, Technology and Philosophy in the Eighteenth Century* (New York: Macmillan, 1939); and Emile Guyénot's *Les sciences de la vie aux XVIIᵉ et XVIIIᵉ siècles* (Paris: A. Michel, 1941). There are a number of excellent special studies of scientific figures, such as Pierre Brunet's *Maupertuis*, 2 vols. (Paris: A. Blanchard, 1929); Jean Torlais's *Réaumur* (Paris: A. Blanchard, 1936); René Taton's *L'oeuvre scientifique de Monge* (Paris: Presses Universitaires de France, 1951); and the short but penetrating studies of French naturalists (Buffon, Daubenton, Cuvier, Lamarck, and Lacépède) by Louis Roule. The literature on Lavoisier is reviewed in my "Lavoisier and His Biographers," *Isis* 45 (May 1954), 51–62.

7. Carl Becker, *The Declaration of Independence* (New York: Harcourt, Brace and Co., 1922), 40–51.

8. If the second chapter of Smith's book is entitled "Newtonian Science," the chapter following (dealing with natural history, geology, and biology) is significantly called "Linnaean Science"; the author makes use of Daniel Mornet's classic *Sciences de la nature en France au XVIIIᵉ siècle* (Paris, 1911) and of the important books of Pierre Brunet which Becker ignored: the basic *Introduction des théories de Newton en France au XVIIIᵉ siècle* (Paris: A. Blanchard, 1931) and the no less important *Physiciens hollandais et la méthode expérimentale en France au XVIIIᵉ siècle* (Paris: A. Blanchard, 1926).

9. See Preserved Smith's chap. 4, "The Place of Science in Eighteenth-Century Thought."

10. Besides the work of Mornet, the reader should consult Edouard Lamy, *Les cabinets d'histoire naturelle en France au XVIIIᵉ siècle et le Cabinet du Roi (1635–1793)* (Paris, 1931?). He will find useful material for the American colonies in Michael Kraus, *The Atlantic Civilization: Eighteenth-Century Origins* (Ithaca, N.Y.: Cornell University Press, 1949), especially in chap. 7, "Scientific Relations Between Europe and America"; and in Brooke Hindle, *The Pursuit of Science in Revolutionary America, 1735–1789* (Chapel Hill, N.C.: University of North Carolina Press, 1956).

11. A. Vartanian, "Trembley's Polyp, La Mettrie, and 18th-Century French Materialism," *Journal of the History of Ideas* 11 (June 1950), 259–86.

12. When Becker writes, "What were most of the scientific academies in France doing but discussing, quarreling about, and having a jolly time over the framing of projects?" (*Heavenly City*, 40), he reveals that he has never perused the solid and technical contents of the *Mémoires* of the Royal Academy of Sciences at Paris. What he says does not even adequately describe the activities of the more amateurish provincial academies, like those of Dijon or Bordeaux. For the Academy of Dijon the reader can consult "Notes et documents pour servir à l'histoire de l'Académie des Sciences, Arts et Belles-Lettres de Dijon," *Mémoires de l'Académie*, 2d series, vol. 16 (1871), as well as Georges Bouchard's *Guyton-Morveau* (Paris: Librairie académique Perrin, 1938). For the Bordeaux Academy there is P. Barrière, *L'Académie de Bordeaux* (Bordeaux: Editions Bière, 1951).

13. I have expanded this point in my paper "Three Eighteenth-Century Social Philosophers: Scientific Influences on Their Thought," *Daedalus*, Winter 1958, pp. 8–24. The reader may consult Louis Bertrand, *La fin du classicisme* (Paris, 1897), for the general problem, though not as applied to science.

14. *Heavenly City*, 17, 20–21. It was precisely Galileo's Platonizing rationalism and his mathematical idealism that his Aristotelian opponents found most difficult to comprehend. On this subject, the standard works are E. A. Burtt, *The Metaphysical Foundations of Modern Physical Science* (New York, 1925), and Alexandre Koyré's *Etudes Galiléenes* (Paris: Hermann, 1940). But see also Koyré's "Galileo and Plato,"

Journal of the History of Ideas 4 (October 1943), 400–428; also Ernst Cassirer, "Galileo's Platonism," in M. F. Ashley Montague, ed., *Studies and Essays in the History of Science Offered in Homage to George Sarton* (New York: Schumann, 1947), 279–97. If Galileo's rationalism is sometimes overstated, as I believe it to be, we can no longer think, if we ever could, that Galileo derived his law of falling bodies from experiment alone or that Newton discovered the laws of motion in the solar system "by looking through a telescope and doing a sum in mathematics," as Becker rather flippantly says (*Heavenly City*, 57).

It seems clearly evident that Becker's opinions concerning the nature of seventeenth-century science are taken largely from Alfred North Whitehead's *Science and the Modern World* (New York, 1925). Professor Whitehead's view that Galileo's science was "objectivist" and "anti-intellectualist" and an expression of the "historical revolt" was not so much a paradox as a sophisticated version of a long-popular view. This view is no longer tenable. See, for example, Alexandre Koyré, "An Experiment in Measurement," *Proceedings of the American Philosophical Society* 97 (April 30, 1953), 222–37.

15. *Heavenly City*, 54, 56–57.
16. Lynn Thorndike, "The True Place of Astrology in the History of Science," *Isis* 46 (September 1955), 223–78.
17. Edgar Zilsel, "The Genesis of the Concept of Physical Law," *Philosophical Review* 51 (May 1942), 245–79.
18. Strongly emphasized by E. A. Burtt and Alexandre Koyré, this view is supported by Hermann Weyl in his *Philosophy of Mathematics and Natural Science* (Princeton: Princeton University Press, 1949). Wilhelm Dilthey, in *Gesammelte Schriften*, 12 vols. (Leipzig, 1914–58), II, 260, seems to have perceived much earlier the "constructive character" and the "Pythagoreanism" of Galileo's thought. The reader can find this interpretation summarized in Herbert Butterfield's *Origins of Modern Science* (London: G. Bell, 1949) and A. R. Hall's more detailed *Scientific Revolution, 1500–1800* (London: Longmans, Green, 1954).
19. A significant example is the use made by Voltaire of these ideas in his assault on the iniquities of French criminal jurisprudence. This I have discussed in the paper referred to above in n. 13.
20. The Baconian influence on the encyclopedists is well known; it permeates d'Alembert's *Discours préliminaire* and, indeed, the whole project; see Herbert Dieckmann, "The Influence of Francis Bacon on Diderot's *Interprétation de la nature*," *Romanic Review* 34 (December 1943), 303–30.
21. See Charles Frankel, *The Faith of Reason* (New York: King's Crown Press, 1948), 143–46.
22. *Correspondance inédite de Condorcet et de Turgot*, ed. Charles Henry (Paris, 1883). Of special interest are letters in which Turgot discusses the problem of the gain in weight of metals on calcination, and advances—before Lavoisier—the theory that air must be the cause. This is discussed in my paper "The Origin of Lavoisier's Work on Combustion." [Subsequently published in *Archives Internationales d'Histoire des Sciences*, no. 47 (1959), 113–35; also treated in my *Lavoisier: The Crucial Year* (Ithaca, N.Y.: Cornell University Press, 1961).]
23. There is little doubt that, as far as Britain and its American colonies were concerned, Newton was, in the words of Cotton Mather, "the perpetual Dictator of the learned World." See, for example, Clinton Rossiter, *Seedtime of the Republic* (New York: Harcourt, Brace and Co., 1953), 133–35. For the influence of Newton's *Opticks* on English poets of the eighteenth century there is, of course, Marjorie Nicolson's *Newton Demands the Muse* (Princeton: Princeton University Press, 1946).
24. Montesquieu's bias toward mechanistic (iatrophysical) physiology is well known. See my chapter "Humanism in Science," in Julian Harris, ed., *The Humanities: An Appraisal* (Madison, Wis.: University of Wisconsin Press, 1950). In his early physical papers, the *Discours sur la cause de la pesanteur des corps* (May 1720) and the *Discours sur la cause de la transparence des corps* (August 1720), the doctrine of attraction is not to be found, nor is there any mention of Newton except for the discovery of the dispersion of light. Two of Montesquieu's chief scientific correspondents were Father Castel and J. J. Dortous de Mairan, both staunch anti-

Newtonians. Montesquieu read with approval Castel's *Vrai système de physique de M. Isaac Newton, exposé et analysé en parallèle avec celui de Descartes* (1743) and remarks that, while Newton is a great geometer, his opinions may be false and that he "auroit souvent été lu sur l'opinion de l'infaillibilité de la géométrie." "Spicilège," *Oeuvres*, édition Pléiade, 2 vols. (Paris: Gallimard, 1949–51), II (1947), 1370. On Mairan and Castel, see P. Brunet, *Introduction des théories de Newton* (Paris, 1931), chaps. 2 and 3; see also Donald S. Schier, *Louis Bertrand Castel, Anti-Newtonian Scientist* (Cedar Rapids, Iowa: Torch Press, 1941). The Cartesianism of Montesquieu was long ago pointed out by Désiré André, *Sur les écrits scientifiques de Montesquieu* (Paris, 1880), 11 and 18–19.

25. John Herman Randall, Jr., *The Making of the Modern Mind*, rev. ed. (Boston: Houghton Mifflin Co., 1940), 262–66.

26. Suggested as early as 1766, a reorganization of the Academy was effected in 1785 to give a greater recognition to sciences like mineralogy, experimental physics, and natural history. See Ernest Maindron, *L'Académie des sciences* (Paris, 1888), 50–57.

27. Some interesting texts have been collected by Gilbert Chinard, *L'apothéose de Benjamin Franklin* (Paris: Librairie orientale et américaine, 1955).

7

Newton et Epicure

❀

Le titre de ma conférence est plus symbolique que précis. Je n'ai pas l'intention de présenter Isaac Newton comme un épicurien au sens ordinaire du mot. Il n'était rien moins que cela, cet homme hautain, solitaire, méfiant, qui avait peur des femmes et qui oubliait de manger les mets que sa servante mettait sur la table.

Je n'ai pas non plus l'intention de chercher sérieusement une influence directe des œuvres d'Epicure ou de Lucrèce sur la pensée de Newton. D'ailleurs, on ne trouve ces auteurs cités qu'une ou deux fois dans ses œuvres. Mon titre doit simplement indiquer que je me propose d'examiner aujourd'hui la théorie de la matière chez Newton,[1] cette théorie corpusculaire ou atomistique qui a—nécessairement—quelque parenté avec la fameuse théorie grecque des atomes et du vide, que nous attribuons à Leucippe, Démocrite et Epicure.

Newton ne développe pas d'une façon explicite sa théorie de la matière dans son grand chef-d'œuvre, les *Principes mathématiques de la philosophie naturelle*; mais des documents inédits, publiés tout récemment pour la première fois, révèlent qu'à cette époque—c'est-à-dire vers 1687—Newton avait déjà au moins esquissé sa théorie.[2] Cependant, ce n'est que vingt ans plus tard, dans son *Traité d'Optique*,[3] que Newton fit paraître pour la première fois les grandes lignes de sa doctrine:

> . . . il me semble trés-probable qu'au commencement Dieu forma la Matiere en particules solides, massives, dures, impénétrables, de telles grandeurs & figures, avec telles autres proprietés, en tel nombre, en telle quantité, & en telle proportion à l'Espace, qui convenoient le mieux à la fin pour laquelle il les formoit. . . .[4]

Ces mots—ne l'oublions pas—ne se trouvent pas dans le texte proprement dit du *Traité d'Optique*, mais dans une des célèbres "Questions" (ou, selon le terme anglais, les "Queries"), ajoutées à la fin du livre dans la traduction latine de 1706. Dans cette "Question 23," Newton applique sa théorie corpusculaire à une variété de problèmes chimiques et physico-chimiques.[5]

Conférence donnée au Palais de la Découverte, le 2 mars 1963. Published by Université de Paris, Palais de la Découverte, Avenue Franklin D. Roosevelt, Paris—VIII.

Cherchant à expliquer comment de ces dures unités pouvait naître à la fois l'architecture et le dynamisme de l'univers, Newton ajoute à sa théorie atomique sa fameuse doctrine de l'*attraction universelle* qu'il applique également (dans les *Principes*) à la physique macrocosmique et à la physique microcosmique ou moléculaire dans cette "Question 23" du *Traité d'Optique*:

> Il me semble d'ailleurs que ces Particules ont non seulement *une force d'inertie*, accompagnée des Loix passives du mouvement, qui résultent naturellement d'une telle *force*; mais qu'elles sont aussi mües par certains Principes actifs, tel qu'est celui de la Gravité, & celui qui produit la fermentation & la cohesion des Corps.[6]

Dans les *Principes*, l'attraction équivaut à la gravitation universelle: cette force mystérieuse par laquelle le soleil et les divers corps de notre système solaire s'attirent mutuellement. Dans le *Traité d'Optique*, Newton considère des forces analogues—"certains principes actifs"—qui obéissent à d'autres lois et qui agissent d'une manière différente:

> ... la Nature se trouvera trés-simple et très-conforme à elle-même, produisant tous les grands mouvements des Corps Celestes par l'attraction d'une pesanteur réciproque entre ces Corps; & presque tous les petits mouvements de leurs particules, par quelques autres Puissances attractives et repoussantes, réciproques entre ces Particules."[7]

Or, bien qu'on admette que Newton ait été tout au moins un corpuscularien —comme le furent, du reste, la plupart des grands savants du XVIIe siècle—sa croyance à l'existence du vide a été récemment mise en doute. Les spécialistes les plus compétents sont en plein désaccord au sujet d'une question fort importante, à savoir: Newton, loin de croire à l'existence d'un vide absolu, était-il un "plé-niste"? Etait-il un partisan de "l'éther"? Suivait-il, sur ce point, non pas Gassendi, mais Descartes? C'est, en substance, ce qu'affirme mon confrère américain, M. Bernard Cohen;[8] c'est ce que nient M. et Mme Hall, dans un article paru en 1960 et reproduit dans le livre qu'ils viennent de publier.[9]

Voici donc un problème qui n'est pas dénué d'intérêt: un problème qui est bien près de toucher au point archimédien de la philosophie newtonienne. En effet, la question du vide et de l'éther ne peut être séparée de la question de l'attraction—ou mieux, de la *cause* de l'attraction. Quelle est véritablement la pensée de Newton à ce sujet? Comment explique-t-il ce mouvement, en apparence spontané, par lequel les corps tendent à se rapprocher les uns des autres? Et de quelle façon faut-il entendre les forces microcosmiques d'attraction et de répulsion imaginées par Newton pour élucider les phénomènes chimiques et physico-chimiques?

Il est certain que l'existence du vide et la réalité de l'attraction furent, pour les disciples de Newton—pour Clairaut, Maupertuis et Voltaire, comme pour la plupart de ses sectateurs anglais—les marques distinctives de la philosophie newtonienne. Chacun connaît la célèbre boutade de Voltaire qui se trouve dans ses *Lettres Philosophiques*:

> Un Français qui arrive à Londres trouve les choses bien changées en philosophie comme dans tout le reste. Il a laissé le monde plein, il le trouve vide. A Paris on voit l'univers rempli de tourbillons de matière subtile; à Londres on ne voit rien de cela.[10]

Et pour ses adversaires—les Cartésiens et Leibniz—il en fut de même. Leibniz, dans sa fameuse polémique épistolaire avec l'ami de Newton, le théologien Samuel Clarke, attaqua vivement, vers 1715, l'idée du vide et l'idée de l'attraction. Dans sa deuxième lettre à Clarke—qui fut, comme on sait, le représentant de Newton—Leibniz n'hésite pas à associer directement Newton à Démocrite et Epicure,[11] ce qui, à cette époque, constituait une véritable provocation, comme nous allons le voir.

Nombreux sont les historiens de Newton qui ont remarqué que dans les *Principes*—où, pour la première fois (en 1687), il utilise sa doctrine de l'attraction macroscopique—Newton évite soigneusement toute affirmation sur la cause de ce phénomène mystérieux. Il lui suffisait de la reconnaître comme un simple fait d'expérience:

> Je vais expliquer les mouvemens produits par ces forces que je nomme *attractions*, quoique peut-être je deusse plutôt les appeler *impulsions*, pour parler le language des Physiciens; mais je laisse à part les disputes qu'on peut élever sur cette dénomination.[12]

Et encore:

> Je me sers ici du mot d'*attraction* pour exprimer d'une manière générale l'effort que font les corps pour s'approcher les uns des autres, soit que cet effort soit l'effet de l'action des corps, qui se cherchent mutuellement, ou qui s'agitent l'un l'autre par des émanations, soit qu'il soit produit par l'action de l'Ether, de l'air, ou de tel autre milieu qu'on voudra, corporel ou incorporel, qui pousse l'un vers l'autre d'une manière quelconque les corps qui y nagent.[13]

Ces citations sont assez caractéristiques de la pensée de Newton dans la plupart de ses œuvres *exotériques*, c'est-à-dire destinées à la publication. De tels passages—on pourrait en trouver d'autres—ont amené plusieurs auteurs à parler hâtivement de Newton comme d'un positiviste. Ce qui n'est d'ailleurs pas totalement faux: en effet, lorsque Newton entreprenait un travail rigoureusement démonstratif, il écartait soigneusement toute spéculation inutile ou superflue. Mais, comme nous le savons fort bien, Newton était doué—et quel savant de premier ordre ne l'est pas?—d'une puissante imagination à la fois physique et métaphysique.

Le deuxième des passages que nous venons de citer montre qu'avant 1687 Newton avait au moins réfléchi au sujet de l'éther. Cette citation est en effet l'écho de quelques tentatives antérieures—ésotériques, si je puis employer ce mot—pour expliquer plusieurs phénomènes au moyen de l'éther. Nous savons qu'à deux reprises, en 1675 et en 1678–79, Newton exposa—d'une façon timide et avec beaucoup de prudence, et même, à en croire les Halls, sans véritable conviction[14]—ses théories d'un éther ambiant.

Dans sa deuxième communication à la Royal Society sur la lumière et les couleurs, Newton imagine un milieu éthéré, ayant quelque ressemblance avec l'air (et peut-être produit par une raréfaction de l'air), mais beaucoup plus subtile et beaucoup plus élastique. Il remplit tout l'espace de notre univers, y compris les pores des corps bruts. C'est précisément l'air, comme le décrivit Robert Boyle, qui lui servit ici de modèle: une matière élastique, mais chimiquement inerte et "flegmatique," remplie de particules actives, d'émanations, etc. De même, dans le corps "flegmatique" de l'éther se trouvent divers "esprits éthérés."

Newton avait ainsi à sa disposition un mécanisme double: par les vibrations du corps de l'éther proprement dit, il expliqua non pas la propagation ou la nature intime de la lumière, mais sa réfraction et ses propriétés périodiques. Il imagina un mécanisme analogue pour expliquer la cohésion, la répulsion entre les corps, et même jusqu'aux mystères de la contraction musculaire. En 1675, il explique la gravitation, non pas par l'action du "corps flegmatique" de l'éther, mais par les courants, pour ainsi dire circulatoires, "d'esprits éthérés."[15] Plus tard, dans sa lettre à Robert Boyle de février 1678–79 (le premier document, d'ailleurs, dans lequel il manifeste ses intérêts chimiques), il utilise le rasoir d'Occam pour simplifier son hypothèse.[16] Dans ce texte, la gravitation, l'attraction moléculaire, la cohésion et même la répulsion sont expliquées par les propriétés du "corps flegmatique" de l'éther, tandis que les "esprits éthérés" semblent tomber dans l'oubli.

Newton avoua franchement le caractère chimérique et fantaisiste de ces spéculations. C'est certainement la raison pour laquelle il ne les a pas publiées. Cependant, il est probable que les études préliminaires qui ont préparé la rédaction des *Principes*, en particulier celle qui a conduit Newton à la grande découverte de la gravitation universelle, ont contribué à lui faire abandonner ces conjectures.

On a raison de croire, aujourd'hui, que lorsque Newton préparait les *Principes*, il était déjà un partisan convaincu du vide épicurien, et qu'il acceptait une certaine théorie atomique qu'on pourrait qualifier d'épicurisme christianisé. Cette théorie est peut-être implicite dans le traitement mathématique de l'espace, des forces centrales, des "points masses," exposé dans les *Principes*. Mais sa position est exprimée en clair dans quelques documents datant de la même époque, publiés récemment par M. et Mme Hall, ainsi que dans quelques autres textes plus connus. Dans un brouillon d'étudiant ou de débutant, le jeune Newton attaque la fameuse identification cartésienne de la matière et de l'extension.[17] L'espace est éternel, immuable, infini et sans bornes, divisible géométriquement en surfaces, lignes et points. Dans cet espace qui est extension pure il n'existe aucune force qui puisse opposer ou favoriser le mouvement des corps. Les corps eux-mêmes sont les créations de Dieu, c'est-à-dire des modifications de l'extension possédant trois propriétés importantes: ils sont étendus, mobiles et impénétrables.

L'espace du jeune Newton, c'est le vide de Démocrite et d'Epicure, mais à une importante différence près, car, dans ce texte de jeunesse, il ne démontre pas seulement que le vide n'est pas (*pace* Descartes) *métaphysiquement* impossible, mais il essaie de prouver—puisqu'il identifie l'espace avec Dieu—que l'espace vide de corps est métaphysiquement *nécessaire*.

Reste à savoir si l'espace est physiquement vide, c'est-à-dire si l'espace, qu'on peut *concevoir* légitimement comme vide de corps, l'est en fait; ou, si, au contraire, un éther, ou une matière subtile, se trouve dans les espaces célestes et remplit les pores des corps.

Il existe un document, court mais précieux, dans la Collection Portsmouth, également publié par les Hall: il date vraisemblablement des années 1690 et constitue, peut-être, le brouillon d'un additif au deuxième livre des *Principes*. Nous y trouvons un exposé très clair de son opinion en faveur de l'existence du vide (avec des arguments *physiques* que nous retrouverons plus tard dans son *Traité d'Optique*). Voici, en substance, ce qu'il y dit: Les corps macroscopiques de tous genres sont composés de particules premières, de *minima* (Newton évite presque toujours le mot "atome"), entre lesquels se trouvent des interstices d'ampleur plus marquée qu'entre les grains de sable sur la plage. Ces pores sont complètement vides. Et nous lisons: "Je ne suis pas du tout troublé par ce sophisme vulgaire [il s'agit de la doctrine de Descartes] suivant lequel les conclusions opposées au concept du vide sont déduites de la caractérisation des corps par l'extension."[18] Les corps sont étendus, mais ils ne sont pas l'extension elle-même, étant entièrement différenciés de l'extension par leur solidité, leur mobilité, la force de résistance et la dureté. Quant au vide, les mouvements réguliers et quasi éternels des planètes—et surtout des comètes qui traversent les cieux dans toutes les directions—prouvent que les espaces célestes sont vides. De même, les différences considérables en gravité spécifique des diverses substances —l'or et l'eau, par exemple—montrent que les corps sont de beaucoup plus rares qu'on a l'habitude de le croire et que les pores des substances sont vides.

En ce qui concerne le rôle architectonique des corpuscules, on peut compléter ces données par d'autres documents de la même période de la carrière de Newton, par exemple un brouillon très important pour la préface des *Principes*, et plus encore la *Conclusio* que Newton n'incorpora pas dans la version imprimée, mais que les Hall viennent de publier. Mieux connu—car il fut publié, avec l'autorisation de l'auteur, au début du XVIIIe siècle—est le *De natura acidorum*.[19] Il date de 1692. En parcourant ces documents, nous voyons que les particules ou *minima* dont parle Newton ont une forte ressemblance avec les atomes classiques: ils sont durs, effectivement insécables, variables en grandeur mais d'une densité présumée égale. Ces particules, en se combinant, produisent une sorte de hiérarchie de particules allant des plus petites aux plus grandes, des plus simples aux plus compliquées, ce que Newton appelle "particules de la première composition," "particules de la deuxième composition," etc., jusqu'aux "particules de la dernière composition" qui forment les substances ordinaires. A chaque niveau, les particules sont douées d'une force attractive, mais qui va en décroissant à mesure qu'on monte dans la hiérarchie. La force attractive est la plus forte pour les substances de la première composition, la plus faible pour les particules de la dernière. Le grand plan ressemble, par conséquent, à une sorte de filet, ou structure réticulaire plus ou moins rigide, composé d'une collection d'assemblages moléculaires. Entre les particules de chaque niveau, il y a le vide. "Les corps," dit Newton, "sont plus rares que l'on ne croit généralement."[20]

❁

Ce résumé des idées de Newton sur la matière suffira, peut-être, à montrer en termes généraux la ressemblance entre la théorie newtonienne et celle d'Epicure ou de Gassendi, même si, dans les détails, on relève des divergences assez importantes. Newton, par exemple, n'insiste point, ou très peu, sur la différence de forme des atomes ou corpuscules; au lieu d'envisager la combinaison des particules par le moyen de crochets ou d'entrelacements, il fait appel à la force mystérieuse d'attraction[21] qui, d'ailleurs, pour beaucoup de personnes, faisait penser au *pondus* inhérent d'Epicure. Malgré les différences qu'on peut relever, l'influence de la philosophie naturelle d'Epicure sur la pensée de Newton—quelle que soit la manière dont elle lui fut transmise et quelles que soient les transformations qu'elle a éprouvées—est hors de doute.[22] Reste à évaluer cette influence; à noter également que l'épicurisme, aussi bien comme philosophie naturelle que comme théorie morale, fut d'une importance capitale en Angleterre, pendant la jeunesse et même pendant l'âge mûr d'Isaac Newton.

❁

L'influence des écrits épicuriens s'est fait sentir beaucoup plus tard en Angleterre qu'en France. Bien qu'on connaisse quelques personnages du règne d'Elisabeth I avec une certaine teinture épicurienne—parmi d'autres, influencés sans doute par la visite de Giordano Bruno, on peut signaler Francis Bacon, qui garda toujours une attitude hésitante concernant la vérité de la théorie atomique, et le mathématicien Thomas Harriot, qui fut peut-être le premier atomiste anglais—il faut noter qu'avant l'année 1640, pas une seule édition de Lucrèce, pas un seul livre sur Epicure ne parut en Angleterre.

A partir de cette année, l'influence croissante de l'épicurisme se fait sentir. A vrai dire, l'influence directe des œuvres mêmes de Lucrèce, Diogène et Epicure ne fut pas très forte au début, vu la rareté des textes publiés. Mais c'est néanmoins à ce courant qu'on doit la plaquette intitulée *Man's Mortalitie*, d'un certain Richard Overton, qui fut un des premiers en Angleterre à oser nier l'immortalité de l'âme et attaquer la distinction entre l'esprit et la matière, tout en invoquant Démocrite et Epicure et leurs théories matérialistes.[23] Membre de la secte des Levellers, Overton, loin d'être un érudit ou un penseur profond, fraya cependant la route à Thomas Hobbes et en même temps mit en garde, sans l'apercevoir, les défenseurs de la foi contre le danger du matérialisme atomistique. Les premiers défenseurs se trouvèrent parmi les théologiens de Cambridge, groupe de tendance libérale et rationaliste qui était quelque peu familiarisé avec ces doctrines. Ce fut un certain John Smith, un "Cambridge man" et élève du célèbre Benjamin Whichcote, qui décocha les premières flèches contre l'épicurisme dans son *Bref Discours d'Athéisme*. Pour Smith, "l'épicurisme n'est que l'Athéisme masqué."[24]

Le courant *indirect*, de beaucoup le plus important, ce fut le courant gassendiste. Mais à quel point, de quelle façon précise et par quelles étapes, les écrivains et savants anglais furent influencés par Gassendi, reste à préciser. Les études, à ce sujet, manquent; il n'existe pas de travaux sur l'influence gassendiste en Angleterre, analogues à ceux que nous avons sur le cartésianisme ou—cela va de soi—sur l'influence de Francis Bacon. On se contente d'affirmations ou de démentis sans preuves. Y a-t-il eu une influence de Gassendi sur Newton? Voltaire l'affirme en précisant que Newton aurait avoué à plusieurs amis français qu'il avait été profondément influencé par les écrits de Gassendi. On a l'habitude d'accepter en France comme un fait établi que Newton était gassendiste, malgré l'absence d'éléments précis confirmant cette opinion.[25] En Angleterre, par contre, les savants ont tendance à nier—ou du moins à minimiser—cette influence. Ils insistent plutôt sur les aspects originaux de la philosophie atomistique de Boyle et par conséquent—puisque Boyle fut peut-être son maître dans ce domaine—de celle de Newton.

Mais c'est Gassendi—en fin de compte—qui fut, pour l'Angleterre comme pour la France, celui qui rétablit la réputation personnelle d'Epicure et ressuscita sa philosophie naturelle. Grâce aux études approfondies de M. Bernard Rochot, on sait que Gassendi s'est consacré à cette tâche dès l'année 1629, mais que jusqu'en 1647—l'année où parut son *De vita et moribus Epicuri*—il ne publia rien sur l'atomisme ou sur Epicure.[26] Avant 1647, il était surtout connu en Angleterre comme astronome et non pas comme philosophe épicurien.[27]

Cependant, bien avant cette date, les travaux philosophiques de Gassendi sur Epicure et l'atomisme étaient connus et appréciés du petit cercle de voyageurs et de réfugiés anglais, amis de Mersenne et d'autres savants français: Sir Charles Cavendish, le mathématician John Pell et le célèbre philosophe Thomas Hobbes.[28] Il est possible—et même probable—que ces savants firent circuler en Angleterre des manuscrits du grand ouvrage de Gassendi, le *Syntagma philosophicum*. C'est donc vraisemblablement par cette voie souterraine, et principalement à travers l'influence de Hobbes, que Walter Charleton, le premier gassendiste anglais, fit son éducation épicurienne.

Le nom de Walter Charleton est tombé dans un oubli injuste et presque total.[29] Immatriculé en 1635 à Oxford, où il avait comme directeur d'études le célèbre John Wilkins, reçu docteur en médecine au moment du déclenchement de la guerre civile, nous le trouvons à Londres vers 1650, pratiquant la médecine, et bientôt membre du Royal College of Physicians. Virtuoso accompli (dans le sens donné à ce mot au XVIIᵉ siècle), ami de Hobbes, qui rentra d'exil en 1651, Charleton fut bientôt mis au courant des travaux de Mersenne, Pascal, Descartes et surtout Gassendi. En même temps, il suivait de près les travaux scientifiques du College of Physicians, ainsi que ceux du groupe d'Oxford ce "collège invisible" qui constitua un premier noyau de la future Société Royale. Charleton nous a laissé, dans un écrit peu connu, une description précieuse des intérêts scientifiques de ces deux groupes.[30]

Au début de sa carrière littéraire et scientifique, Charleton se manifeste comme épicurien et atomiste. En 1652, il publia son *Darknes of Atheism, Dispelled by the Light of Nature*. Ce livre, qu'on peut intituler en français "Les ténèbres de l'Athéisme," est le premier de cette longue liste d'ouvrages physico-théologiques parus en Angleterre à la fin du XVIIe siècle. A mon avis, Charleton peut être considéré comme l'inventeur du genre et comme le précurseur de Bentley, Whiston et les autres commentateurs de Newton, étudiés si savamment par la regrettée Mme Metzger. Comme pour ses successeurs, le dessein de Charleton fut de trouver des preuves de la providence de Dieu en contemplant l'ordre de la nature. Il tirait ses arguments de sources variées: certains étaient traditionnels et livresques, d'autres provenaient de sa propre expérience scientifique; cependant, c'est l'emploi qu'il fit de la théorie atomistique de Gassendi qui nous concerne le plus directement ici.

Pour Charleton, il importait avant tout de montrer que la théorie physique d'Epicure—la théorie atomistique, purgée d'implications athéistes, et christianisée à la façon de Gassendi—loin de conduire au matérialisme, comme John Smith l'avait prétendu, fournit au contraire des preuves irréfutables de l'existence de Dieu. L'atomisme, il en était convaincu, *implique* Dieu. La matière, ne pouvant pas être éternelle, doit être la création de Dieu. La matière, et le système du monde tout entier sont maintenus, comme Newton le pensera plus tard, par "la conservation et la modération constante," suivant l'expression de Charleton, "de la Providence de Dieu."[31]

En outre, l'ordre et la régularité du monde ne sauraient être le résultat d'un hasard aveugle. Il a fallu l'intervention de quelque "principe primitif et actif" qui, par sa puissance infinie, a pu tirer les atomes du néant et, par sa sagesse, a su réglementer ou, pour employer le mot de Charleton, "digérer" ces mêmes atomes dans l'ordre qui constitue aujourd'hui le système du monde.[32] Il est à noter que cette idée constituera un des principaux arguments utilisés par Richard Bentley, quarante ans plus tard, dans ses fameuses Boyle Lectures;[33] ce fut surtout pour renforcer cet argument avec l'autorité de la nouvelle physique newtonienne que Bentley écrivit ses célèbres lettres au grand savant.

Dans le *Darknes of Atheism*, Charleton ne chercha pas à développer toute une philosophie naturelle basée sur l'atomisme gassendiste. Mais il le fit plus tard dans un ouvrage dont le titre résume, en quelque sorte, l'histoire de l'atomisme: la *Physiologia–Epicuro-Gassendo–Charltoniana* (Londres, 1654). Dans cet ouvrage, son but fut moins physico-théologique que purement physique: Charleton voulait démontrer qu'un système complet de philosophie naturelle, expliquant à la fois les faits anciens et les nouvelles découvertes de la science, pourrait être fondé sur le mouvement—l'agrégation et la désagrégation—des atomes dans le vide.

Charleton suit Gassendi de près, mais, plus minutieusement que le savant français et ses précurseurs, il veut montrer comment les qualités des choses—ce que John Locke appellera plus tard les "qualités secondaires,"[34]—c'est-à-dire la couleur, le son, l'odeur, le froid, la chaleur, la dureté, etc.—peuvent résulter des différentes dispositions et arrangements atomiques. C'est, comme nous le savons, le programme de recherche de Robert Boyle.[35] Il est à peu près certain que Boyle connaissait l'œuvre de Gassendi et celle de Charleton, bien qu'il fût aussi loin que possible d'être philosophiquement un épicurien.[36]

On peut néanmoins parler d'un courant, d'une "nouvelle vague" épicurienne, qui date des premiers livres de Charleton. L'année 1656 fut particulièrement significative à cet égard, puisque trois publications épicuriennes parurent cette année-là.[37] Citons d'abord un livre de Charleton lui-même, formé en grande partie d'extraits, sur le caractère et la philosophie morale d'Epicure: *Epicurus, his Morals*.[38] Ensuite une traduction incomplète—mais la première publiée en langue anglaise—du *De rerum natura* de Lucrèce. Cette traduction ainsi que les commentaires qui l'accompagnent sont dûs à l'illustre John Evelyn, autorité en matière de sylviculture, auteur d'un fameux journal intime et un des fondateurs de la Société Royale. Enfin, il y a *The History of Philosophy* de Sir Thomas Stanley, dont les premiers volumes parurent en 1655 et 1656; l'auteur y accorde une grande importance aux premiers atomistes grecs. Dans le dernier volume (1660), il parle longuement d'Épicure et des épicuriens, en s'appuyant sur Gassendi, dont les *Œuvres* parurent en 1658.

Ces années constituent néanmoins une période préparatoire. En effet, en Angleterre, le véritable engouement pour Epicure coïncide assez exactement avec la restauration du roi Charles II et la fin du régime dur et sec des Puritains du Commonwealth. Dans l'entourage du roi, il se trouvait beaucoup de courtisans qui avaient déjà été influencés, pendant leur séjour en France, par les cercles libertins. A Londres, l'épicurisme de salon fournissait une justification philoso-phique au relâchement des mœurs qui marqua la réaction aristocratique de la Restauration. Jusqu'en 1685—l'année de la marée haute du mouvement—la mode était à l'épicurisme. Les coryphées du mouvement sont faciles à identifier: il y avait d'abord l'homme de lettres français exilé, Saint-Evremond, qui exerçait son influence par son exemple plus que par ses écrits; les grands poètes de l'époque: Cowley, Waller et Dryden lui-même; enfin, Sir William Temple, qui publia en 1685 son essai intitulé *Upon the Gardens of Epicurus*, et qui se rangera plus tard du côté des "anciens" dans la fameuse querelle littéraire. Les contempo-rains auraient ajouté sans hésitation à cette liste le nom du grand matérialiste, Thomas Hobbes, bien qu'il ne fût pas corpuscularien, mais pléniste, ne croyant pas à l'existence du vide.

Il est intéressant de trouver dans cet épicurisme de salon des tendances nettement antiscientifiques. Pour les vrais disciples d'Epicure, comme pour le maître lui-même, la philosophie naturelle était l'auxiliaire de la philosophie morale et pratique: elle devait même lui être subordonnée. Sir William Temple ne trouvait rien d'important ni de valable dans "cette partie de la philosophie qu'on appelle 'Naturelle.' " Seule la philosophie morale avait, selon lui, quelque valeur. Pour Saint-Evremond aussi, l'étude des sciences naturelles ne constituait pas une occupation convenable pour un honnête homme.

Cette tendance—à la fois irréligieuse et antiscientifique—aide un peu à expliquer les changements d'attitude et de perspective envers la philosophie d'Epicure qui se manifestèrent au sein de la Société Royale. Au début, c'est-à-dire au moment où l'influence aristocratique et "High Church" était forte, la philosophie épicurienne—et non seulement les spéculations atomistiques—y était très goûtée. N'oublions pas que John Evelyn fut un des membres fondateurs de la Société Royale et que Walter Charleton lui-même fut admis comme "Fellow" en 1661. C'est à cette époque que Robert Boyle publia ses premiers ouvrages corpusculaires qui avaient pour but général d'établir de bons rapports entre les "chymistes" et les "philosophes mécaniciens," c'est-à-dire les corpusculariens.[39]

Mais les savants, comme Boyle, qui considéraient l'hypothèse atomique de plus en plus indispensable pour les sciences, rencontraient une difficulté croissante. En dépit des mesures défensives et prophylactiques prises par Gassendi et Charleton, l'opposition à l'épicurisme se développait dans les milieux universitaires, cléricaux et bien-pensants, et au sein même de la Société Royale, au fur et à mesure que la vogue de cette philosophie de salon se répandait. Déjà en 1665, le maître de Newton, le géomètre et théologien Isaac Barrow, s'en plaignit: "De toutes les sectes et factions qui divisent le monde, celle des railleurs épicuriens est devenue la plus redoutable."[40] Il fallait donc répéter inlassablement que la théorie atomique n'est pas nécessairement matérialiste, qu'elle ne mène pas à l'athéisme et qu'on peut aisément la détacher de la morale irréligieuse des épicuriens. Parmi les savants, c'est Robert Boyle qui s'imposa cette tâche. Dans divers écrits—tout en se disant "corpuscularien"—il se sépare nettement des libertins, des épicuriens et de ces "auteurs atomistiques" qui nient Dieu, se moquent de la religion et croient tout expliquer (et pas seulement en physique) par le vide et les atomes.[41]

Enthousiaste, mais peu méthodique, Boyle se faisait physico-théologien, suivant l'exemple de Gassendi et de Charleton. Comme eux, Boyle était persuadé que l'ordre et l'économie de l'univers—même d'un monde construit mécaniquement d'atomes—témoignent éloquemment de l'existence d'une intelligence créatrice et dirigeante. Selon lui, les atomistes anciens—Leucippe, Démocrite, et Epicure—eurent tort de relier leur théorie physique à l'athéisme. S'ils avaient été des observateurs plus vigilants, ils auraient évité ce piège. Cet argument de Boyle constitue le germe de la théorie développée longuement dans un grand in-folio, illisible et touffu, publié par Ralph Cudworth en 1678 et intitulé *The True Intellectual System of the Universe*.

❁

Ralph Cudworth, le plus sérieux et l'infatigable défenseur de la théorie atomique, était l'un des philosophes et théologiens platonisants de Cambridge, dont nous avons déjà parlé et dont l'influence sur Isaac Newton—surtout en ce qui concerne le cas de Henry More—a été soulignée par M. Koyré.[42] De même que More et Benjamin Whichcote—quoique moins bien peut-être—Cudworth fut connu de Newton comme nous allons le voir.

Dans son *True Intellectual System*, œuvre magistrale qui constitue la défense la plus ambitieuse de la théorie atomique, Cudworth poursuivait un but précis: il se proposait de démontrer—dans la tradition de Gassendi, Charleton, et Boyle—que la physiologie atomique bien comprise, "loin d'être la mère ou la nourrice d'Athéisme," comme on le suppose vulgairement, fournit un rempart solide contre l'incroyance.[43]

Les faits mis en évidence par Cudworth sont de deux sortes: "physiologiques" (c'est-à-dire physiques) d'une part, et "historiques" d'autre part. Considérons d'abord l'évidence "physiologique." La doctrine des qualités secondaires et subjectives—doctrine déduite de l'atomisme et qui amena la chute des formes substantielles des scolastiques—a joué un grand rôle dans la pensée de Gassendi, de Charleton et de Boyle. Pour sa part, Cudworth s'en sert spécifiquement pour prouver l'existence de Dieu. Si les qualités secondaires n'existent pas en réalité dans la matière, elles ne sont que des "idées," des "phantasmes"; d'où peuvent-elles donc venir sinon de Dieu? Dieu les introduit, semble-t-il, dans nos esprits, pour enrichir et embellir un monde triste et morne composé seulement de vide et d'atomes.

Par cette analyse, Cudworth acquiert la conviction que des entités incorporelles et spirituelles existent en même temps que les atomes matériels. Voici son argument: les principes de l'atomisme ayant dissipé le brouillard des formes substantielles et des qualités aristotéliciennes, le mot "corps" peut acquérir une nouvelle signification, beaucoup plus claire. Dorénavant, il désigne une masse étendue, impénétrable, mobile, mais essentiellement inerte et passive. Puisque les corps se meuvent en effet, cette inertie et cette passivité de la matière force à admettre de surcroît l'existence d'une puissance active ou—si l'on veut—de plusieurs puissances actives ou "esprits." Il est évident que la sensibilité, la pensée, la vie même, ne sont pas les propriétés de la matière telle que Cudworth la conçoit: au contraire, ce sont les attributs de quelque chose de distinct, de totalement différent du corps. Et que peut bien être ce quelque chose, sinon—comme l'âme immortelle de l'homme—un principe incorporel?

L'évidence "historique" de Cudworth—et il faut mettre l'adjectif entre guillemets—est ce qu'il y a de plus bizarre, de plus souvent cité et de moins convaincant dans son livre. Selon notre érudit, la théorie atomique n'est pas d'origine grecque; elle fut inventée bien avant Leucippe et Démocrite, auxquels on l'attribue faussement. Elle fut sans doute, selon certains passages des écrits de Strabon et Sextus Empiricus, la découverte d'un certain Moschus, phénicien "qui vécut avant la guerre de Troie."[44] Ce Moschus, dont le nom est cité par Strabon et Sextus d'après Posidonius, pourrait-il être identifié avec Moïse? Cudworth est tenté de le croire. Quoi qu'il en soit, cette doctrine serait passée en *Magna Graecia* où Pythagore et d'autres philosophes de l'école italienne l'ont adoptée. Au début, cet atomisme ne fut rien moins que matérialiste; outre les atomes et le vide, ces premiers philosophes acceptèrent, à en croire Cudworth, l'existence des substances incorporelles et le rôle créateur et directeur de Dieu.

Un changement néfaste fut introduit par Leucippe et Démocrite qui furent, selon les termes de Cudworth, "les premiers athéisants de cette ancienne physiologie des atomes, les inventeurs et les inspirateurs de l'athéisme atomistique."[45] Leucippe et Démocrite furent les premiers—et en ceci ils seront suivis par Epicure—à faire dériver tout le système de l'univers d'atomes inanimés et à supposer la nécessité matérielle de toute chose sans Dieu.

❁

Revenons, maintenant, aux idées de Newton sur la matière et le vide. Newton n'avait pas l'habitude de citer soigneusement dans ses œuvres les autorités auxquelles il se référait, bien qu'il le fît plus souvent que ses contemporains. Nous ne pouvons donc affirmer avec certitude—en dehors de ressemblances frappantes que nous pouvons souligner—ce qu'il doit à Gassendi et à Charleton. Mais la situation n'est pas aussi désespérée dans le cas de Ralph Cudworth. Nous examinerons plus loin un passage du *Traité d'Optique* de Newton où la griffe de Cudworth transparaît avec évidence. Mais pour l'instant Cudworth, et dans une certaine mesure Charleton, nous aideront à commenter quelques passages bien connus, quoique encore quelque peu obscurs, des fameuses lettres échangées entre Newton et Richard Bentley.

Robert Boyle mourut à la fin de l'année 1691. Le pieux savant précisa dans son testament qu'il léguait une certaine somme d'argent destinée à des conférences annuelles sur les évidences de la religion chrétienne: c'est l'origine des fameuses "Boyle Lectures." Le premier conférencier choisi fut le jeune helléniste et théologien Richard Bentley. Suivant la voie tracée par Charleton et Boyle lui-même, Bentley décida de puiser ses arguments contre les athées dans les découvertes scientifiques récentes. Le moment était propice pour une attaque en règle contre les ennemis de la religion. En effet, l'historien du mouvement épicurien en Angleterre date le déclin de l'épicurisme de l'année de la publication des sermons de Bentley, parus sous le titre: la *Confutation de l'Athéisme*.[46] Cette plaquette est, de toute évidence, une défense de la science newtonienne et de la science atomistique et à la fois un réquisitoire contre les hobbistes et les épicuriens.

Le jeune théologien basa le réquisitoire que constituent ses six premiers sermons sur des arguments et des matériaux courants. Comme chez Charleton, ses arguments physicothéologiques étaient d'ordre biologique. Mais dans ses deux derniers sermons, qu'il prononça en novembre et décembre 1692, il eut recours à des arguments tirés des doctrines et découvertes qu'Isaac Newton venait de publier dans ses *Principes*.[47]

Il faut souligner que Bentley reconnaissait en Newton un atomiste et même un partisan d'un épicurisme christianisé. En effet, Bentley explicita, tout en la simplifiant, la doctrine newtonienne de la matière et du vide.[48] Nous n'avons pas besoin de suivre ses raisonnements en détail. Bornons-nous à examiner les diffi-cultés au sujet desquelles Bentley eut soin de consulter Newton. Pendant qu'il préparait son manuscrit—entre décembre 1692 et fin février 1693—Bentley écrivit plusieurs fois à Newton et lui posa diverses questions. Les quatre réponses de Newton sont bien connues.

Dès le début de cette correspondance, une des questions soulevées—question qui, d'ailleurs, avait été examinée auparavant par Charleton—était la suivante: en admettant qu'au début l'univers n'était qu'un chaos d'atomes dispersés uniformé-ment dans l'espace; en admettant aussi que chaque particule était douée "d'une gravité innée vers toutes les autres," notre système solaire aurait-il pu naître spontanément par la simple action des lois de la nature?

Cette question offrit l'occasion d'une réplique fameuse de Newton à Bentley:

> Vous parlez parfois de la gravité comme essentielle et inhérente à la matière. Je vous prie de ne pas m'imputer cette notion-là; car la cause de la gravité est une chose que je n'ai pas la prétention de connaître, et il me faudra plus de temps pour y réfléchir.[49]

Vers la fin de février 1693, Newton reçut de la part de Bentley une longue communication dans laquelle l'auteur considérait les possibilités et les conditions d'une évolution de notre système du monde avec, au départ, un chaos primordial d'atomes. Bentley posait quatre propositions ou, comme il les appelle, quatre "positions." Dans la première, il affirme que:

> supposant un tel chaos, aucune quantité de mouvement (sans attraction) ne pourrait jamais déterminer les atomes ainsi disséminés à se réunir en de grandes masses et à se mouvoir comme dans notre système."

Et voici sa deuxième "position":

> Quant à la gravitation, il est impossible qu'elle soit ou co-éternelle ou essentielle à la matière. . . . Il est inconcevable que la matière inanimée et brute puisse (sans une impression divine) agir sur d'autres matières et les influencer sans contact mutuel, comme il le faudrait si la gravitation en était essentielle et inhérente.[50]

La réponse de Newton a souvent été citée et commentée, mais pas toujours le texte intégral. La voici donc avec la phrase initiale qu'on a l'habitude de négliger:

> La dernière partie [the last clause] de votre deuxième position me plaît beaucoup. Il est inconcevable que la matière inanimée et brute puisse, *sans la médiation de quelque chose d'autre, qui n'est pas matériel,* agir sur une autre matière et l'influencer, comme elle le fait, si la gravitation dans le sens d'Epicure, en est essentielle et inhérente. Et c'est pour cela que je vous ai prié de ne pas m'attribuer l'idée d'une pesanteur innée.[51]

Nous voyons aisément de quoi il s'agissait dans cette correspondance. Dans ses *Principes*, Newton—comme nous l'avons vu—n'offre aucune explication, ni physique, ni métaphysique, de l'attraction. Il la présente toujours comme un fait d'expérience, dont la cause lui échappait, ou ne l'intéressait guère. C'est pourquoi Leibniz n'attaqua Newton que longtemps après la lecture des *Principes*. Néanmoins, une lecture rapide ou superficielle pouvait donner l'impression que Newton faisait de l'attraction une qualité essentielle—au sens littéral du mot—de la matière. Ce fut peut-être là la première impression de Bentley. Mais, comme Newton le savait fort bien, cette doctrine touchait à l'essentiel de la philosophie épicurienne et du matérialisme. Elle sentait terriblement le fagot.

Voici quel en est le danger: si le mouvement (ou la force de l'attraction) est essentiel à la matière, si le mouvement fait partie de notre conception de l'atome, s'il aide à le définir (comme c'est le cas pour l'extension, l'impénétrabilité, etc.), la route menant à l'athéisme est ouverte. En effet, si on explique tout dans le monde physique par la matière en mouvement, on peut se dispenser de la notion de Dieu, quoique à deux conditions seulement: 1° si, en premier lieu, la matière est conçue comme éternelle et incréée, ce que Gassendi et ses disciples ont nié; 2° si, en même temps, le mouvement, ou mieux la tendance au mouvement, le *conatus*, n'est pas seulement inséparable des atomes, mais réellement essentiel à la matière.[52]

Le danger a été plusieurs fois signalé bien avant Newton. Considérer le mouvement comme inhérent ou essentiel à la matière constitua pour John Smith une doctrine néfaste, dont le danger réside dans la possibilité de concevoir un monde entièrement matériel, gouverné par la "nécessité" ou par le "hasard," un monde où Dieu n'a aucune place. Plus tard, Thomas Creech, le traducteur de Lucrèce, résuma cette opinion et reconnut explicitement dans cette doctrine "le fondement de l'athéisme épicurien."[53]

Ainsi, pour pouvoir recourir librement à un atomisme classique, comportant le vide, il était indispensable de le purifier, de le baptiser; et pour effectivement le baptiser, il fallait à tout prix abandonner cette doctrine du mouvement *essentiel*. Sur ce point, tous les défenseurs d'Épicure ou de l'atomisme—Gassendi, Charleton, et Cudworth—étaient d'accord. Par exemple, Cudworth insiste sur le fait que le "fatalisme démocritique" de l'atomisme corrompu—c'est-à-dire démocritique—provient directement de ce que l'on a supposé "que la matière brute et insensible, mue nécessairement, fût la seule origine et le seul principe de toutes choses."[54] Voilà l'erreur fondamentale de ces corrupteurs: Leucippe, Démocrite et Epicure.

Cudworth pensait que c'est à lui qu'il incombait—plus qu'à Gassendi, qui avait en quelque sorte contourné la difficulté—de remettre en vigueur la vieille doctrine phénicienne, tombée depuis des siècles en désuétude. Pour Cudworth, les corpuscules inertes, les atomes, se meuvent dans un espace vide de matière, mus, non pas par Dieu directement, mais par des principes actifs et immatériels spécialement inventés à cette fin. Ce sont les "natures plastiques" qui agissent—comme les agents ou les intermédiaires de Dieu—sur la matière brute, produisant les mouvements des atomes.[55]

A la lumière de ces différentes citations des prédécesseurs de Newton, ce que Newton voulait exprimer dans ses lettres à Bentley se révèle, je l'espère, avec évidence. A cette époque, Newton était loin de croire à un éther. Il était un atomiste classique, un épicurien, au sens assez vague de ce terme, qui croyait à l'existence des atomes et du vide. C'est pour cela précisément qu'il attachait une si grande importance à la question de la cause de l'attraction. Pour ne pas introduire l'*actio in distantia*, qu'il trouvait philosophiquement absurde, et pour éviter le piège épicurien d'un mouvement essentiel à la matière—qui fut, d'ailleurs, en contradiction flagrante avec la première loi du mouvement—il avait recours à un agent immatériel, voire métaphysique.

Je suis persuadé que c'est là la position qu'il préféra pendant la période qui s'étend de l'année de la publication des *Principes*, c'est-à-dire 1687, jusqu'à 1706. En cela, je crois être d'accord avec les premiers disciples et commentateurs de Newton: Bentley, Cheyne, Ray, John Keill, Whiston, Samuel Clarke, et les autres. Tous l'ont interprété de cette façon, et tous ont cru, comme le jeune Voltaire, que chez Newton il n'y avait que les atomes et un espace vide de matière. Par exemple, Samuel Clarke, parlant en disciple de Newton, défendait franchement la doctrine du vide dans ses notes à la *Physica* de Rohault.[56] D'ailleurs, la plupart des "commentateurs newtoniens" ont accepté, comme cause de l'attraction, une activité divine ou une puissance spirituelle émanant de Dieu. Citons en particulier William Whiston. Dans ses *Astronomical Principles of Religion*, il parle de l'impossibilité de toute *actio in distantia*, agissant soit mécaniquement, soit non-mécaniquement:

> Par conséquent, il est évident que la force de la gravitation n'est pas seulement mécanique—c'est-à-dire ne se produit pas par contact corporel ou par choc, mais qu'il n'est pas du tout, à proprement parler, une puissance appartenant aux corps ou à la matière . . . *mais la puissance d'un agent supérieur, mettant tous les corps en mouvement* comme si chaque corps attirait, et fut attiré par tous les autres corps de l'univers, et pas autrement.[57]

❀

Nous approchons, maintenant, au terme de notre enquête. Il nous reste à chercher la confirmation de nos déductions dans le livre où Newton publia enfin, après un délai de plus de vingt ans, ses idées sur la constitution de la lumière, sur la matière et sur les attractions et répulsions microcosmiques. Ce livre, c'est son *Traité d'Optique*, publié en anglais en 1704 et en latin en 1706. C'est avec beaucoup de prudence et de circonspection que Newton y expose ses idées spéculatives, en recourant à l'expédient des "Questions" (*anglice*: "Queries") qui font suite au troisième livre du *Traité*; notre tâche consistera donc à interpréter quelques passages de ces "Questions," surtout de celles qui parurent pour la première fois dans l'édition latine de 1706. Dans la première édition anglaise, en effet, il n'y a rien qui doive spécialement retenir notre attention; les seize "Questions" se rapportent à des problèmes d'optique ou à la théorie de la vision. L'édition latine, par contre, que Newton prépara avec l'aide—pour la traduction —du Dr Samuel Clarke, présente un intérêt tout particulier. Si, dans cette édition latine, la plupart des seize premières "Questions" sont reproduites sans grandes modifications, nous y trouvons sept "Questions" entièrement nouvelles (Questions 17-23), dont les trois dernières offrent un intérêt très vif pour notre problème.

Malheureusement, beaucoup d'historiens des sciences ne connaissent pas de près cette première édition latine et ne voient donc pas à quel point elle diffère des éditions postérieures. Même M. Koyré, qui a été un des premiers à en signaler les principales variantes, ne semble pas en avoir apprécié toute la portée. Quels sont donc les principaux changements intervenus?

Précisons d'abord le point suivant: c'est en 1717 que Newton publia une deuxième édition anglaise, qui est à la base de toutes les éditions ultérieures, anglaises, latines, ou françaises. Dans cette édition de 1717, on trouve la traduction anglaise des sept "Questions" ajoutées à l'édition latine de 1706. On y trouve en outre huit "Questions" nouvelles qui traitent presque exclusivement de l'éther et de la théorie éthérique de la lumière. Elles suivent, dans leurs grandes lignes, la communication que Newton fit à la Royal Society plus de quarante ans auparavant! De ces opinions, il importe de le souligner, il n'y a aucune trace dans l'édition latine de 1706. D'où on peut conclure qu'entre 1706 et 1717, Newton—pour des raisons assez complexes que nous ne pouvons développer ici—a changé d'avis sur les questions qui nous intéressent particulièrement. Il devint plus ou moins partisan de l'éther et décida par conséquent de modifier les opinions qu'il avait avancées systématiquement dix ans plus tôt. L'édition anglaise de 1717 comporte donc plusieurs additifs aux "Questions" de 1706 ainsi que quelques changements verbaux, afin de les rendre conformes évidemment aux idées exprimées dans les nouvelles "Questions."

Par conséquent, si l'on veut connaître la pensée de Newton vers l'année 1706—et avant son revirement subi en faveur de l'existence d'un éther très fin et très délié—il ne suffit pas de consulter les éditions qui dérivent de la seconde édition du *Traité d'Optique* (comme c'est le cas, par exemple, de la traduction française de Coste). Il est indispensable de se reporter à l'édition latine.

C'est dans un cadre physico-théologique—comme on le verra plus loin—que Newton présente ses idées sur la matière dans cette édition de 1706. Et ces idées, qui ont incontestablement un aspect métaphysique, ne sont pas celles d'un mécaniste mûr, d'un partisan de l'éther, mais celles d'un épicurien pour ainsi dire christianisé, dans la tradition de Gassendi, Charleton, Boyle et Cudworth. Pour s'en convaincre, il suffit de prendre la "Question 20" de l'édition latine et de citer quelques passages, en ayant soin de noter en même temps les changements introduits par Newton par la suite. La "Question" débute ainsi:

> Toutes les Hypotheses qui font consister la Lumiere dans une pression ou dans un mouvement continué au travers d'un Milieu fluide, ne sont-elles pas erronées?[58]

Newton a certainement l'air d'y croire, car cette "Question 20"—si on la lit dans la version originale, débarrassée des changements introduits en 1717—*a pour objet principal de réfuter la théorie éthérique de la lumière et l'existence d'un tel milieu fluide.* Recourant, pour réfuter la possibilité d'un éther, à sa théorie de la résistance des fluides, il écrit:

> En plus, il ne peut exister aucun milieu fluide de ce genre, ce que je relie au fait que les planètes et les comètes sont portées par des mouvements réguliers et constants en tout sens au travers des cieux. Car il s'ensuit évidemment de là, que les espaces célestes sont privés de toute résistance sensible et par conséquent de toute matière sensible.[59]

D'autre part, dans la version de 1717 (dans la traduction de Coste), nous trouvons une phrase complètement remaniée: ". . . que les Cieux soient remplis de Milieux fluides, *à moins que ces Milieux ne soient excessivement rares:* c'est ce qu'on ne sçauroit accorder avec les mouvements réguliers & constants des Planetes & des Cometes."[60] Les mots que j'ai mis en italiques ont été ajoutés par Newton, manifestement pour adapter le paragraphe original de 1706 à sa nouvelle doctrine.

Plus bas, dans la même "Question 20" de 1706, Newton donne le résumé suivant:

> Donc pour assûrer les mouvements reguliers & durables des Planetes & des Cometes, il est absolument necessaire que les Cieux soient vuides de toute matiere, excepté peut-être quelques vapeurs très-legeres, ou exhalaisons qui viennent des Atmospheres de la Terre, des Planettes, & des Cometes.[61]

Ici aussi, Newton ajoute dans l'édition de 1717, où la "Question" est devenue Q 28, des modifications pour compléter la phrase précédente: "& un milieu éthéré excessivement rare, tel que nous l'avons décrit ci-dessus," c'est-à-dire dans les nouvelles questions 17–24.

Enfin, nous trouvons en 1706 une allusion tout à fait claire à la théorie quasi-historique de Cudworth, et même à sa doctrine de forces immatérielles opérant dans le vide:

> Ce Milieu [i.e. l'éther] a été rejetté en effet par les plus anciens & les plus celebres Philosophes de Grece *et de Phenicie,* qui établirent pour premiers Principes de leur Philosophie, le Vuide, les Atomes, & la pesanteur de ces Atomes, *attribuant tacitement la pesanteur à quelqu'autre Cause qu'à la Matière.*[62]

Signalons encore une modification significative faite dans l'édition anglaise de 1717: Newton ajoute l'adjectif "dense" pour que la phrase se termine ainsi: "à quelqu'autre Cause qu'à une matière dense." Un peu plus bas, dans le même paragraphe de 1717, nous trouvons cette phrase:

> Qu'est-ce qu'il y a dans les Lieux presque vuides de matiere? D'où vient la pesanteur réciproque des Planetes vers le Soleil, du Soleil vers les Planetes, & des Planetes les unes vers les autres, sans qu'il y ait de la Matiere dense entre eux.[63]

Dans le texte latin, il n'y a pas de "presque" ni de "dense"; nous lisons simplement: *"Quidnam inest in Spatiis Materia vacuis? & Unde est quod Sol & Planetœ ad se invicem gravitent, sine Materia interjecta?"*[64] Voilà encore un exemple d'adaptation à la théorie de l'éther que Newton vient de ressusciter. Il me paraît donc tout à fait évident que, lorsque Newton publia les nouvelles "Questions" de l'édition latine, il rejeta l'idée d'un éther. Comme les "célèbres philosophes de Grèce et de Phénicie," il n'accepta au contraire que les corpuscules durs et insécables et le vide.

Ce qui frappe dans des écrits aussi riches en détails scientifiques que les "Questions" du *Traité d'Optique*—et ce qui manque totalement dans tout ce que Newton publia avant 1706—c'est le ton et le contenu nettement théologique et métaphysique. En effet, la position qu'affirme ouvertement Newton est qu'on n'a pas besoin—qu'on n'a même pas le droit—de séparer la physique de la métaphysique, comme le font les cartésiens. Après avoir souligné que les anciens atomistes (les atomistes imaginés par Cudworth) avaient attribué la gravitation à une autre cause que la matière, il poursuit:

> Les Philosophes modernes ont banni de leurs Speculations Physiques la consideration d'une telle Cause, imaginant des Hypotheses pour expliquer toutes choses méchaniquement, & renvoyant les autres Causes à la Metaphysique; au lieu que la grande & principale affaire qu'on doit se proposer dans la Physique, c'est de raisonner sur les Phenomenes sans le secours d'Hypotheses imaginaires; de déduire les Causes des Effets, jusqu'à ce qu'on soit parvenu à la *Cause Premiere*, que certainement n'est point méchanique.[65]

Que les idées métaphysiques de Newton soient intimement liées à la théorie physique des atomes et du vide, dans la tradition de Cudworth, il est difficile de le nier. On peut citer la réponse de Newton à la question: "Quidnam inest in Spatiis Materia vacuis?" Elle se trouve dans un des passages célèbres, attaqués si vivement par Leibniz, où Newton semble identifier l'espace vide de matière avec l'organe de perception d'un Dieu omniprésent:

> Et ces choses dûëment expliquées, ne paroît-il pas par les Phenomenes, qu'il y a un Etre incorporel, vivant, intelligent, tout-present? qui dans l'Espace infini, comme si c'étoit dans son *Sensorium*, voit intimement les choses en elles-mêmes, les apperçoit, les comprend entiérement & à fond, parce qu'elles lui sont immédiatement présentes.[66]

Dans la fameuse "Question 23" (devenue "Question 31" dans l'édition anglaise de 1717), Newton ne donne aucune explication, ne propose aucune cause spécifique, de l'attraction. Mais le langage qu'il choisit pour exprimer sa pensée en indique cependant assez clairement la direction.

Les corps sont mus par les *principes actifs*, causes de presque tout mouvement observé dans le monde: gravitation, fermentation, cohésion des corps. Au moment de la création du monde, Dieu employa ces *principes* pour mettre les corps en mouvement; d'autres *principes actifs* sont en jeu pour *conserver* le mouvement qui, sans cela, se perdrait, car "le mouvement est beaucoup plus sujet à perir qu'à être produit."[67]

Mais que faut-il entendre par ces "principes actifs," terme par lequel, d'ailleurs, Charleton avait désigné le Dieu créateur? Newton en effet répond de deux façons. Dans un passage bien connu, il emploie ce mot dans un sens plutôt logique ou méthodologique, comme équivalent de "loi de la nature":

> Je ne considere pas ces Principes comme des Qualités occultes, qui soient supposées résulter de la forme specifique des Choses; mais comme des Loix générales de la Nature, par lesquelles les Choses mêmes sont formées; la vérité de ces Principes se montrant à nous par les Phenomenes, quoiqu'on n'en ait pas encore découvert les Causes . . . Nous dire que chaque espece de choses est doüée d'une qualité occulte specifique, par laquelle elle agit & produit des effets sensibles; c'est ne nous rien dire du tout : mais déduire des Phenomenes de la Nature deux ou trois Principes généraux de mouvement, & nous expliquer ensuite comment les proprietés & les actions de toutes les Choses corporelles découlent de ces Principes manifestes; ce seroit faire un progrès très-considerable dans la Philosophie, quoique les causes de ces Principes ne fussent point encore découvertes.[68]

Or, il est évident que, dans d'autres passages, Newton entend par "principes actifs" les *causes* des mouvements. Ce sont des puissances, des *vires* qui se manifestent par les lois du mouvement. Ailleurs, l'équivalence des mots "principe" et "force" apparaît explicitement lorsque Newton traite de l'inertie; la *vis inertiæ* est à la fois une force (*vis*) et un "Principe passif." Les particules de la matière "ont non seulement *une force d'inertie,* accompagnée des Loix passives du mouvement, qui résulte naturellement d'une telle *force*," mais elles sont aussi "muës par certains Principes actifs."[69]

Pour Newton, le mot "principe" signifie tantôt la cause, tantôt le résultat "manifeste" observé. De toute façon, Dieu en est responsable. Il organise et conserve le monde en employant ces "Principes," en mettant en jeu ces forces qui dépendent de sa toute-puissance et de sa volonté. Newton a su se dispenser des agents intermédiaires, comme les natures plastiques de Cudworth et de Grew. Les "forces," les "principes," sont les émanations, les expressions de la toute-puissance de Dieu. Identifié avec l'espace, Dieu est un agent physique; il "est plus capable de mouvoir par sa volonté les Corps dans son *Sensorium* uniforme & infini, & par ce moyen de former & de reformer les parties de l'Univers, que nous ne le sommes par notre Volonté, de mettre en mouvement les parties de notre propre Corps."[70]

❖

Que ce genre d'explication découle en quelque sorte de la tradition intellectuelle dont nous avons suivi les étapes, qu'il dérive—comme le pensait Mme Metzger—aussi bien du milieu intellectuel et moral de l'époque que de considérations purement scientifiques, les doctrines de Whiston, citées plus haut, en témoignent, de même que le livre principal de George Cheyne, les *Principes philosophiques de la religion naturelle et révélée*. Dans ce livre, publié un an avant l'édition latine du *Traité d'Optique*, Cheyne se montre comme celui qui—parmi tous les commentateurs—a le plus apprécié l'aspect mystique et métaphysique de la philosophie newtonienne.[71] Pour le Dr Cheyne, la loi d'attraction universelle ne doit pas être rangée parmi les lois ordinaires de la nature, les "principes mécaniques ordinaires." Elle ne peut pas être expliquée mécaniquement. C'est au contraire le meilleur exemple de ce que Cheyne appelle les "lois de la création." C'est une "grande et primitive loi que l'auteur de la Nature a imprimée sur tous les corps de l'univers."[72]

Nous rejoignons donc ainsi presque la position de Mme Metzger, pour qui "ce pouvoir attractif et irrationnel que nous avons rattaché instinctivement à chaque corpuscule . . . ne dépend pas véritablement d'eux et leur est imposé du dehors par une puissance dominatrice."[73] Mais nous avons vu en outre que cette incursion dans la métaphysique fut dictée à Newton par la nécessité de libérer la vieille doctrine des atomes et du vide de la doctrine dangereuse du mouvement inhérent à la matière.

Notre examen détaillé des "Questions" du *Traité d'Optique* nous a permis de lever la contradiction fondamentale qui a troublé Mme Metzger et qui est à l'origine des divergences d'opinion qui opposent aujourd'hui certains savants. "Dans le système de Newton," écrit-elle, "nous pourrons nettement percevoir des hésitations ou des oscillations entre des manières de voir hétérogènes et peut-être incompatibles; car ou bien l'attraction est un phénomène inexpliqué et premier, résultant immédiatement de la volonté de Dieu s'imposant au monde matériel, ou bien, il se pourrait aussi que cette volonté de Dieu se propage mécaniquement à travers l'esprit universel."[74] D'après ce que nous avons montré, il devient évident qu'il faut parler, non pas d'hésitations et d'oscillations de Newton, mais d'un simple changement d'opinion, après 1706, en faveur de l'existence de l'éther. L'explication de ce revirement, qui dépendait de raisons à la fois psychologiques, métaphysiques et scientifiques, fera l'objet d'un travail ultérieur.

Notes

1. Sur la théorie de la matière chez Newton, voir S. I. Vavilov, "Newton and the Atomic Theory," *Newton Tercentenary Celebrations, 15–19 July, 1946* (Cambridge, 1947), 43–55; et Marie Boas Hall, "Establishment of the Mechanical Philosophy," *Osiris* 10 (1952), 412–541. La communication récente de A. R. Hall et Marie Boas Hall, "Newton's Theory of Matter," *Isis* 51 (1960), 131–44, se trouve reproduite dans leur livre, cité dans la note suivante.

2. A. R. Hall et Marie Boas Hall, *Unpublished Scientific Papers of Isaac Newton: A Selection from the Portsmouth Collection in the University Library* (Cambridge, 1962).

3. En citant cet ouvrage classique en français, je reste aussi fidèle que possible à la traduction de Pierre Coste, faite d'après la seconde édition anglaise de 1717: *Traité*

d'Optique par Isaac Newton, reproduction fac-simile de l'édition de 1722 (Paris: Gauthier-Villars, 1955).

4. *Traité d'Optique*, 486.
5. Cette "Question" est mieux connue comme "Question 31," numéro qu'elle reçut dans la deuxième édition anglaise du *Traité d'Optique* (1717/18). Elle constitue le document le plus important pour comprendre la théorie de la matière et les idées chimiques de Newton. Voir Hélène Metzger, *Newton et l'évolution de la théorie chimique*, extrait de *d'Archeion*, 9 et 10 (Rome, 1928–29); Douglas McKie, "Some Notes on Newton's Chemical Philosophy," *Philosophical Magazine* 33 (1942), 847–70; R. J. Forbes, "Was Newton an Alchemist?" *Chymia* 2 (1929), 27–36; Marie Boas Hall, "Newton's Chemical Papers," in *Isaac Newton's Papers and Letters on Natural Philosophy*, ed. I. B. Cohen (Cambridge, Mass.: Harvard University Press, 1958), 241–48. Pour les "Questions" dans les différentes éditions du *Traité d'Optique*, voir Alexandre Koyré, "Les Queries de l'*Optique*," *Archives Internationales d'Histoire des Sciences* 13 (1960), 15–29.
6. *Traité d'Optique*, 487.
7. *Ibid.*, 482.
8. Pour les opinions de M. Cohen, voir la "General Introduction" à sa collection de documents newtoniens, *Isaac Newton's Papers and Letters*, pp. 4–9, où nous lisons: "Although he always presented his thoughts on the aether with some degree of tentativeness, he did so over so long a period of time that the conclusion is inescapable that a belief in an aetherial medium, penetrating all bodies and filling empty space, was a central pillar of his system of nature." Les mêmes idées sont exprimées par M. Cohen dans son *Franklin and Newton* (Philadelphia: American Philosophical Society, 1956), 166–72, 173–74.
9. Hall and Hall, *Unpublished Scientific Papers*, 183–213. Mme. Metzger trouve que "dans le système de Newton, nous pourrons nettement percevoir des hésitations ou des oscillations entre des manières de voir hétérogènes et peut-être incompatibles." *Attraction universelle et religion naturelle chez quelques commentateurs anglais de Newton* (Paris: Hermann, 1938), 58.
10. *Oeuvres Complètes de Voltaire*, ed. Louis Moland, 52 vols. (Paris, 1877–85), XX (1879), 127.
11. *The Leibniz-Clarke Correspondence*, ed. H. G. Alexander (Manchester: Manchester University Press, 1956), 16.
12. *Principes Mathématiques de la Philosophie Naturelle, par feue Madame la Marquise du Chastellet*, 2 vols. (Paris, 1759), I, 167.
13. *Principes Mathématiques*, I, 200. Cf. Definition 1: "Je ne fais point attention ici au milieu qui passe librement entre les parties des corps, supposé qu'un tel existe." La même prudence est exprimée avec évidence dans le *Traité d'Optique*, 454.
14. Hall and Hall, *Unpublished Scientific Papers*, 184–85.
15. "An Hypothesis explaining the Properties of Light, discoursed of in my several Papers," reproduit en fac-simile dans *Newton's Papers and Letters*, 178–90. Cette communication ne fut jamais publiée du vivant de Newton: elle parut pour la première fois dans Thomas Birch, *History of the Royal Society of London*, 4 vols. (London, 1756–57), III, 248–305.
16. Publié dans *Works of the Honourable Robert Boyle*, 5 vols. (London, 1744), I, 70–74; reproduit en fac-simile dans *Newton's Papers and Letters*, 250–54.
17. Ce document inédit, datant probablement de la période 1664–68, est intitulé *De gravitatione et aequipondio fluidorum*. Il fut découvert par les Hall et publié dans leur collection de documents, *Unpublished Scientific Papers*, 89–156.
18. Hall and Hall, *Unpublished Scientific Papers*, 316.
19. Publié en latin et en anglais dans John Harris, *Lexicon Technicum* (London, 1710); fac-simile dans *Isaac Newton's Papers and Letters*, 256–58. Il est à peu près certain que les copies manuscrites de ce petit ouvrage, ainsi peut-être que d'autres manuscrits chimiques de Newton, circulaient bien avant leur publication. La communication de John Keill, faite à la Royal Society en 1708, a une parenté évidente avec le *De Natura Acidorum*. En 1704, un autre disciple de Newton, John Freind, appliqua, dans des conférences données a Oxford, l'attraction newtonienne à des problèmes de chimie.

20. Hall and Hall, *Unpublished Scientific Papers*, 316.
21. Newton, comme Gassendi (et comme Guglielmini plus tard), notait les formes régulières des cristaux—du nitre et du sel ammoniac par exemple—mais ne croyait pas par là que les formes macroscopiques résultaient des différentes formes des atomes. Il se demandait s'ils n'étaient pas la conséquence de la formation de configurations réticulaires ["net-like figures"] formées par les atomes ["first seeds of things"].
22. Les différences entre l'atomisme de Gassendi et celui de Newton sont mises en évidence par Koyré. L'atomisme classique d'Epicure et de Gassendi "est un atomisme qualitatif et dynamique . . . et les atomes de Newton et de Huygens ne sont plus dynamiques du tout." B. Rochot et al., *Pierre Gassendi*, Centre International de Synthèse (Paris: A. Michel, 1955), 177–78.
23. Thomas Mayo, *Epicurus in England* (Dallas: Southwest Press, 1934), 27. Dans l'étude des courants épicuriens et anti-épicuriens au 17ᵉ siècle, j'ai été beaucoup aidé par les étudiants qui ont participé à mon séminaire à Cornell University. Je dois signaler en particulier les contributions de M. Robert Kargon sur Charleton et Boyle, et celles de M. Joel Rodney sur Cudworth.
24. John Smith, *Select Discourses* (London, 1660), 41.
25. Voir les quelques indications données par Antoine Adam dans Rochot et al., *Pierre Gassendi*, 157–82.
26. Bernard Rochot, *Les Travaux de Gassendi sur Epicure et sur l'Atomisme, 1619–1658* (Paris: J. Vrin, 1944).
27. Voir la lettre de Boyle à Hartlib (8 mai 1647) où nous lisons: ". . . especially Gassendus, a great favorite of mine, I take to be a very profound mathematician, as well as an excellent astronomer, and one, that has collected a very ample treasure of numerous and accurate observations of all, that belongs to the abstruse science of those sublimer bodies." *Works of the Honourable Robert Boyle*, I, 24. Ce passage a été faussement interprété comme montrant que Boyle connaissait les travaux philosophiques de Gassendi. Il n'est pas sans intérêt d'indiquer que la *Institutio Astronomica*, publiée à Paris en 1647, fut le premier des ouvrages de Gassendi imprimé à Londres (1657); cinq éditions anglaises en parurent avant la fin de 1675. Pour les travaux de Gassendi en astronomie, consulter Pierre Humbert, *L'oeuvre astronomique de Gassendi* (Paris, 1936). Une évaluation peu enthousiaste de Gassendi en tant que savant est donnée par Alexandre Koyré dans *Pierre Gassendi*, 59–69.
28. Voir les lettres de Sir Charles Cavendish à John Pell dans James Orchard Halliwell *A Collection of Letters illustrative of the Progress of Science in England* (London, 1841); en particulier la lettre écrite de Hambourg en 1644, où nous lisons: "Mr. Hobbes writes Gassendus his philosophie is not yet printed, but that he hath read it, and that it is as big as Aristotle's philosophie, but much truer and excellent Latin."
29. Sur Charleton et son oeuvre scientifique, il n'existe aucun ouvrage d'ensemble, mais seulement quelques esquisses biographiques: l'article de Norman Moore, M.D., dans le *Dictionary of National Biography*, et celui de Humphrey Rolleston, "Walter Charleton, D.M., F.R.C.P., F.R.S.," *Bulletin of the History of Medicine* 8 (1940), 403 sq. [Since this paper first appeared Robert Kargon has published his *Atomism in England from Hariot to Newton* (Oxford, 1966)].
30. Walter Charleton, *The Immortality of the Human Soul, demonstrated by the Light of Nature. In two dialogues* (London, 1657), 33–48. Je dois cette référence à mon ami et confrère Dean W. R. Keast de Cornell University; c'est lui qui a attiré mon attention sur Charleton.
31. Walter Charleton, *The Darknes of Atheism Dispelled by the Light of Nature: A Physico-Theological Treatise* (London, 1652). Voir "A Preparatory Advertisement to the Reader."
32. *Darknes of Atheism*, 61.
33. Voir *infra*, 93–95.
34. Le philosophe John Locke avait une dette considérable envers Gassendi. Voir H. R. Fox Bourne, *Life of John Locke*, 2 vols. (London, 1876), II, 90–94. Antoine Adam dit que Locke "a gardé sur Gassendi un silence étonnant. . . . Mais dans le *Medical Common Place Book*, où il a donné des indications sur ses études en

1659–1660, on constate qu'il lisait Gassendi et prenait des notes sur lui. Dans sa bibliothèque, à Oates, il y avait Gassendi. Et surtout, pendant un séjour à Paris, après 1675, il a fréquenté Justel, Thoinard et Bernier." *Pierre Gassendi*, 159.

35. Marie Boas Hall, "Establishment of the Mechanical Philosophy." Voir aussi son livre *Robert Boyle and Seventeenth-Century Chemistry* (Cambridge, 1958).

36. Boyle, au début, connaissait Gassendi seulement comme astronome. Voir note 27. Mais vers 1660, quand il écrivit ses premiers ouvrages touchant à la théorie corpusculaire, il était sûrement au courant des ouvrages de Gassendi et de Charleton. Les similarités entre les opinions et même les expressions de Charleton et de Boyle sont soulignées par Robert Kargon dans son article "Walter Charleton, Robert Boyle, and the Acceptance of Epicurean Atomism in England." [Subséquemment publié dans *Isis* 55 (June 1964), 184–92.]

37. Mayo, *Epicurus in England*, 33.

38. Dans l'introduction ("An apologie for Epicurus") à son *Epicurus, his Morals*, Charleton explique qu'il a écrit ce livre pour réparer "les injures faites à la mémoire du modéré, bon et pieux Epicure."

39. "Some specimens of an attempt to make Chymical Experiments useful to illustrate the notions of the Corpuscular Philosophy," *Certain Philosophical Essays* (1661), *Works of the Honourable Robert Boyle*, I, 227–30. Le corpuscularisme de Boyle, visible dans *New Experiments Physico-Mechanical Touching the Spring of the Air* (1660) et dans le *Sceptical Chymist* (1661) se trouve dans sa *Experimental History of Colours* (1663) et particulièrement dans son *Origin of Formes and Qualities* (1666).

40. Cité par Marjorie Hope Nicolson, *Mountain Gloom and Mountain Glory* (Ithaca, N.Y.: Cornell University Press, 1959), 118.

41. Sur Boyle comme physico-théologien, voir Louis Trenchard More, *The Life and Works of the Honourable Robert Boyle* (London: Oxford University Press, 1944); M. S. Fisher, *Robert Boyle, Devout Naturalist* (Philadelphia: Oshiver Studio Press, 1945); et Richard Hunt, *The Place of Religion in the Science of Robert Boyle* (Pittsburgh: University of Pittsburgh Press, 1955).

42. Alexandre Koyré, *From the Closed World to the Infinite Universe* (Baltimore: Johns Hopkins Press, 1957). Sur Cudworth, voir l'esquisse biographique de Thomas Birch dans son édition du *True Intellectual System of the Universe*, new edition, 4 vols. (London, 1820). C'est l'édition que nous avons consultée. De l'ouvrage de Cudworth, un abrégé fut publié à Londres en 1706 (2 vols. in-4°) avec une introduction de Thomas Wise. Des extraits en furent donnés par Le Clerc en 1703 dans sa *Bibliothèque Choisie*. Le nouveau livre de John Passmore, *Ralph Cudworth, an Interpretation* (Cambridge, 1951), ne traite qu'en passant l'atomisme de Cudworth. D'après Antoine Adam, "Cudworth (1617–1688) est gagné à l'atomisme par Boyle. Il publie en 1678 *The True Intellectual System of the Universe*, où il fait maints emprunts au *Syntagma*." *Pierre Gassendi*, 159. Je n'ai rien trouvé que vérifie cette affirmation; Birch nous donne très peu d'informations sur les relations de Boyle et de Cudworth. Voir *Works of the Honourable Robert Boyle*, I, 53, 76.

43. *True Intellectual System*, I, 52–53.

44. Il ne faut pas confondre ce Moschus avec Moschos de Syracuse, le poète bucolique toujours associé avec Théocrite et Bion. L'idée d'attribuer à Moschus l'invention de la théorie atomique n'est pas une innovation de Cudworth. On la trouve en effet dans un passage de Strabon (attribuée à Posidonius) et également dans Sextus Empiricus. La première mention que j'en trouve dans un livre scientifique moderne est de Daniel Sennert, *Hypomnemata physica* (Francfort, 1636), 89. Gassendi y fit allusion plusieurs fois (e.g., *Opera*, I, 257b et VI, 160a), et c'est sûrement de lui que Charleten (*Physiologia*, 87) et Thomas Stanley (*History of Philosophy*, art. "Democritus") la tiennent. Il n'est pas sans intérêt de lire dans le *Sceptical Chymist* de Robert Boyle que "the devising of the atomical hypothesis commonly ascribed to *Leucippus* and his disciple *Democritus* is by learned men attributed to one *Moschus* a Phenician. And possibly the opinion is yet ancienter than so; for it is known, that the Phenicians borrowed most of their learning from the Hebrews." *Works*, I, 315; voir pp. 228, 230. La dernière phrase, aussi bien que la remarque de Boyle citée plus haut, indique, je crois, que la théorie de Cudworth se trouve, pour ainsi dire à l'état

d'embryon, dans les écrits de Boyle. Mais c'est Cudworth qui lui donna son ampleur et en développa toutes les implications.

45. *True Intellectual System*, I, 52–53.
46. Mayo, *Epicurus in England*, 192–95.
47. Les textes essentiels et les indications bibliographiques sont donnés par Perry Miller, "Bentley and Newton," dans *Isaac Newton's Papers and Letters*, 271–394. Pour une analyse des conférences de Bentley, voir Hélène Metzger, *Attraction universelle*, 80–93.
48. Bentley n'eut pas peur d'employer le mot "atome." De plus, à propos de la "propriété constante de la gravitation" qui ressemble au *pondus* d'Epicure, il écrit: "This is the ancient Doctrine of the *Epicurean* physiology, then and since very probably indeed, but yet precariously asserted: But is lately demonstrated and put beyond controversy by that very excellent and divine Theorist Mr. Isaac Newton." *Isaac Newton's Papers and Letters*, 320. Il fait de Newton un partisan du vide: "Now since Gravity is found proportional to the Quantity of Matter, there is a manifest Necessity of admitting a *Vacuum*, another principal Doctrine of the *Atomical* Philosophy." *Ibid.*, 321.
49. *Isaac Newton's Papers and Letters*, 281. Voir *The Correspondence of Isaac Newton*, ed. H. W. Turnbull, 4 vols., others in progress, (Cambridge, 1959–67), III, 234.
50. *Correspondence of Isaac Newton*, III, 249.
51. *Isaac Newton's Papers and Letters*, 302–3. Je ne vois pas comment, vu les mots que j'ai mis en italiques, cette phrase peut être interprétée comme impliquant l'existence de l'éther.
52. Ecoutons Nehemiah Grew: "Motion is not of the Essence of Body; because we may have a Definitive Conception of Body, abstracted from that of Motion." *Cosmologia Sacra* (London, 1710), 5. Roger Cotes était d'accord; en 1713 il écrivit au Dr. Samuel Clarke: "I understand by Essential propertys such propertys without which no others belonging to the same substance can exist." *Correspondence of Sir Isaac Newton and Professor Cotes*, ed. J. Edleston (London, 1850), 158.
53. Mayo, *Epicurus in England*, 68. Charleton appela le mouvement spontané des atomes le "old starting hole, or salley port" des athées. *Darknes of Atheism*, 67.
54. *True Intellectual System*, I, 43.
55. Voir Paul Janet, *Essai sur le médiateur plastique de Cudworth* (Paris, 1860).
56. *Jacobi Rohaulti Physica, latine vertit . . . Samuel Clarke*. 2e édition (London, 1702).
57. William Whiston, *Astronomical Principles of Religion Natural and Reveal'd* (London, 1717), 45–46. Les italiques sont de moi. Et voir plus bas les opinions de Georges Cheyne.
58. *Traité d'Optique*, 434.
59. *Optice: sive de Reflexionibus, Refractionibus, Inflexionibus & Coloribus Lucis Libri Tres. Authore Isaaco Newton, Equite Aurato. Latine reddidit Samuel Clarke, A. M.* (London, 1706), 310. Nous lisons dans l'original latin: "Praeterea, nulla esse omnino istiusmodi Media fluida, inde colligo, quod Planetae & Cometae regulari adeo & diuturno Motu per spatia caelestia undiq; & quaquaversum & in omnes partes ferantur. Inde enim liquet, spatia caelestia omnis sensibili resistentiae, & consequenter omnis sensibilis materiae, expertia esse."
60. *Traité d'Optique*, 438.
61. *Ibid.*, 443.
62. *Ibid.*, 444. La fin de la phrase de la traduction de Coste a été modifiée afin qu'elle soit conforme aux texte latin: "Tacite attribuentes Vim Gravitatis, alii alicui *Causae* a *Materia* diversae." *Optice*, 314. Les italiques sont de moi, sauf dans le texte Latin cité ici.
63. *Traité d'Optique*, 444–45.
64. *Optice*, 314.
65. *Traité d'Optique*, 444.
66. *Traité d'Optique*, 445. Voir Alexandre Koyré et I. B. Cohen, "The Case of the Missing Tanquam," *Isis* 52, (1961), 555–66.
67. *Traité d'Optique*, 483.
68. *Ibid.*, 487–89.
69. *Ibid.*, 439, 487.

70. *Ibid.*, 490–91.
71. *Philosophical Principles of Religion: Natural and Revealed*, 2d edition, corrected and enlarged. By George Cheyne (London, 1715). Voir en particulier pp. 38–44.
72. Metzger, *Attraction universelle*, 143. Pour Cheyne comme pour Newton, les ressorts de la grande machine que constitue l'univers "are an immaterial Principle (if I may so call that of *Gravitation*) which animates the whole and all its Parts: an Original Impress, or a constant efflux from the Divine *Energy*, which enables the whole and all the several Parts, regularly, constantly and harmoniously to attain their destined ends and Purposes."
73. Metzger, *Attraction universelle*, 70.
74. *Ibid.*, 58.

8
Francis Hauksbee: Expérimentateur au Profit de Newton

✿

Cette communication fait en quelque sorte suite à une conférence au *Palais de la Découverte* (sur la théorie de la matière chez Newton et sur sa doctrine d'attraction) et à un travail préliminaire sur le physicien et expérimentateur anglais, Francis Hauksbee.[1]

Sur Hauksbee nous n'avons que de faibles indications biographiques. Nous ne savons quand, ni où, il est né. Il entre mystérieusement en scène à la Société Royale de Londres vers la fin de 1703 comme démonstrateur d'expériences physiques. Le 30 novembre 1705 il est élu "Fellow," mais jusqu'à sa mort—à la fin de mai ou au début de juin 1713—il a continué à jouer le rôle de démonstrateur rémunéré à la Société Royale, sans avoir d'ailleurs le titre officiel dont jouissait avant lui Hooke et Papin, et plus tard Desaguliers. On cite Hauksbee de nos jours surtout pour ses expériences d'électricité; et il est généralement reconnu —peut-être à tort—comme l'inventeur de la première machine électrostatique. Mais on a de lui toute une série de communications sur divers aspects de la physique expérimentale, publiées dans les *Philosophical Transactions* de la Société Royale entre 1704 et 1713.

Les premières furent réunies dans un livre publié en 1709: *Physico-Mechanical Experiments on Various Subjects*, dont une deuxième édition parut en 1719, après la mort de l'auteur, amplifiée avec ses dernières communications. Une traduction italienne de la première édition parut en 1716. Beaucoup plus tard—en 1754—une traduction française, par un certain M. de Brémond, fut publiée avec commentaires par Nicolas Desmarest.[2]

Par conséquent Hauksbee fut un des expérimentateurs anglais les plus discutés par des savants français pendant la deuxième moitié du XVIIIe siècle. Laplace, par exemple, faisait grand cas dans sa *Mécanique céleste* de ses expériences sur la capillarité et la tension superficielle.[3]

✿

Communication faite au Groupe Français d'Historiens des Sciences, à Paris, le 25 avril 1963.

From *Archives Internationales d'Histoire des Sciences* 16 (1963), 1113–28. Revue trimestrielle publiée par la Division d'Histoire des Sciences de l'Union Internationale d'Histoire et de Philosophie des Sciences et avec le concours financier de l'Unesco. Publisher: Editions Hermann, 115 Boulevard Saint-Germain, Paris—6e.

Au XVIIIe siècle, on a remarqué, *grosso modo*, une influence mutuelle de Hauksbee et de Sir Isaac Newton. "M. Hauksbee" écrit Nicolas Desmarest "fournit à M. Newton plusieurs experiences délicates qui appuient très solidement les pensées de ce grand homme."[4] En effet, Hauksbee parle avec vénération de Newton dans son livre; et on peut même dire que les principes de Newton—sur la lumière, sur la matière, sur l'attraction—ont fourni à cet expérimentateur de génie, à ce présumé autodidacte en matière de science théorique, tout son appareil conceptuel.[5] Réciproquement Hauksbee a laissé des traces de son influence dans les derniers travaux de Newton. C'est de cette association féconde du vieux doyen de la science anglaise avec l'expérimentateur plus jeune et plus vigoureux, que je voudrais parler aujourd'hui.

Il est curieux de constater qu'aujourd'hui on a presque oublié—du moins on n'a certainement pas étudié sérieusement—les relations entre Hauksbee et Newton. Le nom de Hauksbee n'est pas cité—pour autant que je sache—dans les biographies réputées de Newton, comme celles de David Brewster et de Louis Trenchard More. Bien qu'on remarque quelquefois le passage de l'*Optique* de Newton où paraît le nom de Hauksbee, et un autre endroit—assez transparent d'ailleurs—où Newton parle de la machine électrique de Hauksbee—personne (si ce n'est W. S. Hardy dans une étude sur un aspect spécial),[6] n'a poursuivi ces indications.

Cependant, il est facile de faire voir l'intérêt exceptionnel que Newton manifestait à l'égard des expériences de Hauksbee. D'un côté nous avons le témoignage de l'*Optique*, et de l'autre celui du *Journal Book*, c'est-à-dire des procès-verbaux inédits des séances de la Société Royale sous la présidence de Newton.

Le *Journal Book* nous apprend que Hauksbee a fait ses débuts devant la Société Royale—le 5 décembre 1703—avec la première expérience d'une série réalisée à l'aide d'une pompe à air particulièrement perfectionée. Il s'agissait de ce qu'on appelait "l'expérience du phosphore mercuriel," une expérience dans laquelle Hauksbee versait le vif-argent dans le vide du récipient de sa pompe à air et montrait la production d'une lumière très vive, très brillante. Ce qu'il est important de souligner c'est que ces débuts de Hauksbee coïncident précisément avec l'arrivée de Newton comme Président de la Société Royale. En fait, Hauksbee produit sa première expérience le jour même où Newton préside pour la première fois. Désormais pendant presque dix ans—depuis la fin de 1703 jusqu'au début de 1713—Hauksbee montre des expériences et des démonstrations à presque chaque réunion de la Société Royale et, presque sans exception, en présence de Newton qui présidait régulièrement les séances. Le *Journal Book* fait mention, à plusieurs reprises, de questions posées par Newton, ou de remarques faites par lui au cours des séances, au sujet des expériences de Hauksbee.

A l'automne de 1705—après avoir démontré une série d'expériences classiques, voire banales, avec sa pompe à air—Hauksbee revient au problème originel: comment expliquer la production de la lumière quand le mercure est agité *in vacuo?*

C'était pour voir ce genre d'expériences, faites par Hauksbee avec sa pompe à air, que Newton a voulu inviter quelques personnages à son appartement de Jermyn Street, au mois de septembre 1705. La démonstration n'a pas eu lieu, car les célébrités en question—Lord Pembroke, Lord Halifax et l'archevêque de Dublin—étaient toutes à la campagne. Mais les deux lettres de Newton à Sir Hans Sloane qui en parlent, témoignent que Newton s'intéressait tout spécialement, et de bonne heure, aux expériences de Hauksbee.[7] Cinq ans plus tard—et il s'agit aussi d'expériences faites en dehors des séances de la Société Royale— Newton a suivi de près les expériences de 1710 dans lesquelles Hauksbee laissait tomber des boules de verre de la coupole de l'église de Saint-Paul pour déterminer l'effet de la résistance de l'air.[8]

❀

Je me bornerai, dans les pages suivantes, à signaler quelques aspects de la pensée de Newton, et quelques endroits dans ses écrits, où il a été manifestement influencé par les expériences de Hauksbee, surtout celles sur la capillarité et celles —plus souvent citées—sur l'électricité et la lumière électrique.

Ces enquêtes si fructueuses de Hauksbee ont eu une origine assez modeste. Chaque fois il a débuté avec une expérience pittoresque, ou bien classique, faite avec sa pompe à air. Ainsi pour son travail sur la capillarité. L'ascension extraordinaire des fluides dans les tubes capillaires avait été remarquée au XVIIe siècle par divers savants y compris ceux de l'Accademia del Cimento.[9]

Les *Saggi* de cette Académie rapportent une expérience dans laquelle il fut démontré que l'effet se produit aussi bien dans le vide—le vide de Torricelli— qu'en plein air. Ce phénomène intéressait également Hooke et Boyle, et ce dernier essaya aussi d'observer l'ascension dans le vide de sa pompe à air. C'est manifestement en répétant l'expérience de Boyle que Hauksbee a débuté dans ce genre de recherches.[10]

Après avoir démontré en avril 1706 que la présence ou l'absence de l'air n'influence en aucune mesure l'ascension des fluides dans les tubes capillaires, il a essayé de généraliser le phénomène et de l'observer dans des conditions analogues. En particulier il a noté l'ascension des fluides entre deux plaques de verre, de marbre ou de métal, employant non seulement l'eau, mais l'alcool et des huiles différentes.[11] En variant ainsi les matériaux employés, il a montré que la montée n'est pas une propriété d'un seul fluide ou d'un seul solide. Dans une expérience de grande portée, il a prouvé que pour des tubes de verre de même calibre l'ascension est toujours indépendante de l'épaisseur du verre, même si un des tubes est dix fois plus épais que l'autre. Par conséquent l'attraction entre le fluide et le solide doit être un phénomène limité à la surface.[12]

Ses premières expériences furent publiées dans les *Philosophical Transactions* et ensuite dans son livre de 1709. Deux ans plus tard, en 1711, Hauksbee reprit ses expériences sur la capillarité, encouragé selon toute probabilité par Newton.

En particulier, dans une série d'expériences, il suivit les mouvements des gouttes de l'huile d'orange entre deux plaques de verre diversement inclinées l'une en face de l'autre.[13] Dans ses dernières communications (de 1712 à 1713) il essaya de déterminer la courbe formée par la surface de l'eau entre les plaques de verre, et tenta d'estimer la force de l'attraction, une force qui s'exerce à très courtes distances et toujours perpendiculaire à la surface intérieure des tubes ou des plaques de verre.[14]

Dans la première édition de son livre—c'est-à-dire, avant d'être beaucoup avancé dans ses expériences sur la capillarité—Hauksbee annonce que pour expliquer ce phénomène il n'invoquera qu'un seul principe: le principe de l'attraction "qui domine partout dans la nature, et par lequel la plupart de ses phénomènes sont explicables." Bien que le mot "attraction" soit un mot "dur et inintelligible" il est tout de même évident qu'une telle puissance existe dans la nature, pas seulement entre les corps du système solaire, "mais aussi entre les corpuscules plus petits et insensibles." La loi d'attraction macroscopique, la loi qui agit entre les grands corps de l'univers, est bien connue; elle est—ce sont les mots de Hauksbee—"parfaitement déterminée et établie." Mais la loi de l'attraction des petites particules de la matière reste à découvrir, bien qu'on sache que dans ce cas les forces d'attraction doivent diminuer plus rapidement que suivant le carré de la distance. Il y aura des difficultés à déterminer cette loi avec exactitude; mais le fait cependant est incontestable, "et les découvertes faites par ce très grand homme, Sir Isaac Newton (l'honneur de notre nation et de la Société Royale) ont jusqu'à présent très clairement mis au jour ces deux lois d'attraction pour ceux qui se serviront de leurs yeux pour les voir."[15]

Tout le monde sait que Newton a démontré la grande loi de la gravitation universelle dans ses *Principes*. Mais à quoi Hauksbee fait-il allusion en parlant de la deuxième "loi" d'attraction? Selon toute apparence, à la fameuse dernière question de l'*Optique*, parue en grande partie pour la première fois dans l'édition latine de 1706. Dans cette question Newton rassemble un tas de faits chimiques et physico-chimiques pour prouver l'existence d'une attraction corpusculaire. Même avant les additions que Newton fit plus tard on pourrait dire avec Fontenelle que "l'attraction domine dans ce plan abrégé de physique."[16]

De son côté Newton s'intéressait vivement aux expériences de Hauksbee sur la capillarité, comme en témoigne clairement une longue addition faite en 1717 à la dernière question (Q. 31). Cette addition—de quatre pages—de beaucoup la plus longue parmi le grand nombre d'additions, et de changements faits par Newton dans les Questions de 1706, n'est qu'un abrégé des expériences de Hauksbee sur l'ascension des fluides dans les tubes capillaires, dans un tuyau de verre "rempli de cendres passées au tamis" et entre des plaques de verre. C'est le 8 janvier 1712–13, à une réunion de la Société Royale, que Newton avait prié Hauksbee de répéter attentivement ses expériences pour estimer la force d'attraction capillaire.[17] Il s'agissait des expériences dans lesquelles Hauksbee suivait le mouvement des gouttes d'huile d'orange ou d'esprit de térébenthine entre deux plaques de verre inclinées l'une en face de l'autre. Newton consacre deux pages de son addition de 1717 à ces expériences, les terminant ainsi avec une mention explicite:

> Or par quelques Expériences de ce genre, faites par feu M. Hauksbee, l'on a trouvé que l'Attraction est presque réciproquement en raison doublée de la distance du milieu de la Goutte au concours des Verres; sçavoir réciproquement en proportion simple à raison de ce que le Goutte se répand d'avantage, & touche chaque Verre par une plus grande Surface; & encore réciproquement en proportion simple à raison de ce que les Attractions deviennent plus fortes, la quantité des Surfaces restant la même. Donc l'attraction que se fait dans la même quantité de Surface attirante, est réciproquement comme la distance entre les Verres; et par conséquent où la distance est excessivement petite, l'attraction doit être excessivement grande. . . . Il y a donc dans la Nature, des Agents capables d'unir ensemble les particules des Corps, par des Attractions très-fortes : c'est à la Philosophie Expérimentale à découvrir ces Agents.[18]

Pour Newton, comme nous allons voir, d'autres expériences de son confrère —surtout celles que Hauksbee a faites sur les phénomènes électrostatiques— indiquaient déjà une voie pleine de promesses pour la découverte de "ces Agents." Mais, en attendant, il aurait pu dire, comme dira beaucoup plus tard l'astronome Lalande: "Les Tubes Capillaires nous mettent entre les mains un indice palpable de la généralité de cette loi qui est la clef de la physique, le plus grand ressort de la nature, et le mobile universel de l'Univers."[19]

✸

Dans l'histoire de l'électricité Hauksbee est considéré, non sans raison, comme un grand précurseur, mais il n'a pas fait exprès ses premières expériences électriques; il y est arrivé progressivement et comme par mégarde. Au cours de ses recherches sur le mercure agité dans le vide, il a constaté que plus il y avait de mouvement et de frottement, plus la lumière produite était vive. Le mouvement était l'essentiel; et il se demandait si d'autres substances—frottées vivement *in vacuo*—donneraient le même résultat. Pour répondre à cette question il inventa un appareil ingénieux où différentes paires de substances pourraient être mécaniquement frottées, l'une contre l'autre, dans le vide du récipient de sa pompe à air.

Avec ce dispositif il a frotté le succin (l'ambre) contre l'étoffe de laine; le silex (pierre à fusil) contre l'acier et—ce qui produisait toujours un effet très spectaculaire—il faisait tourner un petit globe de verre contre de la laine dans le vide du récipient.[20]

Il montra cette dernière expérience avec grand succès devant la Société Royale, le 19 décembre 1705, en présence de Newton; c'était évidemment un succès, car il répéta la démonstration deux fois au mois de janvier 1705–6.[21] Nous avons également une trace, un écho, de cette série d'expériences dans quelques additions que Newton fit cette même année 1706 à l'édition latine de son *Opticks*.

Dans la première édition de ce livre clasique (1704) la Question n° 8 était très courte, consistant en une seule phrase:

> Tous les Corps fixes, lorsqu'ils sont échauffés au delà d'un certain degré, ne jettent-ils pas de la Lumiere, & ne brillent-ils pas? Cette *émission* n'est-elle pas produite par les vibrations de leurs parties?."[22]

Mais dans l'édition latine (dans les *Errata, Corrigenda* et *Addenda*) Newton fait nettement allusion, dans des phrases ajoutées, aux expériences récentes de Hauksbee, en donnant comme exemples de la lumière produite par le mouvement et le frottement:

> . . . le Vif-argent secoüé dans *le Vuide* . . . le Phosphore vulgaire agité par l'attrition de quelque Corps que ce soit . . . des particules d'Acier détachées par le choq d'une pierre à fusil."[23]

Plus loin—dans une des nouvelles Questions de l'édition latine (Q. 22)—il parle de nouveau du fait que le vif-argent "secoüé dans le Vuide . . . brille comme le Feu."[24]

La raison pour laquelle l'électricité n'est pas mentionnée dans l'édition latine de l'*Opticks* de Newton est évidente: Hauksbee n'a commencé les expériences vraiment électrostatiques qu'à la fin de l'année 1706. Cette année-là il avait poursuivi ses expériences sur la mystérieuse lumière produite par le frottement de diverses substances *in vacuo*.

Le 6 novembre il rendit compte à la Société Royale de quelques expériences qu'il venait de faire et qui avaient donné des résultats extraordinaires. Dans une de ces expériences—d'ailleurs souvent citée aujourd'hui—au lieu de faire frotter un globe de verre en le faisant tourner dans le vide—c'est-à-dire un globe solide ou plein d'air—il employait un globe *évacué*, intérieurement vide, qu'il frottait mécaniquement en plein air. Une lumière éclatante se produisait, accompagnée d'effets qu'il allait bientôt reconnaître comme étant électriques.[25]

A cette occasion—le 6 novembre 1706—Newton, le président de la séance, fit une remarque significative au sujet du rapport de Hauksbee. La lumière provoquée par ce frottement mécanique devait être causée, disait-il, par "les émanations des verres, et non pas par celles du corps" qui les frottent.[26]

Newton, en faisant cette observation, pensait peut-être à une expérience qu'il avait faite lui-même, et qu'il rapportait dans sa deuxième communication sur la lumière et les couleurs, envoyée trente ans auparavant à la Société Royale.

Dans cette communication il expliquait, par une théorie d'émanations éthériques, la puissance attractive des petites lentilles de verre, vivement frottées.[27] En tous cas—comme j'ai essayé de le démontrer ailleurs—il se peut que cette observation de Newton ait dirigé l'attention de Hauksbee sur les propriétés électriques du verre, et ait inspiré ses expériences immédiatement postérieures, au cours desquelles il faisait voir les effets électrostatiques produits par le frottement des tubes et des globes de verre. En effet, les premières expériences de Hauksbee sur l'électricité furent montrées à la Société Royale avant la fin de novembre 1706.[28] Les résultats furent publiés dans les numéros 307 et 315 des *Philosophical Transactions*, et ensuite—avec de légères modifications—dans son livre de 1709.

Ces expériences de Hauksbee—si singulières et si impressionnantes—ont beaucoup passionné son auditoire. Il les répétait avec succès dans les cours publics de physique expérimentale qu'il donna chaque année jusqu'à sa mort en 1713.[29] Elles ont fortement séduit Newton: on trouve sans difficulté les traces de cette influence dans les modifications et les additions que Newton a faites aux fameuses Questions en préparant, après la mort de Hauksbee, la seconde édition anglaise publiée en 1717–18.

On peut citer d'abord trois petites allusions aux phénomènes électriques. Dans une des questions totalement neuves que Newton a introduites en 1717, et qui, comme les autres nouvelles questions, se rapporte à l'existence d'un éther, nous lisons:

Si quelqu'un s'avisoit de me demander comment un Milieu peut être si rare; qu'il me dise. . . . comment la friction peut faire évaporer d'un Corps électrique une exhalaison si rare & si subtile (quoique si puissante) qu'elle ne cause aucune diminution sensible dans le poids du Corps électrique; & que se répandant dans une Sphere de plus de deux pieds de diametre, (& quelques fois même de plus de six pieds) elle soit pourtant capable d'agiter et d'élever une feüille de Cuivre ou d'Or, à plus d'un pied de distance du Corps électrique.[30]

Ensuite il y a dans la dernière question (**Q. 31**) l'endroit où Newton parle des attractions à courtes distances:

. . . & peut-être que l'Attraction électrique peut s'étendre à ces sortes de petites distances, sans même être excitée par le frotement.[31]

Plus bas dans la même Q. 31—dans une nouvelle interpolation de 1717 au sujet des "qualités occultes"—Newton fait mention encore une fois, mais seulement en passant, des attractions électriques.[32]

Un écho beaucoup plus important des expériences électriques de Hauksbee se trouve dans une longue addition de 1717—la deuxième faite par Newton, puisque plus haut on en a cité une autre faite par Newton—à la Question n° 8. Je donne maintenant le texte en entier, parce que c'est un résumé assez exact de quelques expériences de Hauksbee, bien que la traduction de Coste ne soit pas d'une clarté limpide:

De même, un Globe de Verre d'environ 8 ou 10 pouces de Diametre étant attaché à une machine, par le moyen de laquelle il puisse tourner rapidement autour de son Axe, venant à tourner, jette de la Lumiere dans l'endroit où il est frotté avec la paume de la main. Si dans le même temps on tient un morceau de Papier blanc, ou de Drap blanc, ou le bout du Doigt, à la distance d'environ un quart de pouce, ou un demi-pouce, de la partie du Verre où le mouvement est le plus grand; la vapeur électrique excitée par la friction du Verre contre la main, venant à donner sur le Papier, sur le Drap, ou sur le Doigt, sera mise dans une telle agitation, que jettant de l'éclat elle rendra le Papier, le Drap,

ou le Doigt aussi lumineux qu'un Vert [*sic*] luisant; & en s'élançant hors du
Verre, elle frappera quelquefois le Doigt si vivement qu'on en sentira le choq.
On a éprouvé la même chose en frottant un long & gros cylindre de Verre ou
d'Ambre avec du Papier qu'on tenoit dans la main, ou tout simplement avec
la main, & en continuant la friction jusqu'à ce que le Verre commençât à
s'échauffer.[33]

Dans ce passage de l'*Optique* de Newton quelques savants ont remarqué
l'allusion à la "machine électrique" de Hauksbee; et le professeur I. Bernard
Cohen a reconnu dans le même passage la découverte par Hauksbee du "vent
électrique." Mais à propos de cette découverte il nous dit: "Isaac Newton répéta
cette expérience et sentit le vent qui soufflait contre une feuille de papier, laquelle, écrivit-il, devint alors lumineux comme un ver luisant."[34] Je ne suis pas
complètement d'accord avec M. Cohen: Newton n'a sûrement pas répété les
expériences de Hauksbee; ce n'était pas nécessaire, Newton les avait suivies de
près à la Société Royale.

❀

De toute façon nous pouvons donner un autre exemple—aussi important que
ceux que nous venons de citer, mais un peu antérieur—de l'influence que les
expériences électrostatiques de Hauksbee ont exercée sur Newton. L'exemple se
trouve à la fin du *Scholium Generale* qui fut—comme on le sait—une des additions principales faites par Newton à la deuxième édition de ses *Principes* en
1713. Le dernier paragraphe, que je cite maintenant dans la traduction de Mme
du Châtelet, a été très discuté:

> Ce seroit ici le lieu d'ajouter quelque chose sur cette espéce d'esprit très subtil
> qui pénétre à travers tous les corps solides, & qui est caché dans leur substance;
> c'est par la force, & l'action de cet esprit que les particules des corps s'attirent
> mutuellement aux plus petites distances, & qu'elles cohérent lorsqu'elles sont
> contigus; ce'st par lui que les corps électriques agissent à de plus grandes
> distances, tant pour attirer que pour repousser les corpuscules voisins : & c'est
> encore par le moyen de cet esprit que la lumiére émane, se réfléchit,
> s'infléchit, se réfracte, & échauffe les corps; toutes les sensations sont excitées,
> & les membres des animaux sont mûs, quand leur volonté l'ordonne, par les
> vibrations de cette substance spiritueuse qui se propage des organes extérieurs
> des sens, par les filets solides des nerfs, jusqu'au cerveau, et ensuite du cerveau
> dans les muscles. Mais ces choses ne peuvent s'expliquer en peu de mots; & on
> n'a pas fait encore un nombre suffisant d'expériences pour pouvoir déterminer
> exactement les lois selon lesquelles agit cet esprit universel.[35]

On a tout récemment signalé que dans la traduction anglaise des *Principia*
de Newton par Andrew Motte (1729) les derniers mots de ce paragraphe sont
modifiés: au lieu de lire "cet esprit universel" nous trouvons l'expression "cet
esprit *électrique* et élastique." D'après M. Koyré et M. Cohen l'auteur de cette
modification fut Newton lui-même.[36]

La question vient d'être éclaircie avec la publication l'année dernière de M. et Mme Hall, des *Unpublished Scientific Papers of Isaac Newton*, qui contiennent des brouillons très intéressants du *Scholium Generale* des *Principes*. Celui que les éditeurs appellent le manuscrit A, affirme sans équivoque l'identité de cet "esprit" avec la force électrique, ou plutôt avec l'émanation ou l'effluve électrique.[37]

Et dans une autre version (le brouillon que M. et Mme Hall désignent par C) nous trouvons une discussion qui est pour ainsi dire condensée dans le dernier paragraphe du *Scholium* tel que Newton l'a publié. Mais le manuscrit donne une esquisse de quelques "propositions" que Newton avait eu un moment l'intention de développer dans la révision de ses *Principes*. Or, les sept premières propositions, intercalées au milieu de C, sont suivies par une allusion à des expériences destinées à les éclairer ou les démontrer. Ces propositions se rapportent aux phénomènes physiques que Newton et Hauksbee ont, pour ainsi dire, étudiés ensemble. La plupart—sinon la totalité—des expériences peuvent facilement être identifiées avec celles de Hauksbee.

> Proposition 1. Que les très petites particules des corps, soit contiguës, soit à très petites distances, s'attirent mutuellement. Expérience 1. Des plaques de verre parallèles. 2. Des plaques de verre inclinées. 3. Des tubes. 4. Des éponges. 5. De l'huile d'oranges.[38]

Bien que M. et Mme Hall reconnaissent que ces expériences sont celles dont Newton fait usage dans la fameuse 31e Question de son *Optique*, le fait que ces expériences sont celles de Hauksbee paraît leur échapper.

Dans la Proposition 2, où Newton propose de démontrer, peut-être dans un *scholium* spécial, "que l'attraction est du genre électrique," l'influence de Hauksbee est également manifeste.

Newton avait suivi de près les expériences de Hauksbee dans lesquelles, quand le verre fut vivement frotté, les pailles et d'autres objets légers sont tantôt attirés, tantôt repoussés, tantôt mus de diverses façons.[39] Et à leur sujet Hauksbee avait écrit: "Quelques-uns de ces phénomènes sont si étranges dans leurs circonstances, que j'avoue que je suis porté à croire qu'il n'y en a pas beaucoup dans la nature de plus surprenants."[40]

La Proposition 3 (sur l'attraction et la cohésion) ne dérive probablement pas des relations de Newton et Hauksbee;[41] mais la Proposition 4, sur la répulsion, devait traiter un phénomène (la répulsion électrostatique), que Hauksbee a, sinon découvert, au moins observé soigneusement pour la première fois.[42]

Par delà toutes les autres, les Propositions suivantes font écho aux démonstrations que Hauksbee réalisait devant la Société Royale:[43]

> Proposition 5. Que l'esprit électrique est un milieu [*medium*] des plus subtils et pénètre très facilement les corps solides. Expérience 7. Il pénètre le verre.
> Proposition 6. Que l'esprit électrique est un milieu très actif et émet la lumière. Expérience 8.

Proposition 7. Que l'esprit électrique est activé par la lumière et reçoit une motion vibratoire, qui donne la chaleur. Expérience 9. Des corps dans la lumière du soleil.

Il ne serait pas difficile de choisir des expériences et des démonstrations faites par Francis Hauksbee et de les employer pour éclairer chacune de ces trois propositions précédentes. C'est évidemment ce que Newton avait l'intention de faire. Par exemple, pour montrer que l'esprit électrique est un milieu subtil qui pénètre le verre, il pouvait citer l'expérience dans laquelle le frottement d'un globe plein d'air excite la production d'une lumière dans un autre globe évacué placé à proximité.[44] Plus convaincantes encore seraient les expériences par lesquelles Hauksbee montra directement la répulsion électrostatique exerçant son influence à travers le verre.[45] Ayant mis un axe de bois à l'intérieur d'un globe, il fixa tout autour du centre de cet axe des fils de laine. Quand le globe fut évacué, et sa surface extérieure frottée dans la machine, les fils de laine devinrent rigides et formèrent des rayons émanants de l'axe. En approchant le doigt du globe sans le toucher, on produisait des mouvements de ces fils qui s'éloignaient. De ce phénomène Hauksbee conclut "qu'il existe une communication entre le milieu à l'extérieur du globe et celui qui est dedans." Et il ajouta: "Pas seulement une *communication,* mais une *continuité,* de la matière qui donne lieu au mouvement des fils."[46]

Pour montrer l'association entre "l'esprit électrique" et la production de la lumière (Proposition 6) Newton avait l'embarras du choix parmi les expériences de Hauksbee. N'importe quelle expérience où les effets électrostatiques et lumineux se produisaient ensemble, aurait pu servir. Mais il avait sûrement l'intention de citer l'expérience avec le globe rotatoire, comme il le fait dans son addition à la Question 8 de l'*Optique.* Par contre il est difficile d'imaginer avec quelle expérience de Hauksbee Newton proposa de compléter sa Proposition 7, car Hauksbee ne croyait pas qu'il fût possible de produire l'électrification comme on produit facilement la lumière, par la seule action de l'effluve d'un globe activé agissant sur un autre. Au contraire, Hauksbee était presque persuadé que l'effluve électrique et l'effluve lumineux étaient des émanations distinctes. Il se peut, pourtant, que Newton n'ait pas fait cette distinction et ait simplement voulu dire que la lumière peut chauffer les corps.

En tous cas Newton abandonna finalement l'idée de traiter en détail, dans sa révision des *Principes,* sa théorie d'un "esprit électrique et élastique." Tout ce qu'il nous a donné en 1713 n'est que l'allusion mystérieuse et provocante que nous trouvons dans le dernier paragraphe du *Scholium Generale.* Mais sa théorie fut développée ailleurs comme tout le monde le sait.

❁

Le lecteur qui s'est familiarisé avec l'*Optique* de Newton dans la version définitive reconnaîtra sans difficulté dans cet "esprit électrique et élastique," ce "milieu très subtil," du *Scholium Generale,* le "milieu éthéré" ou l' "éther" dont Newton parle dans les nouvelles questions (Q. 17-24) qu'il ajouta à son *Traité d'Optique* quand il prépara la deuxième édition anglaise (1717).

Dans une communication récente, j'ai essayé de démontrer qu'en 1706— l'année de la publication de l'édition latine—Newton n'était pas partisan d'une théorie éthérique, et n'eut pas recours à un éther pour éclaircir les propriétés de la lumière et expliquer la gravitation, comme il l'avait proposé, un peu timidement et sans grande conviction, trente ans auparavant.

Toutes les allusions à l'existence d'un éther que l'on trouve dans les Questions de l'*Optique* sont sans exception des additions faites à l'édition anglaise de 1717.

En 1706, et depuis au moins vingt ans, Newton était plutôt un partisan du vide absolu épicurien et en même temps d'une explication métaphysique de la pesanteur.[47]

Le revirement de Newton—sa nouvelle disposition en faveur d'un éther— date des expériences de Hauksbee et de la collaboration scientifique toute particulière entre les deux hommes. La coïncidence est frappante, bien que d'autres influences—comme les critiques des Cartésiens et de Leibniz qui commençaient vers cette époque—ont certainement joué un rôle important. Avec les travaux de Hauksbee l'existence d'une matière très subtile paraissait démontrée. Ces expériences, qui ont donné une si grande ampleur au principe fondamental de l'attraction, ont aussi, pour ainsi dire, rendu visible, même palpable, les effluves matériels, les émanations, semblables à celles que Newton avait autrefois imaginé pour expliquer l'attraction et la répulsion. C'est dans le *Scholium Generale* que nous trouvons la première indication du nouveau point de vue de Newton. C'est dans la revision de son *Optique* qu'il ranima et développa—en forme de Questions, il est vrai—sa vieille théorie de l'éther, l'appliquant aux phénomènes de la lumière et même à la pesanteur.

Notes

1. H. Guerlac, "Sir Isaac and the Ingenious Mr. Hauksbee." Subséquemment publié dans les *Mélanges Alexandre Koyré*, 2 vols. (Paris: Hermann, 1964); et *Newton et Epicure*, Conférence du Palais de la Découverte (Paris, 1963).

2. *Expériences physico-mécaniques sur differens sujets . . .* , traduites de l'anglais de M. Hauksbee, par feu M. de Brémond, 2 vols. (Paris, 1754).

3. *Oeuvres complètes de Laplace*, publiées sous les auspices de l'Académie des Sciences, par MM. les secrétaires perpétuels, 14 vols. (Paris, 1878–1912), IV (1880), 402–11.

   La publication en 1754 de la traduction française du livre de Hauksbee souleva l'opposition des anti-attractionistes comme le P. Gerdil, le P. Abat et le P. Aimé-Henri Paulian. Ce dernier traite Hauksbee sévèrement dans son *Dictionnaire de Physique*, 3 vols. (Avignon, 1761). Pour défendre la doctrine de l'attraction, appliquée à la capillarité, l'astronome Jérôme-François de Lalande écrivit sa "Lettre sur les Tubes capillaires, adressée à Messieurs les Auteurs du Journal des Sçavans," *Journal des Sçavans* (octobre 1765), 723–43.

4. *Expériences physico-méchaniques*, I, p. xxviii.

5. Il est à noter que Hauksbee imite Newton dans l'emploi de l'expédient des "queries" pour préciser les questions soulevées par ses expériences.

6. W. S. Hardy, "Historical Notes upon Surface Energy and Forces of Short Range," *Nature* 109 (1922), 375–78.

7. Ces deux lettres furent publiées par John Nichols, *Illustrations of the Literary History of the Eighteenth Century*, 8 vols. (London, 1817–58), IV (1822), 59. [Et après la publication de cet article, les lettres furent imprimées dans *The Correspondence of Isaac Newton*, ed. H. W. Turnbull, 4 vols. (Cambridge: The Cambridge University Press, 1959–67), IV, 446–47, 448, d'après les lettres originales au British Museum.]

8. On affirme souvent que Newton participait activement à ces expériences, mais l'évidence en faveur de cette opinion n'est jamais citée. Hauksbee a probablement fait les expériences sans l'assistance de Newton, mais Newton présidait à la Société Royale, le 14 juin 1710, quand Hauksbee a fait son rapport. Pour l'importance que Newton attachait à ces expériences voir la lettre de Roger Cotes à Newton (June 9, 1711), dans *Correspondence of Sir Isaac Newton and Professor Cotes*, ed. J. Edleston (London and Cambridge, 1850), 43. Pour les résultats, consulter Hauksbee, "Experiments concerning the time required in the descent of different bodies, of different magnitudes and weights, in common air, from a certain height," *Philosophical Transactions of the Royal Society*, no. 328, pp. 196–98. Nous savons qu'en 1711–12 Newton a proposé à Halley et Hauksbee de faire des expériences magnétiques. Voir Edleston, *Correspondence*, 80.

9. Hardy, "Historical Notes," 315; et E. C. Millington, "Theories of Cohesion in the Seventeenth Century," *Annals of Science* 5, no. 3 (1945), 253–69.

10. "Mr. Hauksbee Showed an Experiment Concerning the Assent [*sic*] of Tinged Liquor in Tubes of Severall Small Diameters in Vacuo," *Journal Book of the Royal Society*, 17 avril 1706. Les résultats se trouvent dans *Phil. Trans.*, no. 305, pp. 2223–24, et dans *Physico-Mechanical Experiments* (London, 1709), 77–81. L'exemple de Boyle, très important dans les travaux de Hauksbee, fut probablement l'influence décisive; mais peut-être les expériences de Carré et Geoffroy, publiées tout dernièrement étaient pour quelque chose. Voir leurs "Expériences sur les tuyaux capillaires," *Mém. Acad. Roy. Sci.* (1705), 241, et le résumé par E. C. Millington, "Studies in Capillarity and Cohesion in the Eighteenth Century," *Annals of Science* 5, no. 4 (1947), 352–53.

11. *Phil. Trans.*, no. 319, pp. 258–65, 265–66; et *Physico-Mechanical Experiments*, 139–50. Pour les expériences de Hauksbee sur la capillarité voir Hardy, "Historical Notes," 375–76; et Millington, "Studies in Capillarity," 353–54.

12. *Physico-Mechanical Experiments*, 151.

13. "Experiment touching the direction of a drop of oil between two glass planes," *Phil. Trans.*, no. 332, pp. 395–96.

14. "Experiment concerning the angle required to suspend a drop of oil at certain stations between two glass planes placed in the form of a wedge," *Phil. Trans.*, no. 334, pp. 473–74. Voir également "Experiment touching the ascent of water between two glass planes in an hyperbolick figure," *ibid.*, no. 336, pp. 539–40 et voir no. 337, pp. 151–56. Brook Taylor fut le premier à insinuer que la courbe de la surface de l'eau entre deux plaques de verre doit être hyperbolique. Dans ses communications de 1712 et 1713 Hauksbee confirmait ce fait par des mesures soigneuses.

15. *Physico-Mechanical Experiments*, 156–58. Je n'ai pas réussi à trouver dans les écrits de Newton un énoncé aussi précis que celui de Hauksbee au sujet de la forme de la loi d'attraction microcosmique.

16. Fontenelle, "Eloge de Newton," dans *Choix d'Eloges*, ed. Paul Janet (Paris, 1888), 329.

17. *Journal Book*, 8 janvier 1712–13.

18. *Traité d'optique . . . Par Monsieur Le Chevalier Newton. Traduit par M. Coste, sur la seconde Edition Angloise, augmentée par l'Auteur. Seconde Edition Françoise* (Paris, 1722), 477–78. Je la cite d'après la reproduction fac-simile (Paris: Gauthier-Villars, 1955).

19. *Journal des Sçavans* (octobre 1768), 124; passage cité (en anglais) par Hardy, "Historical Notes," 376.

20. "Several experiments on the attraction of bodies in Vacuo," *Phil. Trans.*, no. 304, pp. 2165–75; et *Physico-Mechanical Experiments*, 17–36.
21. *Journal Book*, 19 décembre 1705; 2 janvier et 9 janvier 1705–6.
22. *Traité d'Optique*, 405.
23. *Ibid.*, 406.
24. *Ibid.*, 452. Cette "Question" ou "Query" devient Q. 30 dans l'édition anglaise de 1717.
25. *Journal Book*, 6 novembre 1706.
26. *Ibid.*
27. "An Hypothesis explaining the Properties of Light, discoursed of in my several papers," reproduction fac-simile dans *Isaac Newton's Papers and Letters on Natural Philosophy*, ed. I. B. Cohen (Cambridge, Mass.: Harvard University Press, 1958), 178–90. Cette communication ne fut jamais publiée telle quelle du vivant de Newton; elle parut dans Thomas Birch, *History of the Royal Society*, 4 vols. (London, 1756–57).
28. *Journal Book*, 13 novembre 1706.
29. Pour le contenu des cours publics de Hauksbee, voir mon article "Sir Isaac and the Ingenious Mr. Hauksbee," dans les *Mélanges Koyré*; voir Laurence Lewis, *The Advertisements of the Spectator* (London, 1909), 217–18.
30. *Traité d'Optique*, 423.
31. *Ibid.*, 454.
32. ". . . telles que seroient les Causes de la pesanteur, des attractions magnetiques et électriques, & des fermentations," *ibid.*, 488.
33. *Traité d'Optique*, 407.
34. *Benjamin Franklin's Experiments . . .* , ed. with a critical and historical introduction by I. B. Cohen (Cambridge, Mass.: Harvard University Press, 1941), 32–37.
35. Isaac Newton, *Principes Mathématiques de la Philosophie Naturelle*, trad. Mme. du Châtelet, 2 vols. (Paris, 1759), II, 179–80.
36. Marie Boas Hall et A. Rupert Hall, "Newton's Electric Spirit: Four Oddities," *Isis* 50 (1959), 473–76; et la réponse d'Alexandre Koyré et I. B. Cohen, "Newton's 'Electric and Elastic Spirit,' " *Isis* 51 (1960), 337.
37. A. R. Hall et Marie Boas Hall, *Unpublished Scientific Papers of Isaac Newton: A Selection from the Portsmouth Collection in the University Library* (Cambridge, 1962), 349–55.
38. *Unpublished Scientific Papers*, 357, 361.
39. *Physico-Mechanical Experiments*, 42–51.
40. *Ibid.*, 185. La traduction est la mienne.
41. *Unpublished Scientific Papers*, 352, 361.
42. *Ibid.*, 357, 361.
43. *Ibid.*, 352, 362.
44. *Physico-Mechanical Experiments*, 62–68.
45. *Ibid.*, 59–61, 109–14.
46. *Ibid.*, 111.
47. Voir ma conférence, *Newton et Epicure*.

9
Newton's Optical Aether:
His Draft of a Proposed Addition to His Opticks

❖

Students of Newton's physical thought are now in rather general agreement that—contrary to what various writers have asserted—a belief in an aetherial medium was not throughout his life "a central pillar" of Newton's system of nature.[1] By the time he wrote the *Principia*, it seems clear, Newton had not only rejected the Cartesian dense aether but had come to distrust his own youthful aetherial speculations;[2] from the 1680s until some time after 1706, the year the Latin translation of his *Opticks* appeared, an aether played little or no role in Newton's conjectural system of nature. Instead, in the new queries of the Latin *Optice* he speaks of forces of attraction acting at a distance through empty space, and favors, as a possible cause of gravity and of attractions at short range, those immaterial, incorporeal agencies—indebted in all probability to the speculations of More and Cudworth—which he describes as "active principles."[3] Nevertheless by 1713, as the *Scholium Generale* added to the second edition of the *Principia* clearly hints, Newton had reverted to a peculiar sort of effluvial theory, for he refers in the concluding paragraph to the possible role in nature of an "electric and elastic spirit."[4] And when in 1717/18 he published the second English edition of the *Opticks*, inserting more new Queries and touching up the old ones, these additions embodied arguments for the existence of an elastic, tenuous medium: an aether, as we are often reminded, possessed of very peculiar properties, and one which, as I shall also argue, Newton took more seriously as an agency to explain optical effects than as a possible mechanical cause of gravity.

In an earlier paper I called attention to the extent to which the experiments of Francis Hauksbee on electroluminescence, electrical attraction, capillarity, etc. had influenced Newton's scientific thought in this later period of his life, leaving a marked impress on the *Scholium Generale* and on the new material he added to the *Opticks* of 1717/18.[5] And I hazarded the suggestion, which has been received with some skepticism, that it may have been Hauksbee's experiments on electrical attraction and electroluminescence that revived Newton's belief in an aetherial effluvium: he had, as it were, seen it stream from Hauksbee's rotating

From *Notes and Records of the Royal Society* 22 (1967). Published by the Royal Society, Burlington House, Piccadilly, London, W. 1. Edited by Sir Harold Hartley, F.R.S.

globe; he had observed its effects on light objects attracted to rubbed glass tubes; he had even felt it. To be sure, this was simply a conjecture on my part, based upon the concordance in time between Newton's resurrection of the aether and his association with Hauksbee, and upon the fact that the new material in the revised *Opticks* of 1717/18 contained *both* his conjectures about the aether *and* numerous echoes of Hauksbee's work. This was hardly conclusive, for at no point did I find Newton explicitly referring to the aether in print when alluding to Hauksbee's experiments.

Yet it seemed reasonable that Newton could have made just such an inference, as indeed, at about the same time, a French experimenter, Pierre Polinière, did not hesitate to do.[6] That Newton in fact did so, I am now in a position to demonstrate.

Among the unpublished optical manuscripts of Newton in the Portsmouth Collection at Cambridge is a remarkable document that seems to have eluded general notice.[7] It is a much-corrected draft on eight folio sheets in Newton's hand, clearly related to the revisions Newton was contemplating for the second English edition of his *Opticks*. It was written no earlier than 1715, and later portions were probably set down after October 1717. It consists not of Queries but of a number of "Observations," several of which describe those experiments of Hauksbee which eventually found their way into the Queries of 1717/18. The first of these "Observations" sets forth what we may call the "two-thermometer experiment," an experiment by which Newton set great store, but which was not Hauksbee's and about which I shall have something to say in what follows.

Perhaps the most striking thing about this document is the title Newton gave to it. At the top of the first page (fol. 623) we read the caption: "The Third Book of Opticks. Part II. Observations concerning the Medium through which light passes, & the Agent which emits it." Now first of all, this solves a minor but persistent mystery: why the editions of the *Opticks* we commonly consult have the rubric "Part I" in the Third Book, with no "Parts" following, as they do in both the earlier Books. That this designation was added for the edition of 1717/18 is evident: it does not appear in the first edition of 1704 or in the Latin *Optice* of 1706. We have all wondered, I suppose, what Newton had in mind, what he might have treated if he had included a Part II in the Third Book. At least one writer has hazarded the guess that Newton may have planned an axiomatic or deductive exposition of his discoveries on light and color: a synthesis to follow the experimental or analytical presentation of the earlier Books.[8] But this is far from the mark. Our draft of a suppressed Part II of the Book Three had a quite different purpose: it was intended, I believe, to marshal experimental evidence for his aether theory. Yet before the revisions were completed Newton's customary caution won out: he changed his plan at the last minute, and cast this material, or at least a good part of it, in the form of those new or altered Queries with which we are familiar.

Let me briefly describe this document. Two sections of it can be distinguished, both evidently belonging to the years 1715–17, yet one written before the other. The evidence for this division is Newton's numbering, and then his renumbering, of the several "Observations."

Newton's earlier plan did not involve a "Part II." He first proposed simply to add a series of new "Observations," drawn largely from Hauksbee's work, to the eleven in which, in the first part of the Third Book, he had described those experiments on diffraction inspired by the work of Grimaldi. The new "Observations" were to follow immediately after the passage of the *Opticks* in which he explains that he was interrupted in these diffraction experiments and could not pursue them further. Presumably he planned to place the Queries at the end of the new "Observations."[9] These "Observations," at any rate, he proposed to number XII to XVI, and planned to include, besides an account of Hauksbee's experiments, an early experiment of his own, described in his letter to Oldenburg in 1675, almost certainly (from its appropriateness at this point) the experiment showing the attractive and repulsive properties of a rubbed disc of glass. This "Observation" is not written out, but the heading serves, at least approximately, to date this portion of the manuscript, for Newton writes: "Above forty years ago I sent the following Observation to Mr. Oldenburg & have now copied it from one of the Books of the Royal Society."[10] This heading must have been written, at the earliest, late in 1715, and surely, for reasons I shall adduce, before November 1716. After that date, however, Newton adopted another plan. He added a first page recording two new "Observations," to be described in a moment, numbering them Obs. I and Obs. II, and renumbered those he had already written, changing Obs. XII to Obs. III, and so with the rest. To the whole manuscript, thus altered, he added at the top of the first folio the title "The Third Book of Opticks. Part II."

Let us now consider the earlier portion of the manuscript, where Newton described Hauksbee's experiments, and where, to my considerable satisfaction, we find Newton referring to the aether. In Obs. XII (later renumbered as Obs. III) Newton prefaces his account of Hauksbee's experiment with the rotating, evacuated glass globe by remarking

> that gross bodies contain within them *a subtile Aether or Aetherial elastic spirit* wch by friction they can emit to a considerable distance, & which being emitted is found sufficiently Subtile to penetrate & pass through the body of glass, & sufficiently active to emit light at a distance from the gross body if it be there put into a trembling agitation, as is manifest by the following Phaenomena *shewd to the R. Society by Mr. Hawksby.*[11]

Then, speaking of Hauksbee's globe, he writes:

> And if while the glass was in a very swift motion, the Operator held the end of his finger or a piece of white paper or linnen or other convenient body near the globe . . . the body on that side next the globe would shine; the *aetherial spirit or spiritual effluvium* which by friction was excited & emitted

from the glass, dashing upon the finger or paper or linnen or other body, & being thereby put into such an agitation as enabled it to emit light.[12]

Again, with reference to Hauksbee's electrical experiments with rubbed glass tubes, Newton remarks how light objects, like pieces of leaf gold or leaf brass, dart about when approached by a "cilynder of glass newly rubbed" and sometimes "move in curvd lines being agitated & carried about by the *Etherial effluvium* as with a wind. . . ."[13]

There can no longer be any doubt that Newton, having followed Hauksbee's experiments with keen interest, identified the electrical effluvium or electrical spirit with a kind of subtile, penetrating, elastic, and highly active matter pervading all bodies, and was willing to give it the name of "aether." For a brief moment he was sufficiently persuaded by this experimental evidence for the existence of such a tenuous matter to consider introducing the evidence into the Third Book of the revised *Opticks* as a series of "Observations" rather than putting it in the form of Queries, as he ultimately decided to do.

The passages just cited confirm my earlier suspicion that Hauksbee's experiments had turned Newton's mind back to his earlier speculations, focusing his attention on the possible role of a subtile fluid. Yet his remarkably active and inquiring mind was casting about for still more persuasive experimental confirmation. This he hit upon, during the course of 1716, in the form of what I have called the "two-thermometer experiment." His account of this experiment, eventually published as Query 18 in the 1717–18 *Opticks*, is quite familiar. I need only summarize it briefly. If, says Newton, you take two tall cylindrical glass vessels and suspend in each a small thermometer; and if you evacuate one of the vessels with an air pump, leaving the other full of air; then if you carry both vessels from a cold place into a warm one, "the Thermometer *in vacuo* will grow warm as much, and almost as soon" as the thermometer in the vessel filled with air. And he asks his reader: "Is not the Heat of the warm Room convey'd through the *Vacuum* by the Vibrations of a much subtiler Medium than Air, which after the Air was drawn out remained in the *Vacuum*?"[14]

So much for the printed account. Did Newton himself perform the experiment, as is sometimes assumed, or had he read of it or heard an account of it, or perhaps seen it performed by someone else? His words in Query 18 give us no satisfaction. But Obs. I of our manuscript describes this experiment and supplies a clue, for the account begins: "Within two tall & cylindricall Glasses hollow & closed at the upper end, *I caused* two short Thermometers to be hung from the tops of the glasses. . . ."[15] May we gather from this that Newton devised the experiment and "caused" it to be carried out on his behalf? Recalling Newton's close attention to Hauksbee's experiments, and his role in reviving a concern for experiment at the meetings of the Royal Society, this seemed a possibility worth exploring. But who could have performed the experiment? Surely not Hauksbee, who had died in 1713 and in whose papers there is no trace of such an experiment. But if not Hauksbee, why not his successor as Curator of Experiments at the Royal Society? This proved to be the case. The experiment was in fact performed at the Royal Society, in Newton's presence, by Jean-Théophile Desaguliers, whom we all know as a skillful experimenter and a dutiful expositor of Newtonian science. At Oxford, in 1708, Desaguliers had attended a course of lectures in Newtonian experimental philosophy given by John Keill. Two years later, when Keill left for America, Desaguliers succeeded him as lecturer in the same subject at Hart Hall. The year of Hauksbee's death found Desaguliers in London, where he settled in Channel Row, Westminster, and began his course of demonstration lectures "at Mr. Brown's, Bookseller, Temple Bar."[16] His first appearances before the Royal Society date from the early months of 1714; among the earliest experiments he performed for the Society were several inspired by Newton's discoveries, such as experiments to determine the melting points of different metals (in April 1714) or (in July of that year) for "explaining and verifying the President's Treatise of Colours."[17] Newton, it is fair to surmise, was not without influence in inviting Desaguliers to continue at the Royal Society the experimental tradition so brilliantly revived by Hauksbee.

It was on November 15, 1716, with Newton in the Presidential chair, that Desaguliers performed the two-thermometer experiment at the Royal Society. The entry in the *Journal Book* is worth giving in its entirety:

> Mr. Desaguliers Shew'd an Experiment to prove that a Body in a Vacuum of Air was susceptible of heat and cold; by putting two Thermometers of equal Lengths and equally graduated into two tall glass Receivers of equal Contents. The one being exausted [*sic*] of its Air and the other not, upon bringing the said glasses near the fire, the Spirit rose in the Thermometer that was in vacuo almost as much as the Thermometer that was in pleno; only that in pleno rose something faster for it rose 56 degrees or tenths of an Inch, whilst the Thermometer in the exhausted glass rose only 50. When the glasses were removed from the fire, the Liquor in the Thermometer in the Glass full of Air subsided faster than in the other, much in the same proportion.
>
> The President [Newton] said that the Experiments must also be try'd where there is heat without Light, and Desaguliers reported that he had try'd it in a medium so affected namely by holding both the Glasses near a hole where heated Air is convey'd into a Room by means of Iron Cavities round about the fire and the Thermometer in that case rose in vacuo much after the same manner. He further promised to try the Experiment again with a great deal of Care, putting Cement instead of a wet Leather under the exhausted Glass because the wet Leather throws a dew upon the Ball of the Thermometer.[18]

Newton's comment was well taken, for obviously, according to his conception of the nature of light, its corpuscles could easily pass through the vacuum and by their impacts warm the fluid in the thermometer.[19] His keen interest in this experiment is, at all events, clearly manifest, yet it is only from the phrase in Newton's manuscript account ("I caused two short thermometers to be hung, etc. . . .") that we may conclude it was Newton who conceived the experiment and suggested to Desaguliers how it should be carried out.

Fortunately we have a confirmation from Desaguliers himself, though well after the event. In the winter of 1731, Sir Hans Sloane received from a certain J. Brown in Paris a letter in which the writer, a person I have not been able to identify, suggested that by invoking a mechanical aether the chief phenomena demonstrated by Newton in the *Principia*, notably the force of gravitation, could be readily explained. Sloane, evidently doubting his qualifications to deal with such matters, appealed to Desaguliers to supply him with an answer. The letter of Desaguliers to Sloane, dated Channel Row, 4 March 1730/31, begins as follows:

Sir,

As Mr Brown's Letter to you from Paris dated Feb. 21. 1731. was referr'd to me, I take the Liberty to give you my opinion of it. That there is a subtile Medium even finer than Light which serves in the Reflection, Refraction and Inflexion of Light, may be deduced from Phenomena, as is evident to those who read Sir Isaac Newton's Optics with such attention and skill as thoroughly to understand them. But whether there be a subtile Medium call'd Aether which is endued with such Properties as to be the Cause of Gravity by a mechanical Impulse, is only propos'd by Sir Isaac in his last English Edition of his Optics by way of Query. And whether the Medium acting upon Light be the same as the Aether hinted at for the Cause of Gravity, is very cautiously insinuated by our incomparable Philosopher, who mentions an Experiment (*which I made before the Society by his Order*) of a Thermometer in which the Liquor was rais'd as high by Heat *in Vacuo Boyleano* as the Liquor of another equally graduated Thermometer in Pleno, tho' the Heat was not communicated to the first quite so fast; and thence deduces that there must be a Medium to convey the Heat to the Thermometer in the Glass where it was suspended, which Medium or Fluid must penetrate the Glass which is impervious to the Air. . . .[20]

The apparent success of this experiment evidently led Newton to decide, if only momentarily, to revise his plan and to throw the aether "Observations" into prominence by forming them into a Part II of the Third Book of the revised *Opticks*, putting the account of this experiment as the first "Observation" and following it by the other bits of experimental evidence, drawn chiefly from the older work of Francis Hauksbee. This recasting could scarcely have been done before Desaguliers' demonstration of November 15, 1716. There is reason to think, moreover, that the recasting was done even later, indeed several months after Newton had written, and sent to the printer, his "Advertisement II," a new preface to the *Opticks*, dated July 16, 1717. Observation I of our manuscript, the two-thermometer experiment, is followed by a second new "Observation," which describes the venerable experiment of the guinea and the feather let fall in a vacuum, to which he had alluded earlier in Query 20 of the Latin *Optice*. Observation II, however, describes a refinement of the older experiment, and Newton's purpose in introducing it here is to supply conclusive evidence that a vacuum pump can in fact empty a vessel, like that in which one of the thermometers was suspended, effectively of the air. After describing the common version of the experiment, he writes:

> . . . & the Experiment being repeated with two glasses set one upon another so as to make a Vacuum four foot high or with three four or five glasses so as to make a vacuum 6, 8, or 10 foot high the success was very much the same.[21]

This difficult experiment of stacking glasses on one another so as to increase the distance through which the objects must fall in the vacuum was performed by Desaguliers, who demonstrated it to the Royal Society, with Newton presiding, on October 24, 1717. The entry of the *Journal Book* reads as follows:

> Mr. Desaguliers shew'd the experiment of letting fall a bitt of Paper and a Guinea from the height of about 7 foot in a vacuum he had contrived with four glasses set over one another, the junctures being lined with Leather liquored with Oyle so as to exclude the Air with great exactness. It was found that the paper fell very nearly with the same Velocity as the guinea so that it was concluded that if so great a Capacity could have been perfectly exhausted, and the Vacuum preserv'd, there would have been no difference in the time of their fall.[22]

Unless Newton was privy to certain preliminary experiments performed by Desaguliers before the demonstration at the Royal Society, this design of a Part II of the Third Book of the *Opticks* must therefore have been decided upon in the autumn of 1717.[23] At this time, since Newton had already written his "Advertisement II," we may assume that the bulk of the revised manuscript of the *Opticks* was probably at the printers. It is likely, too, that after he had decided upon this plan for a Part II, the Third Book, at least down to the Queries, was sent to the printer with the added rubric "Part I" which has remained to puzzle us. Very soon, however, Newton must have been besieged by doubts; the experimental evidence no longer seemed as persuasive as in his enthusiasm he had first believed. At this point, indeed at the last moment, and in a great hurry, Newton converted these "Observations" into new Queries or used fragments of them to alter some of the older ones. Was there time for this? Evidently so. Even though some rare copies of the second English edition of the *Opticks* bear the date 1717, it should be remembered that Newton and his printer were following the established practice of old-style dating. Accordingly, copies of the book dated 1717 could have appeared not only late in that year, but at any time between January 1718 (according to our method of reckoning) and the end of March, when the old-style New Year began.

Conclusion

The following reconstruction of Newton's changing attitudes towards the aether may be suggested. The experiments of Francis Hauksbee on electrical attraction and repulsion and on electroluminescence had a profoundly stimulating effect on Newton's speculations. They persuaded him, by the time he wrote the concluding paragraph of the *Scholium Generale*, that what he described as a "certain most subtle spirit," an electric and elastic spirit, "lies hid in all gross bodies." This spirit causes the short-range forces acting between the constituent particles of matter; it accounts for electrical attraction at greater distances; it heats bodies; it seems to explain the emission, reflexion, refraction, and diffraction of light; and the vibrations of this spirit, passing along the sensory nerves, yield our sensations, and passing from the brain through the nerves constitute the motor impulses sent to the muscles.[24] Yet by 1715–16 Newton had come to think of this "spirit" as an all-pervading "medium" filling the interstices of all bodies, and indeed all space. And he came to refer to it as the "aether," though by this term he meant something widely different from the dense, mechanical aether of Descartes. The "two-thermometer experiment" was, for Newton, the decisive experiment designed to prove the existence of this rare, all-pervading aetherial medium. But he was less satisfied that this medium could be invoked to explain gravity.[25]

The letter of Desaguliers to Hans Sloane supports this contention, for it does more than show that the experiment was conceived by Newton and performed "by his order." Desaguliers makes it quite clear that Newton little questioned the existence of this medium. But he also makes the point that Newton, while not doubting that it played a part in optical phenomena, was much less confident that the aether "is endued with such Properties as to be the Cause of Gravity by a mechanical impulse." That Newton was convinced of the existence of an optical aether should be evident, Desaguliers insists, to any who read the *Opticks* with "attention and skill," whereas the gravitational role of the aether Newton had merely "propos'd . . . by way of Query." This might seem curious—since ultimately both roles for the aether were presented in the form of Queries—except for what we learned from our manuscript: that at one point, when Desaguliers was working in close collaboration with him, Newton proposed to discuss the optical aether by means of experimental "Observations." Certainly he had no such confidence about the aether as a cause of gravitation, for at the very time he was proposing to present the evidence for the optical aether in the form of "Observations" he had already written out as a new Query his speculations about a gravitational medium. In his "Advertisement II," dated July 16, 1717, Newton tells us that "at the end of the Third Book I have added some Questions, and to shew that I do not take Gravity for an essential Property of Bodies, I have added one Question concerning its Cause, chusing, etc."

It follows that certain of the new Queries added to the *Opticks* in 1717/18 were written before the final decision, made late in 1717, to cast the material on the optical aether in interrogatory form. In the light of this new information, if we take a fresh look at the eight new Queries of 1717/18 (Nos. 17–24), we remark that indeed they fall rather sharply into two groups: the first four Queries, Nos. 17–20, deal with the optical aether, and are obviously related to the "Observations" of the suppressed Part II of the Third Book. On the other hand,

Queries 21–24 are quite different, and some if not all of them were almost certainly written before July 1717. Query 21 is the new "Question" on the cause of gravity, to which Newton refers in "Advertisement II." Query 22 supports Query 21 by arguing that such a medium filling all space need not retard the planets.[26] And the last two Queries seem to expand the hints earlier presented in the *Scholium Generale* about the possible physiological role of the "elastic spirit." Quite clearly these had been planned or written before Newton had in mind to discuss the optical aether in the form of a Part II of the Third Book. On the other hand, the first four Queries, Nos. 17–20, were probably inserted in the revised *Opticks* some time in the autumn of 1717, when the final form of the Third Book was at last decided upon.

Notes

1. *Isaac Newton's Papers and Letters on Natural Philosophy*, ed. I. B. Cohen (Cambridge, Mass.: Harvard University Press, 1958), 7. See also Cohen's *Franklin and Newton* (Philadelphia: American Philosophical Society, 1956), the section headed "Newton's Grand Hypothesis on the Aether," pp. 166–72. Professor Cohen's statement is the most extreme. Among earlier writers, E. T. Whittaker held that "on the whole" Newton favored the aether theory, *History of Theories of Aether and Electricity* (London, 1910), 17; and Philip E. B. Jourdain, in one of his articles of the series entitled "Newton's Hypothesis of Ether and Gravitation," wrote that it "seems impossible to doubt that Newton decidedly leaned towards the hypothesis of a very rare ether." *The Monist* 25 (1915), 439.
2. Neither David Brewster nor Ferdinand Rosenberger took Newton's writings on the aether seriously. Rosenberger seemed to think that Newton abandoned the aether at least as early as the letters to Bentley. *Isaac Newton und seine physikalischen Principien* (Leipzig, 1895), 309. Alexandre Koyré came to much the same conclusion, and Professor Westfall has recently written: "When Newton composed the *Opticks*, he had ceased to believe in an aether; the pulses of earlier years became 'fits' of easy reflection and transmission,' offered as observed phenomena without explanation." Richard S. Westfall, "Isaac Newton's Coloured Circles twixt two Contiguous Glasses," *Archive for History of Exact Sciences* 2 (1965), 190.
3. Henry Guerlac, *Newton et Épicure*, Conférence donnée au Palais de la Découverte (Paris, 1963).
4. The phrase "electric and elastic spirit" appeared for the first time in Andrew Motte's English translation of the *Principia* in 1729; in the second and third editions of 1713 and 1726 Newton spoke only of "a certain spirit" (*spiritus quidam*). But *spiritus electricus* appears in one of the draft variants of the *Scholium Generale*; see *Unpublished Scientific Papers of Isaac Newton*, ed. A. Rupert Hall and Marie Boas Hall (Cambridge, 1962), 357. In Newton's own interlined and annotated copy of the second edition of the *Principia*, Newton wrote in the margin, after the word *Spiritus*, the words "*electrici*" and "*elastici*." See Marie Boas Hall and A. Rupert Hall, "Newton's Electric Spirit—Four Oddities" *Isis* 50 (1959), 473–76; also Alexandre Koyré and I. B. Cohen, "Newton's Electric and Elastic Spirit," *Isis* 51 (1960), 337.
5. Henry Guerlac, "Francis Hauksbee—expérimentateur au profit de Newton," *Archives Internationales d'Histoire des Sciences* 16 (1963), 113–28. See also my "Sir Isaac and the Ingenious Mr. Hauksbee," *Mélanges Alexandre Koyré*, 2 vols. (Paris: Hermann, 1964), I (*L'Aventure de la Science*), 228–53.

6. In his *Expériences de physique* (Paris, 1709), 471–72, Polinière describes an experiment on electroluminescence closely resembling Hauksbee's, and he comments that "the cause of these effects is a sudden motion impressed on a matter much more subtile than the coarse air we breathe. . . . It is the agitation of this matter which makes us perceive the sensation we call light." In an account of the experiments that Polinière had shown to the Académie des Sciences in the late autumn of 1706, he concluded that "the light consisted in a swift vibratory pressure [*une pression subite, tremblante ou trémoussante*] of the aetherial matter, which passes through the glass." *Nouvelles de la République des Lettres*, (January 1707), 107. Such a conclusion was more readily to be expected from a Continental scientist who, like Pardies, D'Ango, and Huygens, identified light with a pulse in a vibrating aether, than from Newton who, in all editions of his *Opticks*, consistently rejects an undulatory theory of light. These experiments of Polinière are discussed by my student, David Corson, in a forthcoming paper. [Subsequently published as "Pierre Polinière, Francis Hauksbee, and Electroluminescence: A Case of Simultaneous Discovery," *Isis* 59 (1968), 402–13.]

7. The University Library, Cambridge, MS. Add. 3970, no. 9, fols. 623–29.

8. Colin Murray Turbayne, *The Myth of Metaphor* (New Haven: Yale University Press, 1962), 44. Turbayne mistakenly writes that "the whole of the *Opticks* is styled 'Part I'." He must have had in mind the Third Book.

9. In the *Opticks* the passage reads: "But I was then interrupted, and cannot now think of taking these things into farther Consideration. And since I have not finish'd this part of my Design, I shall conclude with proposing only some Queries, in order to a farther search to be made by others." In our MS. (fol. 624) Newton alters this to read: "But I was then interrupted & cannot now think of taking these things into further consideration, but must leave this part of my design to be further prosecuted by others who have skill & leisure." The reference to Queries immediately following has significantly been omitted.

10. Add. 3970, no. 9, fol. 627. For the experiment of 1675 see Thomas Birch, *History of the Royal Society of London*, 4 vols. (London, 1756–57), III, 250–51, 270. Newton believed at the time that this experiment suggested that there is a "subtil matter" or "something of an aethereal nature" condensed in bodies, which is rarefied by friction into an "aethereal wind."

11. Fol. 626. My italics.

12. *Ibid.* My italics.

13. Fol. 626 *v*. My italics.

14. *Opticks*, 3d ed., corrected (London, 1721), 323.

15. Add. 3970, no. 9, fol. 623. My italics.

16. For Desaguliers see the article in the *D.N.B.* by Robert Harrison, and, more recently, A. R. Hall in *Dictionary of Scientific Biography*; also Cohen, *Franklin and Newton*, 243–44. Professor Cohen cites a biographical sketch by Jean Torlais, *Un Rochelais grand-maître de la Franc-Maçonnerie et physicien au XVIIIᵉ siècle: Le Révérend J.-T. Desaguliers* (La Rochelle, 1937) which I have been unable to consult. For the relations of Desaguliers with John Keill see the former's *Course in Experimental Philosophy*, 2 vols. (London, 1734–44), II, 404, and the author's Preface to the first volume.

17. Royal Society, Classified Papers, XVIII(2), Experiments of Desaguliers. The earliest reference to Desaguliers in the Society's *Journal Book*, under the date of February 18, 1713/14, establishes that he was initially brought in to construct a thermometer on Newton's design and to carry out experiments proposed by Newton: "It was ordered that Mr. Aguiliers [*sic*] be desired to wait on the President to take his directions as to the Structure of that Instrument and the Experiments to be made with this new thermometer before the Society." The demonstration, made on April 8, 1714, seems to have marked Desaguliers's début before the Royal Society. He was admitted F.R.S. on July 29 of that year.

18. *Royal Society Journal Book*, November 15, 1716.

19. Rosenberger, in commenting upon Newton's account of this experiment in Query 18, found the argument unconvincing, precisely because he saw that, according to Newton's doctrine, light rays should set the particles of matter in motion even in a

vacuum and so the thermometers should be directly heated even in an air-free space. See his *Isaac Newton und seine physikalischen Principien*, 307.

20. British Museum, Sloane MS. 4051, fol. 200. The italics within the parentheses are mine.

21. The University Library, Cambridge, MS. Add. 3970, no. 9, fol. 623.

22. *Royal Society Journal Book*, October 24, 1717. The experiment was repeated at the meeting of December 5.

23. A demonstration was made before the King and the Princess of Wales at Hampton-Court in September 1717, with the King himself pulling the lever to let the bodies fall. See Desaguliers, "An Account of an Experiment to Prove an Interspersed Vacuum," *Philosophical Transactions*, vol. 30, no. 354 (October, November, December, 1717), 717–20. The experiment was repeated before the Royal Society, after the demonstration of October 24, on December 5 of that year and later "in Channel-Rowe, Westminster, before some members of the Royal Society."

24. My friend, Mr. J. E. McGuire, has called my attention to a manuscript in the Portsmouth Collection (MS. Add. 3970, no. 7, fols. 597–604), a Latin draft in Newton's hand, which bears a close relation to the draft materials for the *Scholium Generale* published by the Halls. It was evidently written after Newton had observed Hauksbee's experiments, that is after 1706–7, and probably before 1713. In it Newton speaks of an "electric spirit" whose agitation (rather than the agitation of solid particles) causes bodies to emit light; he imagines it as an effluvium or "spirit" latent in glass, which, when the glass is rubbed, is emitted to short distances. Its presence in the vicinity of bodies accounts for the reflection and refraction of light. In this document Newton does not yet conceive of this "spirit" as a tenuous medium extending any great distance from bodies or filling all space.

   These speculations, of course, closely resemble Newton's earlier views; and these are echoed in the *Opticks* of 1704 (bk. 2, pt. 3, prop. 12), where he introduces what he explicitly calls an "hypothesis" that might seem to explain the "fits of easy reflection and transmission." As I interpret that passage, Newton is not suggesting an aether diffused through all space, but one filling dense bodies and exerting its effects only in close proximity to their surfaces or within them.

25. This evolution of the "electric and elastic spirit" into an all-pervading aetherial medium, and Newton's experiment to test the latter possibility, may perhaps have been primarily motivated by Newton's dissatisfaction with the *pis aller* of "fits of easy reflection and transmission," and even with the earlier form of the aether hypothesis he tentatively invoked to explain these "fits." See above, no. 24.

26. That this Query was written before the autumn of 1717 is suggested by the fact that Newton makes no mention of the very persuasive experiments of Desaguliers on the fall of bodies in an extended vacuum. Yet he took pains to introduce this evidence into the third edition of the *Principia* (1726). Here he adds new sentences at the end of bk. 3, prop. 10, theor. 10, arguing for the lack of resistance in space emptied of air by citing experiments (surely those of Desaguliers) which show that when air is "carefully exhausted by the air pump from under the receiver," bodies of different weights, "though they fall through a space of four, six and eight feet, they will come to the bottom at the same time." *Principia*, ed. Motte-Cajori (Berkeley, Calif.: University of California Press, 1934), 419.

10

Where The Statue Stood:
Divergent Loyalties to Newton in the Eighteenth Century

❖

Since I shall be speaking about Isaac Newton and shall have something to say about the interpretations placed upon his work by some scientific thinkers of the eighteenth century, I shall take my text from Voltaire, his great popularizer in France. In April 1735, Voltaire wrote to his friend Cideville:

> Verses are hardly fashionable any longer in Paris. Everyone begins to play the mathematician and the physicist. Everyone wants to reason. Sentiment, imagination and charm are banished. I am not vexed that philosophy is cultivated, but I should not want it to become a tyrant to exclude everything else.[1]

When Voltaire wrote, science was indeed flourishing in France as never before and was rapidly becoming fashionable among men of letters. The reign of *philosophie*, and of that divinization of nature Carl Becker wrote of, was presaged by this cult of natural science. More particularly, this decade marks the beginning of Newton's ascendancy over the mind of the Enlightenment. His doctrines were at last finding favor in the Academy of Sciences, hitherto wedded to the physics of Descartes and Malebranche; and by Voltaire's efforts—in his *English Letters* of 1734 and in his *Elements of the Newtonian Philosophy*, published four years later—the gospel was brought to a wider audience. Indeed, during the late thirties and the forties the French scientific community was deeply divided by the contest that raged between the partisans of Descartes and Malebranche and the young defenders of Newton's System of the World, men like Maupertuis, Clairaut, and d'Alembert. But the Newtonians soon gained the upper hand; as early as 1743 d'Alembert could call the Cartesians "a sect that in truth is much weakened today." While fifteen years later, in the second edition of his *Traité de dynamique*, he altered this phrase to read "a sect that in truth *hardly exists* today." And this more than anything indicates how swift—after a half century of neglect—was Newton's conquest of France.[2]

For this victory Voltaire—with typical exaggeration—was later to claim the principal credit.[3] In advancing this claim, Voltaire is less than fair to Maupertuis, from whom in fact he learned a good deal, and who, as early as

From Earl R. Wasserman, ed., *Aspects of the Eighteenth Century* (Baltimore: Johns Hopkins University Press, 1965). Copyright © 1965 by The Johns Hopkins Press.

1732, had published an admirably clear and readable popular account of Newton's System of the World, with its theory of universal gravitation.[4]

Yet Voltaire did prepare the principate of Newton, for he was one of the earliest to perceive, or to stress, the vast ideological import of Newton's scheme, and more especially of Newton's method. At the hands of the *philosophes*, when generalized and extended to other spheres, this method served as a critical instrument for the exposure of humbug, prejudice, and intellectual pretension. This "method of analysis," as it was called, was taken to be the necessary and indispensable instrument for all kinds of thinking. "Philosophy," Voltaire wrote, "consists in stopping where the torch of physical science fails us."[5]

All this is by way of introduction. I do not propose to discuss the place of Newtonianism in the philosophy of the Enlightenment, nor ask how well the men who marched under Newton's banner actually understood him. For this you may consult the books of Carl Becker, Preserved Smith, and Ernst Cassirer, where, in the order I have named them, you will find examples of what J. A. Passmore has recently distinguished as the "polemical," the "cultural," and the "problematic" approaches to intellectual history.

Instead I propose to ask, and try to answer, questions of a different sort. What, in fact, was Newton's method, and what was his idea of science? Have we really understood it? How well did scientists of the eighteenth century understand it?

❁

At the start, let me examine a familiar proposition. We usually think of modern physical science as somehow dating from the massive achievement of Newton, especially from the appearance in 1687 of his great work, the *Mathematical Principles of Natural Philosophy*. And we are accustomed to think of the eighteenth century as an age richly ornamented with men of scientific genius who were worthy of following in Newton's footsteps. Their task was to fill out and correct the details of his System of the World, and to extend his method and spirit to illuminate the darker corners of nature.

Not everyone agrees with this interpretation. A short time ago I read—I confess for the first time—that strange book, *La Formation de l'esprit scientifique*, by the late Gaston Bachelard. I was puzzled to find him relegating the eighteenth century to what he called "the pre-scientific age." His condemnation, I discovered, was sweeping: according to Bachelard, the scientists of that age generalized too readily; their explanatory concepts were hasty and vague; their aim was too often to entertain, astound, or edify. When they experimented— like the Abbé Nollet in his famous *expériences de gala*—showmanship took over. Their writings, too, are either stuffed with irrelevant erudition, or they are chatty conversations between the "savant" and the "curieux."

Bachelard may have been a respected philosopher, yet he was certainly a bad historian. To be sure, he gives us many absurd examples of the "prescientific mentality" at work among eighteenth-century scientists, but these examples are taken from some of the cruder popular books, from the works of unregenerate conservatives like Father Castel, from the writings of cranks like the future "Ami du Peuple," J. P. Marat, or of curious, minor figures like the Abbé Bertholon or the Baron de Marivetz. That some of these names are unfamiliar helps to make my point. One would never know from reading Bachelard that in the company of these men there flourished physicists like d'Alembert, Coulomb, or Lagrange; astronomers like Bradley and Herschel; chemists like Joseph Black and Lavoisier; naturalists like Linnaeus and Buffon; or electrical discoverers like Benjamin Franklin. Poor Franklin is mentioned only once, and in a tone of solemn reproof, for the famous "electrical banquet" where, on the banks of the Schuylkill River, "a turkey is to be killed for our dinner by the *electrical shock,* and roasted by the *electrical jack,* before a fire kindled by the *electrified bottle:* when the healths of all the famous electricians in *England, Holland, France* and *Germany* are to be drunk in *electrified bumpers,* under the discharge of guns from the *electrical battery.*"[6] Bachelard is electrically shocked by this bit of whimsy.

Yet there is something in all this to give us pause. The scientific effort of the eighteenth century was incredibly diverse in character and very uneven in quality. All this, I suppose, was the price paid for the often amateur character of science, for democratizing it and making it an instrument of enlightenment.

But the leaders of French science could be as censorious as Bachelard. They did their best to distinguish between cloth of gold and fustian. They snubbed Marat (to their peril, as it turned out) and studiously ignored most of the men in Bachelard's rogues' gallery. Even Buffon was castigated by his fellow Academicians for seeking public applause, departing from scientific rigor, and mingling science with rhetoric.

The Academy of Sciences at Paris, throughout the century, it is generally agreed, represented science at its best; the ablest scientific men in France were its members or at least its correspondents. And just as the Académie Française was supposed to preserve the purity of the French language, so the Royal Academy of Sciences recognized as its prime function to set standards for what was a relatively new kind of intellectual activity.[7]

After about 1750, it was Newton's example and method which set the standard, just as earlier it had been the methodological principles of Descartes and Malebranche. The prolonged resistance to Newton during the first half of the century, and the ensuing debate between the partisans of Descartes and Newton, should perhaps be viewed not so much as a contest between rival theories as a confrontation of two rival standard-bearers, of two different conceptions of science and its proper method.

The delay in accepting Newton's discoveries is otherwise inexplicable, for it was astonishingly long. Some forty years elapsed after Newton published (in 1672) his first experiments on light and color before they were accepted in France. And not until almost sixty years after the appearance of Newton's *Principia* did any scientist in France see in it much more than a display of geometrical ingenuity. These men simply did not understand what Newton was driving at.

The gulf that separated Newton from his contemporaries on the Continent—even from a man of the repute and sagacity of Christiaan Huygens—was far deeper, and more difficult to bridge, than we usually realize. Our usual angle of vision is principally at fault; we have the habit of thinking of Newton as the last and greatest of the mechanical philosophers, as one who synthesized the various ingredients prepared by his predecessors. Instead we ought to think of him as a rebel and a bold innovator, as one who was nurtured by the mechanical philosophy of the seventeenth century but who was able to transcend it. A close comparison of the contrasting viewpoints is well worth making.

☸

The scientific revolution of the seventeenth century, we all know, found expression in two divergent ways of understanding the natural world. One approach —essentially critical, at least at the start, and suspicious of theory—stemmed from Francis Bacon. His disciples of the New Experimental Philosophy urged a continued devotion to observation and experiment: they hoped that by exploring nature the way Bacon had proposed, they might find the "axioms" or "principles" of a new philosophy of nature. Just how this was to be done, nobody was quite sure; meanwhile they observed, experimented, and compiled "histories."

Yet a certain disenchantment with this approach is evident even among the early Fellows of the Royal Society, a body that tirelessly conjured with the name of Verulamus. In 1667 Joseph Glanvill wrote to Lady Margaret of Newcastle: "We have yet no certain theory of nature, and in good earnest, madam, all that we can hope for, as yet, is but the history of things as they are; to raise general axioms, and to make hypotheses must, I think, be the happy privilege of succeeding generations."[8]

In contrast, the mechanical philosophers were less circumspect. By inventing new systems of largely speculative physics, they hastened to fill the vast emptiness left by the collapse of Aristotle's imposing world scheme. Such in essence were the great "hypothetical" systems of René Descartes, Thomas Hobbes, and Pierre Gassendi which dominated men's thinking during the whole second half of the seventeenth century. These men are spoken of as "mechanical philosophers" because they shared the conviction that the real world consists only of material particles, whose motions and combinations could somehow account for the properties, the qualities, of visible things.

To the late seventeenth century this was physics. It was indeed the principal function of the natural philosopher, of the physicist, to suggest plausible mechanisms, or mechanical models, as we should now call them, of the "secret motions of things." Such systems need not be, indeed could not be, true in any real or absolute sense. The mechanical philosophers themselves were explicit about this. Descartes wrote that God has "an infinity of diverse methods" to make things appear as they do. The human mind, he added, cannot really know what method God has employed. Descartes would not have been horrified to learn of Poincaré's proof that if it is possible to represent a phenomenon by a mechanical model, then an infinity of such models is possible.

Hobbes was equally candid. The greatest part of natural philosophy, he wrote, is made up of "things that are not demonstrable," and the most that man can attain to "is to have such opinions as no certayne experience can confute, and from which can be deduced by lawful argumentation no absurdity."[9] Locke advanced similar opinions in his *Essay Concerning Human Understanding*. Natural philosophy cannot lead to certitude, cannot really be a science, for it cannot be demonstrative. And he warns us not to adopt any hypothesis too hastily, or "take doubtful systems for complete sciences." Beware of receiving for unquestionable truths what are really but doubtful conjectures, "such [he adds] as are most (I had almost said all) of the hypotheses in natural philosophy."[10]

Many men in England, like Joseph Glanvill, as we have seen, were despairing enough to fall back upon a crude Baconian empiricism. But others, notably Robert Boyle, sought to combine the experimental and the mechanical philosophies. Boyle, in effect, called upon Bacon—that is, upon experiment— to legitimatize Descartes and Gassendi. In his long career as an investigator he was guided by a single purpose: to illustrate by physical and chemical experiments the plausibility of the corpuscular hypotheses; to use the corpuscular theories in elucidating the rare and the dense, the elasticity of air, the nature of color, and much else. Yet it is difficult to agree with Boyle's admiring *confrères* of the early Royal Society that he was really successful. His method, too, was in some respects "hypothetical."[11]

✷

It was left to Isaac Newton to discover a path to a higher certitude. At the very start of his career, while still a student at Cambridge, he hit upon the method he was to follow all his life. As his early notebooks of this period show, he had been exposed to the works of Descartes and had studied the *Physiologia* of Walter Charleton, a book that seems to have introduced him (and others in England) to the atomistic hypotheses of Gassendi. This latter brand of the mechanical philosophy had a great influence on Newton, but as far as method was concerned he resisted its blandishments. He struck out for himself, narrowed and bridled his speculative ambitions, discarded the hypothetical route, and adopted what he once called his "mathematical way." Though he may have been unaware of the fact, this was the path Galileo had followed when he insisted upon the importance of "geometrical demonstrations founded upon sense experience and very exact observations," upon the necessity of combining "manifest experiences and necessary proofs."[12]

The most often quoted of Newton's scattered statements about his method and his "mathematical way" are to be found in his later writings, in the appendages and asides of his major works: the "Queries" of his *Opticks*, the "Rules of Reasoning" and the "General Scholium" of the later editions of the *Principia*. He describes his method as that of "analysis and composition." Analysis, the upward route, with which one must begin, "consists in making Experiments and Observations, and in drawing conclusions from them by induction." These may serve as "causes" or "principles" or "laws." Composition— the downward synthetic route—consists in drawing out the consequences of these "principles" or "laws" by deductive inference.[13]

There was nothing original about this dual method. Indeed it was formulated as far back as the thirteenth century, if not before; in the sixteenth century it was expounded at great length by Paduan methodologists like Zabarella, and it is echoed in the writings of Galileo. Even the terms in which Newton describes his method are the ancient ones. What is important is the degree of emphasis to be placed on experiment and induction; what is new is the language in which the operations are exclusively to be expressed, the language of mathematics.

"Hypotheses," Newton wrote in the *Opticks*, "are not to be regarded in experimental philosophy." And this assertion is elaborated in the "General Scholium" of the *Principia* by the famous words:

> I frame no hypotheses; for whatever is not deduced from the phenomena is to be called an hypothesis; and hypotheses, whether metaphysical or physical, whether of occult qualities or mechanical, have no place in experimental philosophy.[14]

But how does one arrive, by induction from observations and experiments, at "axioms," "causes," or "principles" which are more reliable than the "hypotheses" of the mechanical philosophers? The answer lies for Newton in his confidence in the simplicity and order of the universe. Nature, he wrote, "is wont to be simple, and always consonant to itself"; the universe is ordered by laws at once uniform and unvarying. We may therefore invoke what writers in the eighteenth century were to call "analogy," and draw forth laws of general applicability by generalizing our observations and the results of even a single well-conducted experiment. This we do with most assurance if we use the language of mathematics, for, as Newton's teacher, Isaac Barrow, said in a lecture Newton probably heard, "Magnitude is the common Affection of all physical Things, it is interwoven in the Nature of Bodies, blended with all corporeal Accidents, and well nigh bears the Principal Part in the Production of every natural Effect."[15] The language of mathematics is therefore the language in which experimental results are to be set down, observations recorded, and the laws of nature best expressed. And once the laws are determined, inferences from them can be drawn, not in the fluid, imprecise language of common speech, but by using the compact symbols and the tested rules of mathematical reasoning. The way followed, therefore, in both analysis and composition is the "mathematical way," and from this the method gained its precision and power. If arguing from experiments and observations be only, as Newton put it, "the best way of arguing which the Nature of Things admits of," yet if done in the mathematical way it gains a degree of clarity and certitude which verbal discourse cannot reach.

❖

Newton's method, "exact, profound, luminous and new"—the words are d'Alembert's—had of course its most dramatic application, and first demonstrated its range and power, in that noble work significantly entitled the *Mathematical Principles of Natural Philosophy*, published when Newton was in the prime of life. But it is not often remembered that all the essentials of that method were embodied in his earliest scientific contribution: the paper on light and color, read on his behalf at the Royal Society in 1672, when Newton was twenty-nine years old. The unfavorable reception accorded this classic paper reveals, as well as anything can, how strange his procedure appeared to contemporaries.

I need only remind you of the following well-known experiments. Newton set up a prism in a darkened room. A beam of sunlight from a small hole in the window shutter passed through the prism and displayed on a screen the "celebrated phenomena of colours," i.e., a solar spectrum. This well-known effect was commonly explained, in Newton's day, by assuming that some sort of essential change, a "modification" or "qualification" of the sun's white light, was produced by passage through the transparent refracting medium. But a striking quantitative disparity led Newton to examine the matter closely: the colored spectrum he obtained formed a band about five times longer than it was wide. This seemed to Newton an "extravagant disproportion" that could not be explained by the well-known law of the bending or refraction of light (Snel's Law). After excluding certain disturbing factors, Newton concluded that the effect was real, and a possible explanation occurred to him. To test it, using a second prism, he carried out what he called his *experimentum crucis.*

Close to and beyond the first prism, he placed a board pierced by a small hole. By rotating the first prism he could make the successive colors of the spectrum pass through the hole, to be refracted again by a second prism, and observe the result projected on a wall or screen. The rays of each given color passed through the second prism with their colors unaltered, and were refracted or bent to the same extent as before. To each colored ray, therefore, a number could be assigned: its characteristic refrangibility. And Newton wrote: "To the same degree of Refrangibility ever belongs the same colour, and to the same colour ever belongs the same degree of Refrangibility."[16] Since the rays

were unaltered by the second prism, he concluded that colors are not "modifications" of white light, but that solar light consists of a bundle or mixture of
rays that are differently refrangible. These rays, each of which produced a
characteristic sensation of color, are merely separated out by the prism. As a
confirmation, he passed the rays of a complete spectrum through a biconcave
lens and showed that as they are brought to a common focus, the colors fade
and whiteness is produced.

Newton's paper produced a controversy that filled the pages of the
Philosophical Transactions for successive issues. His results were criticized by
Robert Hooke; by a young French Jesuit, Father Gaston Pardies; and by the
greatest scientist of the day, Christiaan Huygens. Now, all three of these critics
held to some form of the "modification theory" of the cause of color; and all
of them believed that light was a wave motion, or pulse, in an invisible, all-
pervading aether. For this reason, modern historians have sometimes described
this episode as a dispute between supporters of a wave theory and Newton,
the defender of the theory that light is particulate or corpuscular in nature.
This is not correct, for Newton did not explicitly defend the corpuscularity
of light (he merely hinted at it as a possibility), nor did he attempt to explain
color in terms of it. The dispute turned on something far more fundamental,
as we shall now see.[17]

Both Pardies and Hooke spoke of Newton's explanation of prismatic
colors as his "hypothesis," a word which had many different meanings in the
seventeenth century, but which Newton came to understand in a pejorative
sense. To him it already meant a gratuitous or a priori assumption not scrupulously inferred from experience. This is clear from his reply to Pardies:

> In answer to this, it is to be observed that the doctrine which I explained
> concerning refraction and colours, consists only in certain properties of light,
> without regarding any hypotheses, by which those properties might be ex-
> plained. For the best and safest method of philosophizing seems to be, first to
> inquire diligently into the properties of things, and establishing those properties
> by experiments and then to proceed more slowly to hypotheses for the ex-
> planation of them. For hypotheses should be subservient only in explaining the
> properties of things, but not assumed in determining them; unless so far as
> they may furnish experiments. For if the possibility of hypotheses is to be the
> test of the truth and reality of things, I see not how certainty can be obtained
> in any science; since numerous hypotheses may be devised, which shall seem to
> overcome new difficulties.[18]

Huygens' reaction is notably illuminating. At first he commented that
Newton's theory was "very ingenious," but perhaps not convincingly supported
by experiment. Later he remarked that even if the experiments could be trusted,
there yet remained the great difficulty "of explaining by *mechanical* physics
what causes the colors of the rays." He was unimpressed by Newton's remarkable results precisely because Newton did *not* advance a mechanical hypothesis:

"He hath not taught us, what it is wherein consists the nature and difference of Colours, but only this accident (which certainly is very considerable,) of their *different Refrangibility.*"[19]

Newton's point of view was obviously quite strange, and in a manner suspect, to all three men, Hooke, Pardies, and Huygens. He wished only, he insisted, to "speak of *light* in *general* terms, considering it abstractly," as "something or other propagated every way in streight lines from luminous bodies without determining, what that Thing is."[20] By dealing only with that ancient Greek abstraction, that mathematical entity, the light ray, he was striving for a higher certitude by considering only what he could measure and describe in the language of mathematics. It is apparent that in this early paper he had already imposed a rigid discipline upon himself and adopted the cautionary rule that hasty, gratuitous "explications" can actually block the way to scientific discovery.

The full power of Newton's "mathematical way" could scarcely reveal itself in this brilliant, but restricted, paper on light and color. He had effectively disposed of the "modification theory" of color, though he had done little else of theoretical import. It was a clever performance; but, as Huygens saw it, this was hardly the kind of approach that could lead to an understanding of the System of the World.

Yet this was precisely what Newton was later to achieve in the *Principia*, though once again his goal and his method were misunderstood. In the *Principia*, Newton showed how inadequate was Descartes's physical picture of the planets whirled about in a great solar vortex. This "hypothesis," in Newton's sense of the term, simply could not account mathematically for the elliptical orbits of planets, or for the two other laws of Kepler. On the other hand, Newton's own "mathematical way," after a display of mathematical ingenuity, led him to the observed motions, and yielded a remarkably accurate—although abstract and schematic—account of the System of the World.

But just as in his study of light and color, the underlying mechanism remained hidden. To Newton's universe, the favorite image of a great clock-work mechanism, applicable though it is a Cartesian world, is singularly inappropriate. The springs and wheels operating it are left undescribed. It was enough that a mysterious force like universal gravitation or attraction could be shown by his investigations to work its wonders. To accept its reality, it need not be explained: "What I call Attraction may be perform'd by impulse, or by some other means unknown to me. I use that Word here to signify only in general any Force by which Bodies tend towards one another, whatsoever be the Cause."[21] It was enough for Newton that such forces could be measured and the laws of their action determined, just as he had been content to describe the behavior of colored rays without seeking the cause of their different colors: "For we must learn from the Phaenomena of Nature what Bodies attract one another, and what are the Laws and Properties of the Attraction, before we enquire the Cause by which the Attraction is perform'd."[22]

Once again, content to set forth the quantitative laws of motion and their consequences, Newton was offering an abstract mathematical description in place of the pictorial type of "explication" favored by the mechanical philosophers. This explains why the first criticism elicited by the *Principia*, when it became known on the Continent, was that it was a brilliant display of mathematics, but that it was not *physics* at all. This was the reaction of Christiaan Huygens and of the anonymous reviewer of Newton's book in the *Journal des sçavans*. As physics was then understood, these men were perfectly right. Precisely as the title of his book proclaimed, Newton was setting forth the *mathematical principles* of natural philosophy, not a natural philosophy in the accepted sense.

Newton's use of his "mathematical way" led, therefore, to a peculiar outcome of which he was quite well aware. At the base of his whole system were forces, laws, or principles—chief among them, a universal attraction—the underlying causes of which he did not, perhaps could not, visualize, and about which he was reluctant to speculate in print. Huygens, who understood Newton better than most, had merely chided him for not going far enough. Others felt he had gone too far, perhaps too far backward. They accused him of basing his whole System of the World upon something that was either an absurdity or a mystery; upon the philosophically repellant notion of action-at-a-distance; or upon an occult quality, like those Aristotelian notions that reputable physics had abandoned.

Newton, we know, was doing nothing of the kind. He was not postulating "hypothetically" his Law of Gravitation. Uncomfortable though it was, he had inferred it from phenomena, and he had shown with .what simplicity and precision it explained the observed motions of bodies. Such nonvisualizable entities —if I may call them that—entities at best, or only, representable in equations, are a commonplace to the physicist of today. He is, for the most part, quite happy with them. What—to take an example—is entropy? What, for that matter, is an electron or a photon, entities that display both particle-like and wavelike properties? The modern physicist, being human, may sometimes wonder; but his, too, is the "mathematical way." His trust in it, his reliance upon it, is all but universal. And this reliance dates from the delayed acceptance of Newton's *Principia* and from the stormy burial of the old "hypothetical way" in the third and fourth decades of the eighteenth century. For then, by a typical group of French scientists, a mathematical positivism came to be accepted as the most secure road to the knowledge of physical nature. What Newton had carefully described as the *mathematical principles* of natural philosophy, of physics, became for many— though Newton might not have agreed—the ideal of physics itself.

We can illustrate this interpretation from the words of d'Alembert, one of the earliest of Newton's disciples in France. He set forth his views, about the middle of the eighteenth century, in his *Traité de dynamique* and in his widely read *Preliminary Discourse* to the *Encyclopédie*. Nature, he wrote, is a vast machine whose inner springs are hidden from us; we see this machine only through a veil which hides the workings of its more delicate parts from our view. . . . Doomed as we are to be ignorant of the essence and inner contexture of bodies, the only resource remaining for our sagacity is to try at least to grasp the analogy of phenomena, and to reduce them all to a small number of primitive and fundamental facts. Thus Newton, *without assigning the cause of universal gravitation*, nevertheless demonstrated that the system of the world is uniquely grounded on the laws of this gravitation.[23]

As Newton has shown us we can do, we extend our investigations even to those motions produced by hidden forces and causes, provided that the laws and relationships according to which these causes operate are known to us: "The knowledge or the discovery of these relationships is almost always the only goal we are allowed to reach, and consequently the only one we should have in view."[24] By the application of mathematical calculations to experience, deducing sometimes a great number of consequences from a single observation, one can arrive at truths which closely approximate the certitude of geometrical demonstrations. For contrary to popular belief, d'Alembert reminds his reader, "The most abstract notions, those that ordinary men regard as most inaccessible, are often those that shed the brightest light."[25]

D'Alembert's message is that the frail, imperfect human mind cannot penetrate the reality of nature and is equipped only to observe, compare, and measure. It is also the message of that curious little philosophical tale of Voltaire's entitled *Micromégas*. Perhaps you remember the story. Micromégas, a giant eight leagues high, is a resident of a planet revolving around the bright star Sirius. Banished by the mufti of his planet for scientific heterodoxy, he becomes the first spaceman. Applying his knowledge of the laws of gravitation and hitchhiking on a convenient comet, he makes his way to our solar system. Here he picks up a companion, a resident of Saturn, a mere dwarf some thousand fathoms tall. Together they navigate to the earth. At first sight this ridiculous planet seems uninhabited, but at last they discern a tiny object, a ship in the Baltic sea transporting a group of philosophers and mathematicians. Struck by the ignorance these tiny insects reveal about metaphysical matters, and their lack of agreement among themselves, the two visitors pursue the conversation as follows:

> The traveller was moved to pity for the tiny human race, in which he discovered such astonishing contrasts. "Since you are of the small number of wise men," he said to these gentlemen, tell me, I pray, what you are interested in."
>
> "We dissect flies," said the philosopher, "we measure lines, we gather mathematical data. We are agreed on two or three points we understand, and we argue about two or three thousand we do not."
>
> A fancy forthwith struck the Sirian and the Saturnian to question these thinking atoms, to find out the things on which they were agreed. "What do you reckon to be the distance," asked the Saturnian, "between the Dog-star and the bright star of Gemini?"

"Thirty-two and a half degrees," they all replied at once.

"And what is the distance from here to the moon?"

"In round numbers, sixty semi-diameters of the earth."

"How much does your air weigh?" he continued, thinking to startle them. But they all told him that air weighs about nine hundred times less than an equal volume of the lightest water and nineteen thousand times less than ducat gold. The little dwarf from Saturn, astounded at their replies, was tempted to take for sorcerers these same people to whom a quarter of an hour before he had refused a soul.[26]

The positivistic view set forth by d'Alembert, and advanced in this playful fashion by Voltaire—that physical science at its best can only be a mathematical description of the interconnecting laws of nature—was widely held in eighteenth-century France. Condillac, in some ways the most respected thinker of the age and one whom the scientists were pleased to quote, held a similar opinion. The study of nature, he once wrote, should be limited to discovering the relations (*rapports*) that obtain among the objects of our experience. We cannot construct true systems of nature, for we know nothing of the elements of things, nothing, that is, of nature's underlying mechanism; we can only observe the remote effects. The best causes (*principes*) a physicist can invoke are those phenomena which, like attraction, explain other phenomena, but which themselves depend upon causes that we do not, and perhaps cannot, know.[27]

But was this Newton's own view? Would he have agreed with d'Alembert and Condillac? Was he satisfied with his abstract framework of laws and relations as the only knowledge open to the inquiring mind? I really do not think so. Behind his mathematical demonstrations and his experiments we discern the basic conceptual model of the seventeenth-century thinkers: that the underlying realities in nature are material particles in motion. In all his investigations he was guided by an atomistic philosophy of nature derived from Gassendi and Charleton; it served him invaluably as a psychological prop and a heuristic aid. We may be certain that he believed in it, though he could not demonstrate it. Yet with astonishing consistency and self-discipline he kept those beliefs from obtruding on his "scientificall" demonstrations. These beliefs are set forth for what they were—guesses, probabilities—in the asides and discursive appendices of his serious work. We read them in the *scholia* of the *Principia*, in the famous Queries of the *Opticks*, in some letters and unpublished papers. In short, as far as was humanly possible, Newton kept his science (as we would call it) wholly separate from his speculative natural philosophy.[28]

What, then, did Newton think he was doing, if he did not believe that his "mathematical way" was giving birth to a new physics? I should like to urge a new and perhaps radical interpretation. I suggest that he was preparing the way for, setting the stage for, a new natural philosophy, not wholly unlike those he had attacked. With his abstract, mathematical scheme of laws he believed himself to be marking the boundary conditions for this new natural philosophy, supplying a frame or scaffolding which later generations would fill in and to which they would give substance. The title of Newton's great book, to refer to it once again, is eminently revealing: what he is setting forth are only the *mathematical principles* of natural philosophy. But that philosophy itself is still to come, the work of other hands, though here and there Newton offers hints and suggestions as to what it may contain. This, I believe, is what the famous Queries of the *Opticks* are all about. The speculations and guesses we find there were intended to be so many signposts for later workers. By contrast, what he gave the world in the body of his scientific work was like the steel frame of some great building. The mathematical laws of optics and celestial mechanics are the girders and supporting members; other men will come with the bricks, the mortar, and the cut stone to fill in the walls and lay out the partitions. According to what rules did he feel this should be done? Surely, we must imagine, by following the methodological precepts of "experimental philosophy" as he is at pains to set them forth at the end of the 31st Query. But one thing at least seems certain. The new building, the new Temple of Natural Philosophy, must be erected here, within his framework, inside the boundaries he had marked out by his mathematical laws, and not elsewhere, at some point far afield among the infinite possibilities of which Descartes had written.

We should not be surprised, therefore, that Newton's achievement received various and conflicting interpretations in the course of the eighteenth century, that it was called upon to support quite different methodological and epistemological principles. Cassirer, it seems to me, is much too sweeping when he writes that the ideal of a mechanical philosophy of nature "was gradually superseded until it was finally abandoned entirely by the epistemologists of modern physics."[29] By no means every thinker of the age felt that the physicist "must finally give up trying to explain the mechanism of the universe" and that "he has done enough if he succeeds in establishing definite general relations in nature." Certain men—like some of the odd persons Bachelard mentions in his book, or like the Baron d'Holbach in his *System of Nature*—understood Newton not at all, though they sometimes invoked his name. Others, for example the Dutch methodologists and such English and American scientists as Stephen Hales, Joseph Black, Joseph Priestley, and Benjamin Franklin, saw in Newton the advocate of an experimental philosophy by which men can, with some speculative license, penetrate into the inner workings of material nature. It is this tradition, we might well recall, that culminates, soon after the century's close, in John Dalton's atomic theory. For such men, the hints in the Queries were so many delphic utterances, so many clues to be patiently pursued. Indeed, they seem to have read Newton's program more truly than those we think of as the rigorous Newtonians, than d'Alembert and the other advocates of a positivistic view of science that Newton almost certainly never envisaged.

Notes

1. *Voltaire's Correspondence*, ed. Theodore Besterman, 107 vols. (Geneva: Institut et musée Voltaire, 1953–65), IV (1954), no. 838, pp. 48–49.
2. Jean d'Alembert, *Traité de dynamique*, 1st ed. (Paris, 1748), "Preface," p. v; and 2d ed. (Paris, 1758), "Discours préliminaire," pp. v–vi; my italics.
3. See, for example, the letter to Horace Walpole (July 15, 1768). Besterman, *Voltaire's Correspondence*, no. 14179.
4. Pierre Maupertuis, *Discours sur les différentes figures des astres avec une exposition des systèmes de MM. Descartes et Newton* (Paris, 1732). For Maupertuis' assistance to Voltaire see Pierre Brunet, *Maupertuis*, 2 vols. (Paris, 1929), I, 22–26; E. Sonet, *Voltaire et l'influence anglaise* (Rennes, 1926), 117–18.
5. *The Portable Voltaire*, ed. Ben Ray Redman (New York: Viking Press, 1949), 228. See the passage in Voltaire's *Traité de métaphysique* cited by Ernst Cassirer: "When we cannot utilize the compass of mathematics or the torch of experience and physics, it is certain that we cannot take a single step forward." *The Philosophy of the Enlightenment*, trans. Fritz C. A. Koelln and James P. Pettegrove (Princeton: Princeton University Press, 1951), 12.
6. *Benjamin Franklin's Experiments*, ed. I. B. Cohen (Cambridge, Mass.: Harvard University Press, 1941), 200.
7. See, for example, *Oeuvres de Condorcet*, ed. A. Condorcet O'Connor and F. Arago, 12 vols. (Paris, 1847–49), VII (1847), 295–306, esp. 298–99.
8. Cited by Douglas Grant, *Margaret the First: A Biography of Margaret Cavendish, 1623–1673* (London: Hart-Davis, 1957), 209.
9. Hobbes to Newcastle, July, 1636, *Portland Manuscripts*, II, 128. This passage, like the previous one from Glanvill, I owe to my student, Robert Kargon, whose doctoral dissertation, "Atomism from Hariot to Newton" (Ithaca, 1964), has thrown new light on the mechanical philosophers of the seventeenth century.
10. John Locke, *An Essay Concerning Human Understanding*, chap. 4, secs. 3 and 12.
11. That Boyle shared, and perhaps inspired, Locke's doubts concerning the degree of certainty attainable in natural philosophy is argued by James Gibson, *Locke's Theory of Knowledge and Its Historical Relations* (Cambridge, 1917), 257, 260–65.
12. *Discoveries and Opinions of Galileo*, trans. and with an Introduction and notes by Stillman Drake (Garden City, N.Y.: Doubleday, 1957), 179. Professor Strong has written: "Newton's 'mathematical way' encompassed both experimental investigation and demonstration from principles, that is, from laws or theorems established through investigation." E. W. Strong, "Newton's Mathematical Way," *Journal of the History of Ideas* 12 (1951), 90–110.
13. Sir Isaac Newton, *Opticks, Or a Treatise of the Reflections, Refractions, Inflections and Colors of Light*, reprinted from the 4th ed. of 1728 (London: G. Bell, 1931), 404.
14. Sir Isaac Newton, *Mathematical Principles of Natural Philosophy*, ed. Florian Cajori (Berkeley, Calif.: University of California Press, 1934), 547.
15. Isaac Barrow, *Mathematical Lectures*, trans. John Kirkby (London, 1734), 21.
16. *Isaac Newton's Papers and Letters on Natural Philosophy*, ed. I. B. Cohen (Cambridge, Mass.: Harvard University Press, 1958), 53.
17. That the debate over Newton's paper was not a conflict between the corpuscular and undulatory theories of light has been pointed out by Richard Westfall in his suggestive paper "Newton and His Critics on the Nature of Colors," *Archives Internationales d'Histoire des Sciences* 15 (1962), 47–58. I cannot, however, accept Professor Westfall's interpretation that the debate turned on the nature of qualities, on mechanical versus peripatetic language, or that Newton's first paper "was interpreted by the champions of the mechanical philosophy as a reversion to conceptions associated with the rejected philosophy."
18. *Isaac Newton's Papers and Letters*, 106.
19. *Ibid.*, 136. By "accident" Huygens of course means "property." Newton was by no means alone in having recourse to the old scholastic vocabulary.

20. *Ibid.*, 119. See also p. 106, where, in his reply to Pardies, Newton explains that by "rays of light I understand its least or indefinitely small parts, which are independent of each other; such as are all those rays which lucid bodies emit in right lines, either successively or all together."

21. *Opticks*, 376.

22. *Ibid.* See *Principia*, ed. Cajori, Definition VIII, pp. 5–6, where Newton writes: "I . . . use the words attraction, impulse, or propensity of any sort towards a centre, promiscuously, and indifferently, one for another; considering those forces not physically, but mathematically: wherefore the reader is not to imagine by those words I anywhere take upon me to define the kind, or the manner of any action, the causes or the physical reason thereof. . . ." See also bk. 1, sec. 11, p. 164: "But these Propositions are to be considered as purely mathematical; and therefore, laying aside all physical considerations, I make use of a familiar way of speaking, to make myself the more easily understood by a mathematical reader."

23. D'Alembert, *Mélanges de littérature, d'histoire et de philosophie*, 4th ed., 5 vols. (Amsterdam, 1767), IV, 258–59. My italics.

24. D'Alembert, *Discours préliminaire de l'Encyclopédie*, ed. Louis Ducros (Paris, 1893), 40.

25. D'Alembert, *Discours préliminaire*, 45. Cf. the *Traité de dynamique* (ed. 1758), p. ii, and *Mélanges*, IV, 182–83.

26. My translation is from Ira O. Wade, *Voltaire's "Micromégas," A Story in the Fusion of Science, Myth and Art* (Princeton: Princeton University Press, 1950), 141–42. For other expressions by Voltaire of this view of science, see my article "Three Eighteenth-Century Social Philosophers," in Gerald Holton, ed., *Science and the Modern Mind* (Boston: Beacon Press, 1958), nn. 17, 18.

27. *Oeuvres philosophiques de Condillac*, ed. Georges Le Roy, 3 vols. (Paris, 1947–51), I (1947), 207; III (1951), 439.

28. D'Alembert was fully aware of the speculative dimension of Newton's thought, an aspect which Newton "abstained almost completely from speaking of in his best known writings." *Discours préliminaire*, 107.

29. Cassirer, *Philosophy of the Enlightenment*, 53–54.

11

An Augustan Monument:
The Opticks of Isaac Newton

❖

Scientific books rarely earn a place in histories of literature. So it is not surprising that few if any accounts of the Augustan Age make even a passing mention of one of the most notable works published in the reigns of Queen Anne and of George I. Yet Isaac Newton's *Opticks*, subtitled "A Treatise of the Reflections, Refractions, Inflexions and Colours of Light," first published in 1704, deserves mention, though not a work of imaginative literature, along with Dean Swift's *Gulliver*, the novels of Defoe, and the poetry of Alexander Pope. The extraordinary influence Newton's book exerted during the eighteenth century, not only upon science but upon philosophy, letters, and even—as I shall suggest—upon the arts, marks it as an epoch-making Augustan work. Besides the book itself—its composition, publication, and subsequent revision in later editions—it is the range and spread of its influence that I propose to discuss.

Chronologically Newton was an Augustan. His apotheosis as a national hero, in the last quarter century of his life, almost exactly coincides with the Augustan Age of English literature. He had left Cambridge in 1696 to assume the post of Warden of the Mint, responsible for the great enterprise of the recoinage. Although for some years he kept his professorship at Cambridge, and his Trinity College Fellowship, he resided henceforth in London. Raised to the less exacting and more honorific post of Master of the Mint, a post he held for the rest of his life, he at last severed his connection with his University.[1] The Royal Society of London, whose meetings he had rarely attended up to this time, became thereafter the center of his life. In 1703, the year after the accession of Queen Anne, he was elected President of the Society, and in April of 1705 he was knighted by the Queen. As President, a post that he held until his death in 1727, he was no mere aging figurehead: he concerned himself actively with the affairs of the Society, taking the lead, for example, in finding a new home for that body. He was, as we have come to recognize, intellectually active in science and alert to new discoveries, some of which indeed he actively promoted. Most important, perhaps, he brought out new and revised editions of his great classic, the *Mathe-*

From Peter Hughes and David Williams, eds., *The Varied Pattern: Studies in the 18th Century* (Toronto: A. M. Hakkert Ltd., 1971), 131–63. Copyright © 1971 by A. M. Hakkert Ltd., all rights reserved.

matical Principles of Natural Philosophy (1687), and of his *Opticks*. And the changes he introduced into both these works during the last fifteen years of his life show that his philosophy of nature was capable of far from trivial modifications.[2]

❖

A comparison between Newton's two great masterpieces is perhaps in order. The *Principia,* more often talked about than read, is a difficult work written, if I may put it this way, in two ancient languages: Latin and geometry. The foundation work in rational mechanics, it is more or less rigorously axiomatic and deductive in form. The "First Book" treats bodies as if deprived of all qualities or properties except the "primary" ones of extension and mass; these abstract theorems about mass points or idealized bodies moving under the influence of "central forces" are applied, in the concluding part of the *Principia* ("The System of the World"), to the real bodies of the solar system.

By contrast, the *Opticks* is written in the language of experiment and was first published in English. To be sure, Newton affects a kind of Euclidean form; beginning with definitions (of "rays of light," "refrangibility," and so on) and with a few axioms (certain basic laws of geometrical optics), the work proceeds as a series of "propositions" and "theorems." But these propositions are not statements of abstract mathematical relations; they are assertions of experimental fact, followed by what he calls "proofs by experiment." The axiomatic form is not to be taken seriously; indeed it is systematically followed only in the first of three "books" into which the work is divided. There is little mathematics, and that of the simplest sort. For the most part, the body of the work (exclusive of the famous *Queries* appended to it) consists of a meticulous account of his experiments on light and color. It is no wonder that the book was more widely read, and more accessible to the curious reader, than the formidable *Principia.*

The *Opticks* is in some respects a bibliographical curiosity, and I shall call attention to some of the puzzles it presents to scholars. First of all, contemporaries could have quibbled about the title: the book is not a textbook of optics, but a monograph on color; Newton's purpose is to introduce an exact and quantitative science of color into the ancient and established domain of optics.[3] From his own experiments, and the inferences drawn from them, he explains the colors produced by prisms and droplets of water, the colors of natural bodies, the periodic bands of color ("Newton's rings") observed in thin transparent bodies (bubbles, sheets of mica), and the rainbow.

One puzzle involves the publication date of the *Opticks*. The first edition of 1704 opens with an undated "Advertisement" signed with the initials "I.N." But when, thirteen years later, Newton brought out a second English edition, he added a new preface designated as "Advertisement II"; this too is signed with Newton's initials, and it is dated (July 16, 1717). But the earlier advertisement is reprinted, this time with a date appended. The date—April 1, 1704—is obviously erroneous, for the evidence is clear that the book had been published more than a month earlier, sometime in February 1704.[4] Newton, as we shall see, was indifferent to such minutiae and rather cavalier in revising his own writings.

To answer such questions as "When was the *Opticks* composed?" or "When did Newton put his manuscript in final form?" is not easy. The work is really a conflate text; it embodies experiments and observations that Newton made at various periods of his life; it seems, indeed, pretty much to recapitulate the sequence of his discoveries. The substance of the "First Book" is an expansion of his classic paper, the letter on his "New Theory about Light and Color," published in the *Philosophical Transactions* of the Royal Society in February 1672.[5] This famous paper, you will recall, demonstrated by experiments with prisms that white solar light is a "mixture" of rays "differently refrangible" and suggested, although hardly proving, that there is a one-to-one correspondence between the color of the rays and the degree of refrangibility.[6]

Much of the "Second Book" is almost word for word (with some significant emendations) reprinted from a paper sent to the Royal Society in 1675, describing his observations on the ring phenomena, a paper Newton at the time declined to publish.[7]

The "Third Book" consists of a short section (about nineteen quarto pages) describing experiments on diffraction, that is: the bending and splitting of light when it passes through narrow apertures, or past the sharp edges of bodies, a phenomenon first described by the Jesuit scientist, Francesco Maria Grimaldi.[8] Newton clearly deemed his own experiments on the subject incomplete or inconclusive, for he wrote: "When I made the foregoing Observations, I designed to repeat most of them with more care and exactness, and to make some new ones. . . . But I was then interrupted, and cannot now think of taking these things into further consideration."[9] Then follow immediately the so-called Queries: shorter or longer suggestions or speculations, framed in the interrogative voice, and proposed, as Newton says, "in order to a further search to be made by others." It cannot be determined with certainty when the diffraction experiments were performed, but evidence points to the years just before 1684–85.[10]

The "Third Book" confronts us with still another puzzle. If we consult the later editions of the *Opticks* (including the modern paperback edition in common use) we find that whereas the First Book is divided into two numbered "Parts," and the Second Book has four, the Third Book is labeled "Part I," but there is no "Part II." This designation of a "Part I" is not to be found in the first edition of 1704; it makes its first appearance in the second English edition of 1717–18, a clue that, as we shall see later on, helps unravel the mystery.

When was the *Opticks* actually put together from its diverse elements? How long before its publication in 1704? We know that Newton had in hand a virtually complete manuscript as early as 1694, for in that year David Gregory, the Savilian Professor of Astronomy at Oxford, visited Newton in Cambridge and was shown the manuscript of "Three Books of Opticks." This, Gregory noted, would if printed be the equal of the *Principia*.[11] Later the following summer, doubtless at Gregory's instigation, the Fellows of the Royal Society resolved that a letter be written to Newton urging him "to communicate to the Society in order to be published his Treatise of light and colours, and what other Mathematicall or Physicall Treatises he has ready by him."[12] There was no response from Cambridge; so the next year John Wallis, the venerable Oxford mathematician, wrote to Newton that he had learned of the completion of "a Treatise about Light, Refraction & Colours; which I should be glad to see abroad." And he went on: " 'Tis pitty it was not out long since. If it be in English (as I hear it is) let it, however, come out as it is; & let those who desire to read it, learn English." And Wallis—because of his seniority, eminence, and friendship with Newton he could safely chide that sensitive man—concluded: "You are not so kind to your Reputation (& that of the Nation) as you might be, when you let things of worth ly by you so long, till others carry away the Reputation that is due to you."[13] Newton's reply has been lost, but its burden is clear from the exasperated tone of Wallis's next letter:

> I can by no means admit your excuse for not publishing your Treatise of Light & Colours. You say you dare not *yet* publish it. And why *not yet*? Or, if not now, when then? You adde, lest it create you *some trouble*. What trouble *now*, more then [sic] at another time? Pray consider, how many years this hath lyen upon your hands allready.[14]

Newton's reluctance to publish was, of course, notorious. Only great pressure from the Royal Society, and the direct assistance of Edmond Halley, had brought the *Principia* into being. This reluctance was enhanced by the prolonged controversy—with Robert Hooke and with critics on the Continent—that attended the publication of his first optical paper of 1672. Not long after, Newton wrote to Henry Oldenburg, the Secretary of the Royal Society: "I see I have made my self a slave to Philosophy, but if I get free of Mr. Linus's business I will resolutely bid adew to it eternally, excepting what I do for my privat satisfaction or leave to come out after me."[15]

Evidently the composition of the *Opticks* was just such an activity for his "privat satisfaction," and a good part of the work must have been composed after his resolute vow to Henry Oldenburg. Yet, after his move to London, as we saw, Newton remained deaf to the pleas of his friends. Finally a notable delegation paid him a visit. On Sunday, November 15, 1702 (so David Gregory recorded), Newton "promised Mr. Robarts, Mr. Fatio, Capt. Hally [sic] and me to publish his Quadratures, his treatise of Light, and his treatise of the curves of 2^d genre."[16] Newton kept his promise and must have set to work on the final revisions, and dispatched the manuscript to the printer, before the autumn of 1703. In February of 1704 Newton presented a copy of this handsome quarto, its title page printed in red and black, to the Royal Society. His name nowhere appears, only his initials appended to the undated "Advertisement" where Newton confessed: "To avoid being engaged in Disputes about these Matters, I have hitherto delayed the Printing, and should still have delayed it, had not the importunity of Friends prevailed upon me."[17]

Thus appeared one of the most important scientific books ever published in the English language. Its influence can be attributed to the important discoveries it contained, to the clarity and precision of the writing, to the method of scientific inquiry it exemplified, and—in the remarkable Queries, as Newton expanded them, and added to them, in later editions—to the bold speculations and suggestions for future investigations which Newton set forth.

❖

There was no novelty in publishing a scientific work in English, although Latin was widely favored in the seventeenth century—especially by physicians, who were notably conservative—as the language of scientific communication. Before the middle of the seventeenth century, indeed as early as the late sixteenth, English was employed chiefly in works of a practical nature, such as mathematical works for the surveyor or the navigator, or popular books like John Wilkins's *Discovery of a World in the Moon* (1638). The two earliest classics of English science, William Gilbert's *De magnete* (1600) and William Harvey's epoch-making book on the circulation of the blood, were both written in Latin. But the Royal Society of London, soon after its establishment, profoundly stimulated the use of English, which was the language of its official journal, the *Philosophical Transactions*, and of the majority of the books published under its auspices, for example the *Micrographia* and other writings of Robert Hooke and the *Anatomy of Plants* (1682) of Nehemiah Grew. John Wallis's early mathematical works were published in Latin; but his *Algebra* first appeared in English in 1685 and was only later translated into Latin.[18] This was the common practice of Robert Boyle whose best-known books—his *Spring and Weight of the Air* and his *Skeptical Chemist*—indeed the majority of them, were first composed and published in English but brought out soon after in Latin dress for the benefit of readers on the Continent.[19]

The English style of these early scientific books was often clumsy and cluttered with Anglicized Latin words; none could use them as artfully as Sir Thomas Browne. Boyle is proverbially hard to read; his sentences are long, loose-jointed, and rambling. Yet he made a valiant effort to purify his vocabulary from the influence of scientific Latin, and to find, where possible, equivalents in plain English for the technical terms of medicine and chemistry. At all events, he avoids the gibberish of Walter Charleton, whose *Physiologia* of 1654 exerted in other respects a profound influence upon his thought. Boyle, to say nothing of Newton, would never have written anything like the following description by Charleton of an experiment with a prism:

> As for the Enodation of the *Later Difficulty*, it is comprehended in the Reasons of the Former. . . . This is easily *Experimented* with a piece of narrow black

Ribbon affixt longwise to either side of the Prisme. For, in that case, the light is bipartited into two Borders, or Fringes, the opace part veyled by the Ribbon on each side environed with light, and each border of light environed with two shadows; or, more plainly, between each border of shadows conterminate to each extreme of Light, trajected through the unopacate parts of the Glass: and, therefore, in the commissure of each of the two lights with each of the conterminous shadows, there must be Vermillion to one side, and Caerule on the other.[20]

Compare this, if you please, with Newton's description of one of his early experiments:

In the Sun's beam which was propagated into the Room through the hole in the Window-shut, at the distance of some Feet from the hole, I held the Prism in such a posture that its Axis might be perpendicular to that beam. Then I looked through the Prism upon the hole, and turning the Prism to and fro about its Axis to make the Image of the Hole ascend and descend, when between its two contrary Motions it seemed stationary, I stopt the Prism that the Refractions on both sides of the refracting Angle might be equal to each other. . . . In this Situation of the Prism viewing through it the said hole, I observed the length of its refracted Image to be many times greater than its breadth, and that the most refracted part thereof appeared violet, the least refracted red, and the middle parts blew green and yellow in order.[21]

There are Anglicized Latin terms, to be sure, but no more than Newton deemed unavoidable. The style is crisp and tight; and here, as throughout his *Opticks*, we find that "close, naked, natural way of Speaking; positive Expressions, clear Senses; a native Easiness" that came to be exacted from the Fellows of the Royal Society.[22]

Newton, of course, could not completely free himself from the academic language of his student days. As we leaf through the *Opticks* we find a number of archaic Latinisms: celerity for speed; conduce for contribute; confine for limit or border; intromit for let in; interjacent for lying between. And here and there we find even stranger words, of the kind that flowed so readily from the pen of Walter Charleton: equipollent, consecution, obliquation.

Yet there are words first appearing in the *Opticks* that proved of such utility that they were soon domesticated and added to our language, largely through the efforts of Samuel Johnson. Some twenty years ago, Professor W. K. Wimsatt of Yale published a study of Johnson's *Dictionary* and the *Rambler* and the "philosophic words" drawn from the writings of Francis Bacon, Robert Boyle, and other worthies, including Isaac Newton, which Dr. Johnson naturalized into the literary language. All the Newtonian words were drawn from the *Opticks*, which Johnson owned and had obviously perused with care. They are words now thoroughly at home in modern prose, at least scientific prose, and which writers have used, and still use, metaphorically: accelerate, assimilate, attraction, luminous, medium, rotation, texture, volatile, and many more.[23]

There is one Newtonian word listed by Johnson that deserves our special attention, for it is a key word in modern physics, commonly used, too, for various metaphorical purposes. This is the word *spectrum*, which we usually define as the "band of colors produced by a prism." The *O.E.D.* cites Newton's classic paper of 1672 as the first appearance of the word in English. But if you look at this early paper, or scan the *Opticks* itself, you find something very peculiar. If, says Newton, you let a beam of light through a hole in a window shutter, and project it upon a screen, the "spectrum" will be round and white. If, however, you refract that beam through a prism, and pass the red rays alone through the hole of a second board or screen, projecting it upon the wall, the "spectrum" will be round and red. Newton clearly means by this word any insubstantial, ghostlike optical image.[24] We can see what transpired to give the word its modern meaning: well into the eighteenth century the band of colors produced by a prism (what Newton himself often called the "coloured spectrum," the "solar spectrum," or the "oblong image") is usually called the *prismatic spectrum*. But when the prismatic spectrum became the chief kind of optical image interesting to physicists, it gradually came to be called *the* spectrum, the adjective being simply dropped.[25]

✦

The appeal of Newton's *Opticks* to the nature poets of the English eighteenth century, the best-known literary influence of that book, needs little elaboration from me. Their Newtonian imagery, their preoccupation with color, have been amply described by Marjorie Nicolson in her well-known study, *Newton Demands the Muse*.[26] Miss Nicolson tells us that the keen interest of poets in the *Opticks* began at the time of Newton's death. In the flood of eulogies and elegies published in 1727 and 1728, nearly all the poets mention "Newton's Rainbow" and "Newton's Colours"; yet except for James Thomson's "To the Memory of Newton," and Richard Glover's "Poem on Newton" (from which Miss Nicolson drew the title of her book) she finds these verses crude and undeveloped, "amorphous" is the word she uses. From then on the examples multiply, the quality improves, and poetic allusions to the *Opticks* become more specific, to culminate in Thomson's *Seasons*, Edward Young's *Night Thoughts*, and Mark Akenside's *Pleasures of the Imagination*. To be sure, these poets, while indebted to the *Opticks* for their color language and their prismatic imagery, need not have perused the book itself. They could have read Henry Pemberton's *A View of Sir Isaac Newton's Philosophy* (1728), or a book that dealt chiefly with Newton's discoveries on light and color, Francesco Algarotti's *Il Newtonianismo per le dame*, brought out in English dress (1739) by Elizabeth Carter.

But whether obtained directly or indirectly, the optical lore of Miss Nicolson's poets was derived from the substance of the first two "Books" of the *Opticks*. When Thomson wrote that Newton

Untwisted all the shining robe of day;
And, from the whitening undistinguished blaze,

Collected every ray into his kind,
To the charmed eye educed the gorgeous train
Of parent colours.[27]

he was ennobling in verse the prism experiments set forth in the First Book, Part I, of the *Opticks*. And when the poets describe, in embellished Newtonian language, the rainbow, the iris in a peacock's tail, the permanent colors of natural bodies, it is from Part II of the First Book and the body of the Second Book that the descriptions are drawn. The Third Book, with its marvelous Queries, held its treasures for other and different minds.

✦

If the poets, drawing inspiration from Newton's book, discovered color—color in the natural landscape, color phenomena in the atmosphere—so too did scientists and scientific amateurs. There were efforts to explain in Newtonian terms the colors of the aurora borealis, the blue shadows cast by bodies, and the red color of morning and evening clouds. Others performed experiments on what Buffon called "accidental colors," colors produced by a blow on the eye, or the vivid afterimages seen after strong or prolonged exposure to natural colors, all matters alluded to by Newton.[28]

The keen interest in color for its own sake is attested by the ephemeral success of Father Castel's *clavecin oculaire*, or color organ. A severe critic of Newton's celestial mechanics, the Jesuit Louis-Bertrand Castel, for a time accepted, like most Frenchmen of his day, Newton's discoveries concerning color.[29] He drew powerful encouragement from those passages in the *Opticks* where Newton spoke of the prismatic spectrum as exhibiting seven colors, with intervals corresponding to the notes of a musical octave. Color, then, could provide a kind of music; and in the 1750s Father Castel is said to have entertained audiences with shifting displays of color, color melodies, played silently on his color harpsichord or with musical accompaniment.

Newton's analogy between sound and color was first fully developed in print in the *Opticks*. Here he tells how he projected the spectrum upon a sheet of white paper and asked a friend, whose color perception was keener than his own, to mark the boundaries of the seven colors by lines drawn across the image. To these intervals he assigned ratios corresponding to the notes of an octave: the tonic, major second, minor third, fourth, fifth, major sixth, seventh, and the octave.[30]

Further on, where he reprinted with some changes his early unpublished paper on periodic colored rings, Newton invokes again the musical analogy. He compared the breadths of the rings of different colors to the different thicknesses of the film producing them and found them to be one another "as the Cube-roots of the Squares of the eight lengths of a Chord, which sound the notes in an eighth, *sol, la, fa, sol, la, mi, fa, sol*."[31]

The ideas of the pioneer psychologist David Hartley owe much to the Newtonian passages just quoted, as well as to those in which Newton stated that the periodicity of the rings showed that light rays, for whatever underlying reason, are disposed to display "fits of easy reflexion" alternating with "fits of easy transmission."[32] Hartley's debt to Newton was profound. In his *Observations on Man* (1749) he drew upon the *Opticks* to support his theory of color perception, indeed, as I shall show later on, for his fundamental theory of "vibrations." The sensation of color, Hartley argued, results from vibrations imparted to the retina of the eye. Rays of the seven "primary" colors excite vibrations of different frequencies, and these frequencies are related to each other by the simplest ratios: ratios, he writes that "are also those of the five Tones, and two Semitones, comprehended in the Octave." These ratios are different enough to make at least the five principal colors (red, yellow, green, blue, and violet) "appear distinct from each other to the Mind, for the same reasons, whatever they be, as take place in Sounds." Natural bodies reflect all these colors abundantly "and in sufficient Purity for this Purpose," above all the color green, especially what Newton in grading the colors of successive rings called the green of the third order, i.e., the color of grass and vegetables.[33]

In a later section Hartley takes up our emotional response to color, a strong source of pleasure in young children, especially when colors are combined together in various ways; yet he doubts that there is anything in the relation of colors to each other which corresponds to consonance and dissonance of sounds. Increasingly, as we grow older, our reaction to "mere colours" becomes "very languid," as compared with our response to more sophisticated and intellectual sources of pleasure; yet the pleasures we receive from colors "remain, in a small Degree, to the last"; moreover the feelings transferred to them by association with other sources of pleasure, Hartley says, are "considerable."

> So that our intellectual Pleasures are not only at first generated, but afterwards supported and recruited, in part from the Pleasures affecting the Eye; which holds particularly in respect of the Pleasures afforded by the Beauties of Nature, and by the Imitations of them, which the Arts of Poetry and Painting furnish us with. [34]

And he returns again to the central psychological value of green: the green of Newton's third order:

> It deserves Notice here, that Green, which is the Colour that abounds far more than any other, is the middle one among the primary Colours, and the most universally and permanently agreeable to the Eye of any other.[35]

A generation later the growing interest in color for its own sake is signalized, and was doubtless stimulated, by the publication in 1772 of Joseph Priestley's *History of Vision, Light and Colours*, a work in which discoveries concerning color are carefully recounted from the work of Newton, which he treats at length, until his own time. The book concludes—and not surprisingly when we recall Priestley's debt to David Hartley—with a summary of Hartley's notions about color and the relation of color vibrations to those of musical tones.[36] Whether artists consulted Priestley's book, I do not know; but it is worth noting that two important figures in the history of aesthetics appear among the subscribers to the *History*: Sir Joshua Reynolds and Edmund Burke, who should be remembered not only as orator and political writer but as author of the important essay on *The Sublime and the Beautiful* (1757) with its echoes of Hartley and Newton.[37]

In 1780 Priestley moved to Birmingham, where he became an active member of that informal scientific club, the Lunar Society, whose members included Erasmus Darwin, James Watt, Matthew Boulton, and Josiah Wedgwood, the potter. Among the varied natural phenomena that occupied these men—and Priestley's book or Priestley himself may have been the stimulus—was color, the "accidental colors" first studied by Buffon, and called by them "ocular spectra"; the color of electrical discharges, lightning, and shooting stars; the phenomenon of colored clouds and other atmospheric color effects.[38]

One interesting series of experiments had its origin in the *Opticks*: experiments with a color wheel or color disc, a device (in its later modifications) variously credited to Thomas Young, Helmholtz, or Clerk Maxwell. Yet in its fundamental form it had been suggested by Newton. Divide, Newton wrote, the circumference of a circle into parts "proportional to the seven musical Tones" and paint the pyramidal segments thus drawn so that they represent the successive colors, gradations, and intensities of the prismatic spectrum.[39] The colors can be "mixed" when the disc is rapidly whirled. Such color discs, variously painted, were used by a Lunar Society member, Samuel Galton, to demonstrate the primary colors, the production of white and the complementary pairs. Certain of his discoveries, Erasmus Darwin wrote in summarizing Galton's work, "might be of consequence in the art of painting."[40]

If nothing else, these examples attest to the growing interest in color as the eighteenth century drew to a close, a preoccupation that flowered in one of the most famous and controversial books ever written on the subject: Goethe's *Zur Farbenlehre* (1810).[41] Drawn to the problem of color by his interest in painting, and by his sensitivity to the natural landscape, Goethe, like everyone else, inevitably confronted Newton's *Opticks*, but with a difference. Newton, for Goethe, was the antagonist whose experiments he could not, or at least did not, comprehend. Stubbornly rejecting Newton's discoveries, he devoted a long polemical section of his book to refuting them. But Goethe's "science" was idiosyncratic and his errors were legion: he confused, more than he realized, the physical and the psychological aspects of color. Yet in the latter domain he made acute observations and simple experiments that have proved of great interest to modern physiologists and psychologists.

Mention of Goethe brings us to the moment when, as Professor Gombrich has pointed out, one English painter, John Constable (1776–1837)—who surely had heard more of Newton than of Goethe—dared to brighten his canvases and introduce, albeit timidly, the color green into his landscapes.[42] Constable, al-though always the painter, had a curious scientific bent; his early studies of anatomy enthralled him, and he is said to have been fascinated by astronomy. Clouds and cloud formations obsessed him—their form and their color—and he studied the way in which the ever-changing atmosphere transformed the hues and tones of landscape. In preparation for his Lake District sketches he kept vivid descriptive records of the weather. In a lecture at the Royal Institution in 1836 he remarked, perhaps with his special audience in mind: "Painting is a science, and should be pursued as an inquiry into the laws of nature. Why, then, may not landscape painting be considered a branch of natural philosophy, of which pictures are but the experiments?"[43]

It may be presumptuous, for one not trained in art history, to suggest that at the beginning of the nineteenth century—and owing, if only in part, to the varied reverberations of the *Opticks* of Sir Isaac Newton—we can discern the discovery of color, color for its own sake and in the landscape, by the painters. What we might call the Vitruvian aesthetic, a canon derived from classical sculp-ture and architecture and based on form, mass, and proportion, had long held pride of place. Among the Renaissance painters and their followers there were, to be sure, great colorists; but color, when not used (as in Christian subjects) for its emblematic significance, long tended, as Heinrich Wölfflin has put it, to "subserve form." Especially in landscape painting the brighter colors are often wanting: even the landscapes of Claude Lorrain or the Dutch *paysagistes* were rendered—as were Gainsborough's and the scenic backgrounds of his portraits—in muted brownish tones. Constable took the first step in a movement, which has been with us ever since, towards a chromatic aesthetic.

But among the English painters the first true explorer of color was that great artistic revolutionary, J. M. W. Turner. His sketches, his landscapes in water color and oil—his "pictures of nothing"—were so many excursions into the mysterious world of color: often representations, as Hazlitt put it in a famous essay, "not properly of the objects of nature as of the medium through which they were seen."[44] The chief currents I have been discussing converged upon this extraordinary man: he drew upon, and imitated in his own crude attempts at verse, the color-conscious nature poets: Thomson, Young, and especially Aken-side.[45] He is said to have read and studied Newton's *Opticks*, and he was im-pressed by the Newtonian analogy between color and sound, although his color organ was his painter's palette. He badgered scientific acquaintances for scien-tific facts about light and color, and he experimented tirelessly. He studied the effects produced by light filtering through glass balls filled with colored liquids; he observed the way light is reflected from polished metallic spheres; and he explored the problem of color mixing. He clearly grasped the difference between mixing pigments and mixing spectral colors, and he seems to have anticipated Helmholtz, at least to some extent, in distinguishing between "additive" and "subtractive" color mixing. When, later in life, he read and studied Goethe's *Farbenlehre* (in Eastlake's translation of 1840) and annotated his own copy, his admiration was tempered by sharp criticism; he could hardly have gone along with the great poet's misunderstanding of Newton.[46]

Except for Dr. Johnson and his Newtonian vocabulary, most of the influences I have described, whether direct or indirect, derive from the body of Newton's *Opticks*. Let us now turn to the Queries appended to the Third Book; for these Queries mainly influenced the speculations of philosophers and scientists, and came—in a quite different way—to affect such diverse literary figures as Erasmus Darwin and Shelley.

The Queries did not appear all at once; Newton added new ones, and made changes in some of the earlier ones, in the successive editions of the *Opticks*. They have been much discussed and often misunderstood. In particular, they have been used to characterize the entire *Opticks* to which, in fact and in Newton's mind, they form an appendage. Professor I. Bernard Cohen has written: "The *Opticks* differs from the *Principia, inter alia,* in that Newton freely indulged in hypotheses and speculations."[47] Except for the Queries, this is a poor description of the *Opticks* which, like the *Principia*, in the body of the text is almost wholly without conjectural matter and speculation.[48] Newton is at pains here, as I believe he was in the earlier *Principia*, to exclude anything he is not convinced he can demonstrate mathematically or prove by experiment.

One example should suffice to make my point: throughout the body of the work Newton treats light only as *rays* representable by straight lines. Nowhere does he tell us what the rays physically *are*. Just as nothing in the *Principia* is affected by what we may believe gravitation to be caused by, so in the *Opticks* none of his conclusions, he thought, was in any way dependent upon a theory of the physical nature of light: i.e., whether light is a stream of corpuscles, or a pulse or wave in some ambient medium. In college we were taught that Newton upheld a corpuscular theory of light. But this is simply not the case: Newton makes no such assertions.

Newton had, of course, his private opinions, and the likelihood that light is corpuscular was one of these; but he was largely successful in separating his "scientificall" statements—what he believed he could rigorously prove—from what he deemed possible or even probable. The device he adopted, to separate the two levels or degrees of conviction, was the use of Queries.

The purpose of the Queries of the *Opticks* is clearly set forth in what he says after recounting the experiments on diffraction in Book III and before the first of the Queries:

> When I made the foregoing Observations, I designed to repeat most of them with more care and exactness, and to make some new ones. . . . But I was then interrupted, and cannot now think of taking these things into further consideration. And since I have not finished this part of my Design, I shall conclude, with proposing only some Queries, in order to a further search to be made by others.[49]

The first edition of the *Opticks* in 1704 has only sixteen of these Queries; most of them are quite short, and they set forth notions that occurred to him during the experiments on diffraction. The first, for example, reads: "Do not Bodies act upon Light at a distance, and by their action bend its Rays; and is not this action . . . strongest at the least distance?"[50] From such considerations he was led to questions concerning heat and fire and finally to reveal some of his thoughts about vision. Two years after the publication of the English *Opticks*, Newton brought out the work in a Latin translation, prepared with the help of his disciple, the theologian and philosopher Samuel Clarke. To this Latin *Optice* Newton added seven new Queries, bringing the number up to 23. The first of the new Queries deals with the mysterious phenomenon of double refraction, as described in Iceland Spar (calcite) by Erasmus Bartholinus and more recently by Christiaan Huygens. These matters lead Newton to raise objections in Queries 19 and 20 to the theory, favored by Huygens, that light is a pulse or "pression" in the aether. Here, for example, he writes: "Against filling the Heavens with fluid Mediums, a great Objection arises from the regular and very lasting Motions of the Planets and Comets in all manner of Courses through the Heavens."[51] Such a fluid, he goes on, "can be of no use for explaining the Phaenomena of Nature, the Motions of the Planets and Comets being better explain'd without it. It serves only to disturb and retard the Motions of those great Bodies."[52] And a little later he writes:

> And for rejecting such a Medium, we have the Authority of those oldest and most celebrated Philosophers of *Greece* and *Phoenicia*, who made a *Vacuum*, and Atoms, and the Gravity of Atoms, the first Principles of their Philosophy; tacitly attributing Gravity to some other Cause than Matter.[53]

In the next Query (Q.21) he asks "Are not the Rays of Light very small Bodies emitted from shining substances?" And he gives reasons for believing that this is the case. In this Query and the one that follows we have the nearest thing to an advocacy of a corpuscular theory of light.[54]

The last Query (Q.23) of the *Optice* is the most famous, the most debated, and the longest. It is, indeed, a small essay on a theory of matter, in which much of the supporting evidence is supplied from Newton's extensive knowledge of chemistry. "Have not," he asks, "the small Particles of Bodies certain Powers, Virtues, or Forces, by which they act at a distance, not only upon the Rays of Light . . . but also upon one another for producing a great Part of the Phaenomena of Nature?" And adducing a large array of facts to show that chemical reactions can be understood as the result of preferential attractions between particles, he concludes: "All these things being consider'd, it seems probable to me, that God in the Beginning form'd Matter in solid, massy, hard, impenetrable, moveable Particles, of such Sizes and Figures, and with such other Properties . . . as most conduced to the End for which he form'd them." Nature "will be very conformable to her self and very simple, performing all the great Motions of the heavenly Bodies by the Attraction of Gravity . . . and almost all the small ones of their Particles by some other attractive and repelling Powers which intercede [i.e., act between] the Particles."[55]

It is quite evident that Newton's universe, as he conceived it in 1706, is a Lucretian or Epicurean world, in which atoms of matter move in empty space and interact by means of short-range forces of attraction and repulsion. But in the next few years his opinions underwent a profound change.

In 1717 and 1718 Newton brought out a second English edition of the *Opticks*, and to it he added eight new Queries (numbering them 17–24) and moving the Queries of 1706, now revised and translated into English, to the end, and renumbering them, so that the long chemical Query becomes Query 31.

These new Queries set forth a theory quite at variance with Newton's earlier views of 1706.[56] He describes a universe far from empty; on the contrary it is filled with an aetherial medium "exceedingly more rare and subtile than the Air, and exceedingly more elastick and active." This aether can account for the reflections and refractions of light; by its vibrations it communicates heat to bodies; vision, too, is excited by the vibrations of this medium; and it explains those "fits of easy reflexion and easy transmission," which he had observed in the periodic recurrence of the colored rings.

Newton has returned, with some modifications, to the ideas in a speculative paper which he sent to the Royal Society long before (in 1675) and which he declined to publish.[57] The reasons for this striking reversal of opinion are, I believe, quite evident. Soon after assuming the Presidential chair at the Royal Society he felt the need to reinvigorate the meetings of the Society by appointing a "demonstrator" to perform experiments at the meetings, as Robert Hooke had done in the earlier years. The man chosen was a certain Francis Hauksbee, who carried out this assignment dutifully and brilliantly from 1704 until his death in 1713. The most striking of Hauksbee's experiments—some of them in all likelihood suggested by Newton himself—were electrical demonstrations with a revolving, evacuated globe of glass made to rub against the hand or a piece of cloth. The result was a striking display of electrical discharge under low pressure, the production of a purple electric glow and of remarkable attractive and repulsive effects on nearby light bodies. To Newton, I have argued elsewhere, this demonstrated the existence of a new tenuous kind of matter, "more subtile than the air." He had, in other words, seen the aether, whose existence he had imagined long before, and then rejected.[58] Finally, after Hauksbee's death, the post as the Society's "demonstrator" was taken by Jean-Théophile Desaguliers, who performed a different sort of experiment, one that Newton suggested and described in one of the new Queries. Desaguliers took two tall cylindrical vessels in which he suspended identical thermometers. One of these vessels was evacuated with the air pump, the other remained full of air. When simultaneously exposed to a source of heat, the experiment, as he reported it in 1717, showed that "the Thermometer *in vacuo* will grow warm as much, and almost as soon as the Thermometer which is not *in vacuo*." And Newton asked: "Is not the Heat of the warm Room convey'd through the *Vacuum* by the Vibrations of a much subtiler Medium than Air, which after the Air was drawn out remained in the *Vacuum?*"[59]

You will recall that I mentioned earlier the mysterious indication of a "Part I" of the Third Book of the *Opticks*, but with no subsequent "Part II." A few years ago I turned up in the Cambridge University Library a manuscript draft in Newton's hand of a "Part II." In it are a series of "Observations," among which are references to Hauksbee's experiments and an account of Desaguliers' "two-thermometer experiment." In fact, some of the substance of the so-called aether Queries is given in this draft. It is obvious what happened. Impressed especially by these experiments Newton first believed that he had firm experimental evidence for the existence of a tenuous aether, and proposed to add an account of it to the body of the *Opticks*, as a Part II of Book III in the second English edition. With this in mind, he wrote in pen on the corrected copy of the first edition that he sent to the printer the legend "Part I." But his good sense, or his caution, prevailed; he put the substance of his new ideas in the "aether Queries," but forgot to cancel out the implied promise of a "Part II," where in all subsequent editions it has remained to mystify the reader.[60]

Newton was an indifferent editor. He made a number of rather careless changes in the older Queries to adapt them to his new ideas. Where in 1706 he had argued against filling the Heavens with fluid Mediums, he adds in 1717 the phrase "unless they be exceeding rare." And where he had most strongly advocated the emptiness of space—in referring to the philosophers of Greece and Phoenicia—he now has them attributing "Gravity to some other Cause than *dense* Matter."[61]

It is no wonder that the Queries, in their final form so inherently self-contradictory, have led to quite different interpretations of Newton's theory of matter, both in the eighteenth century and in our day.

By and large the earliest English scientists to be influenced by the *Opticks*, especially by the Queries, ignored Newton's speculations about the aether. This is true of Pemberton and of Desaguliers himself. Nor is the subject of the aether raised in that great textbook of Newtonian optics, Robert Smith's two-volume *Complete System of Optics* (1728). It is Smith, by the way, who is the real advocate of the corpuscular theory of light, and who devotes his efforts to showing how Newton's theories of attraction and repulsion, the interaction between light particles and bodies, can account for all the familiar phenomena of light and color.[62]

The same can be said of the Rev. Stephen Hales, who, in his classic book on plant physiology, the *Vegetable Staticks* (1727), quoted from the Queries many times, and spoke of attraction as "that universal principle which is so operative in all the very different works of nature," yet largely ignored the aether.[63]

As for Newton's French disciples—both popularizers and more proficient men of science—they either passed over Newton's aetherial speculations or treated them with disrespect. And so did David Hume; of all British thinkers none, it can be said, was more influenced by Newton's *Opticks* than Hume. He

knew the book thoroughly before he left college; and Newton, the nonspecula-
tive Newton, was his inspiration and his model. At the close of his great book,
after describing the proper method of scientific inquiry, Newton wrote the lines
that may well have served as Hume's text: "And if natural Philosophy in all its
Parts, by pursuing this Method, shall at length be perfected, the Bounds of Moral
Philosophy will also be enlarged."[64] Hume described his first book, his *Treatise
on Human Nature* (1738), as an attempt to introduce the experimental method
of reasoning into "moral subjects." He had no sympathy for the speculations in
the Queries. Newton, he wrote, the "rarest genius that ever rose for the ornament
and instruction of the species," was particularly to be commended for being
always "cautious in admitting no principles but such as were founded on experiment."[65] Clearly the Queries should be ignored, with their conjectures about
atoms, interparticulate forces, or aetherial mechanisms. And Hume, the rigorous
empiricist, tells us why; in his greatest work, the *Enquiry Concerning the Human
Understanding* (1748), appears this warning:

> It must certainly be allowed, that nature has kept us at a great distance
> from all her secrets, and has afforded us only the knowledge of a few superficial
> qualities of objects; while she conceals from us those powers and principles on
> which the influence of those objects entirely depends.[66]

Hume is not the first, nor the last, to see Newton as what we would call a
positivist; but he was surely one of the few British disciples of Newton to do
so.[67] If the early admirers had ignored the aether Queries, preferring to stress the
Newtonian world of particles in empty space, shortly before the middle of the
eighteenth century we notice a significant shift of focus. In 1743 the Irish physician, Bryan Robinson (1680–1754), published in Dublin his *Dissertation on the
Aether of Sir Isaac Newton.*[68] It opens with a general discussion of this elastic
fluid, and treats in succession—with some display of mathematical apparatus—
the role of the aether in explaining gravity, elasticity, various phenomena of light
and heat, fermentation, and—which was Robinson's chief concern—sensation
and muscular motion.

Robinson's source was, of course, the aether Queries of the *Opticks* of
1717–18. But a year after the appearance of his *Dissertation*, there came to light
an unknown exposition by Newton of his early aether theory: a letter he had
written to Robert Boyle in 1678.[69] Confirmed in his views, Robinson promptly
published a compendium entitled *Sir Isaac Newton's Account of the Aether,
With Some Additions by Way of Appendix*. In it he reprinted the letter to
Boyle, excerpts from the aether Queries, and a further account of his own views
on muscular physiology.

Not long after there appeared an even more important convert to the idea
of an all-pervading aether. If David Hartley, like Hume, was inspired by the
same prognostic passage in the *Opticks*—that the perfection of experimental
philosophy can enlarge the bounds of moral philosophy—he took a strikingly
different tack. For Hartley, whose vibration theory of color I mentioned earlier,

the speculative Queries in the *Opticks* provided him with a physical model, a visualizable mechanism, by which to understand the problems of sensation. He came to his famous principle of the association of ideas from hints he found in the writings of John Locke and the lesser-known John Gay. But the key passage that led to the vibration theory, and gave it physical meaning, came from Newton's Query 23—one of the aether Queries of 1717–18—where Newton wrote: "Is not Vision perform'd chiefly by the Vibrations of this [aetherial] Medium, excited in the bottom of the Eye by the Rays of Light and propagated through the ... optic Nerves into the place of Sensation?"[70]

From this clue Hartley built up his mechanistic psychology in terms of these vibrations: vibrations—as he wrote—excited and propagated by "a very subtle and elastic fluid which Sir Isaac Newton called aether."

It is perhaps significant that this new interest in Newton's aether coincided in time with the discovery of the Leiden jar, the early condenser, and with the dramatic electrical experiments this invention made possible. Newton, we saw, had hinted at the identity of his aether with the mysterious electrical fluid disclosed by Hauksbee's experiments. So it was not long after mid-century that Englishmen, some with impeccable scientific credentials like Benjamin Wilson, F.R.S., and other less qualified persons like Richard Lovett, insisted that the electric matter and the aether were "universally the same thing." Benjamin Martin, a worthy experimenter, held that light, fire, electricity were all different vibratory motions in the aether.[71]

Just as the body of Newton's *Opticks* had inspired the mid-century nature poems studied by Marjorie Nicolson, so the aether Queries left their mark upon such a scientifically curious poet as Erasmus Darwin, and through him upon Shelley in *Prometheus Unbound*, and in the very years when Keats, in *Lamia*, was protesting that Newton had despoiled the rainbow by explaining it. Shelley's notion of a universal fire, of an electrical aether, conceived of as the soul of the world, was not the least of the influences of Newton's remarkable book.[72]

Yet to end on this note would be misleading, for influential philosophers— like David Hume and the Abbé de Condillac in France—and above all the scientists, found the chief value of Newton's *Opticks* in its exemplification of the only proper method of scientific inquiry. The greatest scientists of the later eighteenth century—men like Lavoisier, Laplace, and the Scottish chemists, William Cullen and Joseph Black—were intimately familiar with Newton's discoveries and the compact statement of method with which he ended his last and most famous Query. Rather than repeat, once again, this famous paragraph, I propose to conclude with a more extended statement, a draft of what appeared in Query 31, which has only recently come to light:

> As Mathematicians have two Methods of doing things wch they call Composition & Resolution & in all difficulties have recourse to their method of resolution before they compound so in explaining the Phaenomena of nature the like methods are to be used & he that expects success must resolve before he compounds. For the explications of Phaenomena are Problems much harder

then [sic] those in Mathematicks. The method of Resolution consists in trying experiments & considering all the Phaenomena of nature relating to the subject in hand & drawing conclusions from them & examining the truth of those conclusions by new experiments & drawing new conclusions (if it may be) from those experiments & so proceeding alternately from experiments to conclusions & from conclusions to experiments untill you come to the general properties of things. Then assuming those properties as Principles of Philosophy you may by them explain the causes of such Phaenomena as follow from them: wch is the method of Composition. But if without deriving the properties of things from Phaenomena you feign Hypotheses & think by them to explain all nature you may make a plausible systeme of Philosophy for getting your self a name, but your systeme will be little better than a Romance. To explain all nature is too difficult a task for any one man or even for any one age. Tis much better to do a little with certainty & leave the rest for others that come after you then [sic] to explain all things by conjecture without making sure of any thing.[73]

The main thrust of this paragraph, in the familiar but less spontaneous and more condensed form, appeared in all the later editions of the *Opticks*. Clearly, despite the influences I have described—on poets, philosophers, and artists—this eloquent draft reminds us that we have been dealing with a work of science; and indeed it was, after all, as a model of scientific investigation, that Sir Isaac Newton's *Opticks* exerted its most profound influence.

Notes

1. The standard biographers are David Brewster, *Memoirs of the Life, Writings, and Discoveries of Sir Isaac Newton*, 2 vols. (Edinburgh, 1855), which is still worth consulting, and L. T. More, *Isaac Newton* (New York: Scribner's, 1934). Frank Manuel's *Portrait of Isaac Newton* (Cambridge, Mass.: Belknap Press, 1968) is a psychobiographical study of Newton's complex and often irritating personality.
2. A convenient listing of the editions of these works can be found in George Gray, *Bibliography of the Works of Sir Isaac Newton* (Cambridge, 1907).
3. Geometrical optics, spoken of in the seventeenth century as one of the "mixed mathematics," had a long lineage going back through the Middle Ages to such Greek mathematicians as Euclid, Ptolemy, and Hero of Alexandria. Despite the well-known preoccupation with the rainbow on the part of medieval scholars, color phenomena long resisted precise mathematical treatment and remained largely in the domain of crude experiment and speculation. A good introduction to the early history of the subject is Vascho Ronchi, *Storia della Luce*, 2d ed. (Bologna, 1952); trans. into French by Juliette Taton, *Histoire de la lumière* (Paris, 1956). For the rainbow, and early speculations about color, see Carl B. Boyer, *The Rainbow from Myth to Mathematics* (New York: T. Yoseloff, 1959).
4. In a memorandum of March 1, 1703/4, David Gregory referred to the *Opticks* as already published. See W. G. Hiscock, ed., *David Gregory, Isaac Newton and Their Circle* (Oxford, 1957), 15. The Term Catalogues, ed. Arber, III, 387, list the book under Hilary Term, 1703/4. Books listed for Hilary Term (January 11–31) were licensed for publication in February. I was led to this bit of information by my colleague, Professor Donald Eddy.

5. Experiments 1 and 2 of the First Book record observations not mentioned in Newton's paper of 1672. Probably the first important experiments he made with a prism, they are described in an early notebook (Cambridge University Library, MS. Add. 3975); see A. R. Hall, "Further Optical Experiments of Isaac Newton," *Annals of Science* 11 (1955), 27–43.
6. In facsimile it may be consulted in *Isaac Newton's Papers and Letters on Natural Philosophy*, ed. I. B. Cohen (Cambridge, Mass.: Harvard University Press, 1958), 47–59; also, as printed "with emendations" based on a transcript by Newton's copyist, in *The Correspondence of Isaac Newton*, ed. H. W. Turnbull, 6 vols., continuing (Cambridge, 1959–75), I (1959), 92–107; this will be cited as *Newton Correspondence*.
7. Called the "Discourse on Observations," it was sent to the Royal Society with a letter dated December 7, 1675. The "Discourse" with few changes makes up Parts I and II and half of Part III of the "Second Book" of the *Opticks*. In its original form it was first printed by Thomas Birch in his *History of the Royal Society of London*, 4 vols. (London, 1756–57), III, 272–305; reprinted in facsimile from Birch in *Papers and Letters*, 202–35.
8. In his posthumously published *Physico-Mathesis de Lumine, Coloribus et Iride* (Bologna, 1665).
9. *Opticks* (1704), Third Book, 132. Unlike the later editions, this quarto first edition has separate pagination for the First Book, pp. 1–144, and for the rest of the work: the Second and Third Books and the two Latin mathematical papers appended to the *Opticks*, pp. 1–211. In the Latin *Optice* of 1706, only the mathematical papers have separate pagination. All subsequent references to the *Opticks* will be to the 1704 edition unless noted otherwise.
10. The interruption Newton speaks of was probably occasioned by the writing of the *Principia*, by which time he was familiar with Grimaldi's work, and had himself performed some diffraction experiments. See *Principia* (London, 1687), 231. He probably had not heard of Grimaldi before 1672.
11. Memoranda by David Gregory, *Newton Correspondence*, III, 336.
12. *Journal Book of the Royal Society*, July 4, 1694; *Newton Correspondence*, III, 340, n. 16.
13. *Newton Correspondence*, IV, 100.
14. *Ibid.*, IV, 116–17. Wallis's italics.
15. Letter of November 18, 1676, in *Newton Correspondence*, II, 182–83. Francis Hall, who called himself Line or Linus, was an elderly professor at the College of Jesuits at Liège, and one of those who attacked the findings of Newton's first paper. Linus died in 1675, but the debate continued with Linus's associate, Anthony Lucas, until Newton brought it to an end in the summer of 1678.
16. Hiscock, *David Gregory*, 14. In a later entry, p. 15, Gregory wrote: "Mr. Newton was provoked by Dr. Cheyne's book to publish his Quadratures, and with it, his Light and Colours, etc." George Cheyne, a London physician, published in 1703 his *Fluxionum methodus inversa*, making use of Newton's mathematical discoveries. See Florian Cajori, *History of the Conceptions of Limits and Fluxions in Great Britain* (Chicago and London, 1919), 40.
17. *Opticks* (1704) Advertisement.
18. In the Latin version of his *Algebra* (1693), Wallis made the first public announcement of Newton's fluxional calculus. He referred to it as "some specimen of what we hope Mr. Newton will himself publish in due time." See *Newton Correspondence*, III, 221, n. 1.
19. John F. Fulton, *A Bibliography of the Honourable Robert Boyle*, 2d ed. (Oxford: Clarendon Press, 1961), *passim*.
20. Walter Charleton, *Physiologia Epicuro-Gassendo Charltoniana: Or a Fabrick of Science Natural, upon the Hypothesis of Atoms* (London, 1654), 195–96. A facsimile of this work, with an introduction by Robert Kargon, has been published by the Johnson Reprint Corporation in 1966, The Sources of Science, no. 31. For an account of Charleton, London physician and original Fellow of the Royal Society, see Kargon's introduction to the *Physiologia* and his *Atomism in England from Hariot to Newton* (Oxford: Clarendon Press, 1966), chap. 8.

21. *Opticks* (1704) First Book, Part I, 22–23.
22. Thomas Sprat, *History of the Royal Society* (London, 1667), 113. The Royal Society doctrine was only one contribution to simplification of style in the last third of the seventeenth century. The reform in pulpit oratory, distaste for religious "enthusiasm," and the delayed influence of the King James version of the Bible are all factors emphasized by Louis G. Locke, *Tillotson: A Study in Seventeenth-Century Literature* (Copenhagen: Rosenkilde and Bagger, 1954), chap. 4.
23. W. K. Wimsatt, Jr., *Philosophic Words: A Study of Style and Meaning in the Rambler and Dictionary of Samuel Johnson* (New Haven: Yale University Press, 1948). For a list of words taken from the *Opticks*, see p. 156.
24. Henry Guerlac, "The Word *Spectrum*: A Lexicographic Note with a Query," *Isis* 56 (1965), 206–7. Newton, I now think, borrowed an uncommon classical word used once by Cicero in his *Epistulae ad familiares*; Cicero attributes to one Catius, an Epicurean, the use of the word *spectra* for what Democritus and Epicurus called *eidola*. See *Cicero—The Letters to His Friends*, Loeb Classical Library (London, 1929), III, 296–97. In one of his early notebooks Newton referred to the "Phantome" of colors produced by the prism, a fact pointed out to me by Professor Richard S. Westfall of the University of Indiana. See Hall, "Further Optical Experiments of Newton," 28.
25. The abbreviated form already appears in David Hartley's *Observations on Man* (London, 1749), and in the "Explanation of Technical Terms" in Joseph Priestley's *History and Present State of Discoveries relating to Vision, Light and Colour* (London, 1772).
26. Marjorie Hope Nicolson, *Newton Demands the Muse: Newton's "Opticks" and the Eighteenth Century Poets* (Princeton: Princeton University Press, 1946).
27. *Ibid.*, 12.
28. For Buffon's "Dissertation sur les couleurs accidentelles," see *Histoire et Mémoires de l'Académie des Sciences* for 1743 (Paris, 1746), 147–58. The question of colored shadows and clouds, mentioned by Buffon, is treated by Pierre Bouguer in his posthumous *Traité d'optique sur la gradation de la lumière* (Paris, 1760). For the aurora, see Jean-Baptiste Dortous de Mairan, *Traité physique et historique de l'aurore boréale*, 2d ed. (Paris, 1754), sec. 3, chap. 9, pp. 154–56. Priestley has a general discussion of these problems in his *History of Light and Colour*, 436–49. With due regard for a special point of view, the reader should consult Goethe's *Materialien zur Geschichte der Farbenlehre, Werke*, 14 vols. (Hamburg, 1952–1960), XIV, 7–269.
29. Father Castel attributed his first ideas to an odd passage in the *Musurgia universalis* (Rome, 1650) of Athanasius Kircher, but he refers his reader to the *Opticks* of Newton "pour y voir toutes les couleurs bien diapasonnées avec leurs octaves, quintes, tierces, et septièmes." Later, when he wrote his *Vrai système de physique générale de M. Isaac Newton* (Paris, 1743), he became convinced that Newton's theory of light and color, like his celestial mechanics, was untenable. See Donald S. Schier, *Louis Bertrand Castel, Anti-Newtonian Scientist* (Cedar Rapids, Iowa: Torch Press, 1941), esp. pp. 135–38.
30. *Opticks*, First Book, Part II, 91–93. Newton's earliest reference to the analogy of sound and color occurs in his reply to Robert Hooke in 1672; *Newton Correspondence*, I, 174–75. A good discussion of this aspect of Newton's thought is given by Sigalia Dostrovsky in her "Origins of Vibration Theory: The Scientific Revolution and the Nature of Music," unpublished doctoral dissertation (Princeton, 1969).
31. *Opticks* (1704) Second Book, 17–18 (Obs. 14). This musical analogy does not appear in the original, much shorter, version of Obs. 14 in the "Discourse of Observations" of 1675. For Newton's solmization see Christopher Simpson, *Compendium of Practical Musick*, 3d ed. (London, 1678), 3–4. Here, after describing the old system of Guido d'Arezzo, he writes that four of the old syllables (mi, fa, sol, la) "are necessary assistants to the right Tuning of the Degrees of Sound" but that the other two (ut and re) are superfluous: "We will therefore make use only of Mi, Fa, Sol, La, and apply them to the Seven Letters, which stand for the Degree of Sound." See also John Playford, *An Introduction to the Skill of Musick*, 7th ed. (London, 1674), 1; Playford remarks that the six syllables were used "for many years

past" and in recent times four only are used "being sufficient for expressing the several sounds, and less burdensome for the memory of Practitioners."

32. *Opticks* (1704) Second Book, pt. 3, pp. 78–84. Hartley refers especially to Newton's prop. 16, *ibid.*, 83, where Newton writes that the intervals of the fits of easy reflexion and easy transmission "are either accurately, or very nearly, as the Cube-roots of the Squares of the lengths of a Chord, which sound the notes in an Eight . . . according to the Analogy described in the seventh Experiment of the second Book." See David Hartley, *Observations on Man*, 2 vols. (London, 1749), I, prop. 56, pp. 192–96; I have used the modern facsimile reproduction, with an introduction by Theodore L. Huguelet (Gainesville, Florida: Scholars' Facsimiles & Reprints, 1966).

33. *Observations on Man*, I, 194.

34. *Observations on Man*, I, prop. 60, p. 208.

35. *Ibid.* By "primary colours" Hartley means, like Newton, the unmixed colors produced by the prism, not the pigmentary red, blue, and yellow of the painter. See *Opticks* (1704), First Book, pt. 3, Prop. V., Theor. IV, and Prop. VI, Prob. II.

36. *History of Vision, Light and Colours*, 763–67.

37. Edmund Burke, *A Philosophical Enquiry into the Origin of our Ideas of the Sublime and the Beautiful*, ed. J. T. Boulton (London: Routledge and Kegan Paul, 1958), 73, 138, 159, for Burke's debt to the *Opticks* and to Hartley. The editor of this excellent edition remarks that Burke "clearly owes much" to his study of the *Opticks*; a copy of the 4th English edition (1730) appeared in the sale of Burke's library.

38. Robert E. Schofield, *The Lunar Society of Birmingham* (Oxford, 1963), 189, 272–73; Erasmus Darwin, *Botanic Garden*, 4th ed. (London, 1799), 262–65. Erasmus Darwin's son, Robert Waring Darwin, published a study of psychophysiological color effects (Buffon's "accidental colors") in his "New Experiments on the Ocular Spectra of Light and Colours," *Phil. Trans.* 76 (1786), 313–48. He cites among others, Newton, "the celebrated M. de Buffon," and Joseph Priestley. For Erasmus Darwin's possible role in writing this paper, see Schofield, *Lunar Society*, 272–73.

39. *Opticks* (1704) First Book, pt. 2, p. 115 and fig. 11. Newton does not speak of actually rotating the disc. It may only have been a device to illustrate his experiments of mixing prismatic colors.

40. Schofield, *Lunar Society*, 270–72; and Darwin, *Botanic Garden*, 258–62. Galton may have been led to these experiments, published long after in the *Monthly Magazine* 8 (1799), by reading Priestley's *History*, for Priestley describes Newton's color disc, and reproduces Newton's fig. 11 in his fig. 83 of pl. 12.

41. The *Farbenlehre*, published in two volumes, plus a volume of plates, was the result of Goethe's intensive study of color from about 1790 to 1810. The work is in three parts: the *Didaktischer Teil* (containing Goethe's own observations and theories about color), a *Polemischer Teil*, devoted to attacking Newton, and finally (in the second volume) his long and remarkably detailed *Geschichte der Farbenlehre*, still worth reading. There is a large literature on Goethe as a scientist; the most effective defense is by Rudolf Magnus, *Goethe als Naturforscher* (Leipzig, 1906), English trans., *Goethe as a Scientist*, by Heinz Norden (New York: H. Schumann, 1949). In the same bicentennial year appeared Sir Charles Sherrington's short but highly critical lecture, *Goethe on Nature & on Science* (Cambridge: The University Press, 1949).

42. E. H. Gombrich, *Art and Illusion*, 2d ed. (New York: Pantheon, 1961), 46–48.

43. Cited by Robert C. Leslie, *Life and Letters of John Constable* (London, 1896), 399.

44. Quoted in Lawrence Gowing, *Turner: Imagination and Reality*, The Museum of Modern Art (Garden City, N.Y.: Doubleday, 1966), 13.

45. Jerrold Ziff, "J. M. W. Turner on Poetry and Painting," *Studies in Romanticism* 3 (1964), 193–215.

46. Eastlake's translation was confined to the "Didactic Part" of the *Farbenlehre*, including Goethe's *Einleitung*, with its several deprecatory references to Newton. For Turner's experiments and his reaction to Goethe see Gowing, *Turner*, 21–24. Of considerable importance for Turner's development was his reading of the *Natural System of Color*, written by the entomologist-cum-painter, Moses Harris, published

in 1766 and republished in 1811. For Harris, see Gowing, *Turner*, 23; and Jack Lindsay, *J. M. W. Turner* (London: Cory, Adams, & Mackay, 1966), 208.

47. I. B. Cohen, *Franklin and Newton* (Philadelphia: American Philosophical Society, 1956), 125. Professor Cohen stated this view even more strongly in his preface to the paperback reprint of the fourth edition of the *Opticks*, where he speaks (p. xxvii) of the "progressively conjectural character" of the book and writes that in this work "Newton did not adopt the motto to be found in the *Principia*— Hypotheses non fingo; I frame no hypotheses—but, so to speak, let himself go, allowing his imagination full reign [*sic*], and by far exceeding the bounds of experimental evidence." *Opticks* (Dover Publications, 1952), pp. xxvii, xxiii–xxiv. A similar but more temperate view has been put forward by Alexandre Koyré, "L'Hypothèse et l'expérience chez Newton," *Bulletin de la Société Française de Philosophie* 50 (1956), 59–79; reprinted in trans. in *Newtonian Studies* (London: Chapman & Hall, 1965), 25–52, where it is called "Concept and Experience in Newton's Scientific Thought."

48. The exception occurs at the point where Newton, having described in noncommittal fashion "fits of easy reflexion and easy transmission," offers an explanation: "Those that are averse from assenting to any new Discoveries, but such as they can explain by an Hypothesis, may for the present suppose, that . . . the Rays of Light, by impinging on any refracting or reflecting Surface excite vibrations in the refracting or reflecting Medium or Substance . . . much after the manner that vibrations are propagated in the Air for causing Sound, and move faster than the Rays so as to overtake them." *Opticks*, 4th. ed. (1730), Book Two, Part 3, Prop. 12, p. 280. In his *Newtonian Studies*, 50, n. 1, Koyré says "it is pretty clear that this medium cannot be anything else" but Newton's hypothetical aether. This is by no means clear: Newton is leaving the question open: the medium might be the aether or the material substance of a refracting body. Newton in one of the early Queries speaks of the rays of light as exciting vibrations in the substance of the retina, which vibrations "being propagated along the solid fibres of the optick Nerves into the Brain, cause the sense of seeing." *Opticks* (1704), Third Book, Query 12, p. 135.

49. *Opticks* (1704) Third Book, 132. My view of the matter is that of Thomas Reid, who wrote in 1785:

> Sir Isaac Newton . . . took great care to distinguish his doctrines, which he pretended to prove by just induction, from his conjectures, which were to stand or fall according as future experiments and observations should establish or refute them. His conjectures he has put in the form of queries, that they might not be received as truths, but be inquired into, and determined according to the evidence to be found for or against them. Those who mistake his queries for a part of his doctrine, do him great injustice, and degrade him to the rank of the common herd of philosophers, who have in all ages adulterated philosophy, by mixing conjecture with truth, and their own fancies with the oracles of Nature.

The Works of Thomas Reid, ed. Sir William Hamilton, 8th ed., 2 vols. (Edinburgh, 1880), I, 249. A similar view was earlier expressed by Colin Maclaurin, *An Account of Sir Isaac Newton's Philosophical Discoveries*, 3d ed., (London, 1775), 9–10.

50. Professor Cohen, "Newton's Philosophy of Nature," *Dictionary of the History of Ideas*, argues that the Queries "are all phrased in the negative and are thus purely rhetorical questions rather than genuine interrogations." I cannot see that this tells us more than that Newton thought the propositions likely (in every case, as Cohen remarks, Newton gives evidence in support of his proposition), yet they are matters he cannot demonstrate, and hopes will be investigated by others. [This draft, which I was privileged to read, did not in fact appear in the *Dictionary*.]

51. I have used the translation in a later English edition, *Opticks*, 4th ed. (1730), Third Book, 364–65, modifying it only when Newton himself made small but significant changes. The Latin reads: "Praeterea, nulla esse omnino istiusmodi Media fluida, inde colligo, quod Planetae & Cometae regulari adeo & diuturno Motu per spatia caelestia undiq; & quaquaversum & in omnes partes ferantur." *Optice* (1706), Liber Tertius, 310.

52. *Opticks* (1730), Third Book, 368. See *Optice* (1706), 313, where he speaks of the "materia illa ficta et commentitia" with which people fill the heavens.

53. *Opticks* (1730), Third Book, 369. The italics are Newton's. His Latin reads: "Istiusmodi autem Medium ut rejiciamus, Auctores nobis sunt antiquissimi illi & celeberrimi Graeciae Phaeniciaeq; Philosophi; qui Principia Philosophiae suae, Spatium inane, Atomos, & Gravitatem Atomorum posuerunt; Tacite attribuentes Vim Gravitatis, alii alicui *Causae*, a Materia diversae." *Optice* (1706), 314.

54. *Opticks* (1730), 370–75, and *Optice* (1706), 315–19.

55. *Opticks* (1730), 375–76, 400, 397. *Optice* (1706), 322, 343, 340.

56. That major changes occurred in the Queries between the Latin of 1706 and the later English editions was first pointed out by Samuel Horsley when he edited the *Opticks* for his edition of Newton's collected works, *Isaac Newtoni Opera quae exstant Omnia*, 5 vols. (London, 1779–1785). A list of these alterations was given in F. Rosenberger, *Isaac Newton und seine physikalischen Principien* (Leipzig, 1895). A more detailed analysis is that of Alexandre Koyré, "Les Queries de l'Optique," *Archives Internationales d'Histoire des Sciences* 13 (1960), 15–29. This was actually published after April 1961, and I was unaware of it when I wrote my *Newton et Épicure* (Paris: Palais de la Découverte, 1963), 27–35, where, independently of Koyré, I called attention to the changes but stressed their significance.

57. "An Hypothesis explaining the Properties of Light," enclosed in the letter of December 7, 1675, and read to the Royal Society at meetings of December 9 and 16; first published in Birch, *History of the Royal Society*, III (1757), 248–60, 262–69; facsimile reproduction from Birch in Cohen, *Newton's Papers and Letters*, 178–99; reprinted from the copy in the Royal Society Register Book, supplemented and corrected with the original MS. in *Newton Correspondence*, I, 362–86.

58. Henry Guerlac, "Francis Hauksbee: expérimentateur au profit de Newton," *Archives Internationales d'Histoire des Sciences* 16 (1963), 113–28; also my "Sir Isaac and the Ingenious Mr. Hauksbee," *Mélanges Alexandre Koyré*, 2 vols. (Paris: Hermann, 1964), I (*L'Aventure de la science*), 228–53.

59. *Opticks* (1730), Third Book, 349. The italics are Newton's.

60. Henry Guerlac, "Newton's Optical Aether," *Notes and Records of the Royal Society of London* 22 (1967), 45–57.

61. *Opticks* (1730), Third Book, 364–65, 369. The emphasis is mine.

62. A study of Smith's *System of Optics* has been made by Henry Steffens in "The Development of Newtonian Optics in England, 1738–1831," an unpublished M.A. dissertation (Cornell University, 1965).

63. For Hales's debt to the Queries of the *Opticks* see my "Continental Reputation of Stephen Hales," *Archives Internationales d'Histoire des Sciences*, no. 15 (1951), 393–404; also Cohen, *Franklin and Newton*, 247, 254–55, 266–76. Bishop Berkeley in his *Siris* (1744), that immensely learned but eccentric pamphlet on the virtues of tar-water, several times quoted from the *Opticks* and its Queries. Although he set forth the idea of a universal aether, he identified it with fire or light, conceiving it as a spirit which is neither matter nor mind. In several paragraphs he criticizes Newton's material aether by which "upon later thoughts" Newton explained "all the phenomena and properties of bodies that were before attributed to attraction . . . together with the various attractions themselves." There is no reason "to admit a new medium distinct from light"; to account for the periodic properties of light by vibrations of this medium "seems an uncouth explanation." *The Works of George Berkeley*, ed. A. A. Luce and T. E. Jessop, 9 vols. (London: T. Nelson, 1948–57), V (1953), 107–9.

64. *Opticks* (1730), Third Book, 405. This passage, I think rightly, has recently been interpreted in a manner that would not have appealed to Hume: that a "true natural philosophy must lead to a surer knowledge of God, and thence to a firmly-grounded moral philosophy." See J. E. McGuire and P. M. Rattansi, "Newton and the 'Pipes of Pan,'" *Notes and Records of the Royal Society of London* 21 (1966), 122–23.

65. Cited from Hume's *History of England* by Ernest Campbell Mossner, *The Life of David Hume* (Austin, Texas: University of Texas Press, 1954), 75.

66. *An Enquiry Concerning the Human Understanding*, ed. L. A. Selby-Bigge (Oxford, 1894), 32–33.

67. Burke should be numbered among them, for he wrote:
   When Newton first discovered the property of attraction, and settled its laws, he found it served very well to explain several of the most remarkable phaenomena in nature; but yet with reference to the general system of things, he could consider attraction but an effect, whose cause at that time he did not attempt to trace. But when he afterwards began to account for it by a subtle elastic aether, this great man . . . seemed to have quitted his usual cautious manner of philosophizing. . . .
   Sublime and the Beautiful, ed. Boulton, 129.
68. On Robinson, see Cohen, *Franklin and Newton,* 417–19; also Philip C. Ritterbush, *Overtures to Biology* (New Haven: Yale University Press, 1964), 8. I am also indebted to my student David Corson for a careful study of Robinson.
69. *The Works of the Honourable Robert Boyle,* ed. Thomas Birch, 5 vols. (London, 1744), I, 70–74. This included Birch's *Life of the Honourable Robert Boyle,* published separately in the same year; here the letter of Newton is given on pp. 234–37.
70. *Opticks* (1730), 353.
71. Ritterbush, *Overtures to Biology,* 16–22.
72. Carl Grabo, *A Newton Among Poets* (Chapel Hill: University of North Carolina Press, 1930), esp. chaps. 3 and 8.
73. Cambridge University Library, MS. Add. 3970 (5).

12
Stephen Hales:
A Newtonian Physiologist

✵

Stephen Hales, a clergyman without formal medical training, published his first discoveries in his fiftieth year, yet was soon recognized as the leading English scientist during the second third of the eighteenth century. As the acknowledged founder of plant physiology, he had no worthy successor until Julius von Sachs, a century later. In animal physiology he took "the most important step after Harvey and Malpighi in elucidating the physiology of the circulation."[1] His experiments concerning "fixed air"—and the apparatus he devised—laid the foundations of British pneumatic chemistry and stimulated the discoveries of Joseph Black, Henry Cavendish, and Joseph Priestley. Hales was a primary influence upon the early researches of Lavoisier.

He was born of an old and distinguished Kentish family, but there is no record of his boyhood until, having been "properly instructed in grammar learning," he was sent to Cambridge, where he entered Benet College (now Corpus Christi) in 1692.[2] On receiving the B.A. degree, Hales became a fellow of his college in 1703 and was awarded the M.A. that same year. He was ordained deacon in 1709 and left Cambridge to become "perpetual curate," or minister, of Teddington, a village on the Thames between Twickenham and Hampton Court. He held this position for the rest of his life, and it was at Teddington that most of his scientific work was carried on.

An interest in science was awakened during his years at Cambridge, the university that boasted the great Isaac Newton (who had left for London the year in which Hales entered the university) and the naturalist John Ray, whose earliest book was a catalogue of the plants of Cambridgeshire. Something of a scientific renaissance took place during Hales's last years at the university. William Whiston, Newton's successor as Lucasian professor, was encouraged by Newton's old friend Richard Bentley, who became master of Trinity, the college of Newton and Ray, in 1700. Bentley helped secure the appointment of a gifted young fellow of the college, Roger Cotes, to the newly established Plumian professorship of astronomy and built for him an observatory over the Great Gate of Trinity. When John Francis Vigani became the first professor of

From *Dictionary of Scientific Biography*, vol. 6 (1972), 35–48. Copyright © 1972 American Council of Learned Societies.

chemistry at Cambridge, Bentley provided him with a laboratory "in the mediaeval chambers that look out on the Bowling Green."[3]

In 1703 William Stukeley, the future physician and antiquary, entered Benet College; intent on a medical career, he "began to make a diligent & near inquisition into Anatomy and Botany."[4] He became a close friend of Hales; with Stukeley and other students Hales went "simpling" in the surrounding countryside, Ray's catalogue in hand. In a room that Stukeley's tutor had given him as a sort of laboratory, they performed chemical experiments and dissected frogs and other small animals. Together they devised a method of obtaining a lead cast of the lungs of a dog. It was at this time (about 1706) that Hales carried out his first blood-pressure experiments on dogs. He and Stukeley attended Vigani's chemical lectures and saw his demonstrations in the laboratory at Trinity. Hales, like Stukeley, must have seen the "many Philosophical Experiments in Pneumatic Hydrostatic Engines & Instruments performed at that time" by John Waller, rector of St. Benedict's Church, who later succeeded Vigani as professor of chemistry. Hales knew Waller, for about 1705 the two men "gathered subscriptions to make the cold bath about a mile & a half out of Town."[5] This introduction to pneumatic experiments was probably supplemented by the lectures in experimental physics given by Whiston and Cotes at the observatory in Trinity College. Cotes, in his share of the lectures, demonstrated the experiments of Torricelli, Pascal, Boyle, and Hooke.[6]

Newton's influence was strongly felt at Cambridge. In 1704 appeared his long-delayed *Opticks*, a work that, in its later editions, profoundly influenced Hales. We learn from Stukeley that students at Benet College read the Cartesian *Physics* of Jacques Rohault, but in the edition of Samuel Clarke,[7] who appended Newtonian footnotes to correct the text.

Hales, like Stukeley, doubtless witnessed Newton's arrival in Cambridge in April 1705, when he came to offer himself as the university's candidate for Parliament. On the sixteenth of that month Queen Anne visited Cambridge as the guest of the master of Trinity. As Stukeley recalled it: "The whole University lined both sides of the way from Emanuel college, where the Queen enter'd the Town, to the public Schools. Her Majesty dined at Trinity college where she knighted Sir Isaac, and afterward, went to Evening Service at King's college chapel."[8]

Although doubtless incapable of following the mathematical intricacies of Newton's *Principia*, Hales mastered the main features of the new system of the world. He showed both his mechanical ingenuity and some knowledge of celestial physics by devising a machine to show the motions of the planets. A drawing by Stukeley of Hales's orrery is preserved, along with Stukeley's diary, in the Bodleian Library, Oxford.[9]

At Teddington, Hales was first preoccupied with his parish duties; only several years later did he resume his scientific work; and it was later still before the scientific world heard from him. About 1712 or 1713 he took up again his experiments on animals, this time using as his victims two horses and a fallow doe. But he did not "pursue the Matter any further, being discouraged by the disagreeableness of anatomical Dissections." For several years his scientific endeavors lapsed, yet on March 13, 1717/18 (O.S.), he was elected a fellow of the Royal Society along with his old friend William Stukeley, who was now practicing medicine in London.[10] It was Stukeley, indeed, who brought Hales's name to the attention of the Royal Society.[11] Hales was soon to justify his election.

While conducting his experiments on animal blood pressure, Hales records, "I wished I could have made the like Experiments, to discover the force of the Sap in Vegetables" but "despaired of ever effecting it." Yet early in 1719 "by mere accident I hit upon it, while I was endeavouring by several ways to stop the bleeding of an old stem of a Vine, which was cut too near the bleeding season, which I feared might kill it." His account continues:

> Having, after other means proved ineffectual, tyed a piece of bladder over the transverse cut of the Stem, I found the force of the Sap did greatly extend the bladder; whence I concluded, that if a long glass Tube were fixed there in the same manner, as I had before done to the Arteries of several living Animals, I should thereby obtain the real ascending force of the Sap in that Stem.[12]

Hales was in no hurry to make an appearance at the Royal Society or to contribute to its proceedings; several months elapsed before he appeared in person to sign the required bond.[13] On March 5, 1718/19, perhaps at Stukeley's urging, Hales informed the president, Sir Isaac Newton, "that he had lately made a new Experiment upon the Effect wch ye Suns warmth has in raising ye sap in trees."[14]

Seven years of silence followed, during which, in the free time his parish duties allowed him, Hales followed out this original clue. The experiments on plants were virtually completed, and the *Vegetable Staticks* written out, by the middle of January 1724/25. He submitted his manuscript to the Royal Society, where it was read at successive meetings from January to March.[15] In this form it consisted of six chapters, not the seven of the published book; missing was the long chemical chapter entitled "The Analysis of Air."

Although Hales was urged to publish, two years elapsed before the Royal Society heard from him again. During this time he performed the seventy chemical experiments of Chapter 6 of the *Vegetable Staticks*. In three meetings during February 1726/27 this chemical chapter was read, and at the last of these meetings the book received the imprimatur of the Royal Society, signed by Sir Isaac Newton "Pr. Reg. Soc."[16] The book was already in press, for some of the early sections were read again to the Society in March, probably from advance sheets. On April 13, 1727 (O.S.), a copy of the *Vegetable Staticks*, dedicated to George, prince of Wales (the future George II), was presented to the Society; and the curator of experiments, J.-T. Desaguliers, was asked to prepare an abridgment of it.[17]

Hales next turned to completing and publishing his experiments on animal circulation (mentioned only briefly in the *Vegetable Staticks*) under the title *Haemastaticks*, putting the two works together as the *Statical Essays*. Imbued with the empiricism of John Locke and the principles of Newton's "experimental philosophy," Hales was also influenced by the doctrines of the iatrophysicists—those physicians including Borelli, Baglivi, and the Scottish doctors Archibald Pitcairne and James Keill who insisted, as a certain John Quincy put it, that the application of mechanical principles to "account for all that concerns the Animal Oeconomy" is the best means to "get clear of all suppositions and delusory Hypotheses" and "has appeared to be the only way by which we are fitted to arrive at any satisfactory Knowledge in the Works of Nature."[18]

An immediate stimulus came from James Keill, who, while Hales was still at Cambridge, published a book giving quantitative estimates of the amount of blood in the human body, the velocity of the blood as it left the heart, the amounts of various animal secretions, and so on.[19] Hales may have had Keill's book in mind when he wrote that

> . . . if we reflect upon the discoveries that have been made in the animal œconomy, we shall find that the most considerable and rational accounts of it have been chiefly owing to the statical examination of their fluids, *viz.* by enquiring what quantity of fluids, and solids dissolved into fluids, the animal daily takes in. . . . And with what force and different rapidities those fluids are carried about in their proper channels. . . .[20]

What Hales called the "statical way of inquiry" he deemed the proper way to study living things. For his now obsolete use of the word "staticks" he had ample authority. The usage originated with Nicholas of Cusa; in his *De staticis experimentis*—a work several times reprinted in the fifteenth and sixteenth centuries and translated into English in 1650—Cusa outlined a series of "thought experiments" involving the use of the balance.[21] Statics, for Cusa, meant weighing. Santorio Santorio made the term familiar in medicine with his *De medicina statica aphorismi* (Venice, 1614), a work on "insensible perspiration" often reprinted, and published in English translation in 1712 by John Quincy.[22] In 1718 James Keill published his *Tentamina medico-physica* (a Latin version of his earlier work) and appended to it some studies on perspiration called *Medicina statica britannica*.[23] Hales was familiar with this book, and referred to it several times.

The *Haemastaticks* describes the experiments on blood pressure begun at Cambridge, taken up again at Teddington, laid aside because of the "disagreeableness of the Work," but resumed again after the publication of the *Vegetable Staticks*. At first intended only as an addition to the earlier book, it grew "into the Size of another Volume, so fruitful are the Works of the great Author of Nature in rewarding, by farther Discoveries, the Researches of those *who have Pleasure therein*."[24]

The first Cambridge experiments, Hales tells us, were stimulated by the confusion that existed as to the magnitude of the arterial blood pressure; some maintained that the pressure was enormous, and even that it might be the cause of muscular motion. The results of his series of investigations were of such importance that they have been described as "the most important step in knowledge of the circulation between Malpighi and Poiseuille."[25]

The *Haemastaticks* opens with an account of Hales's most dramatic experiment, a bold and bloody one. He tied a live mare on her back and, ligating one of her femoral arteries, inserted a brass cannula; to this he fixed a glass tube nine feet high; when he untied the ligature, the blood rose to the height of more than eight feet. Detaching the tube at intervals, he allowed a measured quantity of blood to flow out, noting how the pressure changed during exsanguination. He succeeded in recording, by the same method, the venous pressure of a number of animals, including an ox, a sheep, a fallow doe, three horses, and several dogs.

His interest in the mechanics of the circulation now enhanced, Hales turned his attention to the chief factors that must maintain the blood pressure: the output of the heart per minute and the peripheral resistance in the small vessels. He made a rough estimate of cardiac output by multiplying the pulse rate of an animal by the internal volume of its left ventricle, of which he made a cast in wax after the animal had been killed. He noted that the pulse was faster in small animals than in large ones, and that the blood pressure was proportional to the size of the animal.

Hales next studied peripheral resistance with perfusion experiments. Injecting various chemical substances (brandy, decoction of Peruvian bark, various saline solutions), he compared the rate of flow of the perfusate and showed that certain substances had a pronounced effect on the rate at which the blood could flow through an isolated organ. He attributed this to changes in the diameter of the capillaries and so—although he did not observe the phenomenon directly—discovered vasodilatation and vasoconstriction.

Hales's experiments convinced him that the force of arterial blood in the capillaries "can be but very little" and wholly inadequate for "producing so great an Effect, as that of muscular Motion." This "hitherto inexplicable Mystery of Nature must therefore be owing to some more vigorous and active Energy, whose force is regulated by the Nerves." The recent experiments of Stephen Gray suggested to Hales that this energy, these "animal spirits," might be electrical, for Gray had shown that the electrical virtue from rubbed glass

> . . . will not only be conveyed along the Surface of Lines to very great Lengths, but will also be freely conveyed from the Foot to the extended Hand of a human Body suspended by Ropes in . . . the Air; and also from that Hand to a long Fishing Rod held in it, and thence to a String and a Ball suspended by it.[26]

Hales was therefore the first physiologist to suggest, with some evidence to support it, the role of electricity in neuromuscular phenomena.[27]

Despite these achievements, Hales's most original contribution was to apply to the study of plants the "statical method" which had brought such good results with animals. Like his contemporaries he was impressed by the analogies that he perceived between the animal and the vegetable worlds. Perhaps the most obvious was the fundamental similarity of the role of the sap in plants and of the blood in animals. Since the growth of plants "and the preservation of their vegetable life is promoted and maintained, as in animals, by the plentiful and regular motion of their fluids, which are the vehicles ordained by nature, to carry proper nutriment to every part," the same methods ought to be used which had illuminated the animal economy: "the statical examination of their fluids."[28] By an accident, as we saw, he was led to his first attempts to measure the force of the sap in vines and to determine the conditions under which it varied. Although he outstripped his predecessors, Hales was not the first to investigate the flow of sap.

The problem of sap flow had long interested the *virtuosi* of the Royal Society. In 1668 the *Philosophical Transactions* proposed to its readers a long series of "queries" concerning plants, "especially the Motion of the Juyces of Vegetables," asking, for example, whether the "Juyce ascends or descends" by the bark or the pith.[29] Among the responses was a letter from Francis Willughby describing some bleeding experiments on trees performed with John Ray which showed that the sap not only ascended but also seemed to descend and move laterally, and that the rise could not be attributed to capillarity, a common explanation.[30] When the letter was read to the Society, the two naturalists were requested to "try some experiments, to find, whether there be any circulation of the juice of vegetables as there is of the blood of animals."[31] That there might be such a circulation of sap, moving upward through the vessels of woody plants and downward by those between the wood and the bark, was a commonly held view in Hales's day; at late as 1720 Patrick Blair in his *Botanick Essays* tried to prove that this was the case.[32]

Nor was Hales the earliest to apply the "statical way of inquiry" to plants. J. B. van Helmont's famous willow tree experiment, which persuaded him that water was the sole principle or nutrient of plants, is a well-known example.[33] Closer to Hales's time were the quantitative experiments of John Woodward to determine whether water itself, or substances dissolved in it, accounted for the growth of plants. In these experiments Woodward discovered that the phenomenon of transpiration was of considerable magnitude. Growing mint in water, Woodward observed that the plant took up large quantities of water but gave off far more than it retained. He noted that solar heat played a part in the process, but he did not specify or prove that transpiration occurred through the leaves. Much the greatest part of the water imbibed by his plants, he wrote, "does not settle or abide there: but passes through the Pores of them, and exhales up into the Atmosphere."[34]

Hales's experiments on plants, begun in March 1719, were pursued with vigor during the years 1723–25, using the resources of his own garden and plants and trees provided him from the nearby royal garden of Hampton Court.[35] In his early experiments, conducted during the bleeding season, he observed the rise of sap through long glass tubes fastened to the cut end of a branch of a grapevine. In one such experiment Hales joined glass tubes together to a height of thirty-eight feet. The sap was observed to rise in these tubes "according to the different vigor of the bleeding state of the Vine" from one foot up to twenty-five feet. He carefully observed how the sap flow varied with the weather and the time of day.

To measure the sap pressure Hales employed a "mercurial gage," a bent tube filled with mercury which he fixed to the cut branches of the vine, observing again the variations of the pressure at different times of day.

To determine the force with which trees imbibe moisture from the earth, Hales devised what he called "aqueo-mercurial" gauges. He laid bare the root of a small pear tree, cut it, and inserted it into a large glass tube, which in turn was fixed to a narrow tube eight inches long. When the tubes were filled with water and immersed in a vessel of mercury, the root, he reported, "imbibed the water with so much vigor" that in six minutes the mercury rose eight inches. These experiments were carried out in the summer months, when the trees and vines were in leaf. Hales noted that the more the sun shone on the plants, "the faster and higher the mercury rose"; it would subside toward evening and rise the next day. He observed that sometimes the mercury "rose most in the evening about 6 a clock, as the sun came on the Vine-branch." Such results may have suggested to him the role that transpiration—or, as he called it, "perspiration"—might play in causing the sap to rise.

Hales's experiments on transpiration—perhaps the most famous and brilliant of those he performed with plants—were carried out in the summer months of 1724. He grew a large sunflower in a garden pot covered lightly with a thin lead plate pierced by the plant, by a small glass tube to allow some communication with the air, and by another short, stoppered tube through which the plant could be watered. He weighed the pot and plant twice a day for fifteen days, then cut off the plant close to the lead plate, cemented the stump, and by weighing determined that the pot with its earth "perspired" two ounces every twelve hours. Substracting this from his earlier weighings, he found that the plant perspired in that period an average of one pound, four ounces of water.

Hales then stripped off the leaves of the plant and divided them in groups according to their several sizes. Taking a sample leaf from each group, he measured their surface areas by placing over them a grid made of threads, composing quarter-inch squares. By multiplying the area of each sample leaf by the number of leaves in the group and adding his measurements together, he obtained the total surface area of the leaves. His figures for the loss of water from the leaves compare favorably with those obtained long after by Sachs. Hales also attempted to estimate the surface area of the roots to determine the rate of absorption per given area, but these figures are of no value because

Hales "did not know how small a part of the roots is absorbent, nor how enormously the surface of that part is increased by the presence of root-hairs."[36] It was easier to determine the rate of flow in the stem. Always hoping to find analogies between animals and plants, he estimated the total surface area of his sunflower and its weight so as to compare the quantity of water "perspired" by the plant in twenty-four hours with that of an average "well-sized man" over the same period, taking the latter figure from James Keill's *Medicina statica britannica*.[37]

Transpiration could not account for the powerful rise of sap in vines during the bleeding season. Hales devoted a chapter of his book to the experiments which led him to discover root pressure. He cut off a vine, leaving only a short stump with no lateral branches. To it, by means of a brass collar, he fixed a series of glass tubes reaching as high as twenty-five feet. The sap rose gradually nearly to the top of these tubes, both day and night, although much higher in daytime. From this experiment, Hales remarks, "we find a considerable energy in the root to push up sap in the bleeding season."[38]

Similar experiments using his mercurial gauge confirmed that the force of the rising sap was "owing to the energy of the root and stem." Comparing his results with his blood-pressure experiments, Hales concluded that this force was "near five times greater" than that of the blood in the femoral artery of a horse and "seven times greater than the force of the blood in the like artery of a Dog."

Curious whether this force could be detected in vines when the bleeding season was over, Hales performed the same experiment in the month of July and found that the flow of sap ceased when the vine was cut from the stem, thus proving to his satisfaction that after the bleeding season the principal cause of the rise of sap was not root pressure but that which was "taken away, viz. the great perspiration of the leaves." This was evident, too, from a number of experiments which showed that branches stripped of their leaves did not imbibe water "for want of the plentiful perspiration of the leaves."[39]

A series of experiments to discover the direction of the flow of sap, and the portion of the stem through which it moved, were performed by cutting away the bark or slicing off a small section of it. These showed that while there must be some lateral communication, the sap moved upward between the bark and the wood, not downward "as many have thought," and that there is no circulation of the sap. Plants, Hales suggested, make up for the lack of a circulation by the much greater quantity of fluid that passes through them. Nature's "great aim in vegetables being only that the vegetable life be carried on and maintained, there was no occasion to give its sap the rapid motion, which was necessary for the blood of animals."[40]

Hales's explanation of the sap's motion invokes the Newtonian principle of attraction. The chief cause is "the strong attraction of the capillary sap vessels," greatly assisted "by the plentiful perspiration of the leaves, thereby making room for the fine capillary vessels to exert their vastly attracting power."

This "perspiration" results from the sun's warmth acting on the leaves, which are fittingly broad and flat to serve this purpose of absorbing the sun's rays.

An experiment to show the "great force, with which vegetables imbibe moisture" was performed by filling an iron pot nearly to the top with peas and water. Over the peas Hales placed a cover of lead, and on the cover he placed a weight of 180 pounds, which—as the peas swelled with the imbibed water—was lifted up.

The role of attraction, "that universal principle which is so operative in all the very different works of nature; and is most eminently so in vegetables,"[41] was illustrated by Hales's modification of an experiment of Francis Hauksbee, described in Query 31 of Newton's *Opticks*, showing the rise of water through a glass tube firmly packed with sifted wood ashes. Hales measured the imbibing force with his "aqueo-mercurial gage." He quotes Newton's words that "by the same principle, a sponge sucks in water, and the glands in the bodies of animals, according to their several natures and dispositions suck in various juices from the blood." Hales adds:

> And by the same principle it is, that . . . plants imbibe moisture so vigorously up their fine capillary vessels; which moisture, as it is carryed off in perspiration, (by the action of warmth,) thereby gives the sap vessels liberty to be almost continually attracting of fresh supplies.[42]

An influential experiment—it paved the way for some important researches by Sachs—demonstrated the unequal extent of growth in developing shoots and leaves. In the spring, using a comblike device, Hales pricked, with homemade red paint, dots a quarter of an inch apart along a young vine shoot. Several months later, when the shoot was full-grown, he measured the distances between the dots. The shoots, he discovered, had grown chiefly by a longitudinal extension between the nodes; the oldest (basal) internode had grown the least and the youngest (apical) one, the most.[43]

Again, concerned with analogies between plants and animals, this experiment led Hales to see if a similar effect could be observed in the growth of the long bones in animals, with their tubelike cavities. He took a half-grown chick and pierced the thigh and shin bones with a sharp-pointed iron, making small holes half an inch apart. After two months he killed the bird and found that although the bones had grown an inch in length, the marks remained the same distance apart. In contrast with what he had observed in his vine shoots, the growth had occurred not in the shaft but entirely at the junction of the shaft and its two ends, that is, at the symphyses.

In his experiments on plants Hales frequently noticed bubbles of air emerging from the cut stems of vines or rising through the sap, often in such quantity as to produce a froth. This, he remarked, "shews the great quantity of air which is drawn in thro' the roots and stem." The air, he thought for a time, was "perspired off" through the leaves; but an inconclusive experiment led him to suspect that "the leaves of plants do imbibe elastick air."[44] By 1725 he had performed a few experiments to prove that a considerable quantity of air is "inspired" by plants. The problem interested him so much that he deferred publication until he could make "a more particular enquiry into the nature of a Fluid," the air, "which is so absolutely necessary for the support of the life and growth of Animals and Vegetables."[45] These investigations, carried out between 1725 and 1727, were embodied in the long chapter, nearly half the final work, called "Analysis of Air." This chapter was to have momentous consequences for the later development of chemistry.

Since the investigations of Torricelli, Pascal, Otto von Guericke, and, of course, Robert Boyle, the physical properties of air had been pretty well understood: the law of its expansibility, its ability to refract light, its approximate density under standard conditions. But it was no longer thought by most chemists to be an element.[46] Any apparent chemical activity, and its ability to sustain life and support combustion, could be explained by the properties of special substances dispersed through it, such as the nitro-aerial particles imagined by Hooke, John Mayow, and others.[47] Boyle's description of the atmosphere was widely accepted; it was composed, he wrote, of three kinds of particles: the permanently elastic particles making up the air properly speaking, a "thin, diaphanous, compressible and dilatable Body"; vapors and dry exhalations from the earth, water, vegetables, and animals; and, third, "magnetical steams of our terrestrial globe" and particles of light from the sun and stars.[48]

Yet Boyle, Hooke, and other fellows of the Royal Society had shown that "air" ("factitious air") could be produced from solid and liquid bodies in certain chemical reactions: the action of acids on oyster shells or coral, the reaction of dilute acids with iron nails, the explosion of gunpowder.[49] A particularly striking experiment was performed by Frederick Slare in 1694. He poured spirit of nitre (nitric acid) over oil of caraway seeds, and the result was a violent explosion that blew up the glass container. Slare expressed amazement that so much air was produced from small amounts of these liquids.[50] This experiment made a profound impression; Slare's account was read, and the experiment perhaps repeated, by Roger Cotes in his lectures. It was described, too, by Newton in his *Opticks*, although without mentioning Slare by name. Hales was doubtless familiar with this experiment, although when he mentioned Slare in the *Vegetable Staticks* it was for a different experiment.[51] Newton's *Opticks*, to which Hales referred so often in his book, would have been sufficient authority for the existence of "factitious airs." At one point he quotes Newton's words: "Dense Bodies by Fermentation rarify into several sorts of Air, and this Air by Fermentation, and sometimes without it, returns into dense bodies."[52]

Of particular concern for Hales was evidence that air was thought to be of special importance to the plant economy. In France, Guy de La Brosse early in the seventeenth century had argued that plants cannot grow without the air from which they draw "la rosée & la manne."[53] Similar views were advanced by Robert Sharrock, a friend and collaborator of Robert Boyle.[54] This question was taken up in the early meetings of the Royal Society; John Beal suggested in 1663 that it should be determined "what effects would be produced on plants put into the pneumatic engine with the earth about their roots, and flourishing; whether they would not suddenly wither, if the air were totally taken from them."[55] Not long after, Robert Hooke showed that lettuce seed would not sprout and grow, and a thriving plant would wither and die, if kept in a vacuum.[56] In 1669 Beal felt able to conclude that a plant "feeds as well on the Air, as [on] the juice furnish'd through the root."[57] After the discoveries in plant anatomy by Malpighi and Nehemiah Grew, and their description of vessels in plants that appeared, like the trachea of insects, to be tubes for transmitting air, it was suggested that air contributed to the nutrition of plants, or—as John Ray put it—that plants have a kind of respiration.[58]

Except for Malpighi and Grew, whom he cites, Hales may have been unaware of these antecedents. But he was familiar with certain of Boyle's experiments admired by Roger Cotes. By these experiments, published in 1680–82, Boyle showed, as Hales put it in the beginning of his "Analysis of Air," that "a good quantity of Air was producible from Vegetables, by putting Grapes, Plums, Gooseberries, Cheries, Pease, and several other sorts of fruits and grains into exhausted and unexhausted receivers, where they continued for several days emitting great quantities of Air."[59]

In this famous long chapter Hales describes a large number of experiments—some trivial, some confused, but some extremely interesting—performed to discover the amount of air "fixed" in different substances or given off or absorbed under various circumstances. Strictly speaking, Hales was not a chemist, although he had performed some chemical experiments during his Cambridge days, when he had read or consulted George Wilson's practical compendium, *A Compleat Course of Chemistry* (1699).[60] He knew Boyle's work and John Mayow's, and was familiar with Nicolas Lemery's popular textbook.[61] But his approach was more physical than chemical; and it is not surprising—since he thought of air as a unitary substance characterized by its physical property of elasticity—that he failed to note the different chemical properties of the airs he produced.[62]

Hales's true mentor was Newton, whose last query of the *Opticks* (1718) was in fact a monograph on the role of attractive and repulsive forces in chemical processes, and whose short "Thoughts About the Nature of Acids" Hales had also read.[63] He was familiar too with the *Chymical Lectures* in which John Freind attempted to explain chemical reactions in Newtonian terms.[64]

From Newton, Hales derived the fundamental principles by which he explained the effects he observed. Matter is particulate, and the particles are subject to very special laws of attraction and repulsion. In their free state the particles of air exert upon each other strong repulsive forces, which accounts for the air's "elasticity." Yet this elasticity is no immutable property, for

Newton had remarked that "true permanent Air arises by fermentation or heat, from those bodies which the chymists call fixed, whose particles adhere by a strong attraction."[65] When air enters into "dense bodies" and becomes "fixed," its elasticity is lost because strong attractive forces overcome the forces of repulsion between its particles.

Hales's first experiments were distillations in which different substances were strongly heated in a glass or iron retort. The retort was cemented and luted to a globular vessel with a long neck, called a bolthead.[66] This vessel, with a hole cut in the bottom, was immersed in a basin of water; and by means of a siphon the water level was raised in the neck to a point he carefully marked. The amount of air given off or absorbed was determined by allowing the vessel to cool and noting the change in the water level. With this apparatus Hales measured the air produced by weighed amounts of hog's blood, tallow, powdered oyster shell, amber, honey, and a variety of vegetable materials. Whereas he obtained little air from ordinary well water, a considerable quantity was yielded by Pyrmont water, leading Hales to comment that this air "contributes to the briskness of that and many other mineral waters." He distilled iron pyrites, known to be rich in sulphur, and from a cubic inch of this mineral obtained eighty-three cubic inches of air. When he heated minium, or red lead (Pb_3O_4), he obtained a large quantity of air, remarking that this air might account for the increase in weight of lead when it is strongly heated to form minium. This air was doubtless what had "burst the hermetically sealed glasses of the excellent Mr. *Boyle*, when he heated the Minium contained in them by a burning glass."[67]

Two other contrivances were used by Hales to measure the air produced or absorbed in chemical reactions, or, as he put it, in "fermentations." One apparatus consisted of a bolthead placed in a basin of water; over its long neck he inverted a cylindrical vessel, using a siphon to draw up the water a given distance. As in the first apparatus, the amount of air given off or absorbed was determined by the change in the water level.[68] With this apparatus Hales measured the air produced by decomposing sheep's blood, by ale drawn from a fermenting vat, by the fermentation of raisins and apples, and by the action of vinegar on powdered oyster shells. Other experiments showed that salt of tartar (potassium carbonate) treated with acids yielded much air, a discovery that later put Joseph Black on the road to his major chemical discovery.[69] Hales also measured the large amount of air (hydrogen gas) produced from iron filings treated with dilute sulphuric acid. When the iron filings were dissolved in dilute nitric acid, he also obtained much air (in this case, nitric oxide). Of particular interest is Hales's measurement of air produced by the action of oil of vitriol on chalk and his further observation that lime (made from the same chalk) absorbed much air.[70]

His second contrivance has been called his pedestal apparatus.[71] A wooden pedestal is placed upright in a basin of water, and on its expanded top can be placed a candle, a weighed amount of some chemical substance to be ignited, or—in the larger form of this apparatus—a small animal. A glass cylinder is suspended over the pedestal so that its mouth is a few inches under water. As in the other devices, air is withdrawn with a siphon or bellows to raise the water to a convenient level, and a change in the water level indicates the change in the volume of air in the cylinder. With his pedestal apparatus Hales discovered that when phosphorus and sulphur are burned, they absorb air. When he detonated nitre (potassium nitrate) by means of a burning glass, he noted the large amount of air produced but observed that the volume steadily decreased, or as he put it, "the elasticity of this new air daily decreased."[72]

Repeating an experiment of John Mayow's, Hales placed a candle on the pedestal, ignited it with a burning glass, and noted the shrinkage in volume. When he used candles of equal size but in vessels of different capacities, he found that they burned longer in the larger ones and that "there is always more elastic air destroyed in the largest vessel." His burning glass, he found, could not light an extinguished candle "in this infected air."[73] Repeating another of Mayow's experiments, Hales placed a small animal on the pedestal and measured the air absorbed. Here, to be sure, two effects—both unknown to Hales—contributed to the rise of the water level: the intake of oxygen by the animal and its exhalation of carbon dioxide, much of which dissolved in the water. His results led to a series of rebreathing experiments carried out on himself which convinced him that animal respiration "vitiated" the air. His device, a bladder equipped with valves and breathing tube, enabled him to breathe repeatedly his own expired air. He found that he could continue in this fashion only about a minute. In a modification of his device, a series of diaphragms (flannel stretched over thin hoops) was placed in the bladder. When these were soaked with salt of tartar, especially when the salt was calcined (that is, rendered caustic), he found that he could rebreathe for as long as eight and a half minutes. The salt, "a strong imbiber of sulphureous steams," in fact absorbs much carbon dioxide.

In his experiments, when he noticed a decrease in volume of air during certain reactions, Hales always spoke of a loss of elasticity and attributed this to the acid sulphurous fumes which "resorb and fix" the elastic particles of ordinary air.[74] Such fumes, he noted, were produced by burning sulphur, by a lighted candle—indeed, by all "flaming bodies"—and by the expired air of animals and man.

To obviate this effect, Hales devised his most famous apparatus: the first pneumatic trough. Substances were heated in an iron retort; to the long neck of the retort he fixed a bent lead tube which was immersed in a basin of water and projected upward into the open end of an "inverted chymical receiver" filled with water. The released air passing through the bent tube bubbled up through the water and was collected in the top of the glass vessel. Hales's purpose was not to measure the amount of air, as in his other experiments, but to wash the air by passing it through water, to intercept "a good part of the acid spirit and sulphureous fumes." By this means he could collect and store air and ascertain whether its elasticity could be preserved. By separating the generator from the collector, Hales invented the pneumatic trough, later used in modified form by Brownrigg, Cavendish, and Priestley.

With his trough Hales collected air from a variety of substances—horn, human bladder stones, pyrite, saltpeter, minium, salt of tartar, and various vegetable materials—and claimed that the greater part of the air remained for the most part "in a permanently elastick state" and so was true air, not a mere flatulent vapor. He did not explore the different chemical properties of the air produced from different substances—indeed, he had no great reason to believe they could be found. Yet he suspected that there were at least some physical differences. Newton had written of bodies rarefying into "several sorts of air," an opinion that Hales seems to have shared, for he suggested that since air arises from a great variety of "dense" bodies, it is probable that airs from different sources may differ in the size and density of their constituent particles and may have "very different degrees of elasticity." But his crude attempts to see if common air and the air produced by salt of tartar (carbon dioxide) differed in density and compressibility disclosed no difference.

Hales's explanation of combustion was a physical one.[75] He rejected the notion that fire is "a particular distinct kind of body inherent in sulphur," as the chemists Willem (or Guillaume) Homberg and Louis Lemery believed. Instead, he followed Newton in distinguishing between heat and fire: heat is the rapid intestine motion of particles; fire is merely "a Body heated so hot as to emit Light copiously," and flame is only a "Vapour, Fume or Exhalation heated red hot." Hales owed much also to the speculations of John Mayow, but he did not believe that combustion results from the activity of some nitro-aerial spirit. Candles and matches cease to burn not because they have rendered the air "effete, by having consumed its *vivifying spirit*," but because of "acid fuliginous vapours" that destroy the air's elasticity. A continual supply of fresh elastic air is necessary to produce the rapid intestine motion of the fuel; this motion is the result of the "action and re-action" of acid sulphurous particles and the elastic particles of air. "Air cannot burn without sulphur, so neither can sulphur burn without air."

Despite the limitations of his achievement—he had prepared a number of gases without recognizing their differences—Hales passed on to the eighteenth century the conviction that there was such a thing as "fixed air" and that it abounds in all sorts of animal, vegetable, and mineral substances. Air is "very instrumental in the production and growth of animals and vegetables," serving in its fixed state as the bond of union "and firm connection of the several constituent parts" of bodies, that is, the chief elements or principles of which things are made: "their water, salt, sulphur and earth." He concluded that air should take the place of "mercury" or "spirit" as a fifth element:

> Since then air is found so manifestly to abound in almost all natural bodies; since we find it so operative and active a principle in every chymical opera-
> tion . . . may we not with good reason adopt this now fixt, now volatile

Proteus among the chymical principles, and that a very active one, as well as acid sulphur; notwithstanding it has hitherto been overlooked and rejected by Chymists, as no way intitled to that denomination?[76]

For Hales, science was more than the avocation of a country minister: it was a natural extension of his religious life. If he was a devotee of the mechanistic world view and held that the living organism was a self-regulating machine, this was in no way incompatible with his faith. For him, as for many other "physical theologians," nature testified to the wisdom, power, and goodness of the all-wise Creator "in framing for us so beautiful and well regulated a world."[77]

But Hales never doubted what Robert Boyle called "the usefulness of experimental philosophy." Hales's study of plants would, he was confident, improve man's skills in "those innocent, delightful and beneficial arts" of agriculture and gardening. He was well aware, too, that his studies of the animal vascular system and respiration would prove of medical value. Like Benjamin Franklin, one of the many who read the *Statical Essays* and were influenced by them, he was constantly alert to the practical possibilities of his discoveries. In describing his perfusion experiments on animals, Hales took occasion to warn the heavy imbibers of alcoholic liquors of the consequences of their vice. Indeed, he soon directed two pamphlets against the growing evil, "the Bane of the Nation," and, according to Gilbert White, was instrumental, under the patronage of Sir Joseph Jeckyll, in securing the passage of the Gin Act of 1736 "and stopping that profusion of spiritous liquors which threatened to ruin the morals and the constitution of the common people."

In the *Vegetable Staticks* he had described an ingenious mercury gauge used to determine the pressure exerted by peas expanding in water, and this led him to imagine its adaptation as a "sea gage" to measure the depths of the ocean. He applied his chemical knowledge to suggesting ways of keeping water sweet during long sea voyages and exploring the obstinate problem of distilling fresh from salt water.

With the publication of his *Haemastaticks*, Hales's career in pure science came to a close. From 1733 to the end of his life he devoted himself to applying scientific knowledge, technical skill, and his rich inventiveness to alleviating human problems, both medical and social. But even earlier he had turned his attention to a problem that had long challenged the resources of the medical profession: the painful affliction of kidney and bladder stones.

Early in 1727, while the *Vegetable Staticks* was in press, he obtained a specimen of such a human calculus from a friend, the famous surgeon John Ranby. On distilling this stone, Hales collected a much greater proportion of air than he had obtained from any other substance. Since various chemical agents were known to release this "strongly attracting, unelastic air," he thought it at last possible to find a solvent to dissolve the calculi and obviate the painful operation of being "cut for the stone." He carried out a number of experiments and published the results with his *Haemastaticks*. His attempts to find a useful solvent failed, and the paper is noteworthy chiefly for his success in perfusing a dog's bladder with one of his solutions and for his invention of a surgical forceps, which Ranby and other surgeons promptly used with success to remove stones from the human urethra. Ironically, it was for this largely useless work on human calculi—not for the remarkable experiments on plants and animals and on air published in the *Statical Essays*—that Hales was awarded the Royal Society's Copley Medal in 1739.

His newly acquired expertise entangled Hales in a rather notorious episode.[78] A Mrs. Joanna Stephens had for some years been treating victims of the stone with a secret proprietary remedy, supposedly with some success. Attempts to persuade her to divulge her secret led Parliament to vote a substantial reward and to set up a group of trustees to receive her disclosure and evaluate the effectiveness of her nostrum. Hales was one of the trustees, and he set to work to determine the effective ingredients in the odd mixture. Experiments convinced him that it was the lye used in soapmaking, and lime from eggshells used in her formula, that seemed to have the desired property of dissolving the stone. The result was Hales's suggestion—destined to be taken up by others—that lime water might prove an effective if somewhat corrosive remedy.

Hales's experiments on air and respiration were the stimulus for the invention that more than any other contributed to his contemporary fame: the ventilators he contrived in order to remove fetid air from prisons, hospitals, and slave ships. His experiments—especially the rebreathing experiments—had convinced him that "elastic" air, free from noxious fumes, was necessary for respiration, for there was great danger in respiring "vitiated air." These theories fitted well with the current belief that many diseases were attributable to bad air and "miasmata." After a victory over a rival inventor, Hales's ventilators were installed in His Majesty's ships, as well as in merchant vessels, slave ships, hospitals and prisons. The ventilators did not, of course, eliminate airborne bacterial or viral diseases, but they seem to have markedly reduced mortality rates. As one of the first to call attention to the importance of fresh air, Hales deserves his reputation as a pioneer in the field of public health.

These varied activities did not interfere with his parish duties. He preached regularly and presided with some severity over the morals of his village; he enlarged the churchyard and virtually rebuilt the old church. In 1754 Hales engineered a new water supply for the village and, as Francis Darwin remarks, "characteristically records, in the parish register, that the outflow was such as to fill a two-quart vessel in 3 swings of a pendulum, beating seconds, which pendulum was 39 + 2/10 inches long from the suspending nail to the middle of the plumbet or bob."[79]

Hales's later years were graced with honors. Oxford conferred on him the doctorate of divinity in 1733. He was one of the trustees of the Georgia colony; and John Ellis, the merchant-naturalist who was governor of the colony and a correspondent of Linnaeus, named after him a genus of American flowering shrubs (*Halesia*). Hales was one of the founders of what is now the Royal

Society of Arts and became one of its vice-presidents in 1755. In 1753 he was chosen a foreign associate of the Paris Academy of Sciences, replacing Hans Sloane, who had died earlier that year. Hales's portrait was painted by Francis Cotes and by his neighbor at Twickenham, the popular Thomas Hudson.

He had many acquaintances in the neighborhood, among them Alexander Pope (he was one of the witnesses to Pope's will) and Horace Walpole (who called him "a poor, good, primitive creature"). He was patronized by Frederick Louis and Augusta, the prince and princess of Wales, who lived not far distant at Kew. The prince, it is said, enjoyed surprising Hales in his laboratory at Teddington.

Walpole's unflattering description bears out the opinion of contemporaries, who spoke of Hales's native innocence and simplicity of manner. Peter Collinson testified to "his constant serenity and cheerfulness of mind." He died after a brief illness and was buried under the tower of his beloved church. A monument in Westminster Abbey was erected to his memory by the princess of Wales, with a bas-relief of "the old philosopher" in profile. If there is anything in the church at Teddington recalling Hales to memory, the guidebooks make no mention of it. Instead, they single out a monument to Hales's most famous (and notorious) parishioner, the actress Peg Woffington.

✸

[A visit to Hales's church in Teddington, after this article was written, disclosed the recent addition of a stained-glass window with the figures of Hales and the minor poet Thomas Traherne. Visitors are also shown a ventilator at the entrance to the church, which doubtless blew a cold draft on the necks of his parishioners.]

Notes

1. J. F. Fulton, *Selected Readings in the History of Physiology*, 2d ed. rev. and enl. (Springfield, Ill.: C. C. Thomas, 1966).
2. *Gentleman's Magazine* 34 (1764), 273. This article by Peter Collinson seems to have been based on information supplied by Stukeley. It is reproduced in *Annual Register* (1764), "Characters," 42–49.
3. G. M. Trevelyan, *Trinity College* (Cambridge, 1946), 55. See James Henry Monk, *Life of Richard Bentley*, 2d ed., 2 vols. (London, 1833), I, 204.
4. *Family Memoirs of the Rev. William Stukeley, M.D.*, 3 vols. (London-Edinburgh, 1882–1887, I, 21. Henceforth referred to as *Family Memoirs*.
5. *Ibid.*, 21–22.
6. *Hydrostatical and Pneumatical Lectures by Roger Cotes*, ed. Roger Smith (Cambridge, 1738; 2d ed., 1747). These posthumously published lectures, as we have them, were delivered after 1706 (Cotes refers to Newton's Latin *Optice* of that year) and perhaps before 1710. Cotes died in 1716. Whiston's lectures on this subject were never published.

7. *Family Memoirs*, I, 21, where we read: "Mr. Danny read to us . . . Pardies Geometry, Tacquets Geometry by Whiston, Harris's use of the Globes, Rohaults Physics by Clark. He read to us Clarks 2 Volumes of Sermons at Boyles Lectures, Varenius Geography put out by Sr. Isaac Newton & many other occasional peices [*sic*] of Philosophy, & the Sciences subservient thereto."

8. A. Hastings White, ed., *Memoirs of Sir Isaac Newton's Life by William Stukeley* (London, 1936), 9. See also *Family Memoirs*, I, 23–24.

9. The sketch is reproduced in A. E. Clark-Kennedy, *Stephen Hales, D.D., F.R.S.: An Eighteenth Century Biography* (Cambridge, 1929), (pl. 4), and in R. T. Gunther, *Early Science in Cambridge* (Oxford, 1937), 160. Stukeley says Hales "first projected, & gave the idea of horarys." *Family Memoirs*, I, 21. The name "orrery" was attached to such devices after the one later built by John Rowley for his patron, the fourth earl of Orrery.

10. After earning the degree of bachelor of medicine from Cambridge, Stukeley studied "the practical part of physick" under Richard Meade at St. Thomas's Hospital; early in 1717 he opened his own London practice.

11. On March 6, 1717/18; see *Royal Society Journal Book*, V (1714–1720), 235. Stukeley, although formally elected the same day as Hales, had been nominated much earlier by Edmond Halley, and his nomination evidently approved by the Council.

12. *Vegetable Staticks* (London, 1727), p. iii. Hales, probably writing his preface late in 1726, states that this accidental observation occurred "about seven years since." Unless otherwise noted, all future references to *Vegetable Staticks* will be to the 1727 edition.

13. "Mr. Hale [*sic*] having been formerly Elected, and lapsed the time of his admission, the same was dispensed with by the Society, and he Subscribed the Obligation and was admitted accordingly." *Journal Book*, V, entry of November 20, 1718 (O.S.), 250–51.

14. *Ibid.*, 289. A good summary of the experiment is here transcribed.

15. *Ibid.*, VI (1720–1726), 438–40 ff.

16. *Ibid.*, VII (1726–27), 44–45, 48–50.

17. *Philosophical Transactions of the Royal Society*, vol. 35, no. 398, pp. 264–91; no. 399, pp. 323–31. In April and May 1727, Desaguliers repeated before the Royal Society certain of Hales's experiments. *Journal Book*, VII, 74, 83.

18. "Of Mechanical Knowledge, and the Grounds of Certainty in Physick," in his translation of Santorio's *Medicina Statica*, 2d ed. (London, 1720), 1.

19. *An Account of Animal Secretion, the Quantity of Blood in the Human Body, and Muscular Motion* (London, 1708). James Keill was strongly influenced by his older brother, the mathematician and Newtonian disciple John Keill.

20. *Vegetable Staticks*, 2–3.

21. Cusa's *De staticis*, one of the *Idiota* dialogues, appeared in many editions, sometimes appended to editions of the *De architectura* of Vitruvius. The first English translation is *The Idiot in Four Books; the First and Second of Wisdome, the Third of the Minde, the Fourth of Statick Experiments, or Experiments of the Ballance. By the Famous and Learned C. Cusanus* (London, 1650).

22. *Medicina Statica: Being the Aphorisms of Sanctorius, Translated into English with Large Explanations*, trans. John Quincy (London, 1712). This popular work was, of course, known in a number of Latin editions, some with commentaries by Giorgio Baglivi and Martin Lister. In a preface to the volume of *Philosophical Transactions* for 1669 we read "The Ingenious Sanctorius hath not exhausted all the results of Statical indications." *Philosophical Transactions*, vol. 4, no. 45, p. 897. The *Oxford English Dictionary* gives as the earliest example of the word in English: Sir Thomas Browne's reference in the *Pseudodoxia Epidemica* (1646) to "the statick aphorisms of Sanctorius." Quincy's was not the earliest English version; a translation had been published by J. Davis (London, 1676).

23. James Keill, *Tentamina medico-physica quibus accessit medicina statica britannica* (London, 1718).

24. *Haemastaticks* (1733), preface.

25. Arturo Castiglioni, *History of Medicine*, ed. and trans. E. B. Krumbaar (New York:

Knopf, 1941), 614. Malpighi was the first to observe the capillaries; J. L. M. Poiseuille studied blood viscosity and rate of flow and introduced the mercury manometer for the measurement of blood pressure.

26. *Haemastaticks* (1733), 58–59.

27. Hales doubtless knew the passage in Francis Hauksbee's preface to his *Physico-Mechanical Experiments* (1709), in which Hauksbee wrote that electricity may possibly explain "the Production and Determination even of *Involuntary Motion* in the *Parts of Animals*," for he quotes Hauksbee three times, but on other matters, in the "Analysis of Air." He surely also knew the concluding passage of the General Scholium of Newton's *Principia*, 2d ed. (1713), in which Newton hints that "an electric and elastic spirit" may account for sensation and cause "the members of animal bodies [to] move at the command of the will, namely, by the vibrations of this spirit, mutually propagated along the solid filaments of the nerves, from the outward organs of sense to the brain, and from the brain to the muscles."

28. *Vegetable Staticks*, 2–3.

29. *Philosophical Transactions*, vol. 3, no. 40 (1668), pp. 787–801.

30. The letter was communicated June 10, 1669 (O.S.) and published in *Philosophical Transactions*, vol. 4, no. 48, pp. 963–65. See also Charles Raven, *John Ray* (Cambridge, 1950), 187–88.

31. Thomas Birch, *History of the Royal Society of London*, 4 vols. (London, 1756–57), II, 382.

32. The theory of sap circulation was advanced by Christopher Merret in 1664 and by Johann Daniel Major a year later. See Julius von Sachs, *History of Botany* (Oxford, 1890), 456; and J. Reynolds Green, *History of Botany in the United Kingdom* (London, 1914), 76. This theory is clearly set forth by John Locke. See *Philosophical Works of John Locke*, ed. J. A. St. John, 2 vols. (London, 1705–1706), II, 487. Even later than Blair were the claims of a Mr. Fairchild in 1724 to have proved by experiments "a constant Circulation of the Sap in Trees and Plants." *Journal Book*, VI, 377.

33. For the background of this experiment, suggested by Cusa in his *De staticis*, see Herbert M. Howe, "A Root of van Helmont's Tree," *Isis* 56 (1965), 408–19. See also A. D. Krikorian and F. C. Steward, "Water and Solutes in Plant Nutrition," *BioScience* 18 (1968), 286–92.

34. "Some Thoughts and Experiments Concerning Vegetation," *Philosophical Transactions*, vol. 21, no. 253 (1699), pp. 193–227. The quotation is from p. 208. Hales, when describing an experiment on the imbibition by a spearmint plant growing in water, wrote: "I pursued this Experiment no farther, Dr. *Woodward* having long since . . . given an account . . . of the plentiful perspirations of this plant." *Vegetable Staticks*, 28.

35. Hales writes "by the favour of the eminent Mr. *Wise*." *Vegetable Staticks*, 17–18. Hales owed something to his relations with "the skilful and ingenious Mr. Philip Miller" of the Chelsea Physic Garden and author of the popular *Gardener's Dictionary* (1724). On Miller see Green, *History of Botany*, 156–57, and *passim*.

36. Francis Darwin, *Rustic Sounds* (London, 1917), 126.

37. *Vegetable Staticks*, 10.

38. *Ibid.*, 103.

39. See, for example, Hales's experiments 7 and 28. *Vegetable Staticks*, 28–29, 90.

40. *Ibid.*, 136. See also pp. 13–14, where he writes that the sap has "probably only a progressive and not a circulating motion as in animals."

41. *Ibid.*, 96.

42. *Ibid.*, 100. (The quotation from Newton is on p. 99.)

43. *Ibid.*, 329–37. He was struck, as Sachs observes, by the fact that the longitudinal growth allows the capillary vessels to retain their hollowness, as when glass tubes are drawn out to fine threads.

44. *Ibid.*, 102–3, 148. For the inconclusive experiment see experiment 122 in the chapter "Of Vegetation." After the publication of the *Vegetable Staticks* Hales repeated the experiment and convinced himself that leaves imbibe air. He informed Desaguliers of these results by June 1727; see Desaguliers's postscript to his abstract of Hales's book in *Philosophical Transactions*, vol. 35, no. 399, p. 331.

45. *Vegetable Staticks*, 155–56.
46. The prevailing view in the seventeenth century (of men like Jean Beguin, Lemery, and Homberg) was that there are five elements or principles: three active principles (variously described as spirit, oil, and salt or as mercury, sulphur, and salt) and two passive ones, water and earth. This was clearly a compromise between the Aristotelian theory of the four elements and the *tria prima* of the Paracelsians. For a clear statement of this view in Hales's day, see John Harris, *Lexicon technicum* (1704), article "Principle."
47. Henry Guerlac, "John Mayow and the Aerial Nitre," in *Actes du Septième Congrès d'Histoire des Sciences* (Jerusalem, 1953), 332–49; and "The Poet's Nitre," *Isis* 45 (1954), 243–55.
48. Robert Boyle, *General History of Air* (London, 1692), 1. See also Harris, *Lexicon technicum*, article "Air"; and Cotton Mather, *The Christian Philosopher* (London, 1721), 65.
49. When the experiment on powdered oyster shells was shown to the Society on March 15, 1664/65, the air was collected in a deflated bladder. But when it was repeated a short time later, a large glass filled with water was inverted ("whelmed") over the reactants; and when the reaction was over, it was found that the "whelmed glass" was about a quarter full of an aerial substance. Birch, *History of the Royal Society*, II, 22, 27. This early anticipation of the principle underlying the pneumatic trough seems to have escaped notice; it is not mentioned in John Parascandola and Aaron J. Ihde, "History of the Pneumatic Trough," *Isis* 60 (1969), 351–61.
50. *Philosophical Transactions*, vol. 18, no. 212, pp. 212–13.
51. *Hydrostatical and Pneumatical Lectures by Roger Cotes* (1747 ed.), 220–23; and Isaac Newton, *Optice* (1706), 325, and *Opticks* (1718), 353. Hales quotes the experiment in which Slare distilled or "calcined" an animal calculus and found that the greatest part of this stone "evaporated in the open fire." *Vegetable Staticks* (1727), 188–89.
52. Query 30 of *Opticks* (1718), 349–50; and *Vegetable Staticks*, 312. Hales also quotes (*Ibid.*, 165) from another long passage of the *Opticks* in which Newton speaks of airs formed from those bodies "which Chymists call fix'd, and being rarefied by Fermentation, become true permanent Air." Query 31, *Opticks* (1718), 372. Newton and Hales both used the word "fermentation" to mean chemical reactions that are accompanied by the production of heat and ebullition. The term seems to have originated with Thomas Willis in his *De fermentatione sive De motu intestino particularum in quovis corpore* (London, 1659).
53. *De la nature, vertu, et utilité des plantes* (Paris, 1628), 75; see also pp. 94–95.
54. *The History of the Propagation & Improvement of Vegetables* (Oxford, 1660), 40–42, 84–85. Robert Sharrock, an Oxford graduate who became archdeacon of Winchester, supplied prefaces to three of Boyle's works. His book is dedicated to Boyle.
55. Birch, *History of the Royal Society*, I, 304.
56. *Ibid.*, II, 54, 164; III, 418, 420–21.
57. *Philosophical Transactions*, vol. 3, no. 42, p. 854.
58. *The Wisdom of God Manifested in the Works of the Creation*, 8th ed. (London, 1722), 72. Mather, *Christian Philosopher*, 69, was clearly paraphrasing Ray when he wrote: "Yea, *Malpighius* has discovered and demonstrated, that the *Plants* themselves have a kind of respiration, being furnished with a Plenty of Vessels for the Derivation of *Air* to all their Parts."
59. *Vegetable Staticks*, 156. Boyle's experiments, carried out with Denis Papin, using the latter's improved air pump, were published in Boyle's *A Continuation of New Experiments . . .*, which appeared in Latin in 1680 and in English in 1682.
60. For one such Cambridge experiment see *Vegetable Staticks*, 195.
61. Hales also cited Hermann Boerhaave's *New Method of Chemistry*. An unauthorized version of Boerhaave's lectures had appeared in Latin in 1724; Hales seems to have used the English translation by Peter Shaw and E. Chambers, dated 1727; if so, his references to it in the *Vegetable Staticks* were obviously added while his book was in press.
62. Although he records the combustibility of the gases produced by distilling peas, he

failed to note the same property in coal gas. In describing the air produced by the action of dilute acid on iron filings (that is, hydrogen), he does not remark that it is inflammable.

63. *Vegetable Staticks*, 291. Newton's paper was published by John Harris in the introduction to his *Lexicon technicum*, II (1710), where Hales consulted it.

64. "And Dr. Freind has from the same principles [as Newton] given a very ingenious Rationale of the chief operations in Chymistry." *Vegetable Staticks*, preface, p. v.

65. *Ibid.*, 165. See Newton, *Opticks* (1718), 372.

66. The bolthead was a chemist's globular flask with a long cylindrical neck, what Boyle called a "glass egg with a long neck." It was named for its resemblance to the head of a bolt or arrow.

67. *Vegetable Staticks*, 287.

68. The method of collecting air by the displacement of water, used in all Hales's devices, was not original with him. It had been used as early as 1665, probably by Robert Hooke, in an experiment performed at the Royal Society. But it was doubtless from John Mayow that Hales learned of this method; Mayow used it extensively in his *Tractatus Quinque Medicophysici* (1674) and illustrated several modifications in an accompanying plate.

69. Henry Guerlac, "Joseph Black and Fixed Air," *Isis* 48 (1957), 435, and n. 141.

70. *Vegetable Staticks*, 223.

71. Henry Guerlac, "The Continental Reputation of Stephen Hales," *Archives Internationales d'Histoire des Sciences* 4, no. 15 (1915), 396–97. See also Parascandola and Ihde, "History of the Pneumatic Trough," 355.

72. Hales, *Vegetable Staticks*, 187, compared his results with the observations of Francis Hauksbee, who had noted the same effect. See Hauksbee's *Physico-Mechanical Experiments on Various Subjects* (London, 1709), 83.

73. *Vegetable Staticks*, 231.

74. *Ibid.*, 183.

75. *Ibid.*, 272–75, 278–85.

76. *Ibid.*, 315–16.

77. For Hales's "argument from design" to justify his scientific work, see in particular his eloquent preface to the *Haemastaticks* (1733).

78. For a detailed account of this episode, and its later influence on the work of Joseph Black, see Henry Guerlac, "Joseph Black and Fixed Air," 137–51.

79. "Hales, Stephen," in *Dictionary of National Biography*.

Bibliography

I. Original Works

Hales's major writings in English are *Vegetable Staticks: Or, an Account of Some Statical Experiments on the Sap in Vegetables . . . Also, a Specimen of an Attempt to Analyze the Air . . .* (London, 1727), also reprinted with a useful foreword by M. A. Hoskin (London: Scientific Book Guild, 1961); *Statical Essays: Containing Vegetable Staticks* (London, 1731), the 2d ed., "with amendments"; *Statical Essays*, 2 vols. (London, 1733): vol. I is the 3d ed. of *Vegetable Staticks*, and vol. II is the 1st ed. of *Haemastaticks: Or an Account of Some Hydraulic and Hydrostatical Experiments Made on the Blood and Blood-Vessels of Animals*, with a separate preface, Hales's "Account of Some Experiments on Stones in the Kidnies and Bladder," an appendix with nine "Observations" relating to the motion of fluids in plants, seven additional experiments on air, and "Description of a Sea-gage, Wherewith to Measure Unfathomable Depths of the Sea." Vol. II is reprinted in facsimile as no. 22 in History of Medicine Series of the Library of the New York Academy of Medicine (New York: Hafner, 1964), with a short introduction by André Cournand, M.D.

Statical Essays, 2 vols. (London, 1738–1740): Vol. I is 3d ed. of *Vegetable Staticks* and vol. II is 2d ed., "corrected," of *Haemastaticks*; and *Statical Essays*, 2 vols. (London, 1769): vol. I is 4th ed. of *Vegetable Staticks*, and vol. II is 3d ed. of *Haemastaticks*.

Translations of Hales's major works are *La statique des végétaux, et l'Analyse de l'air* . . . , trans. G. L. L. Buffon (Paris, 1735), an influential French translation which has the famous "Préface du traducteur," in which Buffon praises the experimental method, and includes Hales's appendix of 1733; *Haemastatique, ou la statique des animaux* (Geneva, 1744), the first French version of the *Haemastaticks*, translated by the physician and botanist François Boissier de Sauvages; *Statique des végétaux, et celle des animaux*, 2 pts. (Paris, 1779–1780); pt. 1 is Buffon's translation of the *Vegetable Staticks* "revue par M. Sigaud de la Fond," and pt. 2 is Boissier de Sauvages's translation of the *Haemastaticks; Statick der Gewächse* (Halle, 1747), translated, with a preface, by the philosopher Christian von Wolff; *Statick des Geblüts*, 2 pts. (Halle, 1748), pt. 1 is the *Haemastaticks*, and pt. 2 is Wolff's translation of the *Vegetable Staticks; Emastatica, ossia statica degli animali*, 2 vols. (Naples, 1750–1752), Italian translation from the French of Boissier de Sauvages, vol. 2 has a translation of Hales's work on bladder and kidney stones and two medical dissertations by Boissier de Sauvages; and *Statica de'vegetabili ed analisi dell' ari*, trans. D. M. A. Ardinghelli (Naples, 1756, 1776) with commentary.

Hales's minor works were *A Sermon Preached Before the Trustees for Establishing the Colony of Georgia in America* (London, 1734); *A Friendly Admonition to the Drinkers of Brandy and Other Distilled Spirituous Liquors* (London, 1734), anonymous, but attributed to Hales; *Distilled Spirituous Liquors the Bane of the Nation; Being Some Considerations Humbly Offered to the Hon. the House of Commons* (London, 1736); *Philosophical Experiments: Containing Useful and Necessary Instructions for Such as Undertake Long Voyages at Sea* . . . (London, 1739); *An Account of Some Experiments and Observations on Mrs. Stephen's Medicines for Dissolving the Stone* (London, 1740), also translated into French (Paris, 1742); *A Description of Ventilators* . . . (London, 1743), French translation by P. Demours (Paris, 1744); *An Account of Some Experiments and Observations on Tar-Water* (London, 1745); *Some Considerations on the Causes of Earthquakes* . . . (London, 1750), French translation by G. Mazeas (Paris, 1751), with the letter of the bishop of London, Thomas Sherlock, on the moral causes of the London earthquakes of 1750; *A Sermon Before Physicians, on the Wisdom and Goodness of God in the Formation of Man* (London, 1751), the annual Croonian sermon of the Royal College of Physicians, *not* the Croonian lecture of the Royal Society; *An Account of a Useful Discovery to Distill Double the Usual Quantity of Sea-water* . . . *and an Account of the Great Benefit of Ventilators* . . . (London, 1756); and *A Treatise on Ventilators* . . . (London, 1758).

II. Secondary Literature

General and biographical sources include the following (listed chronologically) : "Some Account of the Life of the late excellent and eminent STEPHEN HALES, D.D., F.R.S. chiefly from Materials communicated by P. Collinson, F.R.S.," in *Gentleman's Magazine* 34 (1764), 273–78, see also *Annual Register of World Events* (1764), 42–49; Jean-Paul Grandjean de Fouchy, "Eloge de M. Hales," in *Histoire de l'Académie Royale des Sciences* for 1762 (Paris, 1764), 213–30; Robert Watt, *Bibliographia Britannica*, 4 vols. (London, 1824), I, col. 457; F. D. [Francis Darwin], "Hales, Stephen," in *Dictionary of National Biography*, an excellent summary; Francis Darwin, *Rustic Sounds* (London, 1917), 115–39, a useful essay of a distinguished botanist; G. E. Burget, "Stephen Hales," *Annals of Medical History* 7 (1925), 109–16; A. E. Clark-Kennedy, "Stephen Hales: Physiologist and Botanist," *Nature* 120 (1927), 228–31; George Sarton, "Stephen Hales's Library," *Isis* 14 (1930), 422–23; A. E. Clark-Kennedy, *Stephen Hales, D.D., F.R.S.: An Eighteenth Century Biography* (Cambridge and New York, 1929; reprinted Ridgewood, N.J., 1965), the only full-length biography; Jocelyn Thorpe, "Stephen Hales," in *Notes and Records of the Royal Society of London* 3 (1940), 53–63; and Lesley Hanks, *Buffon avant "L'histoire naturelle"* (Paris: Presses Universitaires de France, 1966), 72–101, which discusses Buffon's translation of the *Vegetable Staticks* and the Newtonianism of Hales and Buffon.

On his work in public health, see D. Fraser Harris, "Stephen Hales, the Pioneer in the Hygiene of Ventilation," *Scientific Monthly* 3 (1916), 440–54.

On animal physiology see the following (listed chronologically) : John F. Fulton, *Selected Readings in the History of Physiology* (Springfield, Ill., and Baltimore: C. C. Thomas, 1930), 57–60, 75–79, 235. See also the greatly enlarged edition with material supplied by Leonard Wilson (Springfield, Ill.: C. C. Thomas, 1966); and *Physiology*, in the Clio Medica series (New York, 1931), 35–36, 42–43; Thomas S. Hall, *A Source Book in Animal Biology* (New York: McGraw-Hill, 1951), 164–71, which reprints Hales's preface to the *Haemastaticks*, without the concluding acknowledgment, and experiment I; and Diana Long Hall, "From Mayow to Haller: A History of Respiratory Physiology in the Early Eighteenth Century," Ph.D. dissertation (Yale University, 1966), 118–21.

Hales's work in plant physiology is discussed in the following (listed chronologically): Julius von Sachs *Geschichte der Botanik* (Munich, 1875), 514–21, 582–83, English translation by Henry E. G. Garnsey, revised by I. B. Balfour, *History of Botany (1530–1860)* (Oxford, 1906), 476–82, 539; J. Reynolds Green, *A History of Botany in the United Kingdom* (London, 1914), 198–206 and *passim*; R. J. Harvey-Gibson, *Outlines of the History of Botany* (London, 1919), 46–50 and *passim*; and Ellison Hawks and G. S. Boulger, *Pioneers of Plant Study* (London: Sheldon Press, 1928), 228–30.

Hales's chemistry is discussed in the following (listed chronologically): Hermann Kopp, *Geschichte der Chemie*, 4 vols. (Brunswick, 1843–1847), III, 182–83 and *passim*; Ferdinand Hoefer, *Histoire de la chimie*, 2d ed., 2 vols. (Paris, 1866–1869), II, 338–42; Henry Guerlac, "The Continental Reputation of Stephen Hales," *Archives Internationales d'Histoire des Sciences* 4, no. 15 (1951), 393–404; Milton Kerker, "Hermann Boerhaave and the Development of Pneumatic Chemistry," *Isis* 46 (1955), 36–49; Henry Guerlac, "Joseph Black and Fixed Air, A Bicentenary Retrospective," *Isis* 48 (1957), 124–51, 433–56; Rhoda Rappaport, "G.-F. Rouelle: An Eighteenth-Century Chemist and Teacher," *Chymia* 6 (1960), 94: Henry Guerlac, *Lavoisier—The Crucial Year* (Ithaca, N.Y.: Cornell University Press, 1961), *passim*, for Hales's influence upon Lavoisier; J. R. Partington, *History of Chemistry*, 4 vols. (London: Macmillan & Co., 1962), III, 122–23; and John Parascandola and Aaron J. Ihde, "History of the Pneumatic Trough," *Isis* 60 (1969), 351–61.

13
Newton and the Method of Analysis

❁

Isaac Newton's disciples in the eighteenth century were impressed not only by his discoveries in optics and celestial mechanics and by his admirably ordered System of the World, but also by the method he employed. A variety of writers mention him as the inventor of the only proper way of investigating nature: of a method that d'Alembert called "exact, profound, luminous and new." Laplace found his method "happily applied" in the *Principia* and the *Opticks*, works valuable not just for the discoveries contained in them but as the best models to be emulated, as the embodiments of this method.[1] Moreover, Newton's method—as understood by men of letters like Voltaire and Condillac—was thought to be, besides a technique for investigating physical nature, "a new method of philosophizing" applicable to all areas of human knowledge. What Newton called his "Experimental Philosophy" had wide application; it set the bounds to human presumption; it was systematic, yet as Condillac pointed out, was opposed to the *esprit de système*; it rejected unsupported or gratuitous *hypotheses*.

Analysis, the dissection of nature, men in the eighteenth century took to be the key to knowledge, the great and novel intellectual tool, indeed the essence of Newton's method. Yet if we read carefully the important Newtonian passages, or the best of Newton's expositors, we discover that Newton's methodological prescriptions were by no means confined to this "dissection of nature," although he lays great stress upon it. Men who possess Newton's experimental philosophy, wrote Roger Cotes,

> . . . proceed in a twofold method, synthetical and analytical. From some select phenomena they deduce by analysis the forces of Nature and the more simple laws of forces; and *from thence* by synthesis show the constitution of the rest. This is that incomparably best way of philosophizing, which our renowned author most justly embraced in preference to the rest.[2]

Later, Colin Maclaurin wrote, "In order to proceed with perfect security, and to put an end for ever to disputes, [Newton] proposed that, in our inquiries into nature, the methods of *analysis* and *synthesis* should be both employed in a proper order."[3]

From *Dictionary of the History of Ideas*, vol. III (1973), 378–91. Copyright © 1973 Charles Scribner's Sons.

Newton first referred in print to his methodological principles, at least in
a major work, in the new Queries added to the Latin version of his *Opticks*
which appeared in 1706. In one of these Queries (Q. 20/28) he reproves those
"later philosophers" (*physici recentiores*) who invoke mechanical hypotheses
to explain all things, "whereas the main Business of natural Philosophy is to
argue from Phaenomena without feigning Hypotheses, and to deduce Causes
from Effects."[4] He is even more explicit in a passage towards the close of the
last new Query (Q. 23/31), a passage that was greatly expanded when Newton
in 1717–18 brought out a second English edition of his *Opticks*. Since this
English text is more familiar, to say nothing of being more complete, and
indeed is the *locus classicus* for any study of Newton's theory of method, it
deserves to be printed here in full, with the passages that had earlier appeared
in the much shorter Latin statement of 1706 given in italics:

> *As in Mathematics, so in Natural Philosophy, the Investigation of difficult
> Things by the Method of Analysis, ought ever to precede the Method of Com-
> position [synthesis].* This Analysis consists in making Experiments and Observa-
> tions, and in drawing general Conclusions from them by Induction, and
> admitting of no Objections against the Conclusions, but such as are taken from
> Experiments, or other certain Truths. For Hypotheses are not to be regarded in
> experimental Philosophy. And although the arguing from Experiments and
> Observations by Induction be no Demonstration of general Conclusions; yet
> it is the best way of arguing which the Nature of Things admits of, and may
> be looked upon as so much the stronger, by how much the Induction is more
> general. And if no Exception occur from Phaenomena, the Conclusion may be
> pronounced generally. But if at any time afterwards any Exception shall occur
> from Experiments, it may then begin to be pronounced with such Exceptions
> as occur. By this way of Analysis *we may proceed from Compounds to In-
> gredients, and from Motions to the Forces producing them; and in general,
> from Effects to their Causes, and from particular Causes to more general ones,
> till the Argument end in the most general.* This is the Method of Analysis: And
> *the Synthesis consists in assuming the Causes discover'd, and establish'd as
> Principles, and by them explaining the Phaenomena proceeding from them,
> and proving the Explanations.*[5]

Several points in this text should be noted. First, Newton advocates a
single procedure, made up of two "methods," both of which must be employed,
but one of which (the analytic) must be carried out before the other (the
synthetic or compositional). Second, that although he describes these two
methods by terms used for analogous methods in mathematics, by analysis he
means the making of observations or the performing of experiments, and
deriving conclusions from them by induction. Experiments are among the
certain truths, yet inductions from them do not "demonstrate" the conclusions
drawn; still this is "the best way of arguing which the Nature of Things
admits of."

If this is Newton's most important statement of his scientific method,
what of the famous Rules of Reasoning in Philosophy which Newton placed

at the beginning of Book III of the *Principia?* With the single exception of Rule IV, they do not deal with method, at least as the term was commonly used in Newton's day, and will be used in this article. The first two rules— statements of the law of parsimony and of the principle of the analogy and uniformity of nature—can perhaps be described as basic articles of scientific faith, as metascientific principles. Rule III, which has been much discussed and variously interpreted, can be characterized, at least in a loose way, as an analogical rule. Its manifest purpose is to justify extending observations and measurements concerning gravity on the earth and in the solar system so as to "allow that all bodies whatsoever are endowed with a principle of mutual gravitation." Rule IV was added to the third edition of the *Principia* (1726) and clearly echoes what he had written some eight years before in the expanded Query 23/31 of the *Opticks*. This Rule reads as follows:

> In experimental philosophy we are to look upon propositions inferred by general induction from phenomena as accurately or very nearly true, notwithstanding any contrary hypotheses that may be imagined, till such time as other phenomena occur, by which they may either be made more accurate, or liable to exceptions.

This is clearly a methodological rule which, Newton adds, "we must follow, that the argument of induction may not be evaded by hypotheses." Newton in his *Universal Arithmetick*, a work significantly subtitled "A Treatise of Arithmetical Composition and Resolution," describes arithmetic as synthetic, since we "proceed from given Quantities to the Quantities sought," whereas algebra is analytic because it "proceeds in a retrograde order" assuming the quantities sought "as if they were given." And he comments that "after this Way the most difficult Problems are resolv'd, the Resolutions whereof would be sought in vain from only common Arithmetick."[6] These remarks cast light on what Newton had in mind when, in Query 31, he compares the method of investigation in natural philosophy with those in mathematics.

Method versus Logic

At this point, let us agree to an important distinction: that between logic and method. Both, of course, are concerned with how our mind should operate in order to arrive at reliable knowledge, but there are differences. The word *method* seems to have been a coinage of Plato; it first appears in his *Phaedrus*, where Socrates is advocating an art or technic of rhetoric as opposed to the devices of the Sophists. The word suggests a "path" or "route," being derived from *meta* and *odos*, indicating a movement according to a road.

To think or argue clearly and effectively it is necessary to understand the route along which we conduct our thoughts. He who has such a route, such a direction, possesses method. Logic and method are not the same thing. Logic is, of course, indispensable to method: it is the *inner machinery* conducting us along the path; it provides us with the tactics we employ. If method indicates the grand strategy, the road (or roads) we should follow, logic in turn provides the means of transportation, together with the *code de la route*.

Aristotle was not primarily concerned with the problem of method as the word is defined here. But there are passages in the *Organon* which testify that he believed there were two modes or directions of conducting our reasoning: the deductive or syllogistic and the inductive. Both lead to understanding, and understanding requires knowledge of the reason for the fact. But the deductive or syllogistic mode of demonstration assumes that we know the cause or principle from which the consequences can be drawn. In the jargon of the medieval philosophers, it is demonstration *propter quid* (demonstration *wherefore*), by which we start from what is prior in the order of nature and end up with what is "prior in the order of our knowing," that is, what is directly accessible to us. Some inquiries properly move in the opposite direction: from what is more knowable and obvious to us we proceed to those things that are "more knowable by nature." In the *Physics* Aristotle makes it clear that in this branch of philosophy we must follow this inductive path and advance from what is more obscure by nature, but clearer to us, towards what is more clear and more knowable by nature.[7]

There is, however, a passage in the *Nicomachean Ethics*[8] where Aristotle contrasts men who deliberate with those who analyze in geometry, and men who act upon those deliberations with those who engage in geometrical synthesis. This is perhaps the earliest explicit echo of those two *directional* ways of conducting thought in mathematics to which Newton referred in our chief text.[9] In all likelihood the two mathematical methods were known in the time of Plato and Eudoxus. Nevertheless the classical account is that given much later in the *Mathematical Collections* of Pappus, who attributes the elaboration of two methods to the work of Euclid, Apollonius of Perga, and Aristaeus the Elder.[10] In analysis, Pappus writes, the mathematician assumes what is sought as if it were true, and by a succession of operations arrives at something known to be true. Synthesis, on the other hand, reverses the process: the geometer starts with what is known (axioms, definitions, theorems previously proved) and by a series of deductive steps arrives at what he has set out to prove. This synthetic method is, of course, characteristic of the most familiar proofs of Euclid's geometry. The two methods may be thought of as alternative paths to be chosen according to the demands of a particular inquiry. Yet it appears likely that the analytic method is one used for purposes of investigation and discovery (as Aristotle implies), and that this is then followed, for formal demonstration, by the method of composition or synthesis which, unlike analysis, follows the "normal" direction of logical consequence.[11]

Pappus's text, unknown to the Middle Ages, was rediscovered in the Renaissance, and it gained currency especially through Commandino's Latin translation of 1589. The two procedures or methods in mathematics were obviously common knowledge by Newton's day. Algebra, in which the sixteenth and the seventeenth centuries made such notable progress, was seen to explore problems analytically, although quite differently from the "geometrical analysis of the ancients." It is in this sense that the two procedures are discussed by Newton's friend Edmond Halley in the preface to his edition of Apollonius. Algebraic analysis Halley describes as *brevissima simul perspicua* ("very short and clear"); synthesis, by contrast is *concinna et minime operosa* ("elegant and the least laborious").

There is—and it surely deserves mention—a text with which Newton must have been familiar. In his *Mathematical Lectures*, delivered in 1664–65, Newton's friend and teacher, Isaac Barrow, equates the mathematical and philosophical uses of these two terms. Barrow explains why, in enumerating the parts of mathematics, he is "wholly silent about that which is called Algebra or the Analytic Art."

> I answer, this was not done unadvisedly. Because indeed *Analysis* . . . seems to belong no more to *Mathematics* than to *Physics, Ethics* or any other Science. For this is only . . . a certain Manner of using Reason in the Solution of Questions, and the Invention or Probation of Conclusions, which is often made use of in all other Sciences. Wherefore it is not a Part or Species of, but rather an Instrument subservient to the Mathematics: No more is *Synthesis*, which is the manner of demonstrating Theorems in Contradiction to *Analysis*.[12]

The relation between mathematical procedures and general intellectual method had, we saw, been rather casually invoked by Aristotle. Yet he devoted most of his attention to elaborating his demonstrative logic and his theory of the syllogism. Early in our era—by the second and third centuries A.D.—philosophers and commentators began to use the mathematical terms in writing about method. For example Alexander of Aphrodisias, in his commentaries on Aristotle, mentions geometrical analysis as one of nine different senses of the word analysis used by philosophers.

A very important figure is surely Galen, for he brings the subject out of the realm of pure dialectic into the practical world of the physician. His concern is with method (or perhaps with methods) and with the proper way the doctor should learn about and teach his art. Galen, we know, wrote a major work called *Concerning Demonstration*, but it was subsequently lost. Almost certainly it did not deal with logic in the narrow sense, but with the problem of method, a problem that arose for him because of the different approaches to medicine of the chief rival schools into which his contemporaries and rival physicians were divided: the Empiric and the Dogmatic. What he sought was a middle way between those who relied wholly on accumulated experience, and those who based their procedures upon medical theory. In a treatise called *On Medical Experience* Galen wrote "The art of healing was originally invented and discovered by the logos [reason] in conjunction with experience. And to-day also it can only be practised excellently and done well by one who employs both of these methods."[13] But how are these methods to be employed?

References to method are scattered through Galen's major works; but his small work, the *Ars parva* (or *Microtegni*)—one of the first Greek medical writings to be made available in Latin—was the chief vehicle for transmitting Galen's thoughts on method. It was translated from an Arabic version into Latin by Constantine the African in the eleventh century, and later by Gerard of Cremona, who rendered it along with the remarks of its Arab commentator, Haly Rodohan. In this form it was printed as an early medical incunabulum. Galen's introductory paragraph is very short, yet it provided the basis for much subsequent discussion of method in medicine. Galen says there are three ways of teaching or demonstrating the art of medicine: these are by *analysis*, by *synthesis*, and by *definition*.[14] The use of the mathematicians' terms "analysis" and "synthesis" may have come to Galen from philosophical writings, yet it calls to mind his great admiration for the demonstrative procedures of the Greek geometers.

Perhaps too much has been made of Galen's methodology, but it should be emphasized that he is talking about methods of teaching, of leading the thought of the learner in teaching him medicine. There is nothing to suggest that the two procedures are to be used together, or that they are supplementary aspects of a single method. Rather, medicine can be taught analytically, by rising from the facts of observation (as, for example, in anatomy or pathology) to the principles or causes of health and disease. Or it can be presented in reverse fashion by starting with the principles—i.e., with medical theory—descending thence to the observed facts. But Haly in his preface identifies these two methods of teaching with the two directions of reasoning Aristotle presents in the *Posterior Analytics*: reasoning that moves from causes to effects, and that which proceeds from effects up to causes.[15] (The Latin translation gives *conversio* and *solutio* for "analysis" and *compositio* for "synthesis.")

The earliest text in the Latin West involving a discussion of method is related not to medicine but to philosophy. It is a passage in the commentary of Chalcidius, a fourth-century Christian Neo-Platonist, on Plato's *Timaeus*. Chalcidius is discussing the number and nature of the principles or elements of things; there is, he says, a *duplex probatio* for dealing with such matters, a double method of demonstration, the two parts of which are called *resolutio* and *compositio*, terms which correspond respectively to *analysis* and *synthesis*. *Resolutio* is a method of inquiry that begins with things sensible, prior in the order of understanding (that is, more known to us), from which we infer the principles of things, principles which are prior "in the order of nature." *Compositio* (or synthesis) is the method of syllogistic inference from the principles. The historian of science, Alistair Crombie, asserts that Chalcidius "defined the combined *resolutio-compositio* as the proper method of philosophical research."[16] But an examination of the Chalcidius text shows that the business is more complex: the two procedures, while in some sense supplementary, do not seem to form a single method, but are alternative methods. *Resolutio* is the method used in arriving at the material principles of things; *compositio* has a wider application: by means of it we demonstrate the formal relationships (*genera, qualitates, figuras*) from which we are led to grasp God's harmonious order and his providential role.[17]

Chalcidius's commentary was widely read in the early Middle Ages. For example we find the terms *resolutio* and *compositio* used by the ninth-century thinker John Scotus Erigena in his mystical *De divisione naturae*, where the aim is metaphysical understanding. Clearly, then, *resolutio* need not be an analysis of natural phenomena, but an analysis of thought.

The philosophers of the twelfth century had less interest in problems of method than in logic as they first found it in the old logic (*logica vetus*) of Aristotle. The rationalism, moreover, of an Anselm, a Gilbert de la Porrée, a Richard of St. Victor, even an Abelard, led them to deprecate the evidence of the senses as leading only to "opinion," not truth, and as unsuitable for handling the questions that really interested them. The only appropriate procedure was deduction from necessary and indemonstrable first principles. With the recovery of the later books of Aristotle's *Organon*, the *logica nova*, men of the thirteenth century focused upon syllogistic logic, and paid scant attention to the problem of method. With a single important exception, discussion of method was confined to the medical centers of Italy. And even this remarkable person was almost certainly influenced by what he knew of Galen. The exception, of course, is Robert Grosseteste, to whose role as scientific methodologist and scientist Professor Crombie has devoted a major book. For Grosseteste scientific knowledge is knowledge, as it was for Aristotle, of the causes of things, knowledge *propter quid*. The natural procedure in any science is to proceed from those particulars and whole objects known to us, directly but confusedly through the senses, up to the principles or causes, and then by a deductive chain to show the dependence of the particulars upon the principles or causes. But Grosseteste's method is primarily dialectical; its aim is the discovery of a *definition*, a generalized verbal characterization, and it is perhaps not surprising that he discusses composition before taking up resolution, for in certain sciences (notably mathematics) the synthetic or compositive method is all that seems to be needed in most cases. A different approach is needed in physics, which is uncertain, because, as Crombie paraphrases him, there can be only "probable knowledge of changeable natural things."[18] Causal definitions in physics could not be arrived at *a priori* (or *simpliciter*) like the axioms of geometry; they had to be reached by analysis or resolution of experimental objects, a process involving first, a dissecting of the object or phenomenon, and then, an inductive leap. But "the special merit" of Grosseteste's methodology, as Crombie points out, was to recognize that the induction is not probative, not a demonstration. What is necessary is verification or falsification of the principles or definitions arrived at by analysis (*resolutio*). The procedure of deducing the consequences of the definition, cause or principle, is of course the *compositio*: it serves to confirm (or falsify) the analytic results. Together both procedures—*resolutio* and *compositio*—constitute a single method. In the study of nature the two procedures must be used together.

Speculations on method never flourished in Paris or Grosseteste's Oxford, but attracted much attention in the universities of northern Italy, in the late fifteenth century and more especially in the sixteenth century. Two writers had a particularly great influence, Agostino Nifo (c. 1473–1545) and Jacopo Zabarella (1533–89). Both men, arguing much like Grosseteste, asserted that the object of a science is to discover the causes, the *propter quid*, of observed effects. To discover the causes, one must first proceed *a posteriori*, inferring causes from effects, i.e., using first the method of resolution or analysis; then the demonstrative or compositive method can be used to develop the consequences. The double procedure constitutes the method.

Ernst Cassirer was among the first to call attention to Zabarella, to bring him to light once more, and to see him as an influence on Galileo's scientific method, a position taken later, and even more strongly, by J. H. Randall, Jr.[19] But Neal Gilbert in a recent book, *Renaissance Concepts of Method*, casts doubt on this interpretation (see especially Chapter 7). As with Grosseteste, the emphasis of these Renaissance philosophers is on method for its own sake, on method as prescriptive for all areas of knowledge; the concern is not with its application to natural science alone but to all disciplines, metaphysical, moral, dialectical. The empirical element, as Gilbert has pointed out, is weak. While it is true that in analysis or resolution we pass from what is better known to what is more remote from us, the better known "experiences" may not be observations of scientific fact; they can as well be "clear and distinct ideas" resulting from the analysis of thoughts and concepts, and the principles or causes are verbal definitions. Even if this is somewhat unjust to Zabarella, there is an important difference between his method and that of Galileo, and perhaps an even greater gap between Galileo and Newton.

In the *probatio duplex*—the double method of Grosseteste and Zabarella— the really probative element is supplied by the synthetic, deductive arm. The analytic or resolutive procedure is merely suggestive or conjectural. This, it would appear, is also characteristic of Galileo, but with him the empirical element which Gilbert finds lacking in Zabarella is, of course, much more important.

In many passages, notably in the *Letter to the Grand Duchess Christina*, Galileo, in phrases that are reminiscent of Galen's injunctions, insists that the proper approach to natural philosophy is to employ jointly "manifest experiences and necessary proofs"; "direct experience and necessary demonstrations"; "experiments, long observation, and rigorous demonstration."[20] In such phrases he seems to be implying a double method of resolution and composition, of analysis and synthesis. But how does one carry this out?

The Third Day of the *Discorsi* throws light on the matter. Galileo opens with the famous statement of purpose: that he intends "to set forth a very new science dealing with a very ancient subject," the subject of motion in nature. Concerning this, he remarks, "I have discovered some properties of it which . . . have not hitherto been either *observed* or *demonstrated*."[21]

After several pages describing the kinematics of uniform motion, Galileo enters upon the subject of accelerated motion:

> And first of all it seems desirable to find and explain a definition best fitting natural phenomena. For anyone may invent an arbitrary type of motion and discuss its properties . . . but we have decided to consider the phenomenon of bodies falling with an acceleration such as actually occurs in nature and to make this definition of accelerated motion exhibit the essential features of observed accelerated motions.[22]

This suggests, if his earlier statement about discovering his results did not satisfy us, that he has observed and perhaps crudely determined the acceleration of falling bodies in order to arrive at his definition. Thus at least a crude *analysis* of experience led him to the rule nature might follow, led "by the hand, as it were, in following the habit and custom of nature herself, in all her various other processes." "And this, at last," he says in the same paragraph, ". . . after repeated efforts we trust we have succeeded in doing. In this belief we are confirmed mainly by the consideration that experimental results are seen to agree with and exactly correspond with those properties which have been, one after another, demonstrated by us."[23] Such an experimental confirmation completes the synthesis, or compositional phase, of Galileo's double method. It is notable that the language used (as later with Newton) is that of mathematics: kinematic descriptions, measurable and representable (as his pages show) by numbers and geometry. Galileo's "definitions" are mathematically symbolized "laws."

A passage in the *Dialogue on the Two Great World Systems* seems to confirm our inference; it contains, also, an explicit reference to the method of resolution:

> *Simplicio*   Aristotle first laid the basis of his argument *a priori*, showing the necessity of the inalterability of heaven by means of natural, evident, and clear principles. He afterwards supported the same *a posteriori*, by the senses and by the traditions of the ancients.
> *Salviati*   What you refer to is the method he uses in writing his doctrine, but I do not believe it to be that with which he investigated it. Rather, I think it certain that he first obtained it by means of the senses, experiments, and observations, to assure himself as much as possible of his conclusions. *Afterward* he sought means to make them demonstrable. This is what is done for the most part in the demonstrative sciences; this comes about because when the conclusion is true, one may by making use of the analytical methods [*methodo resolutivo*] hit upon some proposition which is already demonstrated, or arrive at some axiomatic principle. . . . And you may be sure that Pythagoras, long before he discovered the proof for which he sacrificed a hecatomb, was sure that the square on the side opposite the right angle of a right triangle was equal to

the squares on the other two sides. *The certainty of a conclusion assists not a little in the discovery of its proof.*[24]

What Galileo seems to be saying is that exploratory methods, based on trial and experiment, can lead to a degree of certainty, or probability, in the conclusion. This conclusion can then be *demonstrated,* either because (as in the mathematical analysis) it leads to something already known, or because it leads to something that can be tested experimentally. A point worth emphasizing is the stress that Galileo places on the *analytic* or *resolutive* procedure as strongly suggestive, though falling far short of probative demonstration. The analytic procedure, as he makes clear, is the method of discovery (in natural philosophy the only method of discovery or invention); the synthetic procedure rounds out the process, and is the method of final demonstration and formal presentation. In the seventeenth century the problem of method, as distinct from logic, became of paramount concern. Indeed—as the Kneales point out in their *Development of Logic* (1962)—this led to a neglect or an impoverishment of logical studies in this century. The new concern, an attempt to formulate a doctrine of method in natural science, scientific method, is first encountered in Francis Bacon.

Bacon has suffered at the hands of many historians of science and of philosophy, and he has often been grossly misinterpreted. His self-appointed role was to stress experience and experiment, and to do so with all the rich resources of rhetoric. His aim, as he put it, is to restore "the commerce between the mind of men and the nature of things." In the study of nature we cannot succeed if we rely excessively or exclusively on the human reason, "if we arrogantly searched for the sciences in the narrow cells of the human understanding, and not submissively in the wider world."[25] But any careful reading of Bacon reveals that his goal is the discovery of axioms and principles from which a demonstrative science can be constructed. In any case, these axioms and principles should not be ad hoc or gratuitous; they should not be "hypotheses" in Newton's pejorative sense: they must somehow be rooted in, derived from, Nature herself. What Bacon wrestles with, if not too successfully, is the problem of induction, in other words the problem of increasing the probative value of the analytical arm of the double method; since the synthetic arm had been thoroughly investigated from Aristotle to his own time, it could be momentarily left aside. To arrive at axioms we must learn how to analyze and dissect nature, *dissecare naturam*: "Now what the sciences stand in need of is a form of induction which shall analyse experience and take it to pieces, and by a due process of exclusion and rejection lead to an inevitable conclusion [that is, to axioms and causes]."[26] In his effort to strengthen the upward procedure perhaps Bacon helped to distract attention away from the *double* method. Interest in the double method, and an appreciation of its power as a scientific instrument, waned in mid-century. But for this Descartes is perhaps as much at fault as Bacon; at his hands the "method" is distorted in a very interesting way. It is Newton who has the honor of restoring and sharpening it as a tool of what he called "Experimental Philosophy."

It has been claimed that the double method of analysis and synthesis, of resolution and composition, is the central feature of Descartes's famous method. As everyone knows, his doctrine of method is set forth in the readable, but here and there oddly cryptic, *Discourse on Method* (1637). The method is summed up in the famous four rules which Descartes introduces in Part II of his book, but is more completely set forth in the twenty-one rules of his post-humous *Regulae ad directionem ingenii.*[27]

The *Discourse*, as we commonly encounter it, was only a preface, an introduction, to those illustrations of the "method" which were published in the original book, and which have ever since been generally omitted from modern editions: the *Dioptrique*, the *Météores*, the *Géométrie*, intended to-gether to illustrate the range of application of the method. When one examines the two physical essays as examples of Descartes's method, there is certainly no trace of a double way, a *probatio duplex*. The *Dioptrique* begins with a dis-cussion of the nature of light and the phenomena of refraction, presented in synthetic fashion. The *Météores* is an even better example of hypothetical reasoning, taking its departure from a purely conjectural picture of the shapes of particles.

It would seem that for Descartes analysis and synthesis are simply two alternative directions in which one can conduct one's thoughts in orderly fashion. Analysis, to be sure, is the road to first principles, leading to the clear and distinct ideas; but it is the analysis of concepts, of thought, not the analysis of sense experience or experiments. In either direction one follows long chains of reasoning in which the validity of each step involves the spontaneous opera-tion of the *vis cognoscens*, the power of the mind to grasp directly the "simple natures," the "atoms of evidence" which are the links of the chain. This power, or rather the action of the mind at each of these elementary steps, is what Descartes calls *intuitus*.

All reasoning, for Descartes, is thus a series of intuitive steps. And what men need, instead of the rules of formal logic which may be dispensed with, are the practical injunctions of his Four Rules. Method, for Descartes, is merely *order* in thought, order that will permit the natural intellect to operate unim-peded. This order can be ensured by observing the simple rules of intellectual behavior which Leibniz found so absurdly obvious yet so vague. Descartes's famous rules are perhaps best described as propaedeutic, or even as prophylactic, injunctions. When they are scrupulously observed, the power of the mind, the *vis cognoscens*, operates reliably and surely. There seems, accordingly, to be little justification for finding in Descartes's method the dual procedure we have been describing. The truth value does not come from the mutual support of the two limbs of a dual method, but from perceiving clear, distinct, and irrefutable ideas.

Whether or not it is adequate to present Descartes's "method" in this fashion, one thing at least is certain: those of his followers who discuss *analysis* and *synthesis* clearly see these as *two* sorts of method, not as jointly constituting, when used one after the other, a single method. Arnauld and Nicole, the authors of the book called, in its English versions, the *Art of Thinking*, a work first published in French in 1662 and often called the *Port Royal Logic*, write as follows: "We distinguish two kinds of method: the one for the discovery of truth is called *analysis* or the *method of resolution* or the *method of invention*; the second, used to make others understand the truth, is called *synthesis* or the *method of composition* or the *method of instruction*."[28] Analysis, they remark farther on, is used "to investigate a specific thing rather than to investigate more general things as is done in the method of instruction [i.e., composition]." And they add that this analysis "consists more of discernment and acumen than of particular procedures," a statement that reminds us not only of Descartes but also of Bacon's remark that analysis by experiment, which he calls the Chase of Pan, is really a kind of sagacity. And the Port Royal logicians bluntly state that "the more important of the two methods" is the method of composition "in that composition is used for explanation in many disciplines."[29]

A similar distinction is made by Pierre-Sylvain Régis, a Cartesian natural philosopher. In his *Système de philosophie*, published in 1690, Régis speaks of *two* methods: "of which one serves to instruct ourselves and is called *analysis*, or the method of division, and the other which is used to instruct others is called *synthesis*, or the method of composition."[30]

A much earlier work than those just discussed, Robert Sanderson's *Logicae artis compendium*, published in Oxford in 1615 and frequently reprinted in the seventeenth century, was written to restore Aristotle's prestige and make some compromises with the popular doctrine of Ramus.[31] It is of particular interest since we know that Newton studied it either just before, or soon after, he entered Cambridge. Written well before the Cartesian works we have mentioned, the book had a clear exposition of a theory of method that involved two contrasting procedures, one which he termed the "method of invention," the other the "method of doctrine":

> Each proceeds from that which is more known by us to that which by us is less known. . . . For we discover precepts by ascending, that is, by progressing from the concrete and the particular, which to us is more directly known, towards the intelligible and universal, which are more known by nature. But we transmit precepts by descending, that is, by progressing from the universal and intelligible, which are more known by nature, and more clearly known by us also, to that which is less universal, and closer to the senses, and as it were less known.[32]

The same ideas are expressed in W. J. 'sGravesande's *Introductio ad philosophiam, metaphysicam et logicam continens* (Leiden, 1736).[33] Book II is devoted to logic, and the third part of this is called "On Method." The opening words are as follows:

It now remains to indicate the route that the person . . . should follow to reach a true understanding of the things he has set out to examine.

The method should be different, according to the different circumstances.

First I shall treat the method for discovering truth, and then the method that we use to explain to others that which we know.

The first method is called *analytic*, or the method of resolution; the other is *synthesis* or the method of composition.

The general difference between the two methods consists in this: that in the first method one passes from the complex to the simple by resolution; and in the second one goes from the simple to the compounded.[34]

We see how different from Newton's these statements are, which is curious when we recall that 'sGravesande is chiefly remembered for his exposition of the Newtonian philosophy. In his pages on method the Dutch scientist owes much, it would seem, to the later Cartesians and perhaps more immediately to the *Port Royal Logic*.

To return to Sanderson's *Compendium*: especially worth noting is his treatment of the second of his alternative methods, his "method of invention," which should have caught the eye of men like the young Newton:

The method of invention has four means, and as it were four stages through which we ascend. First is perception, by the help of which we assemble some notion of individual things. Second is observation or seeing accurately, in the course of which we collect and arrange what we have assimilated at different times by perception. Third is proof by experiment, wherein we subject the multitude of assembled observations to fixed tests. Fourth and last is induction, in which we summon the multitude of collected and tested proofs so as to make up a universal conclusion.[35]

Newton's Scientific Method

Before trying to assess Newton's method of analysis and synthesis, comparing it with the twofold scheme so long and so variously elaborated by his predecessors, it might be well to consider a longer and more relaxed exposition that Newton never published, and which is closely related to the famous methodological section of Query 23/31 cited at the beginning of this article:

As Mathematicians have two Methods of doing things wch they call Composition & Resolution & in all difficulties have recourse to their method of resolution before they compound so in explaining the Phaenomena of nature the like methods are to be used & he that expects success must resolve before he compounds. For the explications of Phaenomena are Problems much harder then [*sic*] those in Mathematics. The method of Resolution consists in trying experiments & considering all the Phaenomena of nature relating to the subject in hand & drawing conclusions from them & examining the truth of those conclusions by new experiments & drawing new conclusions (if it may be) from those experiments & so proceeding alternately from experiments to conclusions & from conclusions to experiments untill you come to the general properties of things. Then assuming those properties as Principles of Philosophy you may by them explain the causes of such Phaenomena as follow from them: wch is the method of Composition. But if without deriving the properties of things from Phaenomena you feign Hypotheses & think by them to explain all nature, you may make a plausible systeme of Philosophy for getting your self a name, but your systeme will be little better than a Romance. To explain all nature is too difficult a task for any one man or even for any one age. . . . Tis much better to do a little with certainty & leave the rest for others that come after you then [sic] to explain all things by conjecture without making sure of any thing.[36]

Others before Newton used the word "romance" to describe fanciful hypotheses. See, for example, Henry Power in his *Experimental Philosophy* (1664), who speaks on page 186 of those "that daily stuff our Libraries with their Philosophical Romances."

In contrast to Descartes, the logicians of Port Royal, and 'sGravesande, Newton sees the two methods as constituting a single procedure, in which one begins by analysis or resolution, and follows this by a synthetic demonstration. Formally, this is the double way, the *probatio duplex*, of Grosseteste, Nifo, Zabarella, and the other early methodologists. Unlike them, however, Newton— like Galileo—would have us analyze not so much our ideas about things as the *phenomena*. But in turn Newton's double method differs from that of Galileo in a subtle but important way. With Galileo, as we saw, the analysis by experiment and observations is merely suggestive or indicative. The real cogency of the method depends on the demonstration: on synthesis or mathematical deduction. With Newton, however, the stress is on the *analysis* which "consists," as he says in the *Opticks*, "in making experiments and observations and in drawing general Conclusions from them by Induction." For Newton the analytic procedure is independently probative, although falling short of strict demonstration. Indeed (like Bacon before him) he feels it necessary to stress this analytic procedure, as he does in the *Opticks* by devoting more space to it than to the synthetic arm. Like Descartes and the Port Royal Logicians, he too sees analysis as the true method of discovery, of "invention." We must, he wrote, admit of "no Objections against the Conclusions, but such as are taken from Experiments, or other certain Truths. For hypotheses are not to be regarded in Experimental Philosophy." Although this experimental and inductive process does not lead to demonstration, "yet it is the best way of arguing which the Nature of Things admits of."[37]

If the force of Newton's dual method does not wholly depend (as it seems to have mainly done with Galileo) upon the synthetic procedure, what does this deductive limb of the double method actually contribute? Newton does not restrict it to a confirmatory role; still less does he limit its use to presenting or teaching what has already been discovered. The deductive limb can also be a means of prediction and discovery for, as he points out in the draft from which we have quoted above, one can deduce unexpected consequences. Having discovered "from Phaenomena" the inverse square law of universal gravitational force, and then using this force as a Principle of Philosophy, he writes (in the same manuscript cited above):

> I derived from it all the motions of the heavenly bodies & the flux & reflux of the sea, shewing by mathematical demonstrations that this force alone was sufficient to produce all those Phaenomena, & deriving from it (a priori) some new motions wch Astronomers had not then observed but since appeare to be true, as that Saturn & Jupiter draw one another, that the variation of the Moon is bigger in winter then in summer, that there is an equation of the Moons meane motion amounting to almost 5 minutes wch depends upon the position of her Apoge to the Sun.

The later history of science has again and again confirmed the power of a well-founded theory to predict new phenomena, and to explicate other facts which had not been considered when the theory was elaborated.

In Query 31, immediately after describing his method, Newton tells us how the two procedures are exemplified in the foregoing books of the *Opticks*. In the greater part of the First Book, Newton sets forth his classic experiments showing that light is a heterogeneous mixture of rays of different refrangibility, and that rays of different refrangibility differ also in color. Although in this book Newton affects a kind of axiomatic presentation beginning with definitions and axioms, and enunciating a series of propositions, the procedures are really analytic in his sense, as he tells us they are: the propositions are not abstract mathematical statements, but affirmations of physical or experimental fact, and they are justified, not by mathematical deduction, but by what he calls "proofs by experiment." These discoveries being proved, Newton writes, "they may be assumed in the Method of Composition for explaining the Phaenomena arising from them." An example "of which Method I gave at the end of the first Book." He does not specify what propositions he means, but it is clear that he is referring us to those propositions he designates as "problems" rather than as "theorems" and which we encounter in Book I, Part II (1730 ed., pp. 161–85): to explain the colors produced by a prism (Prop. VIII. Prob. III); to elucidate the colors of the rainbow; (Prop. IX. Prob. IV); and to explain the permanent colors of natural bodies (Prop. X. Prob. V).

Newton and Experiment

At first sight it may seem odd to find Newton equating the analytic procedure with experimentation. Yet if we think about it for a moment, Newton's reasons are quite clear. A convincing and well-designed experiment involves a sort of dissection or analysis of nature, an isolation of the phenomenon to be examined, and the elimination of disturbing factors. As Lavoisier wrote long after Newton's time: "One of the principles one should never lose sight of in the art of conducting experiments is to simplify them as much as possible, and to exclude [*écarter*] from them all the circumstances that can complicate their results."[38] Experiment, indeed, is usually necessary to determine which factors can safely be eliminated or at least must be held constant, and which are those that primarily determine the phenomenon. As W. Stanley Jevons wrote: "The great method of experiment consists in removing one at a time, each of those conditions which may be imagined to have an influence on the result."[39] Physical nature does not readily reveal its secrets to the phenomenologist, but only to those who analyze.

Newton, in any event, profoundly altered that conception of experiment which Bacon had advocated and which in his spirit was accepted by so many *virtuosi* of the early Royal Society. Newton's Experimental Philosophy is not what Thomas Sprat or Henry Power, or even Robert Boyle, called by that name. Newton would not have agreed that experiment merely serves to render plausible the great sweeping "hypotheses" of the mechanical philosophers. Nor, at the other extreme, could he have agreed with Samuel Parker that probably "we must at last rest satisfied with true and exact Histories of Nature,"[40] or with Locke, who argued that improving knowledge of substances by "experiences and history . . . is all that the weakness of our faculties in this state of mediocrity which we are in in this world can attain to."[41]

On the title page of his *Experimental Philosophy* (1664), a miscellany of microscopic observations and experiments with the Torricellian tube and with the magnet, Henry Power described them as providing "some Deductions, and Probable Hypotheses . . . in Avouchment and Illustration of the now famous Atomical Hypothesis." With greater experimental gifts and a richer scientific imagination, Robert Boyle can be said to have guided his own investigation in the same spirit.

For Newton, on the other hand, experiment is essentially a device for problem solving, for determining with precision the properties of things, and rising from these carefully observed "effects" to the "causes." More clearly than Bacon was able to do, Newton showed by his method that experimentation could lead with at least "moral certainty" to axioms, principles, or laws. In two ways Newton's method must be distinguished from that of the majority of his predecessors and nearly all his contemporaries. He insisted upon the cogency of a single, well-contrived experiment to answer a specific question, as opposed to the Baconian procedure of collecting and comparing innumerable "instances" of a phenomenon. Perhaps even more significant, Newton's experiments, whenever it is possible, are quantitative.

Robert Hooke, to be sure, was fully capable of designing and carrying out experiments to test a conjecture or working hypothesis. This he did in his "Noble Experiment" in which, by dissecting away the diaphragm of a dog and blowing air through the immobilized lungs, he showed that the animal could be kept alive, and in this way verified "my own *Hypothesis* of this Matter," namely that it is the air passing into the blood, not the motion of the lungs, that was necessary for life. Yet in his dispute with Newton over the latter's first paper on light and color Hooke's arguments are often Baconian.

He argues that Newton's famous prismatic experiment, what Newton called his *experimentum crucis*, being a single isolated experiment, is unpersuasive, compared to "all the experiments and observations," and the "many hundreds of trials" he (Hooke) had made.[42]

But it was upon this lone experiment, Newton replied, that "I chose to lay the whole stress of my discourse." By the Baconian term *experimentum crucis*—a phrase he borrowed from Hooke's *Micrographia* (1665)—Newton means an experiment designed to decide between two alternative outcomes, in other words an experiment designed (like Hooke's "Noble Experiment") to answer a clearly formulated question posed to Nature. In adopting this point of view, Newton of course had distinguished forerunners in Galileo, William Gilbert, William Harvey, and Blaise Pascal, among others. But it is interesting to cite some words of Isaac Barrow, a man to whom Newton was greatly indebted:

> The Truth of Principles [Barrow wrote] does not solely depend on *Induction*, or a perpetual Observation of Particulars, as *Aristotle* seems to have thought; since only one Experiment will suffice (provided it be sufficiently clear and indubitable) to establish a true Hypothesis, to form a true Definition; and consequently to constitute true Principles. I own the Perfection of Sense is in some Measure required to establish the Truth of Hypotheses, but the Universality or Frequency of Observation is not so.[43]

Newton's Mathematical Way

For his notion that scientific investigation should consist in the solving of discrete, well-defined problems Newton surely owed much, as the name of Isaac Barrow suggests, to the mathematical tradition, for that is how mathematicians of necessity proceed. We should remember that among students of physical nature were men like Galileo, Torricelli, and Pascal—all of them mathematicians more than natural philosophers—who pointed the way and demonstrated by their achievements that this modest, piecemeal approach was the most fruitful way of studying not only mathematical problems but also nature. Few if any of these men would have described what they were doing as "physics," for in the seventeenth century "physics" meant natural philosophy, which was sharply set apart from mathematics, as it had been since the time of Aristotle. Subjects like optics, mechanics, music (acoustics and harmonics), which we now consider branches of physics, were described as belonging to the "mixed or concrete mathematics."[44] They were subjects that treated mathematically things perceived by the senses, whereas pure mathematics dealt only with things "conceived by the mind." But they were not parts of physics.[45]

Physics in Newton's day was exemplified by those all-embracing, all-encompassing systems of nature devised by the so-called mechanical philosophers: Pierre Gassendi, Thomas Hobbes, Descartes, and their lesser followers. These men all shared the view—in opposition to Aristotle with his "substantial forms" and "occult qualities" and to Paracelsus with his spiritual agencies—that the underlying principles of physical nature were to be found in matter and its motions, and they built their different systems on an all-embracing mechanism. Common to all their systems, despite their rejection of Aristotle, was the conviction that the purpose of physics—a science of nature in a sense that is almost Aristotle's—was to explain the visible world in terms of particulate matter: the sizes, shapes, motions, and mechanical interaction of invisible particles, or what Francis Bacon had called "the secret motions of things." Physics, a branch of philosophy, was a dialectical science that imparted knowledge, derived from "first principles," about the whole material universe. As Descartes put it, physics was that second branch of philosophy (the first being metaphysics) "in which, after having found the true principles of material things, one examines in general how the whole universe is composed; then in particular what is the nature of this earth and of all the bodies that are commonly found around her, like air, water, fire, the magnet and other minerals" (*Oeuvres*, IX, 14). Even more self-confident and succinct is the definition of Descartes's disciple, Jacques Rohault. Physics, he wrote in 1671, is the science "that teaches us the reasons and causes of all the effects that nature produces."[46]

The logical model for the builders of these systems was mathematics, and their method of presentation was, in general, synthetic and deductive. Yet the language and syntax are verbal, not mathematical. Mathematizable in principle, Descartes's *Principles of Philosophy* has no trace of mathematics. Indeed because of this widely accepted separation between the disciplines of mathematics and physics, a mathematical physics appeared to most men to be a contradiction in terms. But there are important exceptions.

One of the earliest is Galileo, who wrote in his *Saggiatore* a passage that has often been quoted:

> Philosophy is written in that great book which ever lies before our eyes—I mean the universe—but we cannot understand it if we do not first learn the language, and grasp the symbols, in which it is written. This book is written in the mathematical language, and the symbols are triangles, circles, and other geometrical figures, without whose help it is impossible to comprehend a single word of it; without which one wanders in vain through a dark labyrinth.[47]

Even more eloquent in opposing the conventional split between mathematics and natural philosophy was Isaac Barrow, one of that small number of Englishmen who had mastered Galileo's work, and from whom, in all likelihood, Newton was led to understand the thought and achievement of the great Italian scientist. Barrow, in discussing those "Sciences termed *Mixed Mathematics*," commented:

> I suppose they ought all to be taken as Parts of *Natural Science*, being the same in Number with the Branches of *Physics*. . . . For these mixed Sciences are stiled Mathematical for no other Reason, but because the Consideration of

Quantity intervenes with them, and because they require Conclusions to be demonstrated in Geometry, applying them to their own particular Matter. And, according to the same Reason, there is no Branch of natural Science that may not arrogate the Title to itself; since there is really none, from which the Consideration of Quantity is wholly excluded, and consequently to which some Light or Assistance may not be fetched from Geometry.

And he goes on, in what is almost a paraphrase of the famous Galilean passage:

> For Magnitude is the common Affection of all physical Things, it is interwoven in the Nature of Bodies, blended with all corporeal Accidents, and well nigh bears the principal Part in the Production of every natural Effect.[48]

Elsewhere Barrow wrote that no one can expect to understand or unlock the hidden meanings of nature without the "Help of a Mathematical Key":

> For who can play well on *Aristotle's* Instrument but with a Mathematical Quill; or not be altogether deaf to the Lessons of Natural *Philosophy*, while ignorant of *Geometry?*[49]

We need hardly stress the essentially mathematical character of Newton's major work, *The Mathematical Principles of Natural Philosophy*. If a glance at the book were not enough to convince us, Newton makes sure that we understand what he is about, and how he has bridged the gulf between mathematics and physics. In his Preface he writes that like "the moderns" he has in his treatise "cultivated mathematics as far as it relates to philosophy" and offers his book "as the mathematical principles of philosophy, for the whole burden of philosophy seems to consist in this—from the phenomena of motions to investigate the forces of nature, and then from these forces to demonstrate the other phenomena."[50]

By "philosophy" Newton means, of course, natural philosophy, or "physics." And he seems in this passage to refer to the traditional distinction between mathematics and physics. Yet he makes clear that these "principles"— the laws and conditions of certain motions and of powers or forces—are the things "we may build our reasonings upon in philosophical inquiries." One passes indeed without difficulty from one domain to the other:

> In mathematics we are to investigate the quantities of forces with their proportions consequent upon any conditions supposed; then, *when we enter into physics*, we compare those proportions with the phenomena of Nature.[51]

One question immediately confronts us: did Newton conceive of the famous method—his double procedure of analysis and synthesis set forth and exemplified in the *Opticks*, and where the analytic arm is identified with experiment and observation—as applying equally well to the *Principia*? This has recently been denied, yet the answer is surely in the affirmative. To be sure, the two works offer a striking contrast; they treat not only different aspects of nature, but at first glance seem to treat them in different ways. The *Principia* is, at least in the first two books, a work of abstract rational mechanics, strictly mathematical and presented in axiomatic fashion. A chain of propositions treats of mass points or idealized spherical bodies subject to certain imagined forces. Yet even when the results are applied to "physics"—to the real bodies of the solar system—these are viewed as bodies qualitatively similar, deprived of what John Locke would have called their "secondary qualities" and differing only in such quantifiable properties as mass, extension, impenetrability, and state of rest or motion.

Abstract and mathematical though it appears throughout, the *Principia* was deemed by Newton to be as firmly rooted in observation and experiment as the *Opticks*. In the *Scholium* to the axioms or laws of motion Newton wrote: "I have laid down such principles as have been received by mathematicians, and are confirmed by abundance of experiments" (*Principia*, 21). This, he felt, need not be insisted upon for the first two laws of motion. But his third law, the law of equality of action and reaction, he saw to be a novel assertion requiring further justification. To this end he invoked at some length the experiments on elastic impact carried out some years before independently by Christopher Wren, John Wallis, and Christiaan Huygens; and he concludes that "so far as it regards percussions and reflections [the third law] is proved by a theory exactly agreeing with experience" (*Principia*, 25). To show further that the law can be extended to attractions, he cites an experiment he has made on the mutual attraction of a lodestone and iron. And elsewhere throughout the work we find scholia serving the same purpose of supporting important propositions by experimental evidence. Many years later, in the unpublished discussion of his method of analysis and composition (the first part of which was quoted above) he wrote:

> Thus in the Mathematical Principles of Philosophy I *first shewed from Phae-nomena* that all bodies endeavoured by a certain force proportional to their matter to approach one another, that this force in receding from that body grows less & less in reciprocal proportion to the square of the distance from it & that it is equal to gravity & therefore is one and the same force with gravity.[52]

Having tried to persuade us that the famous principle and law of universal gravitation was discovered through analysis, he describes in the passage quoted earlier his subsequent use of the synthetic method.

Newton's *Opticks*, by contrast, deals with the "secondary qualities" of things: chiefly color and—if we take the famous Queries into consideration—those attributes which differentiate various kinds of bodies: chemical behavior, phenomena associated with heat, and such physical properties as cohesion, surface tension, and capillary rise.[53]

Yet it is wrong to insist, as one scholar has done, that there are two kinds of Newtonianism: the mathematical Newtonianism of the *Principia* and the "experimental Newtonianism" of the *Opticks*. To remove any reasonable doubt as to what Newton himself thought, we may quote from an anonymous review what are generally acknowledged to be his own words: "The Philosophy which

Mr. *Newton* in his *Principles* and *Optiques* has pursued is Experimental; and it is not the Business of Experimental Philosophy to teach the Causes of things any further than they can be proved by Experiments."[54] The *Opticks*, unlike the *Principia*, consists largely of a meticulous account of experiments. Yet it can hardly be called nonmathematical, although little more than some simple geometry and arithmetic is needed to understand it. In spirit it is as good an example of Newton's "mathematical way" as the *Principia*: light is treated as a mathematical entity, as *rays* that can be represented by lines; the axioms with which he begins are the accepted laws of optics; and numbers—the different refrangibilities—serve as precise tags to distinguish the rays of different colors and to compare their behavior in reflection, refraction, and diffraction. Wherever appropriate, and this is most of the time, his language of experimental description is the language of number and measure. It is this which gives Newton's experiments their particular cogency.

Observation is not merely looking and seeing; it is a kind of reporting. We report to ourselves or to others some aspect of an object that has aroused our interest. In this broad sense a painting or a poem is a kind of report; some aspect of visual or auditory or tactile experience is singled out from the flux of nature to be attended to. But not all reports, as we know to our sorrow, are really observations. Any observation deserving the name, certainly any observation we might call "scientific," involves a comparison with something else. And the most precise and unambiguous comparisons are those expressed in the language of number and measure. When we measure we do not simply contrast two objects with one another. We do not just report that object A appears bigger, heavier, brighter, or faster than object B; we report *how much* they differ from each other. What is required is some way of attaching a more precise meaning to "bigger," "heavier," and so on. This we do by comparing both objects with some standard. Just as in counting we compare a set of objects with that abstract standard or scale we call the system of natural numbers, so when we *measure* we physically compare the objects at hand with a unit or standard of measure, which in turn involves comparing both the object and the standard with our abstract numerical scale. When we perform this operation of comparison, using the language of numbers—that is, when we measure—we are reporting this relationship of the objects as *ratios*. This, indeed, is the meaning that Newton attaches to the word "measure."

Newton interprets the numbers themselves as ratios or measures. Thus he writes: "By *Number* we understand not so much a Multitude of Unities, as the abstracted Ratio of any Quantity, to another Quantity of the same Kind, which we take for Unity. And this is threefold; integer, fracted, and surd: An *Integer* is what is measured by Unity, a *Fraction*, that which a submultiple Part of Unity measures, and a *Surd*, to which Unity is incommensurable."[55]

An experiment is, of course, only a contrived observation, and all the advantages of precision and lack of ambiguity that accrue to observations by being cast in the language of number and measure must necessarily be found in what we call "quantitative experiments," which Newton's almost always are. There is no better instance of Newton's quantitative approach to his experiments than the following undated manuscript page describing things "To be tryed" to elucidate the phenomenon of diffraction, that is the bending of light, and the production of colored fringes (*fasciae* to Newton), when light passes through a tiny hole or past a knife edge:

1. *What are the numbers limits and dimensions* of the shadow & fasciae of a hair illuminated from a point at several distances.

2. *How far* a hair in the edge of light casts light into the shadow surrounding the light.

3. Whether in the approach of a hair to the shaddow the fasciae *encreas* & wch fasciae vanish first.

4. *How many* fasciae can be seen through a Prism.

5. *At what distances* each fascia begins to appear.

6. *What alteration* is made by the bluntness & shapness [*sic*] of the edge or by the density of the matter.

7. *How much* the shadow of a pin or slender wiar is broader then that of a hair.

8. Whether one hair behind another makes a broader shadow & *how much.*

9. *At what distances* from one another two hairs, two backs or edges of knives or raisors, two wiars or pins & two larger iron cilinders make their fasciae meet.

How the same or other bodies make their fasciae go into one anothers shadows.

In what order the fasciae begin to appear or disappear *increase or decreas* [*sic*] in going into or out of any well defined shadow.[56]

From a series of observations men habitually are impelled to generalize. To generalize is to report and sum up in some tidy way the results of a series of comparisons. The pitfalls of ordinary language compound the dangers of the generalizations we make in everyday life. But even the murky business of generalizing—of making an inductive inference—gains precision through the use of numbers, of mathematical rather than verbal language. The end product is a mathematically expressed "rule," or "law," or—to employ a favorite word of Newton's day—a "principle." Thus Newton opens the *Opticks* with what he called "Axioms," which are simply the well-established laws of geometrical optics. When, on the other hand, he enunciates a law that he has himself discovered, a generalization that he has reached by an inductive inference and which is quantitatively expressed, he often employs the word *rule*. Thus after reporting a series of detailed measurements on the colored rings produced when light passes through thin, transparent bodies, he concludes: "And from these Measures, I seem to gather this Rule: That the thickness of the Air is proportional to the secant of an angle, whose Sine is a certain mean proportional between the Sines of Incidence and Refraction."[57]

Clearly Newton's extreme confidence in his Method of Analysis, in the probative power of inductive inferences from his experiments, depends not a little on the fact that the ascending chain of comparisons by which he reaches these "rules" or "laws" is expressed in the language of mathematics. This use of number, one hardly needs to add, strengthens the deductive synthetic limb of his double method, for the syntax of mathematical demonstration is at his disposal, in pursuing the downward path from "principles" and "laws" back to the phenomena. It is a syntax well understood and devoid of the ambiguities and traps of purely verbal deduction. E. W. Strong, in his "Newton's Mathematical Way" (in *Roots of Scientific Thought*, ed. Wiener and Noland), summed the matter up when he wrote: "Newton's 'mathematical way' encompasses both experimental investigation and demonstration from principles, that is, from laws or theorems established through investigation" (p. 413), and this procedure "requires measures for the formulation of principles in optics and mechanics— principles that incorporate a rule of measure. Were there not mathematical determinations in the experiment, there would be no subsequent determination in the demonstration" (p. 421).

Notes

1. P. S. de Laplace, *Exposition du système du monde*, 6th ed. (Paris, 1835), 430–31.
2. Sir Isaac Newton, *Mathematical Principles of Natural Philosophy*, ed. F. Cajori (Berkeley, Calif.: University of California Press, 1934) xx–xxi. My italics.
3. Colin Maclaurin, *Account of Sir Isaac Newton's Philosophical Discoveries*, 3d ed. (London, 1775), 9.
4. *Opticks*, 4th ed. (London, 1730), 369.
5. *Ibid.*, 404–5.
6. *Universal Arithmetick* (London, 1728), 1.
7. See *Physics*, I, 1, 184a–184b, 10.
8. Bk. III, chap. 1, 1112a–1113a, 14.
9. Thomas L. Heath, *Mathematics in Aristotle* (Oxford, 1949), 270–72.
10. M. R. Cohen and I. Drabkin, eds., *A Source Book in Greek Science* (New York: McGraw-Hill, 1948), 38–39.
11. Jaako Hintikka, "Kant and the Tradition of Analysis," *Deskription, Analytizität und Existenz, 3–4 Forschungsgespräch des internationalen Forschungszentrum für Grundfragen der Wissenschaften Salzburg*, ed. Paul Weingartner (Salzburg and Munich: Pustet-Verlag, 1966).
12. Isaac Barrow, *Mathematical Lectures Read in the Publick Schools at the University of Cambridge*, trans. John Kirkby (London, 1734), 28.
13. *Galen on Medical Experience*, ed. and trans. R. Walzer (London: Oxford University Press, 1944), 85.
14. Galen, *Opera omnia*, ed. C. G. Kühn, 20 vols. (Leipzig, 1821–33), I, 305–7.
15. For Haly's prologue see A. Crombie, *Robert Grosseteste and the Origins of Experimental Science* (Oxford, 1953), 77–78.
16. *Ibid.*, 59.
17. *Timaeus a calcidio translatus commentarioque instructus*, ed. J. H. Waszink (London and Leiden: Warburg Institute and J. Brill, 1962), secs. 302–5, pp. 303–6: vol. 4 of *Plato Latinus*, ed. Raymond Klibansky, in the *Corpus Philosophorum Medii Aevi*. See also, *Platonis Timaeus interprete Chalcidio cum eiusdem commentario*, ed. Johannis Wrobel (Leipzig, 1876; reprinted 1963), 330–34.
18. *Robert Grosseteste*, 59.

19. Ernst Cassirer, *Das Erkenntnisproblem*, 3 vols. (Berlin, 1922–23), I, 136–44.
20. *Discoveries and Opinions of Galileo*, trans. Stillman Drake (Garden City, N.Y.: Doubleday, 1957), 179, 183–84, 186, 197.
21. *Dialogues Concerning Two New Sciences*, trans. H. Crew and A. de Salvio (Evanston and Chicago: Northwestern University Press, 1939), 153. My italics.
22. *Ibid.*, 160.
23. *Ibid.*
24. *Dialogue Concerning the Two Chief World Systems*, trans. S. Drake (Berkeley, Calif.: University of California Press, 1953), 50–51. The italics in Salviati's reply are mine.
25. Cited by Basil Willey, *The Seventeenth Century Background* (London: Chatto & Windus, 1953), 36.
26. *The Philosophical Works of Francis Bacon*, ed. J. M. Robertson (London: George Routledge and Sons, 1905), 249.
27. *Oeuvres de Descartes*, ed. C. Adam and P. Tannery, 13 vols. (Paris, 1891–1912), X, 359, 469.
28. Arnauld and Nicolle, *The Art of Thinking* (London, 1685), 302.
29. *Ibid.*, 309.
30. Cited by Paul Mouy, *Le Développement de la physique cartésienne, 1646–1712* (Paris: J. Vrin, 1934), 148.
31. Wilbur Samuel Howell, *Eighteenth-Century British Logic and Rhetoric* (Princeton: Princeton University Press, 1971), 16–21.
32. *Ibid.*, 19.
33. Translated into French as *Oeuvres philosophiques et mathématiques de Mr. G. J. 'sGravesande*, ed. J. N. S. Allamand, 2 parts in one vol. (Amsterdam, 1774).
34. *Ibid.*, pt. 2, p. 120. Translated from the French version.
35. *Eighteenth-Century British Logic and Rhetoric*, 20. For Howell's earlier discussion of Sanderson see his *Logic and Rhetoric in England, 1500–1700* (Princeton: Princeton University Press, 1956), 299–308.
36. Cambridge University Library, MS. Add. 3970 (5).
37. *Opticks*, 4th ed. (London, 1730), 404.
38. *Traité élémentaire de chimie* (Paris, 1789), 57.
39. W. S. Jevons, *The Principles of Science* (London, 1905), 417.
40. Cited by Henry G. Van Leeuwen, *The Problem of Certainty in English Thought* (The Hague: Martinus Nijhoff, 1963), 75, n. 74.
41. *Essay Concerning Human Understanding*, bk. 4, chap. 12, sect. 10; compare chap. 3 sect. 29.
42. *Isaac Newton's Papers and Letters on Natural Philosophy*, ed. I. B. Cohen (Cambridge, Mass.: Harvard University Press, 1958), 110–11.
43. Barrow, *Mathematical Lectures*, 116.
44. *Ibid.*, 16–20. See also, Proclus's *Commentary on Euclid's Elements* in Cohen and Drabkin, *A Source Book in Greek Science*, 2–5.
45. *A Source Book in Greek Science*, 90–91.
46. *Traité de physique* (Paris, 1671), 1.
47. *Opere di Galileo Galilei*, ed. A. Favaro, 20 vols. (Florence, 1890–1909; reprint 1929–1939), vol. VI, sec. 6, p. 232.
48. Barrow, *Mathematical Lectures*, 21.
49. *Ibid.*, pp. xxvi–xxvii.
50. *Principia*, ed. Cajori, p. xvii–xviii.
51. *Ibid.*, 192; my italics.
52. Cambridge University Library, MS. Add. 3970. My italics.
53. I. B. Cohen, *Franklin and Newton: An Inquiry into Speculative Newtonian Experimental Science* (Philadelphia: The American Philosophical Society, 1956), 115–17.
54. "Account of the Booke entituled Commercium Epistolicum, etc.," *Philosophical Transactions of the Royal Society*, vol. 19, no. 342 (1717), 222.
55. Newton, *Universal Arithmetick*, 2.
56. Cambridge University Library, MS. Add. 3970, fol. 643. My italics.
57. *Opticks* (1704 ed.) bk. 2, pt. 1, p. 12. See also *Opticks*, 4th ed. (1730), 205.

14
The Background to Dalton's Atomic Theory

✿

We are often reminded of John Dalton's debt to the atomistic speculations of Sir Isaac Newton. And various contributors to this conference have made it clear that a corpuscular conception of matter was a commonplace in the two centuries that preceded Dalton's successful application of the atomic theory to a quantitative chemistry. What I shall attempt to do is to contrast the different approaches to the corpuscular view that we encounter in Newton's immediate predecessors, in Newton himself, and in some of those, both in France and in Britain, who were influenced by him and who in differing ways thought of themselves as following the path he had marked out.

Newton is, of course, the culminating figure of what we call the Scientific Revolution of the seventeenth century, an historic series of events, immensely complex, but whose main outlines are surely familiar to everyone. If called upon to single out the enduring discoveries made before Newton set to work, any of us would include without hesitation in his list a series of famous achievements: Galileo's demonstration of the law of falling bodies, Kepler's determination of his three planetary laws, Harvey's painstaking proof of the circulation of the blood, the experiments of Torricelli and Pascal on the properties of air, Snel's discovery of the law of refraction, and similar well-known discoveries by other men.

Now these all have certain features in common: they were investigations into a strictly limited area of experience; and, where appropriate, they were carried out, and their results expressed, in the language of mathematics. But what I wish to stress, however obvious it may seem, is that these enduring discoveries all concerned the *macroscopic* world, or at least the world of the immediately visible and palpable. They dealt with phenomena directly impinging on the senses. They were *not* carried out to support or illuminate some great system of natural knowledge; they were, in other words, largely independent (except on some deeper psychological level) of any theory about the invisible world, about —for example—the structure of matter, the intimate nature of light, the cause of motion, and so on. These men agreed on one thing: they distrusted the world

From *John Dalton and the Progress of Science* (papers presented to a conference to mark the bicentenary of Dalton's birth) (Manchester: The University Press; New York: Barnes & Noble, 1968), 57–91. © 1968 Manchester University Press.

view of Aristotle, with its occult doctrines of form and matter, substance and accident, and all the rest. If, as many of them did, they shared the conviction of the mechanical philosophy—that the valid objects of study are material bodies in motion—they held their speculations in check, and confined themselves to what they could observe and measure.

There were, however, others of a different cast of mind for whom this strictly limited, piecemeal investigation of the objects of perception was not enough, and who were anxious to substitute for Aristotle's great ordered picture of nature a new and more satisfying, but equally ambitious, world image. Influenced by the Renaissance revival of early Greek speculation, by the classical atomic doctrines in particular, they shared with many of the more circumspect scientists the conviction that the real world must ultimately consist in material particles whose motions, shapes, combinations, and separations could best account for the differing properties, and the curious behavior, of the objects perceptible to the senses. Upon this conviction the so-called mechanical philosophers—the French priest, Pierre Gassendi, England's controversial Thomas Hobbes, and the French philosopher René Descartes—erected great systems of natural knowledge, divergent in detail, of course, but having the common object of deducing observed phenomena from the hypothetical behavior of invisible corpuscles. This, after the middle of the century, was what men commonly meant when they spoke of *physics* or *physiologia*; and it was seen as the task of the natural philosopher, the true physicist, to suggest plausible mechanisms, models—as we should say— of corpuscular behavior, pictures of what Francis Bacon called "the secret motions of things." As Newton put it, in natural philosophy there was "no end of fancying."[1]

These great conjectural systems were the subject of keen debate in England when Newton entered Cambridge as a student, and their influence on his first speculations is beyond question. His earliest commonplace book of that period shows his reading of Descartes, of Hobbes, of Walter Charleton, a disciple of Gassendi, and of course of Robert Boyle.[2] Several pages of this notebook are devoted to musings about atoms, and to similar speculations about the nature of light. As early as 1664–66 Newton conceived of light rays as a stream of corpuscles or, as he calls them, of "globules," and even drew a sketch of one. This view of the corpuscularity of light he never abandoned, any more than he did his firm belief that matter is ultimately composed of "solid, massy, hard, impenetrable and moveable" particles; though whether the atoms had their abode in empty space, or in a tenuous aether, was a question on which his opinion seems to have varied at different times of his life.

Yet—and this is a point I now must emphasize—although Newton was manifestly influenced by this mechanistic view, it would scarcely be correct to call him a partisan of the mechanical philosophy. He well knew how much of it was sheer "fancying," and would have agreed with John Locke, who wrote that "most (I had almost said all) of the hypotheses of natural philosophy" are really but doubtful conjectures.[3] Quite early in his career Newton seems to have set himself the task of proving to himself and his critics that natural philosophy, or what we should call physical science, could and should be made, as far as possible, probative and demonstrative. It was against mechanical philosophers of all stripes—as well as the latter-day Aristotelians—that Newton was protesting when, late in life, he wrote the famous words: "I feign no hypotheses; for whatever is not deduced from the phenomena is to be called an hypothesis; and hypotheses, whether metaphysical or physical, whether of occult qualities or mechanical, have no place in experimental philosophy."[4] By physical and mechanical "hypotheses" Newton meant precisely those gratuitous, imaginary "principles"—the figured atoms of Gassendi, the whirling particles of Descartes—when they are taken as the unsubstantiated premises of scientific argument. Instead, under the title of "Experimental Philosophy," he invoked Galileo's more cautious method. Even as a very young man he saw the necessity of proceeding, as Galileo had urged (though it is doubtful if Newton was well acquainted with Galileo's books), "by geometrical demonstrations founded upon sense experience and very exact observations,"[5] even when this procedure led to principles, like that of universal gravitation, for which no indisputable cause could be found.

As early as his first scientific paper—the famous memoir of 1672 in which he showed that sunlight is composed of diversely colored rays, each with its different refrangibility—Newton clearly stated his determination not to "mingle conjectures with certainties."[6] In this early paper, though he hinted at it, he did not defend his privately held opinion that light is a stream of corpuscles. For the business at hand he treated light rays, as he said, only in general terms, abstractly, as "something or other propagated every way in straight lines from luminous bodies, without determining what that thing is."[7] This point of view his older contemporaries, Robert Hooke and Christiaan Huygens, for example, found difficult to accept, if indeed they understood what he was about.

The same caution pervades Newton's great *Mathematical Principles of Natural Philosophy*. Though the atomic doctrine surely lurks behind the theorems of this historic work, notably in Rule III of the *Regulae Philosophandi* of 1713, he does not discuss it and nothing that he demonstrates stands or falls as a result of one's believing or disbelieving in the philosophical doctrine of atoms and the void. Nor, to the confusion of his readers, did he try to account for the novel doctrine of universal attraction, insisting instead that it must be understood, mysterious though it seemed, simply as a fact of nature. The same discretion is evident later in the main body of his *Opticks*, where the experimental results do not in any way depend upon what we may believe the intimate nature of light to be.[8] To keep this distinction before us, I shall speak for convenience of Newton's "Natural Philosophy" when I mean his conjectures about the nature of matter or of light and the cause of attraction, conjectures that he kept clearly distinct from his "demonstrative" or "scientificall" work. On those speculative matters he was for a long time silent in print, and when later he spoke of them, he deliberately set them apart, in scholia added to the *Principia* (notably in the second edition of 1713), or in the famous new Queries added to the Latin *Optice* of 1706.

The surprising consistency with which Newton prevented his Natural Philosophy—his conjectures—from intruding on his more rigorous demonstrations does not in the least imply a lack of concern about the deeper problems. Yet the first order of business was to do what must be done: to give direction and precision to the scientific enterprise, to show the power and scope of a method of inquiry—a blend of observation, experiment, and mathematics—which Newton was at pains to describe and advocate. The element of self-discipline and self-restraint is well expressed by Newton in one of his unpublished manuscripts:

> But if without deriving the properties of things from Phaenomena you feign Hypotheses & think by them to explain all nature, you may make a plausible systeme of Philosophy for getting your self a name, but your systeme will be little better than a Romance. To explain all nature is too difficult a task for any one man or even for any one age. . . . Tis much better to do a little with certainty & leave the rest for others that come after you then [*sic*] to explain all things by conjecture without making sure of any thing.[9]

At Newton's hands these canons of scientific procedure yielded remarkable results. But since they depended, in principle as well as in fact, upon sense experience, he could apply them only at the visible or the macroscopic level. Yet he was obsessed by the possibility that this method ought somehow to be applicable for exploring the submicroscopic world with the same rigor and reliability as in the study of the visible universe. We should recall that telling, if cryptic, passage in Newton's Preface to the *Principia* where, after informing the reader that the central problem of science "seems to consist in this—from the phenomena of motions to investigate the forces of nature, and then from these forces to demonstrate the other phenomena," he adds the significant words: "I wish we could derive the rest of the phenomena of Nature by the same kind of reasoning from mechanical principles."[10] If this were to prove possible, then the hidden, invisible world which had so preoccupied the mechanical philosophers, and which held the deepest secrets of nature, might eventually become part of science as Newton conceived science should be. He brooded over this problem throughout his life, for many years confining his thoughts to letters and personal memoranda, or to papers he was reluctant to publish. And when, late in his career, he appended to his *Opticks* the famous Query 31, setting forth his conjectures about the atomic theory of matter (and at the same time reminding his readers of the strict method that must be followed in experimental philosophy), he was, in effect, bequeathing this problem to those who came after him as the great unfinished business of science.[11]

Let us now ask what Newton's disciples and followers made of these pronouncements, what interpretations they placed upon his scientific work, his speculations (so cautiously phrased for the most part in the interrogative voice), and his doctrines of method. The interpretations of his eighteenth-century admirers are surprisingly divergent; they fall rather sharply into two patterns, one best illustrated by his British disciples, the other, at least until late in the century, by the majority of Frenchmen.

In Britain, quite understandably, Newton was the revered sage. His vision penetrated—no one doubted—into the inner reality of nature. One had only to follow him faithfully: in mathematics (by adhering to his fluxional calculus), in celestial mechanics, of course, in optics and in chemistry. Robert Smith, in his massive *System of Optics*, published in 1738, made no distinction between Newton's conjectures and his more rigorous discoveries. He adopted and elaborated without hesitation Newton's suggestion that light was corpuscular, that its particles were acted upon by attractive forces in a way that could explain reflection, refraction, and diffraction. It is not too much to say that this Master of Trinity was, if not the real author of the corpuscular theory of light, at least its most influential advocate.[12] As for chemistry, to a degree that can scarcely be exaggerated, Newton's Queries left their mark on that science in Britain for more than a century. From the early writings of Keill, Freind, Stephen Hales, and Peter Shaw down to Dalton himself, English writers unhesitatingly invoked Newton's atomic explanation of chemical processes.[13]

By contrast, Newton's followers in France were many of them skeptical of his speculations in natural philosophy. On the other hand, they were captivated by his statements concerning method, citing repeatedly his rejection of hypotheses and his advocacy of the "Method of Analysis." They interpreted him as concerned only to observe, measure, and determine the laws according to which the objects of our experience behave. The physicists, especially, were embarrassed by Newton's talk of attractions, and tended to avoid discussions about atoms and the intimate structure of matter. This was the opinion that d'Alembert expressed in his *Treatise on Dynamics* and elsewhere in his writings. Nature, he wrote:

> is a vast machine whose inner springs are hidden from us; we see this machine only through a veil which hides the workings of its more delicate parts from our view. . . . Doomed as we are to be ignorant of the essence and inner contexture of bodies, the only recourse remaining for our sagacity is to try at least to grasp the analogy of phenomena, and to reduce them all to a small number of primitive and fundamental facts. Thus Newton, without assigning the cause of universal gravitation, nevertheless demonstrated that the system of the world is uniquely grounded on the laws of this gravitation.[14]

Just like Newton discussing gravity, d'Alembert insisted that we can learn something about the hidden causes in nature only if the laws and relationships by which they operate can be discovered: "The knowledge or the discovery of these relationships [d'Alembert wrote in another place] is almost always the only goal we are allowed to reach, and consequently the only one we should have in view."[15] Such a positivistic approach to Newtonian science, while not universal, was widespread among the more influential French thinkers. It was set forth by the Abbé de Condillac in his philosophical writings, and popularized for a still wider audience by Voltaire. For example, in discussing ever so briefly Newton's speculations about an all-pervading, elastic aether, Voltaire concluded that even an hypothesis that can explain nearly everything, as he admitted Newton's aether might do, ought not to be accepted. We should build nothing on conjectures, he wrote, not even on those of Newton himself.[16]

Neither school—neither the British nor the French—fully grasped, I think, the essentially disjunctive character of Newton's view of natural knowledge: his cautious separation of demonstrative science from suggestive conjecture. Nor did they quite understand Newton's attitude towards his own atomistic theories: that, for the moment at least, such theories could not be firmly incorporated into the structure of experimental philosophy, but that eventually this might be done, if the proper route could be discovered, the key experiments performed and rightly interpreted.

The procedure of the mechanical philosophers had been, we saw, to derive the qualities and behavior of visible or macroscopic objects from the properties and behavior assigned to hypothetical corpuscles. They sought to travel, as it were, from the inside of nature outward. But how, with any assurance or rigor, could this be done? This question sums up what some recent American philosophers have termed, somewhat barbarously, the "problem of transdiction."[17] Just as one *pre*-dicts when one makes inferences forward in time; just as geologists and astronomers (and I suppose historians) who argue from present evidence to the past can be said to *retro*dict; so *trans*diction is the term some philosophers have applied to the process of inference from the phenomenal world to that deeper level of reality that the senses cannot perceive. Could this be done by some credible procedure? If not, the limits of truly scientific inquiry must be severely restricted. Both Descartes and Newton were sensitive to this problem. Descartes, for example, wrote in his *Principles of Philosophy*: "Someone could ask how I have learned what are the shapes, sizes and motions of the tiny particles of each body, several of which I have determined just as if I had seen them, although it is certain that I could not observe them with the aid of the senses, since I have admitted that they are imperceptible." And he explains that he has done so through an intellectual analysis of common experience:

> I have considered in general all the clear and distinct notions we may have in our minds concerning material things, and having found no others except those ideas we have of shapes, sizes and motions . . . I judged that it must necessarily follow that all the knowledge that men can have of nature must be drawn from them alone.[18]

As to the hidden mechanism of nature, Descartes goes on to explain that having examined *in thought* the possible shapes, sizes, and motions of these tiny bodies, and the "perceptible effects that could be produced by the different ways in which they mingle together" he was pleased to encounter similar effects in those vastly larger bodies our senses can perceive: "Then I believed that they must infallibly have been so, when it seemed to me impossible to find, in the whole extent of nature, any other cause capable of producing them."[19] Infallibly? Well, not quite. Descartes is well aware that, as he says, "God has an infinity of different means" by which He could make things in the world appear as they do. Perhaps it is enough if the causes that are postulated can reasonably produce the observed effects; yet Descartes is quite explicit that the mechanisms he has imagined convey what he calls a "moral certainty," a certainty short of demonstration, to be sure, but with what we would call a high degree of probability.

Newton's other mentor, Robert Boyle—influenced as he was by both Descartes and Gassendi—little doubted that some version of the corpuscular hypothesis would best explain the divergent properties of particular substances. His wide range of experiments—on the elasticity of air, on the colors of bodies, on chemistry—seem designed to show how plausibly the corpuscular model can account for the diverse "forms and qualities" of things. By broadening the experimental evidence he hoped to increase—extensively, rather than intensively, if I may put it that way—the "moral certainty" of such corpuscular explanations.

Such reasoning did not satisfy Isaac Newton. The "moral certainty" of Descartes was delusive, as Newton clearly demonstrated in refuting the Cartesian vortices in the second Book of the *Principia*. Something better, too, was needed than the illustrative empiricism, the multitude of suggestive experiments, and the cautious hints of Robert Boyle. And Newton believed he had hit upon it— "the best method of arguing which the Nature of Things admits of"—in his dual method of "Analysis and Synthesis," or of "Resolution and Composition," a method of ancient lineage, to which he gave a very special twist. He describes it compactly towards the end of Query 31, but presents it more freely, and less schematically, in a manuscript draft only recently discovered:

> The method of Resolution consists in trying experiments & considering all the Phaenomena of nature relating to the subject in hand & drawing conclusions from them & examining the truth of those conclusions by new experiments & . . . so proceeding alternately from experiments to conclusions & from conclusions to experiments untill you come to the general properties of things. Then assuming those properties as Principles of Philosophy you may by them explain the causes of such Phaenomena as follow from them: wch is the method of Composition.[20]

Analysis through observation and experiment then—not, as with Descartes, a mere *conceptual* analysis—is the path that must be followed if one is to build up a persuasive and valid science: an analysis, too, that whenever possible should be cast in quantitative, in mathematical, language. It is the Method of the *Principia* and the body of *Opticks*. But can this method reach down into the realm of the nonperceptible? Is there a kind of experiential or experimental evidence that can supply a key to the invisible world? Newton was confident that such evidence might be found; two areas of investigation particularly impressed him as offering alluring possibilities: the study of light and color; and, more pertinent to our present discussion, chemistry.

In one of his earliest papers Newton described his observations on the colors produced when light passes through thin transparent bodies, the phenomenon of what came to be called Newton's rings.[21] The brilliant feature of these experiments was his precise measurement of the varying thickness of thin films that yielded the different spectral colors and the different "orders" of colors. Now Newton was very early convinced, as he wrote to Robert Boyle, that "the colours of all natural bodies whatever seem to depend on nothing but the various sizes and densities of their particles."[22] Colors, then, might provide an indication of particle size. And since his measurements of the rings supplied him with a correlation of film thickness with color, it was possible, if one could accept the "analogy" between spectral colors and those produced by reflection or transmission through translucent media, to infer from the colors of natural bodies (and the "order" to which their colors belonged) the dimensions, or more precisely the relative dimensions, of their particles. Newton was sufficiently persuaded by his argument to retain his faith in this early paper and include it in his *Opticks*, where, slightly modified, it constitutes the Second Book.

But it was chemistry that offered a more likely way of penetrating—by inference from experiment—into the invisible realm. At least by the 1670s Newton had begun to devote a considerable part of his energy to chemical experimentation and wide reading in the chemical and alchemical literature. These interests are clearly expressed in the famous letter he wrote to Robert Boyle in February of 1679, a letter often cited for Newton's speculations about the mechanical role of a subtle aether. Here Newton frankly indulges in some "fancying" of his own. He invokes the aether as a hypothetical agent to explain such phenomena as surface tension, capillary rise, and the refraction and reflection of light: that is, to explain *physical* processes. But he invokes something very like his later notions of "attraction" and "repulsion" when he discusses *chemical* phenomena:

> When any metal is put into common water [he writes], the water cannot enter into its pores, to act on it and dissolve it. Not that water consists of too gross parts for this purpose, but because it is unsociable to metal. For there is a *certain secret principle* in nature, by which liquors are sociable to some things, and unsociable to others.[23]

By the time Newton composed the *Principia*, say in 1686–87, his opinions in one important respect had profoundly altered. The aether has been virtually discarded and attractive *forces* have taken its place, not only for explaining the motions of large bodies but also (though he is cautious about such speculations in print) to account for phenomena, both chemical and physical, explicable in molecular terms. The "certain secret principle" in nature that accounts for sociableness and unsociableness has been transformed into forces of attraction and repulsion.

> I am induced by many reasons [his Preface to the *Principia* continues] to suspect that [the rest of the phaenomena of nature] may all depend upon certain forces by which the particles of bodies, by some causes hitherto un-

known, are either mutually impelled towards one another, and cohere in regular figures, or are repelled and recede from one another. These forces being unknown, philosophers have hitherto attempted the search of nature in vain.[24]

This is all, for a time, that he said in print. But Rupert and Marie Hall brought to light a few years ago an extended draft of this same preface, as well as a suppressed "Conclusio" to the *Principia*, in both of which Newton spells out his ideas in considerable detail.[25] In each of these precious documents we discover pretty much the same argument, and many of the same bits of supporting chemical evidence, that we find in the *De natura acidorum* of 1692 and in Query 31 of the *Opticks*. Clearly, by 1687 the main features of Newton's theory of matter had been sketched out; already he foresaw the area of investigation—chemistry and those phenomena we treat under physical chemistry—which eventually would lead with some assurance into the molecular and atomic realm. In the Queries of the *Opticks* he indicates at least the general direction scientists would have to follow if they were to come to grips with the "problem of transdiction" and turn their plausible conjectures about the atomic structure of matter into a true science. But the special route that he appears to be advocating in Query 31 was less fortunate, though by a circuitous route it may be said, at the very least, to have contributed to the achievement of Dalton. But before we see what Newton's English followers made of his suggestions, I must return, for a brief moment, to the transdictive problem in general.

We should remind ourselves of what was involved when scientists of the nineteenth century presumed to speak about atoms and molecules, when they invented mathematically conceived molecular and atomic theories to which they imputed a degree of credibility. For this is what Newton was searching for. These theories, it seems to me, embody a definable procedure. Certain properties are assigned in thought to the hypothetical particles, properties which are in principle quantifiable. Then certain precise configurations or interactions are imagined which are in turn susceptible to mathematical description. But these properties, configurations, and so on must possess a further characteristic. They must be such as to yield a statistical or aggregate behavior that manifests itself in some observable and measurable *molar* properties: the behavior of these aggregates must be predictable from the assumed characteristics of the particles that compose them. Perhaps you will agree that, at least in a general way, this is what was involved in such varied nineteenth-century achievements as the kinetic theory of gases, electrochemistry, chemical solution theory, and van't Hoff's work on stereoisomerism.

Professor Hall in his paper at this conference has already mentioned two recognizable examples of this sort of procedure in the *Principia*: Newton's derivation of Boyle's law (though Newton does not mention Boyle) on the assumption of a fluid composed of mutually repellent particles; and his derivation of a proposition that bore, as Newton contented himself with saying, "a great resemblance" to the Snel-Descartes law of refraction. But Newton is oddly diffident about these results, taking refuge in his claim that his inquiry is mathematical, not physical: in the first case he remarks: "But whether elastic fluids do really consist of particles so repelling each other, is a physical question."[26] The power of this sort of method seems to have eluded him, or perhaps he distrusted it.

In retrospect, a more compelling example—and a striking foreshadowing of nineteenth-century procedures—was published by Daniel Bernoulli in 1738. Bernoulli imagined a fluid composed of "very tiny corpuscles moving in every direction" and confined in a cylindrical vessel closed by a pistonlike lid held down by a variable weight. If the force balancing this weight be thought of as caused by the impacts of the innumerable particles on the lid, then as the weight is decreased the volume must increase, and conversely. In effect, Bernoulli shows that the product of pressure and volume is constant ($PV = C$), as Newton had shown starting from different assumptions. Bernoulli, however, admitted only impacts. The elasticity of air can only result from what we call the translational kinetic energy of the particles. Accepting the mechanical theory of heat—as indeed did Newton—he was therefore led on to inquire into the effects of a change of temperature on the elasticity of air; he came to the remarkable conclusion that the increased pressure when a vessel of air is heated at constant volume must be proportional to the square of the velocity of the particles. This, in essence, was the theoretical derivation, on molecular assumptions, of the law of Charles and Gay-Lussac, which had been anticipated very early in the eighteenth century by the experiments of Amontons.

Bernoulli's achievement is a rare example of this sort of inquiry until late, very late, in the eighteenth century.[27] Not until the last years of the eighteenth century and the opening years of the nineteenth did the molecular and atomic realm open some of its secrets to men like Coulomb, Laplace, Avogadro, and—of course—John Dalton. Only then do we discern the opening of a Second Scientific Revolution, a Molecular Revolution that, at least dimly, Newton had foreseen.

The principal line of attack that Newton envisaged as a solution of the transdictive problem was, as I have suggested, derived from his chemical investigations. Though more esoteric motives may have encouraged him in his dogged perusal of the alchemical literature, it is at least clear that he was confident that chemistry could supply empirical evidence, not only for confirming the existence of atoms and of corpuscles of higher complexity formed from them, but also for providing some insight into the "secret motions of things," and the forces producing them. Here, as well as when exploring the larger universe, the chief task of the scientist, as Newton described it, was "to investigate the forces of nature, and then from these forces to demonstrate the other phenomena." Chemistry—taken widely enough to embrace a number of physical phenomena—was to provide the key. One of his disciples, Peter Shaw, wrote shortly before Newton's death that when the great man discusses the laws, actions, and the powers of bodies, he always produces chymical experiments for his vouchers; and when, to solve other phenomena, he makes use of these powers, his refuge is to chemistry; whence he manifestly shows, that without the assistance of this art, even he could hardly have explained the peculiar nature and properties of particular bodies.[28]

Shaw, of course, has especially in mind the Queries of the *Opticks*, where, as Priestley put it later, Newton "indulged bold and excentric thoughts," but he could equally as well have pointed to the little tract *De natura acidorum* first printed by John Harris in his scientific dictionary, the *Lexicon Technicum*, in 1710.[29] The main burden of both works is to present his corpuscular theory and—as in the preface of the *Principia* he announced he hoped to do—to offer evidence that there must exist short-range forces of attraction and repulsion, analogous to, yet different from, the gravitational force between bodies. In a rather vague, and largely qualitative, fashion Newton suggested in Query 31 how attractions and repulsions could explain such physical effects as cohesion, surface tension, and capillarity, and a multitude of strictly chemical phenomena: he speaks not only of solution and precipitation but also of a wide range of different chemical reactions.[30] The extent of Newton's chemical information, and its precision, is remarkable; most of the reactions he describes can be readily interpreted in modern terms. None of them, I suspect, was original with him, though he may have followed nearly all of them with his own eyes. Most are described in the writings of such seventeenth-century authorities as Robert Boyle, Nicolas Lemery, Angelus Sala, John Mayow, and Thomas Willis. It is how Newton interprets this information—in corpuscular and attractionist terms—that is significant. For example when "salt of tartar" (potassium carbonate) deliquesces in moist air, this is because the particles of the salt *attract* particles of water from the air. This attraction for water, which occurs also in the case of "oil of vitriol," but not in common salt or saltpeter, has a definite limit or saturation value, an observation of central significance.

Attraction too, seemed capable of explaining in a qualitative way common displacement reactions: the release of spirit of salt (HCl) when common salt is treated with sulphuric acid; or the precipitation of insoluble salts from solution when salt of tartar is added.

The most interesting passage—for it most strikingly illustrates the preferential attraction of certain substances for each other—describes the elementary experiment of showing that iron displaces copper from acid solution, copper displaces silver, and so on for a number of metals, in an order that corresponds to our electromotive series. "Does not this argue," Newton asks, "that the acid Particles of the *Aqua fortis* are attracted more strongly . . . by Iron than by Copper, and more strongly by Copper than by Silver, and more strongly by Iron, Copper, Tin, and Lead, than by Mercury?"[31] Clearly, Newton thought, the attractive forces, acting at minute distances between the particles, must vary in strength from one chemical individual to another. In this striking passage, together with others in Query 31, Newton set forth, in attractionist terms, what came to be called in the eighteenth century the doctrine of *elective affinities*. That eventually this famous Query exerted a profound effect cannot be denied, but whether it was the immediate stimulus to, and the preponderant influence upon, the century's preoccupation with affinities is not so certain.

In 1718, not long after Newton's *Opticks* appeared, a leading chemist of the French Academy of Sciences, Etienne-François Geoffroy, presented a now classic memoir entitled *Sur les rapports des différentes substances en chimie*.[32] In it Geoffroy published the first table of chemical affinities. Each of a series of parallel columns, headed by the conventional symbol for a familiar substance, displays one above another the symbols of those reactants with which the substance was known to combine. These follow one another in such an order that each reactant displaces all others lower in the column from union with the substance at the head of the column.

Nowhere in his paper does Geoffroy speculate about the cause of these phenomena; he says nothing about atoms or attractions, and speaks only of "rapports" or relationships; he seems to have the purely practical objective of summarizing in tidy and convenient fashion a body of chemical fact. Had he read Newton's *Opticks* and been stimulated by it? We cannot be sure, and indeed it is not necessary to assume that this is the case. Much, if not most, of the information in the table he could have drawn from the seventeenth-century chemical tradition, as indeed Newton himself had done in accumulating the chemical facts that he set forth in Query 31.

Chemists had long been aware of the property of specific reactivity; they well knew that some substances react together with ease and often with violence, others more quietly or with difficulty; still others were found to be almost or totally inert with respect to common reagents. And it was early observed that certain pairs of substances are particularly disposed to combine together and, once combined, to resist separation. Replacement reactions, too, were described by a number of different chemists. Glauber, for example, explained that when zinc oxide is heated with sal ammoniac, the zinc displaces the ammonia from combination with the acid, setting it free.[33] Otto Tachenius (1666) knew not only that acids differed in strength, but also that a stronger acid could displace a weaker one from its salt.[34] Indeed both Glauber and Robert Boyle had described, well before Newton published his *Opticks*, the displacement series for the common metals which we cited above.[35]

Yet Newton's was the first text—one is tempted to call it the first short monograph—in which the well-known instances of chemical affinity were described in one place and in brief compass: the earliest general treatment of a subject to which earlier writers had referred only in passing and in widely separated places. Geoffroy could well have read the important passages of Query 31. He had close contacts with British science; in 1698, on a visit to England, he had been elected, along with other Europeans, a Fellow of the Royal Society, and he kept in close touch with the Society as a faithful correspondent of Sir Hans Sloane. He was sent a complimentary copy of Newton's *Opticks* (1704), which is now in the library of my university, and it is hard to believe that he did not also receive, or at least consult, the Latin *Optice* of 1706, in which the famous Query first appeared.[36] Of this, however, we cannot be certain. At all events, Geoffroy's paper makes no mention of Newton, and he says nothing of those attractions between particles in terms of which Newton sought to explain the selective reactivities of chemical substances. Geoffroy's unwillingness to indulge in speculations earned the approval of Fontenelle, who wrote, in his account of the year's activity at the Academy:

> C'est ici que les sympathies & les attractions viendroient bien à propos, si elles étoient quelque chose. Mais enfin en laissant pour inconnu ce qui l'est, & en se tenant aux faits certains, toutes les expériences de la chimie prouvent qu'un même corps a plus de dispositions à s'unir à l'un qu'à l'autre. . . . Ce sont ces dispositions, quelqu'en soit le principe, & leurs degrés que M. Geoffroy appelle *Rapports*, & une plus grande disposition est un plus grand rapport.[37]

Geoffroy's table elicited some criticism but more interest and approval; one writer, for example, remarked in 1723 that by his table alone Geoffroy "a rendu plus de service à la Chymie qu'une infinité d'Auteurs par des volumes de raisonnemens physiques."[38] There were, of course, doubters, and it was several years before attempts were made to devise improved tables: among the earliest of these revisions and elaborations were those of Grosse (1730), Gellert (1750), and Rudiger (1756).[39] In France, after about 1740, G.-F. Rouelle, Lavoisier's teacher—and the teacher of a whole generation of French chemists—gave special attention in his lectures to Geoffroy's table, and may have been chiefly responsible for the interest a number of his pupils—among them J.-F. Demachy and G.-F. Venel—later displayed in this aspect of chemistry.[40] None of these men, it should be emphasized, indulged in what Rouelle deplored as "vains raisonnemens" about the cause of the relationships disclosed in the tables; all, indeed, had practical considerations chiefly in mind, and they avoided not only the unpopular idea of "attraction," but for the most part also the sort of explanations that had been current in the previous century.

At this point we should recall that early chemists had various expressions to describe, and perhaps even to elucidate, the mystery of selective reactivity. A few used the analogy of magnetic attraction; but the majority appealed to an "animate model" and readily used metaphorical images derived from human relationships. The tendency is well illustrated by Francis Bacon, who wrote:

> It is certain that all bodies whatsoever, though they have no sense, yet have perception: for when one body is applied to another, there is a kind of election to embrace that which is agreeable, and to exclude or expel that which is ingrate: and whether the body be alterant or altered, evermore a perception precedeth operation; for else all bodies would be alike one to another.[41]

In the same spirit Joachim Jungius, a generation later, spoke of a "power" or "appetite" in natural bodies by which they are mutually drawn together and combine. Glauber, the greatest practical chemist of the seventeenth century, remarked in his *Furni novi philosophici* (1648) that "sand and its like have a great community [Gemeinschaft] with the salt of tartar and they love each other very much, so that neither of them willingly parts from the other."[42] Newton himself, we saw, spoke in 1679 of chemical substances as being "sociable" or "unsociable."

The word *affinity* (*adfinitas* or *affinitas*) was often used, and had animistic overtones we are inclined to forget. Originally, in classical Latin, it denoted a human relationship by marriage, as distinguished from consanguinity, and the Catholic Church later extended it to such spiritual and sacramental relationships as that between godparents and godchildren. The metaphor came easily to be broadened to imply any kinship or close resemblance.

In chemistry, "affinity" made an early appearance in the writings of Albertus Magnus in the thirteenth century; we find the term used by the Latin Geber; by Bernard Palissy in the sixteenth century; and in Newton's day by Sylvius de le Boë (1659), Hooke (1665), Mayow (1674), and Barchusen (1698).[43] The "animate model" is everywhere evident, and the word carried with it the notion of a fundamental similarity, a close kinship, even an identity of natures, between chemical substances that react together. John Mayow made the marriage metaphor quite explicit (as Barchusen did the image of friendship or kinship) when he wrote: "Salt has as great affinity and relationship with nitro-aërial spirit and also with sulphur; for these very active elements are by turns married to salt as to a fitting bride, and are fixed in its embrace."[44]

The early makers of affinity tables doubtless were as unwilling to become entangled in such outmoded conceptions as they were skeptical about invoking Newton's short-range attractive forces. Just as they avoided the word "attraction," so at first they were even reluctant to speak of "affinity."

Two chemists of outstanding repute in the eighteenth century can be said to have restored respectability to the term: Hermann Boerhaave, author of the widely read *Elementa chemiae* (1732); and with considerably greater enthusiasm, P.-J. Macquer, the acknowledged leader of chemical thought in France in the generation before Lavoisier.

Let us first consider Boerhaave, for he has been credited with having "impressed on the term *affinity* the meaning which it has retained since his time," that of the force which brings and holds together chemically *dissimilar* substances.[45] Boerhaave may well have resuscitated the word, but he certainly did not envision affinity in this way. Like his Dutch contemporaries, he was more or less under the Newtonian spell, and in his early inaugural discourses praised Newton's experimental method in science and spoke favorably of the evidence Newton supplied for the role of attractive forces in chemistry. Nevertheless the older views had left their mark, for when he discusses in his famous textbook the action of solvents (*menstrua*), in the passages to which scholars are wont to refer, Boerhaave invokes *both* Newtonian attractions *and* the older notion of affinity, distinguishing sharply between them. Attractions, which he calls in one place an "appetite of union," account for the coming together of *unlike* substances, in particular for the dissolving action of solvents upon solutes. Affinity, however, he uses in the traditional sense: it is that mysterious power which acts between *identical* or *like* substances, which brings together particles of the same nature into homogeneous masses, as when particles of a solute come together during precipitation and crystallization.[46]

If, as most historians of chemistry would agree, the theory of affinity, or of elective attractions, was the boldest attempt made in the eighteenth century to give unity to the newly emerging science of chemistry, much credit must be accorded P.-J. Macquer. In his *Elémens de chymie théorique* (1749) he was the first to place a discussion of affinities at the very center of an exposition of chemical theory, and his views were later set forth in the article "Affinité" of his great *Dictionnaire de chymie* (1766). Macquer adopts a theory of the constitution of chemical substances which in some degree resembles Newton's. The underlying entities are of two kinds: constituent particles (*parties constituantes*) and molecules or integral particles (*parties intégrantes*); constituent particles combine and separate acording to the laws of chemical affinity. Affinity, Macquer insists, is not a word devoid of sense "but a truly physical, very real, very general" property of bodies.[47] It is responsible for all chemical combination, not merely the combination of "like" substances, and is subject to invariable laws. In an elaborate discussion, he distinguishes affinities of several kinds ("les affinités simples, les composées, les réciproques, les doubles," etc.) and gives examples of each. Yet that diffidence towards unsupported conjecture, that positivistic tendency among French scientists, is clearly evident in Macquer: he declines to speculate about the cause of these affinities. In his early work he mentions attraction only once, and in a deprecatory way.[48] In the *Dictionnaire* he wrote:

> On ne cherche point ici la cause de ce grand effet, qui est si général qu'il peut être regardé lui-même comme cause de toutes les combinaisons, & servir à en rendre raison. Il est peut être une propriété aussi essentielle de la matière que son étendue & son impénétrabilité, & dont on ne peut dire autre chose, sinon, qu'elle est ainsi.[49]

And the reader, he goes on, may satisfy his curiosity by consulting the works of Newton, Freind, Keill, and the Abbé Marcuzzi "qui ont essayé de porter la lumière du calcul sur ces objets obscurs." One senses perhaps a cautious change in Macquer's views between 1749 and 1766, and indeed this may point to a growing inclination in France to give a greater credence to Newton's chemical speculations, or at least to a hope that a physical explanation for the phenomenon of chemical affinity could be discovered.

Maupertuis noted this change of attitude among the chemists. In his *Dissertation physique à l'occasion du Nègre Blanc* (1744), the short preliminary version of his *Vénus physique*, he devotes a paragraph to Geoffroy and his "rapports." A few years later he added a new paragraph to the *Vénus physique*, identifying the "rapports" with attractions, and remarking that, though the astronomers were the first to employ attractions, "La Chymie en a depuis reconnu la nécessité; & les chymistes les plus fameux aujourd'hui, admettent l'Attraction, & l'étendent plus loin que n'ont fait les astronomes" (*Œuvres de Mr de Maupertuis*, Dresden, 1752, 247).

It was to this end, and perhaps stimulated by the works of Macquer, that the Academy of Rouen proposed as a prize essay subject for 1758 to: "Déterminer les Affinités qui se trouvent entre les principaux Mixtes, ainsi que l'a commencé Mr Geoffroy; et trouver un Système Physico-méchanique de ces Affinités." The prize was divided between George Louis Lesage, the Genevan physicist and mathematician who died the following year, and Jean Philippe de Limbourg, a physician from the neighborhood of Liège. Lesage, who produced an elaborate mechanistic theory to account for attractions, was deemed to have treated in superior fashion the second part of the proposed question, whereas Limbourg, a pupil of Rouelle, devoted special attention to the chemical evidence. The second chapter of Limbourg's *Dissertation*, entitled "De la nature & des causes des affinités," is especially relevant to our discussion. Affinity cannot be explained, he insists, by an "identité de principes" in the substances that combine, nor by the action of some mechanical instrumentality. The principal cause is the force of attraction, and similitude enters in only insofar as a "similitude de parties"— some correspondence, we are to suppose, in corpuscular size or shape, or in disposition of parts—permits bodies to approach near enough together for the short-range forces of attraction to be exerted. These forces are identical with the attraction that physicists speak of, and differ only in degree. In any case, the cause of these chemical attractions or affinities is as obscure as that of the physicists' gravitational attraction; to attempt to explain either kind is to indulge in purely imaginary suppositions.[50]

The influence of Limbourg's *Dissertation* is hard to assess; there was little in it to induce chemists to discard the conveniently vague term "affinity" and speak instead of Newtonian "attractions." Not long after its publication, two of Rouelle's other pupils made this point clear. Demachy commented in 1765:

> Tirons néanmoins avec les Chymistes tout le parti qu'ils ont raisonnablement tiré des affinités & de leur application aux travaux chymiques, & ne regardons ce mot affinité que comme une acception, un terme qui désigne un effet, mais qui rarement explique la cause.[51]

The same attitude is expressed in the same year by Venel in his article "Rapport ou Affinité" in Diderot's *Encyclopédie*. Chemists understand by these *words* (Venel italicizes the word "*mots*") only the aptitude of certain substances to unite chemically with certain other substances, whatever may be the cause. "Les Chymistes sagement circonspects," he writes, "se gardent bien de théoriser sur le formel, le mécanisme, les causes de l'affinité chymique."[52]

In the spirit of their master, Rouelle, both Demachy and Venel are defending the autonomy of chemistry; chemistry is not physics, as Venel, in particular, had strongly and picturesquely argued.[53] The times, however, were changing; in physics, of course, Newton had triumphed in France; but even chemistry was not quite what it had been: men with a different background and training, whose formation was not derived from medicine and pharmacy, were becoming interested in it: men who knew, or thought they knew, something about physics.[54] Buffon was one of these, and so too was the Dijon lawyer, Guyton de Morveau, whom Buffon came to know in 1762, and with whom he corresponded and, for a brief period, collaborated in experiments destined for the *Histoire des miné-raux*. In the same year 1765 in which Demachy and Venel published the opinions we have just cited, Buffon defended the opposing view:

> Les lois d'affinité par lesquelles les parties constituantes de ces différentes substances se séparent des autres pour se réunir entre elles, et former des matières homogènes, sont les mêmes que la loi générale par laquelle tous les corps célestes agissent les uns sur les autres; elles s'exercent également et dans les mêmes rapports des masses et des distances; un globule d'eau, de sable ou de métal, agit sur un autre globule comme le globe de la terre agit sur celui de la lune; et si jusqu'à ce jour l'on a regardé ces lois d'affinité comme différentes de la pesanteur, c'est faute de les avoir bien conçues, bien saisies, c'est faute d'avoir embrassé cet objet dans toute son étendue.[55]

To Guyton, Buffon is the "Newton of France"; he has seen farther than Newton into what lies behind chemical affinity; he has restored unity in nature; and his insight into the universality of the law of attraction "will henceforth be the compass for chemical theory."[56] In 1772 Guyton published his remarkable *Digressions académiques*, in which he applied Buffon's attractionist theory to problems of solution, crystallization, and other chemical matters.[57] For French chemistry, a turning point had been reached; and the publication three years later of Torbern Bergman's *Disquisitio de attractionibus electivis* confirmed it. Guyton became the correspondent and translator of Bergman, and soon emerged as the acknowledged expert on affinity theory in France.[58]

Yet before Bergman, and before the French chemists, the British had taken the important step of interpreting the affinity tables of the Continental writers in attractionist, and Newtonian, terms. They were the earliest to do so. The first man in England to call attention to Geoffroy's table was the largely forgotten physician and chemist, William Lewis. He reproduced the table in a pharmaceutical work, his *New Dispensatory* of 1753, but confined himself to calling attention to its practical importance for "officinal processes," that is, for the preparations of the pharmacist.[59]

The chief innovators were men of the Scottish school—the chemists of Glasgow and Edinburgh—who deserve to be remembered for this as for so much else. The first published reference to "elective attractions" appeared in Joseph Black's *Experiments upon Magnesia Alba* (1756). Black, in interpreting his results, makes frequent use of the concept of differing forces of attraction between substances, and at one point remarks that "a few alterations may be made in the column of acids in Mr Geoffroy's table of elective attractions."[60] Black's use of the phrase "elective attractions," a phrase that Bergman later made popular, is at least suggestive. Perhaps it was from Scotland that Bergman received his inspiration.

The doctrine of elective attractions, and the phrase, were almost certainly derived by Black from his teacher at Glasgow, Dr. William Cullen. Cullen launched his course of chemistry at Glasgow in 1747, and according to his biographer began at once to discuss chemical affinities with his students.[61] This is hardly surprising, for he was aware of Geoffroy's table as presented by Lewis, and recommended Boerhaave to his students "or Macquer to those that understand French."[62] Indeed it is not too much to say that Cullen's doctrine is Newtonian atomism and attractionism combined with Macquer's elaborate picture of the several kinds of affinity. Of attraction, Cullen remarked:

> The combination of bodies depends on that inclination or tendency which they have to approach one another and in certain circumstances to remain coherent. To this general fact the English philosophers have given the name of *Attraction*, a word which has given rise to endless cavils among foreigners, because it seems to imply a power in bodies by which they can run together. If this were really what were meant by it, if it were intended to express an original primary principle, such cavils would not be groundless, as such a supposition would stop any farther researches into the vast chain of causes and effects. . . . But the great NEWTON who first brought this term into general use, has often put us on our guard against this by telling us that it was merely intended to express a general fact, and it is in this sense that we use it. The foreigners who have not embraced this term have been under the necessity of using others in the same acceptation.[63]

Cullen, then, would seem to have been the earliest to develop a full-blown chemical theory that combined Newton's attractionist atomism with those chemical facts of affinity assembled and sifted by chemists on the Continent. Cullen's doctrine was picked up by his pupil, Joseph Black, who employed it not only in his famous doctoral dissertation but in his teaching as well. Both men taught that chemical combination takes place between atoms, forming the larger aggregates we call (as indeed, on occasion, did Macquer) "molecules." And both men used diagrams to illustrate what they believed to take place in reactions of double decomposition.[64] These diagrams may have been, first and foremost, a pedagogical device, a "visual aid." But they disclose crude attempts to quantify the doctrine of "elective attractions," and to determine—as Newton had long ago hoped would prove possible—the relative strength of these forces. As an example, Cullen used the following diagram to show what happens when silver nitrate reacts with mercuric chloride, using letters to indicate the relative strength of attractions:

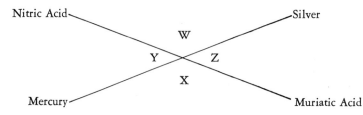

From the table of affinities $y > x$ and $z > w$, therefore $y + z > x + w$. Thus when silver nitrate and mercuric chloride (corrosive sublimate) are mixed together, the partners will be exchanged and silver chloride and mercuric nitrate will result. Joseph Black, in his turn, actually assigned numbers, instead of letters, to these relative affinities. They are arbitrary, if you will, but were intended to reflect the information derived from the tables of affinity.

This sort of thing was soon carried to great lengths, for example by William Higgins in his *Comparative View of the Phlogistic and Antiphlogistic Hypotheses* (1789). A thoroughgoing atomist, Higgins later claimed to have anticipated John Dalton; but except for having perhaps recognized the important fact of multiple proportions, though not its full significance, his pretensions are surely unjustified. Yet he is interesting for his exhaustive efforts to assign numbers (after the manner of Cullen and Black) to indicate the relative strength of interparticulate attractions. "Chemical philosophy," he wrote in words that Newton would doubtless have approved, "will never reach its meridian splendour except by means of such principles."[65]

Newtonian atomism—with its attendant doctrine of attractions—was, it should be apparent, a widespread chemical doctrine when John Dalton entered upon his scientific career. In the last decades of the eighteenth century it flourished not only in Britain, but—through the influence of Buffon, Guyton de Morveau, and especially Torbern Bergman—on the Continent, where its popularity is attested by the title that Goethe gave to a famous novel: *Elective Affinities*.

But this specific path to a solution of the transdictive problem—this key to finding the underlying forces of corpuscular nature—was far from leading chemistry to its "meridian splendour." Indeed, for reasons which were becoming increasingly evident, it was destined to prove illusory. Yet it may have contributed, though indirectly and almost accidentally, to Dalton's very special answer of the problem of transdiction. At least, as an intruder into Daltonian scholarship, I argued that this possibility, once widely held, deserves re-examination.[66]

At this point an important question merits brief consideration. It may be protested that the new habit of speaking of "attractions" instead of "rapports" or "affinities" was merely a matter of words, and that it can hardly have had any real significance for the work of chemists of the late eighteenth century. This, I believe, was not the case: men were led, as a result, to think of interparticulate *forces*, and therefore of entities that, in principle, could be measured. We noted this tendency in the case of Cullen, Black, and Higgins. It was characteristic, too, of Bergman, about whom Guyton wrote in 1786: "il a prévu tout l'avantage que la Chymie pouvoit retirer de l'expression des affinités en nombre & de la détermination des figures des molécules qui s'unissent."[67] And Bergman's example undoubtedly stimulated Wenzel and Kirwan. Wenzel sought to obtain numerical values for affinity from the speed of reactions; Kirwan, for his part, in a careful quantitative study of the solution of various metals in mineral acids, wrote:

The advantages resulting from these inquiries are very considerable, not only in promoting chymical science, which, being a physical analysis of bodies, essentially requires an exact determination, as well of the quantity and proportion, as of the quality of the constituent parts of bodies.

And further on:

But the end which of late I had principally in view, was to ascertain and measure the degrees of affinity or attraction that subsist betwixt mineral acids, and the various bases with which they may be combined, a subject of the greatest importance, as it is upon this foundation that Chymistry, considered as a science, must finally rest.[68]

Even more elaborately and somewhat later a similar investigation was undertaken by the German chemist J. B. Richter. In search of some sort of mathematical law, he embarked on a long series of experiments on the common acids and bases to discover whether the weights with which they combined to form neutral salts could be some measure of their "affinities" or "attractions." He ascertained the weight of each base (lime, potash, magnesia, etc.) necessary to neutralize a fixed amount of each of the common acids, and conversely the amount of each acid that would neutralize a fixed quantity of a chosen base. His figures were scattered through long, prolix works, filled with extravagant computations. The results only became usable, and more widely known, when another chemist, E. G. Fischer, reduced them to a simple table. This table, which was included by Berthollet in his *Essai de statique chimique* of 1803, revealed the combining weights, or rather the combining weight ratios, of the acids and bases.[69] Now if you were a chemist in the British tradition, for whom the existence of Newtonian atoms admitted of no possible doubt, and who unquestioningly believed that the process of chemical combination is a joining of atom with atom, it might occur to you that these gross combining weights must reflect, in the case of pure substances, the relative weights of the constituent atoms themselves.

This was the interpretation that John Dalton placed on Richter's work when he learned of it, though precisely when is hard to determine.[70] It is now well established that Dalton's great theory was originally inspired by his interest in meteorology and the gases of the atmosphere. In studying the physical behavior of mixed gases, he adopted Newton's model that air is an elastic fluid constituted of small particles or atoms of matter, which repel each other by a force increasing as their distance from each other diminishes. Yet Dalton's experiments showed that in a mixture of two gases each constituent behaves as if the other were not present: the particles of one species neither attract nor repel the particles of the other kind, and he sought to confirm this by experiments on gaseous diffusion and solubility. Obviously the atoms of different gases must be different. But what property set them apart? It occurred to him that the distinguishing property of the differing atomic species might be not the shape or volume of the particle, but *its weight*. But how could one be sure? How could one actually determine the relative weights of single *atoms*?

The answer seemed to lie in chemical evidence. So, I have suggested, he combed the chemical literature, in the course of which he encountered Richter's table of equivalents. Here was a method that, if applied to reactions between elementary substances—between gases, or between gases and substances like carbon or the metals—could yield the weights he sought. It was in the course of this reading, or in conducting his experiments, that he hit upon the great confirmation of his speculations: the *Law of Multiple Proportions*, which he announced at the same time as his atomic theory. This "law" states that if two elements combine to form two or more compounds, the fixed proportions in which two elements combine together are simple integral ratios. Such a "law" makes sense only if the elements combine as separate discrete units, each unit possessing a characteristic weight. As Berzelius, the great Swedish chemist, later wrote in a letter to Dalton: "The theory of multiple proportions is a mystery without the atomic theory."

This attribution of a definite weight to the atoms of each individual kind of chemical element is, we all know, the cardinal feature of Dalton's work. Though Dalton would not have thought of it in this way, it is obvious that it satisfied those transdictive criteria of which I spoke at the beginning of this paper: (1) a quantifiable property—the relative weight of the atoms—is assigned to the ultimate particles of the elements or simple substances; (2) when substances combine, atom with atom, the weight of the resulting molecule will be obtained by simple addition, if the molecule is formed by the elective attraction of atom to atom; (3) and lastly—if only pure substances are involved—the relative weights of molar or aggregate quantities must be as the weights of the atoms of the substances being compared. Dalton's was surely the most important step ever taken in the quantification of chemical theory. And it is the first successful example of the kind of transdictive procedure which Newton was hoping for. There are, of course, other ways of proceeding. But with Dalton's work the atom, long-invoked and much talked about, had at last a claim to enter the inner shrine of science.

Notes

1. Letter to Robert Boyle, 28 February 1678/9. *The Works of the Honourable Robert Boyle*, ed. Thomas Birch, 5 vols. (London, 1744), I, 70.
2. The University Library, Cambridge, MS. Add. 3996. See Richard Westfall's careful study of this notebook in *The British Journal for the History of Science* 1 (1962), 171–82.
3. *An Essay Concerning Human Understanding*, bk. 4, chap. 12, art. 13.
4. *Mathematical Principles of Natural Philosophy*, ed. Florian Cajori (Berkeley, Calif.: University of California Press, 1934), 547; hereafter cited as "Motte-Cajori," the

English translation being that of Andrew Motte (1729). In deference to current pedantry I have altered Motte's "frame" to "feign," which is probably a better rendering of *fingere*.

5. "Letter to the Grand Duchess of Tuscany," in Stillman Drake, *Discoveries and Opinions of Galileo* (Garden City, N.Y.: Doubleday, 1957), 179.

6. I. B. Cohen, ed., *Isaac Newton's Papers and Letters on Natural Philosophy* (Cambridge, Mass.: Harvard University Press, 1958), 57. Reply to Hooke.

7. *Ibid.*, 119.

8. Newton's famous doctrine of "Fits of easy Reflexion and easy Transmission" (*Opticks*, bk. 2, pt. 3, prop. 12–20) may seem to refute this statement, but it is not an hypothesis in Newton's sense of the term; he intends it rather as a noncommittal description of the effects he observed in studying refraction phenomena of thin transparent media.

9. The University Library, Cambridge, MS. Add. 3970, fol. 478 *v*. This was apparently an early version of the section on method near the end of Query 23/31 of the *Opticks*.

10. Motte-Cajori, p. xviii.

11. In the Latin *Optice* of 1706 this appears as Q. 23. Modern scholars have only recently realized how drastically the Queries of the Latin edition of the *Opticks* differ from the new and altered ones of the second English edition of 1717/18. See Alexandre Koyré, "Les Queries de *l'Optique*," *Archives Internationales d'Histoire des Sciences* 13 (1960), 15–29. But Koyré overlooked the fact that Samuel Horsley, in his *Isaaci Newtoni Opera quae exstant omnia* (1779–85), had indicated all the important changes.

12. Smith was nevertheless selective: he ignored completely those speculations about the aether Newton added to the later Queries of the *Opticks*. See Henry John Steffens, "The Development of Newtonian Optics in England, 1738–1831," a Thesis presented to the Faculty of the Graduate School of Cornell University for the Degree of Master of Arts (Ithaca, N.Y., 1965).

13. Shaw's most influential contributions to chemical Newtonianism were his notes to Boerhaave in *A New Method of Chemistry*, 2d ed., 2 vols. (London, 1741).

14. D'Alembert, *Mélanges de littérature, d'histoire, et de philosophie*, 4th ed., 5 vols. (Amsterdam, 1767), IV, 258–59.

15. *Discours préliminaire de l'Encyclopédie*, ed. Louis Ducros (Paris, 1893), 40. For a fuller discussion, see my paper, "Where the Statue Stood: Divergent Loyalties to Newton in the Eighteenth Century," in Earl R. Wasserman, ed., *Aspects of the Eighteenth Century* (Baltimore: Johns Hopkins Press, 1965), 317–34.

16. *Elémens de la philosophie de Neuton* (Amsterdam, 1738), 176.

17. Maurice Mandelbaum, *Philosophy, Science and Sense Perception* (Baltimore: Johns Hopkins Press, 1964), 61–62.

18. *Les Principes de la philosophie*, pt. 4, art. 203. This argument is echoed in Newton's famous Rule III of the *Regulae Philosophandi* added to the second edition of his *Principia* in 1713.

19. *Ibid.*

20. The University Library, Cambridge, MS. Add. 3970, fol. 478 *v*.

21. What has been called Newton's "Second Paper on Light and Colours," the results of which long remained unpublished, was sent to the Royal Society as a letter to Henry Oldenburg, dated December 7, 1675. See Cohen, *Newton's Papers and Letters*, 177–200; and *The Correspondence of Isaac Newton*, ed. H. W. Turnbull, 6 vols., continuing (Cambridge, 1959–76), I, 362–86.

22. *Works of the Honourable Robert Boyle*, I, 71.

23. *Ibid.* My italics. For Newton's chemical researches see Douglas McKie, "Some Notes on Newton's Chemical Philosophy," *Philosophical Magazine*, ser. 7, 33 (1942), 847–70; also Marie Boas and A. Rupert Hall, "Newton's Chemical Experiments," *Archives Internationales d'Histoire des Sciences* 11 (1958), 113–52.

24. Motte-Cajori, p. xviii.

25. A. Rupert Hall and Marie Boas Hall, *Unpublished Scientific Papers of Isaac Newton* (Cambridge, 1962), 302–8, 320–47.

26. Motte-Cajori, 302. Note Halley's comment in his review of the *Principia*: "Next the

density and compression of Fluids is considered . . . and here 'tis proposed . . . whether the surprizing *Phenomena* of the Elasticity of the Air and some other Fluids may not arise from their being composed of Particles which flie each other; which being rather a Physical than Mathematical Inquiry, our Author forbears to Discuss." Cohen, *Newton's Papers and Letters*, 409.

27. Historians of science have been puzzled to account for the long oblivion which fell upon Bernoulli's work, but I think it can be readily explained. Bernoulli was pretty much of a Cartesian; his work appeared at precisely the moment when French physicists were becoming adherents of the attractionist physics of Newton; and attractions had no place in Bernoulli's scheme of things. But more important, perhaps, is a fact to which I have already alluded: that the scientists in France, the physicists who were mathematically equipped to cope with such matters, had a persistent doubt that anything meaningful or demonstrable could be asserted about the invisible world of atoms and molecules.

28. *The Philosophical Works of the Honourable Robert Boyle . . . by Peter Shaw, M.D.*, 3 vols. (London, 1725), I, 260. Compare Boerhaave's similar comments in his *Sermo academicus de chemia* (Leiden, 1718), 39–40.

29. John Harris, *Lexicon Technicum: Or, an Universal English Dictionary of Arts and Sciences*, 2 vols. (London, 1704–1710), II. Newton's paper is here printed in the Introduction; it has been reproduced in facsimile in Cohen, *Newton's Papers and Letters*, 256–58.

30. *Opticks: Or a Treatise of the Reflections, Refractions, Inflections and Colours of Light by Sir Isaac Newton*, reprinted from the 4th ed. (London: G. Bell, 1931), 375–406.

31. *Ibid.*, 381.

32. *Mém. Acad. Roy. Sci.*, 1718 (Paris, 1719), 202–12. Geoffroy's paper is mentioned in every history of chemistry covering the period. For recent treatments see J. R. Partington, *A History of Chemistry*, 4 vols. (London: Macmillan & Co., 1962), III, 49–55; also Maurice Crosland, "The Development of Chemistry in the Eighteenth Century," *Studies on Voltaire and the Eighteenth Century* (Geneva, 1963), 382–90, a most useful summary.

33. Partington, *History of Chemistry*, II, 355.

34. *Ibid.*, 293.

35. Glauber, in his *Furni novi philosophici* (Amsterdam, 1648), described the relative facility with which the several metals amalgamate with mercury. Boyle in his *Mechanical Qualities* (1675) explained the order in which metals replace one another from acid solutions of their salts. *Works of the Honourable Robert Boyle* (London, 1744), III, 640. Before either man, Sylvius de le Boë explained the replacement of one metal in acid solution by another as the result of the varying "affinities" of the different metals for the acid. See Partington, *History of Chemistry*, II, 288.

36. This volume is described in my note, "Newton in France—Two Minor Episodes," *Isis* 53 (1962), 219–21. Geoffroy did not receive the book until late in 1705 or early 1706, for he wrote to Sloane from Paris on January 20, 1706: "J'ay reçu tout ce que vous m'avés envoyé depuis le commencement de la guerre, sçavoir les volumes des transactions philosophiques des annés 1702 et 1703 avec L'excellent traitté d'Optique de Mʳ Newton que le père de Fontenays ma remis entre les mains de votre part." British Museum, Sloane MSS. 4040, fol. 114.

37. *Hist. Acad. Roy. Sci.*, 1718 (Paris, 1719), 36. In his *éloge* of Geoffroy, Fontenelle wrote later in 1731: "Il donna en 1718 un système singulier et une table des affinités ou rapports des différentes substances en chimie. Ces affinités firent de la peine à quelques-uns, qui craignirent que ce ne fussent des attractions déguisées, d'autant plus dangereuses que d'habiles gens ont déjà su leur donner des formes séduisantes: mais enfin, on reconnut qu'on pouvait passer par-dessus ce scrupule et admettre la table de Geoffroy, qui, bien entendue et amenée à toute la précision nécessaire, pouvait devenir une loi fondamentale des opérations de chimie et guider avec succès ceux qui travaillent." *Oeuvres de Fontenelle*, new ed., 8 vols. (Paris, 1792), VII, 400; *Hist. Acad. Roy. Sci.*, 1731 (Paris, 1733), 99–100.

38. *Nouveau cours de chymie suivant les principes de Newton & de Sthall* (Paris, 1723), p. lxvii. Sénac, author of this anonymous book, was familiar with Newton's *Opticks*,

and discusses the work of Keill and Freind; far from avoiding "raisonnemens physiques," this is what his book is about; instead of "attractions" in the Newtonian fashion, he speaks of the "magnétisme des corps." His speculations were virtually without influence; they are ignored by Baron who cites Sénac frequently in his edition of Lemery's *Cours de Chimie* (Paris, 1756).

39. The earliest discussion of these affinity tables is Demachy's *Recueil de dissertations physico-chymiques* (Amsterdam, 1774); a later, and more valuable, source is Guyton de Morveau's article "Affinité" in the *Dictionnaire de chimie* of the *Encyclopédie méthodique* (Paris, 1786), I, 535–613. As Guyton points out, Demachy was responsible for resurrecting the table of Grosse, a chemist about whom almost nothing is known except what he tells us, and assigning it the date 1730. In 1737 Voltaire sought chemical information from Grosse whom he describes as an associate of Boulduc. See *Voltaire's Correspondence*, ed. Theodore Besterman, 107 vols. (Geneva: Institut et Musée Voltaire, 1953–65), VI (1954), nos. 1207 and 1289.

40. For Rouelle see Rhoda Rappaport, "G-F. Rouelle: An Eighteenth-Century Chemist and Teacher," *Chymia* 6 (1960), 68–101. Pierre Duhem found the school of Rouelle opposed to theories in chemistry, and stresses their "empirisme chimique." See his *Le mixte et la combinaison chimique* (Paris, 1902), 43.

41. *The Works of Francis Bacon*, ed. Spedding, Ellis, and Heath, 15 vols. (Cambridge, Mass., 1863), V, 63.

42. For Jungius, see Partington, *History of Chemistry*, II, 415–22. The citation from Glauber is given, without page reference, by John Maxson Stillman, *The Story of Early Chemistry* (New York and London, 1924), 499.

43. For the early history of affinity theory, see Hermann Kopp, *Geschichte der Chemie*, 4 vols. in 2. (Braunschweig, 1843–47), Part II, 285–90. The work of Albertus Magnus, *De rebus metallicis et mineralibus libri V*, is analysed by Ferdinand Hoefer, *Histoire de la chimie*, 2d ed., 2 vols. (Paris, 1866–69), I, 384–85; Albert wrote that sulphur darkens silver and generally burns metals by its affinity to these bodies (*propter affinitatem naturae metalla adurit*). The relevant work of Geber is the *Summa perfectionis*, which may be consulted in the translation of Richard Russell (London, 1678). For Palissy, see Partington, *History of Chemistry*, II, 75; for Barchusen, see Kopp, *Geschichte der Chemie*, 287–88, and Raoul Jagnaux, *Histoire de la chimie*, 2 vols. (Paris, 1891), I, 300–301. For the later history of affinity consult Jagnaux, 301–60, and M. M. Pattison Muir, *A History of Chemical Theories and Laws* (New York, 1907), 379–430. Short accounts are given by Maurice Daumas in *Histoire générale des sciences*, ed. René Taton, 4 vols. (1958–66), II, *La Science moderne* (Paris: Presses Universitaires de France, 1958), 559–69; by Daumas in *Histoire de la science* (Paris: Encyclopédie de la Pléiade, 1957), 934–44; and by Maurice Crosland, "Development of Chemistry," 382–90.

44. John Mayow, *Medico-physical Works*, Alembic Club Reprints, no. 17 (Edinburgh and Chicago, 1908), 35. Robert Boyle objected to this sort of explanation: "I look upon amity and enmity as affections of intelligent beings, and I have not yet found it explained by any, how those appetites can be placed in bodies inanimate and devoid of knowledge or of so much as sense." Cited by Pattison Muir, *Chemical Theories and Laws*, 380. Nicolas Lemery (who, like Boyle, explained differential reactivity in terms of the *shapes* of the particles) had no patience with the "animate model." In his *Cours de chimie*, ed. Baron (Paris, 1759), 49, he uses the word "affinité" only once, and pejoratively, in referring to the astrologer's doctrine of the correspondence of the seven planets and the seven metals.

45. Pattison Muir, *Chemical Theories and Laws*, 381. The error originated with Jagnaux, who only says, however, that Boerhaave "redonna au mot affinité le sens qu'il a maintenant en chimie." Pattison Muir's elaboration of this statement is echoed by Stillman and by Daumas.

46. Peter Shaw, *A New Method of Chemistry*, 2d ed., 2 vols. (London, 1741), I, 492–93. Boerhaave's use of attraction in the case of *dissimilar* substances may have been only a way-station, but it was immensely significant. The rival theory of Stahl and his disciples still involved affinities in the other sense: all substances react because of a fundamental resemblance, or a sharing of common qualities. See Hélène Metzger, *Newton, Stahl, Boerhaave et la doctrine chimique* (Paris: F. Alcan, 1930),

139–48. Stahl's peculiar system of *latus*, an application of this principle, is discussed by Demachy, *Recueil de dissertations*, 100–105.

47. *Plan d'un cours de chymie expérimentale et raisonnée, avec un discours sur la chymie. Par. M. Macquer . . . et M. Baumé* (Paris, 1757).

48. Between different bodies, Macquer wrote, there is "une convenance, rapport, affinité, ou attraction si l'on veut, qui fait que certains corps sont disposés à s'unir ensemble. . . . C'est cet effet, quelle qu'en soit la cause qui nous servira à rendre raison de tous les phénoménes que fournit la Chymie, & les lier ensemble." *Elémens de chymie théorique* (Paris, 1749), 20. Macquer has completely discarded the notion of identity or similarity as a cause of combination, and unlike Boerhaave, discerns a single effect, his "affinité," in all chemical reactions.

49. *Dictionnaire de Chymie*, 2 vols. (Paris, 1766), I, 48.

50. *Dissertation de Jean Philippe de Limbourg, Docteur en Médecine, sur les affinités chimiques, qui a remporté le prix de physique de l'an 1758, quant à la partie chymique* (Liège, 1761). See also Fortunato Felice, *De Newtoniana attractione unica cohaerentiae naturalis causa* (Berne, 1757), *passim*; and Jacob Reinbold Spielmann, *Instituts de chymie*, trans. Cadet, 2 vols. (Paris, 1770), I, 22–24.

51. *Instituts de chymie*, I, 96–97.

52. *Encyclopédie, ou dictionnaire raisonné des sciences, des arts et des métiers*, 35 vols. (Paris, 1751–80), XIII (1765).

53. *Ibid.*, art. "Chymie," III (1753). See Charles Coulston Gillispie, *The Edge of Objectivity* (Princeton: Princeton University Press, 1960), 184–87.

54. This subject is taken up in my paper, "Some French Antecedents of the Chemical Revolution," *Chymia* 5 (1959), 73–112.

55. *Histoire naturelle*, XIII (1765), seconde vue; *Oeuvres Complètes de Buffon*, ed. Pierre Flourens, nouvelle ed., 12 vols. (Paris, n.d.), III, 414–24. Buffon argued that whereas the shape of celestial bodies has no effect on the law of gravitation because the distances are so vast, when the distances are small the effect of differing shape may be very great. Accordingly, he believed it possible to calculate the shape of "parties élémentaires" or "parties constituantes" of homogeneous bodies if the law of attraction for a particular substance could be determined by experiment.

56. *Elémens de chymie, théorique et pratique*, 3 vols. (Dijon, 1777), I, 57. For the influence of Buffon on Guyton de Morveau see Crosland, "Development of Chemistry," 385–87.

57. *Digressions académiques, ou essais sur quelques sujets de physique, de chymie & d'histoire naturelle* (Dijon, 1762). The book was actually published in 1772, but the incorrect date appears on the majority of surviving copies. See W. A. Smeaton, "L. B. Guyton de Morveau (1737–1816)," *Ambix* 6 (1957), 18–34. The *Digressions* has two principal parts, a "Dissertation sur le Phlogistique," the influence of which I have discussed in my *Lavoisier—The Crucial Year* (Ithaca, N.Y.: Cornell University Press, 1961), and an "Essai physico-chymique sur la Dissolution et la Crystallization, Pour parvenir à l'explication des affinités par la figure des parties constituantes des Corps."

58. In 1780–85 Guyton translated the first two volumes of Bergman's *Opuscula physica et chemica* (Moström nos. 150 and 165). These did not include the important "Disquisitio de attractionibus electivis," *Nova Acta Regiae Societatis Scientarum* 2 (1775), 161–250, Moström no. 97. Guyton does not refer to Bergman in the section on affinity in his *Elémens de chymie* (1777), but discusses his work at length in the *Encyclopédie méthodique*. See above, note 39. Guyton surely had a hand in the French translation of Bergman's "Disquisitio," *Traité des affinités chymiques ou attractions électives* (Paris, 1788).

59. Nathan Sivin, "William Lewis (1708–1781) as a Chemist," *Chymia* 8 (1962), 67–73. Sivin discusses the modifications Lewis made in Geoffroy's table.

60. *Experiments upon Magnesia Alba, Quick-lime and Other Alcaline Substances*, Alembic Club Reprints, no. 1 (Edinburgh, 1893), 46.

61. J. Thomson, *An Account of the Life, Lectures, and Writings of William Cullen, M.D.*, 2 vols. (Edinburgh and London, 1859), I, 23–31, 44–45. This may not be exact. W. P. D. Wightman has tried to reconstruct from surviving fragments Cullen's Glasgow lectures of 1748–49, the first year in which he took sole responsibility for

the new course in chemistry. "William Cullen and the Teaching of Chemistry—II," *Annals of Science* 12 (1956), 192–205. Wightman's identification of the phrase "habits of mixts" with "elective attractions" is not convincing; Cullen's chemical theories at this time were evidently primitive and confused, derived partly from Boerhaave and Stahl. See Cullen's "Reflections on the Study of Chemistry," in Leonard Dobbin, "A Cullen Chemical Manuscript of 1753," *Annals of Science* 1 (1936), 138–56.

62. "Notes of Dr Cullen's Lectures on Chemistry made by Dr John White of Paisley." I am indebted to Catherine R. McEwan, Librarian of the Corporation of Paisley, for providing me with a microfilm of this manuscript. It bears the date 1754, but the lectures were probably delivered at Edinburgh in 1757–58, and certainly before Cullen could recommend Andrew Reid's translation of Macquer, *Elements of the Theory and Practice of Chemistry*, 2 vols. (London, 1758).

63. "Lectures on Chymistry by Dr Willm Cullen." Clifton College Falconer MS., fols. 13–14. This dates from the 1760s, for Cullen refers to "the late Dr Alston" (Charles Alston, the physician and botanist, who died in November 1760) and mentions "some late experiments at Petersburgh" on the freezing of mercury, a clear reference to the experiments of Braun and Lomonosov, which were carried out in December 1759.

64. M. P. Crosland, "The Use of Diagrams as Chemical 'Equations' in the Lecture Notes of William Cullen and Joseph Black," *Annals of Science* 15 (1959), 75–90. This paper was actually not published until 1961; earlier I called attention to Black's use of what Crosland calls the "double-circle" diagrams, and emphasized that Cullen and Black had preceded Bergman in discussing double elective attractions. See Marshall Clagett, ed., *Critical Problems in the History of Science* (Madison, Wis.: University of Wisconsin Press, 1959), 517–18.

65. For a careful evaluation of Higgins see Partington, *History of Chemistry*, III, 736–54. In the same year (in the "Discours Préliminaire" of his *Traité élémentaire de chimie*) Lavoisier spoke with some diffidence of chemical affinities and elective attractions. He called them the part of chemistry "la plus susceptible, peut-être, de devenir un jour une science exacte," but added that "les données principales manquent, ou du moins celles que nous avons ne sont encore ni assez precises ni assez certaines pour devenir la base fondamentale sur laquelle doit reposer une partie aussi importante de la chimie." *Oeuvres de Lavoisier*, 6 vols. (Paris, 1862–93), I, 5–6. For the subsequent history of affinities, notably the work of Berthollet, see Frederic L. Holmes, "From Elective Affinities to Chemical Equilibria: Berthollet's Law of Mass Action," *Chymia* 8 (1962), 105–45.

66. "Some Daltonian Doubts," *Isis* 52 (1961), 544–54. A re-examination seems indeed to be under way, my critics still favoring the view that Dalton's achievement was independent of Richter. See the generous but skeptical remarks of Frank Greenaway, *John Dalton and the Atom* (London: Heinemann, 1966), 234–35; and Arnold Thackray's recent paper, optimistically entitled "The Origin of Dalton's Chemical Atomic Theory: Daltonian Doubts Resolved," *Isis* 57 (1966), 35–55.

67. In the *Encyclopédie méthodique*, see the *Dictionnaire de chimie*, 6 vols. (Paris, An IV), I, 539. Guyton describes in his *Elémens de chymie*, I, 60–67, experiments to measure affinities by determining the attractive forces between mercury and various metals. See Crosland, "The Development of Chemistry," 387.

68. *Conclusion of the Experiments and Observations Concerning the Attractive Powers of the Mineral Acids* (London, 1783), 20–22.

69. See my "Quantification in Chemistry," *Isis* 52 (1961), 205–6.

70. "Some Daltonian Doubts," 548.

III
THE CHEMICAL REVOLUTION AND LAVOISIER

15

John Mayow and the Aerial Nitre: Studies on the Chemistry of John Mayow—I

✿

In 1674 there was published at Oxford a work entitled *Tractatus Quinque Medico-Physici*, written by a young medical practitioner of Bath, Dr.[1] John Mayow. He was then only thirty-three, a man of some local repute, but no very high scientific standing.[2] This work led directly to Mayow's election to the Royal Society of London in November 1678, less than a year before his premature death. It is doubtful if any scientific work of the seventeenth century has been the subject of more thoroughly conflicting interpretations. It has been praised as one of the great classics[3] of chemistry and physiology, and more recently damned without mercy as absolute nonsense.

The debate and disagreement have centered about the theory of combustion and respiration which Mayow developed in this treatise and which he supported by numerous ingenious experiments. This theory has a strikingly—and it should be quickly added, a deceptively—modern ring. Air, according to Mayow, is made up of two parts: a large inert mass (the air itself): and a smaller proportion of highly active, "fermentative" particles, which he sometimes described as nitre-like, nitrous, or igneo-aerial, but which he usually preferred to call a nitro-aerial spirit (*spiritus nitro-aereus*). This nitro-aerial spirit, like the oxygen of present-day chemistry, rendered the air fit for respiration, and when absorbed by the lungs imparted the bright-red color to arterial blood. It was also the chief agent of combustion; it explained the necessity of air for burning, and the increase in weight of metallic antimony when it is calcined. This mysterious substance was thought by Mayow to be similar to the active ingredient in nitre or saltpeter, for the presence of nitre permits substances to be burned *in vacuo*, and therefore must contain within itself that which is ordinarily supplied by the air.

Despite the honor bestowed on its author, this theory had little immediate success. With a single important exception, his results were ignored by the chemists and physiologists of the eighteenth century.[4] Only after the discovery of oxygen by Priestley and Scheele, and Lavoisier's unraveling of the mystery of combustion and respiration, was its significance suddenly appreciated. In conse-

From *Actes du Septième Congrès d'Histoire des Sciences* (Jerusalem, 1953), 332–49. All rights reserved to: Académie Internationale d'Histoire des Sciences, 12 rue Colbert, Paris—2ᵉ.

quence, various publications appeared late in the eighteenth century in which Mayow's nitro-aerial spirit was described as tantamount to the discovery of oxygen and an anticipation of the antiphlogistic chemistry. Writers of the nineteenth century contributed, virtually without dissent, to develop the myth that Mayow was an unappreciated precursor of the Chemical Revolution until, at last, accounts of his work lost all resemblance to the original. One twentieth-century critic has called Mayow "one of the greatest scientific men of the seventeenth, or indeed any century,"[5] while another referred in 1928 to Mayow's forgotten "discovery of Oxygen in the seventeenth century, which was repeated by Priestley a hundred years later."[6]

In 1931 an opposite tide suddenly set in when the late T. S. Patterson of Glasgow published a long study of Mayow's work. In something over 90 pages he dismembered one by one the Mayow mythmakers, and contended that the *Tractatus Quinque* was merely a mass of speculations drawn almost entirely from the work of others.[7] Patterson quite convincingly disposes of the idea that Mayow's aerial nitre was a gas identifiable with oxygen, and that his views anticipated to any extent the New Chemistry of Priestley, Scheele, and Lavoisier. The burden of Dr. Patterson's paper, as he put it, was to show:

> that if the modern commentators had read Mayow's *Tractatus quinque* with an open mind, instead of quoting and amplifying the quite unjustified eulogies of Beddoes and Yeats, Mayow, long ago, would quietly have dropped back into the oblivion from which there was never any real necessity to recall him."[8]

That Patterson's evaluation was excessively one-sided, despite its very considerable worth, has not escaped notice; but instead of retracing the ground, most historians of chemistry have contented themselves with taking an unsubstantiated middle position. I believe there is need to reopen the controverted Mayow question, largely because of the important influence which Mayow's work had upon the theories and experiments of Stephen Hales, and therefore upon the British pneumatic school and, by this path, upon Lavoisier.[9]

Many aspects of the problem need careful sifting. But in the present study I shall confine myself to only one: the meaning and origin of Mayow's theory of the nitro-aerial spirit, and its presumed derivation from the work of Robert Boyle, Malachia Thruston, and Robert Hooke. I shall bring forward evidence of an unsuspected common source from which both Hooke and Mayow may have derived their theory.

※

Mayow's theory of the nitro-aerial agent in the air was first published six years before the appearance of the *Tractatus Quinque*. In 1668 he had brought out a small book called the *Tractatus Duo*, which contained, besides a treatise on rickets, a work called the *De Respiratione*. The *Tractatus Duo* was reprinted the following year without change. In both printings, this work of his early years—he was twenty-seven at the time—is excessively rare.[10] The two tracts were reprinted later in the *Tractatus Quinque*,[11] but with significant changes, so that Mayow's early views on respiration cannot be reliably judged from the later work.[12]

The *De Respiratione* makes no mention of combustion; but in it Mayow states his belief that there is something in the air that is necessary to life, and which passes into the blood during respiration. Air is rendered unfit for living things when this something (*quicquid sit*) is totally consumed. And he conjectures that certain subtle and vital particles that abound in the air are communicated to the blood.[13] Mayow refers to this mysterious substance as "this aerial nitre" (*nitrum hoc aereum*) or simply as "this nitre" (*nitrum hoc*),[14] and he clearly puts it in the class of saline particles.[15]

Mayow then invokes this aerial nitre to explain other physiological processes. Fermenting with the sulphureous parts of the blood, not only in the heart but also in the bloodstream, it accounts for the production of the animal heat. Still more boldly Mayow explains muscular contraction as due to an "explosion" that results when the nitrous particles carried by arterial blood meet the animal spirits or the "volatile spirit of blood." This explosive effect is enhanced by the presence in blood of sulphureous as well as nitrous particles, which render the mixture nitro-sulphureous.[16]

Patterson makes short work of this treatise. There is in it, he feels, "nothing new"; it is merely an account of a subject "developed by others"; the theory of the expansion and contraction of the muscles by a nitrous "explosion" is "at the best, extremely crude."[17] There is no novelty even in the hypothesis that something is removed from the air in respiration, for something of the sort had been suggested by Robert Boyle, who in turn cited Paracelsus as his authority.[18] The theory of the aerial nitre, Patterson contends, was derived from Robert Hooke's theory of combustion set forth in his *Micrographia* (1665). Mayow's application of the nitre theory to respiration may even owe something to one Malachia Thruston, who, in 1664, had presented a thesis at Cambridge in which he asserted that air is laden with nitrous particles which are taken into the lungs during respiration, leaving the expired air vapid and effete.[19]

Let us examine these claims in some detail, especially the dependence of Mayow's views upon those of Hooke and Thruston.

Patterson recognizes that Thruston's thesis, though publicly defended at Cambridge early in 1664, was not published until 1670.[20] But he seems to feel that Mayow knew of it and used it in the *De Respiratione* of 1668–69, for he speaks of Mayow's reference to Thruston in the *Tractatus Quinque* of 1674 as "rather belated." But Mayow clearly gives the impression that he did not know of Thruston's thesis until he was at work on the *Tractatus Quinque*; and unless we have better reason than at present to impugn Mayow's honesty, we must take this at face value. Indeed it is likely that Mayow first encountered Thruston's work, not in the first edition of 1670, but in a reprint of 1671 where it forms part of a collection that included Mayow's own *Tractatus Duo*, and which, therefore, he is more likely to have seen.[21] To Thruston we shall return, for there is good evidence that he drew upon the same common source for his theory of nitrous particles as did Hooke and Mayow.

The earliest instance of Hooke's speculations on the nitrous agent dates from January 4, 1664/5 (O.S.), when according to the records of the Royal Society he attempted to show "that air is the universal dissolvent of all sulphureous bodies" and to prove that this was caused by a "nitrous substance inherent and mixt with the air."[22] Soon after, at the suggestion of Robert Boyle, Hooke used the vacuum pump to demonstrate that gunpowder could burn *in vacuo*, and that sulphur dropped upon nitre in a red-hot iron crucible in an exhausted receiver flamed "as freely, as if it had been in the open air."[23] Late in 1665, Hooke published his *Micrographia*, where he developed at somewhat greater length this theory of combustion. "Fire," he writes, "is nothing else but a dissolution of the burning body by the most universal menstruum of all sulphureous bodies, namely the Air." This dissolution "is made by a substance inherent, and mixt with the Air, that is like, if not the very same, with that which is fixed in Salt-peter."[24] Now it is one of Patterson's chief points that this cautious statement is far superior to Mayow's references to nitrous particles—or, as Patterson believes, to actual nitre in the air. Patterson insists that Mayow's later abandonment of this terminology and his substitution of the phrase "nitro-aerial spirit" in the *Tractatus Quinque* is an important improvement since it no longer assumes that actual nitre is the agent in the air.

Now it can, I think, be shown that the difference between Hooke's and Mayow's theories has been greatly exaggerated. Mayow nowhere flatly states that particles of actual solid nitre are present in the air. Even his statement that these particles are "saline," as we shall see, proves little. What is more, in a passage of the *Micrographia* which Patterson has overlooked, Hooke compares air with its dissolvent particles to "*saline* menstruums, or spirits,"[25] and in a later reference to the problem he stated his belief that alkalis exposed to the air "would arrest the *volatile salt*, which is the air, and turn it into nitre."[26]

Now to the modern chemist, who thinks of a salt as a solid compound substance, produced by the reaction of an acid with a base, which is more or less soluble in water, but only in exceptional cases volatile, it is difficult to understand this use of the word. To chemists of the late seventeenth century an essential mark of many salts was that they produced upon distillation an acid, corrosive spirit, which in aqueous solution provided the most important acid menstruums or solvents: spirit of salt (HCl), spirit of vinegar (acetic acid), spirit of nitre (aqua fortis), and so on.[27] These *acid spirits*, Robert Boyle remarks at one point, "seem to belong to the family of salts."[28] And elsewhere he speaks of the volatility of a substance as proving "how saline a nature it is." Thus, far from implying the solid state, the saline characteristics of a substance were assumed to give it volatility; or, put in another way, it is the volatile ingredient in solid substances like nitre, vitriol, or sea salt which, to these early chemists, represented the true salt. It seems clear, therefore, that when Hooke speaks of a substance in the air, "that is like, if not the very same, with that which is fixed in Saltpeter" he is thinking of a saline spirit, perhaps not the same as, but closely resembling, the well-known spirit of nitre. It is also evident that when Mayow alters his terminology, in the *Tractatus Quinque*, to speak of the nitro-aerial spirit, he is not, as Patterson supposes, abandoning an earlier view that common nitre was found in the air. This he never held. What he and Hooke both believed, with no very great clarity on either side, was that a volatile salt, or saline ingredient, of common nitre was to be found in the air. Mayow's later choice of the current term "spirit" was therefore no great terminological advance, and need hardly be attributed solely to the intervening use of the term by Lower.[29]

Having stressed sufficiently the similarity of the theories of Hooke and Mayow, let us examine the contention that Mayow borrowed Hooke's theory without acknowledgment.[30] Now Patterson is correct when he says that "there is no reason to suppose, either that Mayow intended or that Hooke felt any plagiarism," although Mayow nowhere mentions Hooke in his work.[31] It is significant that Hooke, a notoriously jealous, sensitive, and combative man, never took umbrage at Mayow's failure to refer to him.[32] Far from it; for after the publication of the *Tractatus Quinque*, which seems to have brought the two men together, they met pleasantly on several occasions;[33] and it was Hooke who, in fact, nominated Mayow to the Royal Society.[34] One recent writer, not without justice, has described Hooke as viewing Mayow as an ally, sympathetic to his theory of combustion.[35] If neither man had reason to feel that he could claim any special credit for the basic idea that a nitrous substance in the air played an important role in natural phenomena, an important source of potential disagreement would vanish.[36] In the following pages, I shall try to show that this central idea of a nitrous agent, and even an intimation of the use made of it by Thruston, Hooke, and Mayow, was widely current in the seventeenth century.

❖

As Patterson points out, Robert Boyle in his first scientific publication was led to conjecture "that there is some use of the Air which we do not yet so well understand, that makes it so continually needful to the life of animals," and that there might be "in the air a little vital Quintescence . . . which served to the refreshment and restauration of our vital Spirits."[37] Boyle supports his conjecture by quoting Paracelsus "that as the Stomack concocts Meat, and makes part of it useful to the Body, rejecting the other part, so the Lungs consume part of the Air, and proscribe the rest."[38]

To this work of Boyle, and to his growing prestige as well, we can safely attribute much of the interest early manifested by the Royal Society *virtuosi* in the problem of air and its role in respiration and combustion. This subject was repeatedly discussed in the early meetings. One of the first to raise a problem of this sort before the Society was that most occult-minded and eccentric of the early Fellows, Sir Kenelm Digby. On January 23, 1660/61 (O.S.), Digby delivered at Gresham College a wordy, rambling, but nonetheless interesting *Discourse Concerning the Vegetation of Plants*.[39] In this tractate, published shortly afterward,[40] there are some extremely significant passages.

For example, Digby described, from his own experiments and those of others, the astonishing influence of solutions of saltpeter upon plant growth, and advanced the theory that plants receive their fecundity not only from a nitrous salt, a balsamic saline juice in the earth which causes a plant to "Swell, Germinate, and Augment it self," but also from a similar substance in the air. Indeed, Digby suggests that the saltpeter in the earth must attract to itself, like a magnet, a similar salt found in the air. "And this," he added in a cryptic and tantalizing aside, "gave cause to the Cosmopolite to say, *there is in the Air a hidden food of life*." Airs, he goes on, "as are most impregnated with this benign fire, are healthful to live in," while others which have little balsamick Salt in them—"the food of the Lungs, and the nourishment of the Spirits"—are, as he put it, "unsound."[41]

Here, I believe, is one of the principal direct sources of the views held by Thruston, Hooke, and Mayow. Thruston, for one, credits Digby, along with Gassendi and George Ent, for his ideas about nitrous particles.[42] While Mayow does not quote Digby, at one point in his *De Respiratione* (1668–69) he writes that the aerial nitre is necessary for all living things, including plants, which cannot grow and flourish without it, for the air provides this fertilizing salt.[43] This seems to be a clear echo of Digby. Hooke also shows Digby's influence. On June 7, 1665, he described to the Royal Society an experiment that in all probability was designed to test Digby's theory that air is necessary for the growth of plants. He divided a parcel of lettuce seed and sowed some in earth exposed to air and the rest in earth placed in the receiver of a vacuum pump he then exhausted of its air. The seed which had been exposed to the air germinated and sprouted in eight days, to the height of an inch and a half; but the seeds in the exhausted receiver remained dormant.[44] Patterson hazards the opinion that Mayow's statement, just quoted, was based on this experiment of Hooke's;[45] but it is more likely that Digby's *Discourse* is the common source of both men.

Let us pursue the matter further and inquire into Digby's source for his theory of an aerial, nitrous salt necessary for life. In the passage cited above, Digby mentions, besides the mysterious *Cosmopolite*, to whom we shall shortly return, the name of Cornelis Drebbel (1572–1633), the famous scientific adventurer, inventor, and alchemist, who had spent the last twenty years of his life in England and whose exploits were well known. Drebbel, like Paracelsus, to whom he may have been indebted, believed that air was composed of two parts, a quintessence fit for respiration, and a "carcass" that is inert. His theory seemed to be confirmed, or at least was widely publicized, by the tale of his famous submarine demonstrated early in the century in the presence of King James I. The method by which he claimed to "revive his languishing guests, in his straight house under water,"[46] as Digby puts it, by releasing an aerial substance prepared from saltpeter, gave strong support to the theory of a life-giving nitrous substance in the air. This episode, mentioned only briefly by Digby, was described in detail by Robert Boyle in his *New Experiments* (1660).[47] I do not propose to answer the question whether, as I think entirely possible, Drebbel actually prepared oxygen—over a century and a half before Scheele and Priestley—by the method favored by Scheele, namely by the thermal decomposition of saltpeter; I only wish to emphasize that the widely repeated story of Drebbel's submarine was a powerful force in dramatizing the idea of a nitrous agent in the air.[48]

It is not difficult to identify Digby's *Cosmopolite*. We must begin with a Scottish alchemist, Alexander Seton by name, who in 1602–3 was plying his dubious trade on the Continent, first in Holland, then in Zurich, Strasbourg, and other cities of the Empire.[49] Arriving at last in Dresden, he undertook to effect a transmutation for the skeptical Elector of Saxony, Christian II. When Seton failed in his attempt, he was threatened by Christian and, so the story goes, was imprisoned and tortured. A certain Michael Sendivogius, a Polish gentleman at that moment in Dresden, who was an ardent devotee of the hermetic arts, determined to rescue Seton and extract his secret from him. With the connivance of Seton's wife, Sendivogius was able to free the suffering alchemist from prison and to spirit him out of the country. Seton died soon after at Cracow, early in 1604, and Sendivogius fell heir both to the widow and to the manuscript. This work, the *Novum Lumen Chymicum*, Sendivogius promptly published at Prague in 1604, putting his own name, in the form of an anagram, on the title page.[50] This treatise was many times reprinted during the seventeenth and eighteenth centuries, often accompanied by one or more of the treatises ascribed to Sendivogius himself, such as the important *Tractatus de Sulphure*. A French translation by De Bosnay, called the *Cosmopolite ou Nouvelle Lumiere de la Phisique Naturelle,* appeared in 1609 and was several times reprinted; an English translation, called *A New Light of Alchymie*, was published in 1650.[51] The works of Sendivogius and Seton, especially the *Novum Lumen Chymicum*, were widely read and extremely influential in their day. They are included in such collections as Nathan Albineus's *Bibliotheca Chemica Contracta* (1673), Manget's *Bibliotheca Chemica Curiosa* (1702), and the *Musaeum Hermeticum* (1749).[52] Sendivogius and Seton are mentioned by Pierre Borel[53] and by Borrichius.[54] The tract *De Sulphure* is cited by Sir Thomas Browne in the *Pseudodoxia*,[55] and Athanasius Kircher refers to Sendivogius in his widely read *Mundus Subterraneus* (1664).[56] The dramatic events of Seton's life were well known from accounts published by Hoghelande (1604), Dienheim (1610), and Morhof (1673). As late as 1732, Hermann Boerhaave cites the name of Sendivogius; but Peter Shaw warns, in his translation of Boerhaave's *Elementa*, that Sendivogius' writing "should be read with caution; being full of vain promises."[57] Yet barely an echo of Sendivogius' great reputation in his own century is to be found in the works of modern historians of chemistry.[58]

The passage which Digby cites is readily found in the epilogue to the *Novum Lumen Chymicum*. In the somewhat biblical style of the English translator, it reads as follows:

> Man was created of the Earth, and lives by vertue of the Aire; for there is in the Aire a secret food of life, which in the night wee call dew; and in the day rarified water, whose invisible congealed spirit is better than the whole earth.[59]

There are other, somewhat less confusing, passages which allude to the *Cosmopolite*'s belief—to be traced back to its source in Paracelsus—that there is a mysterious and efficacious ingredient in the air. But by all odds the most impressive statement is not in the *Novum Lumen Chymicum*, but its accompanying tract, *De Sulphure*:

> In [air] also is the vitall spirit of every creature . . . it nourishes them, makes them conceive, and preserveth them; and this daily experience teacheth, that in this element not only Minerals, Animals, or Vegetables live but also other Elements. For wee see that all Waters become putrefied, and filthy if they have not fresh Aire: The Fire also is extinguished, if the Aire be taken from it. . . . In briefe the Whole structure of the world is preserved by Aire. Also in Animals, Man dies if you take Aire from him, &c. Nothing would grow in the world, if there were not a power of the Aire, penetrating and altering, bringing with it selfe nutriment that multiplies.[60]

In the light of the later speculations of Boyle, Hooke, and Mayow, this is indeed a surprising passage to have escaped the notice of historians of science. And the importance of these words, imbedded though they are in elaborate passages phrased in the calculatedly high-flown and evasive style of later alchemy, is enhanced from the point of view of our problem when we find the *Cosmopolite*, like his contemporary Drebbel, identifying the "secret food of life" and the "power of the Aire" with nitre and a nitrous substance. In one otherwise obscure passage the *Cosmopolite* speaks of the "waters of our dew out of which is extracted the Salt Petre of Philosophers, by which all things grow and are nourished."[61] In still another place he talks of the fertilizing power of rain and attributes it to the same mysterious agency:

> Therefore when there is rain made, it receives from the aire that power of life, and joins it to the salt-nitre of the earth . . . and by consequence the greater plenty of Corn grows, and is increased, and this is done daily.[62]

These passages are echoed by Digby when he says that it is "well-digested dew" that "makes all Plants luxuriate and prosper most" and when he concludes that rain and dew owe their "prolifick virtue" to the nitrous salt dissolved within them, which "gives foecundity to all things."[63]

Lastly the *Cosmopolite* is clearly Digby's source for the theory that saltpeter in the earth is generated from the nitre in the air by an attraction of like to like.

He likens this to the attraction of the lodestone for iron, a comparison that Digby borrows, and to the hygroscopic properties of calcined tartar, drawing water from the air: "such attractive power," he writes, "hath the salt-nitre of the earth, which also was aire."[64]

❖

Views similar to Digby's—and perhaps also derived from the source we have just identified—had been published as early as 1641 by Dr. George Ent (1604–89), a friend and disciple of William Harvey and one of the founders of the Royal Society of London.[65] Ent's chief work, and almost certainly the source referred to by Thruston, was his *Apologia pro circulatione sanguinis* (1641), written to defend the Harveian doctrine of the circulation against the attacks of an Italian physician, Emilio Parigiano (1567–1643). In this book, written in clear and precise Latin, Ent sought to overwhelm his antagonist by close reasoning and a weighty arsenal of classical and contemporary references.[66] Though it contains little that is original, and added no new experimental evidence, the book affords an illuminating summary of contemporary physiological theories. Here we find the first sustained application of the nitre theory to problems of respiration. In Ent's book the earliest significant reference to the subject appears in a discussion of the respiration of fish, where he offers it as his opinion that air "in the state we know it" is not contained in water, but that the "nitrous quality . . . for which we chiefly breathe the air, is likewise contained in water, and fish live upon that."[67] In another section, which in its main lines resembles the later writings of Thruston, Mayow, and Lower, Ent explains that the blood absorbs air during its passage through the lungs in higher animals and is carried to the left ventricle where—as in ordinary combustion—it maintains the fire in the heart. Deprived of air, this vital heat is extinguished like any fire when the air is cut off.[68] The role of a nitrous substance in respiration is expounded in the following passage, which is perhaps the most significant of the book:

> It is well known to everyone how [fire is extinguished] in medical cupping glasses, in a closed furnace, and elsewhere. And this happens, because the fire is fostered by the nitre of the air (I have previously called it dew). And the more nitre is contained in the air, the more strongly this flame burns. . . . That is the power (*virtus*) for which we breathe; without it, air would only do us harm. Thus from rotten swamps and caverns redolent of Mephitis, pestilential air, deprived of this nitre, breathes forth and brings certain death to living creatures.[69]

Whatever opinions we may harbor about these speculations, it is interesting to find the nitre theory of respiration applied also to combustion, as in the *Tractatus de Sulphure*. Ent makes no mention of Sendivogius, nor have I been able as yet to determine what his sources were or how much he drew from them. But a reference of the *Palladium Spagyricum* (1624) of Pierre-Jean Fabre suggests that the alchemical and iatrochemical literature in which the nitre theory originates was not unknown to him. The popularity of the *Novum Lumen Chymicum* and the *Tractatus de Sulphure*, and the numerous editions of them already available, makes it highly probable that Ent was, directly or indirectly, influenced by these writings.

Only those unacquainted with the writings of Paracelsus or of his followers will be surprised to find the words "balsam" and "dew" used for this mysterious, life-preserving substance. These terms are common in the late alchemical literature from the time of Paracelsus onward. In Paracelsus we may accordingly expect to find some anticipations of the views of the *Cosmopolite* or Sendivogius, and we will not be deceived. In his famous theory of the Elements, the *salt* of the *tria prima* is described as a natural balsam of the living body and its protection against decay.[70] And although he stresses the importance of the earth, and the nutrients of plants and animals drawn from the earth, as a chief source of this balsamic salt,[71] Paracelsus does not weary of repeating that in some manner the air plays a deeply significant role in the phenomena of life. "The air lives of itself, and gives life to all other things."[72] The life of fire, for example, is air "for the air makes the fire blaze more strongly and with greater impetuosity. Some air proceeds from all fire, sufficient to extinguish a candle or to lift a light feather, as is evident to the eyes. All live fire, therefore, if it be shut up or deprived of the power to send forth its air, must be suffocated."[73] Above all, air contributes in some mysterious way to the maintenance of the life of men, as the passage that Boyle cited clearly suggested, some part of the air being utilized, and the rest rejected. And at one point we find Paracelsus saying:

> The life, then, of all men is none other than a certain astral balsam, a balsamic impression, a celestial and invisible fire, an included air, and a spirit of salt which tinges. I am unable to name it more clearly, although it could be put forward under many distinctive titles.[74]

I have found nothing to suggest that Paracelsus identifies this balsamic salt in the air with *nitre*.[75] It seems to have been his followers, like the *Cosmopolite* or Sendivogius, who sought to "name it more clearly." Until more evidence is forthcoming—and I do not claim to have combed the works of Paracelsus and all his disciples thoroughly enough to say that there is no such evidence—I shall be tempted to believe that the theory of a life-preserving balsamic nitre, or nitrous spirit, in the air, was the work of the *Cosmopolite* and Sendivogius, developing their Paracelsian legacy. Whether or not this will turn out to be the final explanation, it was almost certainly through the *Novum Lumen Chymicum* and the *Tractatus de Sulphure* that this physiological theory of an aerial nitre became known in the seventeenth century, passed to Ent and Digby, and from them to Hooke, Thruston, Lower, and John Mayow, who found it a useful conceptual scheme by which to interpret the steadily growing mass of experimental evidence concerning combustion and respiration.

✿

In this paper I have confined myself to a small group of chemical and medical authors in whom we find, well before the time of Thruston, Hooke, and Mayow, references to the life-sustaining properties of a nitrelike ingredient in the air. I have said little about Hooke's theory of combustion, beyond trying to show that his view of the nitrous particles differs hardly at all from that presented by Mayow in his *De Respiratione* of 1668. I have elucidated neither Thruston's reference to Gassendi, nor Mayow's peculiar theory of muscular contraction as due to the violent explosion of nitrous and sulphureous ingredients in the blood and tissues. In a subsequent paper, I hope to fill in the background of this widely accepted idea of an aerial nitre by taking up an all-but-forgotten meteorological theory, widely current in the seventeenth century, which gave strong support to the physiological theories I have just set forth. But I have, I trust, produced enough evidence up to this point to suggest that we need not, with Dr. Patterson, assume that Mayow in the *De Respiratione* of 1668 derived his ideas from either Hooke or Thruston. Thruston alone records his debt to men like Digby and Ent, and we have traced back the ideas of these men to Drebbel and the *Cosmopolite*. We can only assume that Hooke and Mayow, who were not kind enough to steer us to their sources as Thruston and Digby did, drew their nitrous spirit from the same alembics.

Notes

1. Mayow was LL.D., not M.D. On the title page of his chief work he refers to himself as "LL.D. & Medicus."
2. It has recently been shown that Anthony à Wood's biographical assertions must be revised, and that Mayow was born in Cornwall (Manor of Bray) in 1641 not, as generally said, in 1643. See Douglas McKie's note in *Nature* 148 (1941), 728, and his "The Birth and Descent of John Mayow: A Tercentenary Note," *Philosophical Magazine* 32 (1942), 51.
3. Extracts of the work in German translation are included in Ostwald's *Klassiker der exacten Wissenschaften*, no. 125; and the whole has been done into English in Alembic Club Reprints, no. 17 (Edinburgh, 1907).
4. The exception is the Rev. Stephen Hales. Ferdinand Hoefer, however, lists a half dozen followers of Mayow, mostly obscure, all of whom published their works between 1676 and 1689. See his *Histoire de la chimie*, 2 vols. (Paris, 1843), II, 271. But his theories were criticized by Borelli, Boyle, and Haller, and he is ignored by Stahl.
5. Francis Gotch, *Two Oxford Physiologists* (Oxford, 1908), 5.
6. Charles Singer, *A Short History of Medicine* (New York: Oxford University Press, 1928), 189.

7. T. S. Patterson, "John Mayow in Contemporary Setting, A Contribution to the History of Respiration and Combustion," *Isis* 15 (1931), 47–96, 504–46.
8. Patterson, "John Mayow," 55, n. 10.
9. See the present writer's "The Continental Reputation of Stephen Hales," *Archives Internationales d'Histoire des Sciences*, vol. 4, no. 15 (1951), 393–404.
10. I have used, on microfilm, a copy of the 1669 reprint, kindly supplied by the John Crerar Library of Chicago.
11. For full bibliographical details on Mayow's writings, the reader should consult John F. Fulton, "A Bibliography of . . . Richard Lower and John Mayow," Oxford Bibliographical Society, *Proceedings and Papers* 4 (1935), 1–62.
12. Patterson, "John Mayow," 81 and 88.
13. *De Respiratione* (1669), 43. Mayow's Latin reads: "Et verisimile est *tenuiores esse, & nitrosas particulas*, quibus abundat aer, quae quae [*sic*] per pulmones sanguini communicantur."
14. *Ibid.*, 44, 49.
15. *Ibid.*, 50–51.
16. The Latin of this passage is given by Patterson, "John Mayow," 76, n. 47. Nevertheless he omits a word in the next to the last sentence, and does not give the original of the very significant last sentence, though he gives its substance in his text; these sentences read: "*Spiritus* hi, inquam, quoties motus obeundi gratiâ à nervoso genere amandati, prioribus illis *nitro salinis*, & diversi generis, particulis occurrunt, ex eorum mixturâ veluti *ex spiritu sanguinis volatili*, & *liquore salino* unitis subitam illam explosionem, & per consequens musculorum inflationem simul, & contractionem fieri verisimile est. Et forte etiam *sanguis* ad ebullitionem hanc nonnihil conducit; cujus particulae sulphureae, nitro inspirato conjunctae, liquorem *nitro-sulphureum* & magis adhuc explosivum efficiunt." *De Respiratione* (1669), 51–52.
17. Patterson, "John Mayow," 76–78.
18. *Ibid.*, 62–63, 77.
19. *Ibid.*, 82–84.
20. It is entitled *De Respirationis Usu Primario, Diatriba*. I have consulted it in Daniel Le Clerc and Jean Jacques Manget, *Bibliotheca Anatomica*, 2 vols. (Geneva, 1699), II, 166–85.
21. Fulton, "Bibliography," 44; and Patterson, "John Mayow," 82.
22. Thomas Birch, *The History of the Royal Society of London*, 4 vols. (London, 1756–57), II, 2.
23. *Ibid.*, 15 and 19.
24. Robert Hooke, *Micrographia*, 103.
25. *Ibid.*, 104.
26. Birch, *History of the Royal Society*, II, 307.
27. Robert Boyle, "Of the Producibleness of Spirits," in *Works*, ed. Thomas Birch, 6 vols. (London, 1772 ed.), I, 609.
28. Robert Boyle, *Skeptical Chemist*, in *Works*, I, 540.
29. Patterson, "John Mayow," 88, does concede that the term "may have been in common use at the time." What Lower says in his *Tractatus de Corde* (1669) is that the change in color from venous to arterial blood is due to a "nitrous spirit of the air"; see p. 169.
30. This charge was first made by the British chemist Thomas Thomson in his *System of Chemistry*, 4 vols. (Edinburgh, 1802), I, 347. Patterson seems to know only the briefer statement in Thomson's *History of the Royal Society* (London, 1812), 467. Oddly enough, Thomson has almost nothing to say about Mayow in his *History of Chemistry*. For Patterson's views on the relationship of Hooke and Mayow, see his paper "John Mayow," 51, 69, 77.
31. Patterson, "John Mayow," 51, n. 8.
32. It should be recalled that Hooke does not apply his theory to respiration, but confines himself to promising elsewhere "to shew the use of air in respiration," but as Patterson admits, he never seems to have fulfilled his promise. At the Royal Society, Hooke seems to have hinted on one occasion that respiration depends upon a "nitrous quality" in the air (June 20, 1688), and in June 1672 he suggested that "by the air something essential to life might be conveyed into the blood; and something that was

noisome to it, be discharged back into the air." See Birch, *History of the Royal Society*, II, 184; and III, 55–56.

33. *The Diary of Robert Hooke*, ed. H. W. Robinson and Walter Adams (London: Taylor & Francis, 1935), 130 and *passim*.

34. Douglas McKie, *Phil. Mag.* 32 (1942), 57; and see Birch, *History*, III, 384.

35. D. J. Lysaght, "Hooke's Theory of Combustion," *Ambix* 1 (1938), 94.

36. This does not imply that Hooke did not take full credit for his elaboration of his theory of combustion. See his *Lampas: Or, Descriptions of some Mechanical Improvements of Lamps and Waterpoises, etc.* (London, 1677), where he speaks of "The Hypothesis of Fire and Flame I did about eleven years since publish in my Micrographia, which hath so far obtained, that many Authors have since made use of it, and asserted it." This would seem clearly to include Mayow's *Tractatus Quinque*.

37. Cited here from Patterson, "John Mayow," 62. See also Boyle, *Works*, I, 107.

38. Patterson, "John Mayow," 62, n. 20, has identified the passage in Paracelsus that Boyle probably had in mind. It is from the *De Morbis Metallicis*.

39. Birch, *History of the Royal Society*, I, 13.

40. *A Discourse Concerning the Vegetation of Plants: spoken by Sir K. D. at Gresham College, on the 23 January, 1660, at a meeting of the society for promoting knowledge by experiments* (London, 1661). A French translation by P. de Trehan was published in Paris in 1667; a Latin translation by O. Dapper appeared in Amsterdam in 1669. Also in 1669 the discourse was republished in English in conjunction with Digby's earlier *Of Bodies, and of Man's Soul* and his well-known *Discourse of the Powder of Sympathy*. This is the edition consulted.

41. Digby, *Of Bodies, and of Man's Soul* (London, 1669), 222–23.

42. Le Clerc and Manget, *Bibliotheca*, II, 173.

43. Mayow, *De Respiratione* (1669), 44. The Latin reads: "Adeo enim ad vitam quamcunque nitrum hoc aereum necessarium est; ut ne plantae quidem in terra eodem privata crescant, quae tamen si aeri exposita, sale hoc foecundante denuo impregnetur, plantis demum alendis rursus idonea evadit: plane ut vel *ipsae plantae qualemque respirationem, aerisque necessitatem* habere videantur."

44. Birch, *History of the Royal Society*, II, 54.

45. Patterson, "John Mayow," 75–76, n. 45.

46. Digby, *Of Bodies*, 223.

47. Boyle, *Works*, I, 107. This passage, which follows directly his reference to Paracelsus, was unaccountably overlooked by Patterson. Boyle elsewhere cites Mersenne's testimony, and that of Drebbel's son-in-law, in support of the same episode. *Ibid.*, 453.

48. For a full account of Drebbel's activities, and the numerous references to his "nitrous air" and the submarine by Oldenburg, Huygens, Leibniz, and others, see Gerrit Tierie, *Cornelis Drebbel (1572–1633)* (Amsterdam, 1932), 64–71. The present writer has found the story repeated by Isaac Newton in one of his unpublished notebooks.

49. A brief summary of the facts concerning Seton and Sendivogius can be found in John Ferguson, *Bibliotheca Chemica*, 2 vols. (Glasgow, 1906), II, 368–70, 374–76, with extensive references. See also A. E. Waite, *Lives of the Alchemystical Philosophers* (London, 1888); and Louis Figuier, *L'Alchimie et les alchimistes* (Paris, 1856) and later editions.

50. Divi Leschi Genus Amo (Michael Sendivogius). A second (?) edition was published by the chemist Jean Beguin in Paris in 1608; what seems to be the third edition appeared in Frankfurt in 1611 under the title *De Lapide Philosophorum*. This is the earliest Latin version I have consulted. No detailed bibliographical study of the Seton-Sendivogius writings seems to have been made, but see Ferguson, *Bibliotheca Chemica*, II, 364–70, 374–77.

51. *A New Light of Alchymie: Taken out of the fountaine of Nature, and Manuall Experience. To which is added a Treatise of Sulphur . . . by J. F., M.D.* (London, 1650). A second edition of this appeared in 1674. As the title indicates, this includes a translation of the *Tractatus de Sulphure*, a French version of which, by F. Guiraud, appeared in 1628, in the same year and with the same publisher as an edition of De Bosnay's translation of the *Novum Lumen Chymicum*. These were bound together in the French version I was able to consult.

52. Ferguson, *Bibliotheca Chemica*, II, 367–68.
53. Pierre Borel, *Bibliotheca Chimica* (Paris, 1654), 210; for a fuller account see the same author's *Tresor de Recherches et Antiquitez Gauloises et Françoises* (Paris, 1655), 474–89, 581–86.
54. Olaus Borrichius, *De Ortu et Progressu Chemiae Dissertatio* (Copenhagen, 1668), 144.
55. Sir Thomas Browne, *Pseudodoxia Epidemica* in *Works*, ed. Charles Sayle, 3 vols. (London, 1904), I, 240.
56. Athanasius Kircher, *Mundus Subterraneus*, 2 vols. in one (Amsterdam, 1664), II, 266.
57. Peter Shaw, *A New Method of Chemistry*, 2 vols. (London, 1741), I, 55, which refers to the 1674 edition of the English translation. See also p. 419.
58. Hoefer and Kopp devote space to him; but Stillman, otherwise so informative on the chemists of the period, merely lists him among the alchemists. Partington, in *A Short History of Chemistry*, 2d ed. (London: Macmillan Co., 1948), omits him entirely.
59. *A New Light of Alchymie* (1650), 40. The Latin reads: "Creatus homo de terra, ex aere vivit; est enim in aere ocultus vitae, cibus, quem nos rorem de nocte, de die aquam vocamus rarefactum, cuius spiritus invisibilis, congelatus melior est, quam terra universa." *De Lapide Philosophorum* (1611), 47–48.
60. *A New Light of Alchymie*, 96.
61. *Ibid.*, 41.
62. *Ibid.*, 43–44.
63. Digby, *Of Bodies*, 222. Since the designation "Cosmopolite" does not appear in the English translation—or, as far as I have been able to ascertain, in the title of the various Latin versions—it would seem that Digby encountered this work in the French edition. However that may be, it is likely that Digby is not quoting from the English translation by "J. F., M.D.," for this translator renders the crucial phrase as "a secret food of life," whereas Digby says "a hidden food of life."
64. *A New Light of Alchymie*, 43.
65. For facts about Ent see the *Dictionary of National Biography*. His importance as a founder of the Royal Society is described in Martha Ornstein, *The Role of Scientific Societies in the Seventeenth Century*, 3d ed. (Chicago: University of Chicago Press, 1938), 93 ff. For a brief reference to his defense of Harvey against Parigiano see H. P. Bayon, "William Harvey, Physician and Biologist: His Precursors, Opponents and Successors," Parts I and II, *Annals of Science* 3 (1938), 86.
66. George Ent, *Apologia Pro Circulatione Sanguinis: Qua respondetur Aemilio Parisano Medico Veneto* (London, 1641). I was generously lent a copy of this rare book by the Yale Medical Library. This copy had been presented to Dr. J. F. Fulton by his friend, the distinguished American surgeon, Harvey Cushing.
67. Ent, *Apologia*, 18. For the transcription and translation of these passages, as well as for much assistance in rendering other passages I have quoted, I am greatly indebted to my wife, Rita Guerlac.
68. *Ibid.*, 96–97.
69. *Ibid.*, 97–99.
70. *The Hermetic and Alchemical Writings of Paracelsus*, ed. Arthur Edward Waite, 2 vols. (London, 1894), I, 259. See also pp. 257–58.
71. *Ibid.*, I, 258 and 263. The idea of a universal magnetic property in nature is also Paracelsian. See *ibid.*, I, 132. Waite defined the word "magnet" as alchemically speaking the "dew of the philosophers." *Writings of Paracelsus*, II, 373.
72. *Ibid.*, I, 137. See also p. 134: "None can deny that the air gives life to all corporeal and substantial things which are born and generated from the earth."
73. *Ibid.*, I, 137.
74. Waite, *Writings of Paracelsus*, I, 136. It is of interest to find Helmont's term "gas" defined in terms of these Paracelsian ideas by a late-seventeenth-century writer: "Gas, a term used by Helmont, and signifies a Spirit that will not coagulate, or the Spirit of Life, a Balsam preserving the Body from Corruption." See Stephen Blancard, *The Physical Dictionary*, English trans., 3d ed. (London, 1697).

75. Yet Paracelsus stresses the great use made of nitre or saltpeter in alchemical operations and says, "By means of this salt many of the Arcana in Alchemy are brought about. . . ." *Writings*, I, 263. There is, however, the interesting statement that nitre "is an essential spirit and excrement of all salts, possessing a hermaphroditic nature." See his *De Pestilitate*, cited by Waite, *Writings*, I, 100 note.

[My suggestion that the nitre theory might have a Paracelsian origin was soon followed up by Allen Debus "The Aerial Niter in the 16th and early 17th Centuries," *Proceedings of the Tenth International Congress of the History of Science*, 2 vols. (Paris: Hermann, 1964), II, 835–39; and by W. Hubicki "Michael Sendivogius's Theory, Its Origin, and Significance in the History of Chemistry," *ibid.*, II, 829–33.]

16

The Poets' Nitre:
Studies in the Chemistry of John Mayow—II

✿

The results reported here grew out of an inquiry into the scientific achievement
and present reputation of John Mayow, whose contributions and originality
seemed in need of serious reappraisal. Historical investigations notoriously refuse
to confine themselves to preordained paths, and I have found myself intruding
upon that distinctly underpopulated fraternity, the historians of meteorological
theory. I have been led to exhume a theory of atmospheric phenomena which,
though it flourished for more than a century, leaving its imprint upon English
literature as well as upon various branches of scientific speculation, seems to have
been overlooked by historians of science as completely as though it had never
existed. Yet it tells us of ingenious ideas concerning the nature of our atmosphere, and the mysterious changes that take place in it, and supplements what I
have adduced in a previous paper concerning the chemical and physiological
background of Hooke's and John Mayow's famous theory of nitro-aerial particles.

✿

In his *Tractatus Quinque Medico-Physici* published in 1674, John Mayow outlined a theory of respiration and combustion which, while it had little immediate
success and was virtually forgotten during the eighteenth century, has deeply
impressed modern writers by its striking resemblance to Lavoisier's oxygen
theory. Mayow believed that air was composed essentially of two parts, an inert
mass which provided a solvent or *menstruum* for numerous effluvia and impurities, and a small proportion of highly reactive particles which he referred to as
nitrelike or nitrous, and more often as the "nitro-aerial spirit." Like the oxygen
of modern science, these particles alone made air suitable for respiration, and
when absorbed through the lungs they imparted the bright-red color to arterial
blood. The nitro-aerial particles, also, explained the necessity of air for combus-

From *Isis* 45 (1954). Copyright 1954 by the History of Science Society, Inc.

A preliminary version of this paper was read before a joint session of the History of
Science Society and the American Historical Association in New York, December 1951.
In this research I was greatly assisted by a grant from the Rockefeller Foundation and by
my membership during 1953–54 at the Institute for Advanced Study.

tion, as well as the increase in weight of certain metals when they are calcined.[1] Ever since the late T. S. Patterson of Glasgow published his classic paper on Mayow[2] it has been amply clear in what fundamental respects Mayow's hypothetical particles differed from oxygen, and equally clear that he owed much to such predecessors as Boyle, Hooke, and Lower, the last two of whom made use of the same concept of a nitrelike agency in the air. I have tried to prove in a recent paper that this theory had wide currency in the seventeenth century; that views similar to those of Thruston, Hooke, and Mayow on the role of aerial nitre in combustion and respiration are encountered in various chemical and medical writings of the earlier part of the century, and that they seem to have had their origins in the *Novum Lumen Chymicum* first published by Michael Sendivogius in 1604.[3]

Yet there are numerous aspects of Mayow's fully developed theory which are not wholly attributable to the iatrochemical tradition. As is well known, Mayow applied his theory to a ridiculously wide range of phenomena, and has been severely criticized for so doing. He invoked his nitro-aerial particles, on little or no experimental evidence, to explain the cohesion of solids and the elasticity of air. But what is especially important, Mayow advanced the theory that all combustions and fermentations—apparently all chemical transformations whatsoever—involve not only the nitro-aerial particles but also their interaction, more or less violently, with sulphureous ones which he took to be the characteristic ingredient of all combustible substances. What we may describe as his sulphur-nitre theory of chemical change, or a sulphur-nitre theory of the elements, which he substitutes for the *tria prima* of the Paracelsians, is vividly set forth by Mayow in the following words:

> The nitro-aerial spirit and sulphur are engaged in perpetual hostilities with each other, and indeed from their mutual struggle when they meet and from their diverse state when they succumb by turns all the changes of things seem to arise.[4]

This theory was given a physiological application, which called forth the scorn of Professor Patterson, when Mayow sought to explain muscular contraction as resulting from an "explosion" of sulphurous and nitrous particles reacting together.[5] It will emerge, I trust, from what sources this modification of the nitre theory was in all likelihood drawn.

❁

Some measure of the currency of a scientific theory, though hardly of its precise standing at any given time, can reliably be sought in its popular vogue and its echoes in polite literature. Such is the case with the theory of a "nitre in the air." That it had become a commonplace before the time of Hooke and Mayow can best be shown by the literary allusions made to it. That it outlasted its scientific popularity, and as a debased coinage of literary convention was frequently invoked in metaphor and poetic description during the eighteenth century, does no damage to our argument, though it may be of some importance to the literary historian to have its origins revealed. As I shall try to show, it was from another scientific tradition, besides the alchemical or iatrochemical one, that the literary folk mainly drew. The poetic references lead us without hesitation to the meteorological theory to which I alluded in my introduction.

For Aristotle, the chief phenomena of the atmosphere were explained by the peculiar behavior of two kinds of exhalation, a watery vapor and a smoky exhalation. Lightning was caused when contracting clouds forcibly ejected the dry exhalation, which ignited as it escaped.[6] Pliny in some degree echoed the opinion of Aristotle.[7] Seneca, however, believed that lightning was produced by the friction or collision of the storm clouds—just as sparks fly from stones that are struck together—and Lucretius attributed the heat and light to "seeds of fire" or fire atoms extracted in like fashion.[8] There was greater unanimity among the ancients as to the cause of earthquakes, which were quite generally assumed to be due—according to a humble physiological analogy—to entrapped winds rumbling in the bowels of the earth, a feature which made it possible to treat them appropriately in meteorological discussions.[9]

It may fairly be asked what happened to these quasi-scientific theories—especially those of Aristotle and Pliny, which were specially favored as late as the sixteenth century—after men lost faith in the infallibility of the ancients and before more truly scientific explanations were to be had.[10] Now the evolution of meteorological theory has yet to be accurately reconstructed, and the student may be forgiven if he turns in despair to Andrew D. White's pithy *A History of the Warfare of Science with Theology in Christendom*, where some space is devoted to these problems. If he does so he will gather the impression that meteorology in the seventeenth and early eighteenth centuries was a churchly monopoly, and that a pious meteorology, invoking supernatural agencies, either supplanted or contended with the ancient theories. Attempts were indeed made to reconcile Aristotelian explanations with the Psalmist's references to "the arrows of the thunder"; but, so we are to believe, lightning and thunderbolts were generally deemed the expressions of divine wrath, or credited to Satan and his diabolical agents, until Benjamin Franklin's demonstration of the electrical nature of lightning emancipated human reason with a scientific explanation.[11]

This picture by no means suffices. For more than a hundred years—during a sort of scientific interregnum—a theory of great simplicity, chemical rather than physical, was widely in vogue to explain by one and the same agency the frightening mystery of lightning and thunder, the origin of earthquakes, the formation of snow, as well as certain imagined properties of wintry air.

The new theory of the cause of lightning and earthquakes was the child of technological progress, specifically of the discovery and improvement of gunpowder in the later Middle Ages. We may call it for convenience the sulphur-nitre theory, or with greater brevity the gunpowder theory, for it assumed that the obvious similarity between the effects of chemical explosions and thunderstorms was not purely coincidental. This resemblance was at least metaphorically suggested by those first familiar with gunpowder. The author of the *Liber Ignium*, which contains one of the earliest formulas for an explosive mixture, describes the result as making "thunder";[12] and Roger Bacon, or whoever may have been the author of the *Letter Concerning the Marvelous Power of Art and Nature*, promised the adept who could decipher his cryptic account of the powerful mixture of charcoal, sulphur, and saltpeter, that he would forthwith "make thunder and lightning."[13] Such references are multiplied in the course of the sixteenth century soon after the full impact of this new mode of destruction became widely recognized.[14] Georgius Agricola spoke for many of his contemporaries when he complained in 1556 that "there has been sent from the infernal regions to the earth this force for the destruction of men, so that Death may snatch to himself as many as possible by one stroke." After showing the close resemblance of a gunpowder explosion to the effects of a thunderstorm, he wrote that it could be better said "of impious men of our age than of Salmoneus of ancient days, that they had snatched lightning from Jupiter and wrested it from his hands."[15]

It was no great step to assume that the "material causes" of lightning, thunder, and earthquakes were in fact identical with, or very similar to, the materials that went into gunpowder. Just as the essential ingredients of gunpowder (abstraction made of the charcoal that was included) were the sulphur and that "master-ingredient,"[16] saltpeter or nitre, so these same substances reacting together in the clouds or beneath the surface of the earth could account for these familiar yet terrifying natural phenomena. The sulphureous smell associated with volcanic eruptions, and the sulphurlike odor (actually due to the momentary occurrence of ozone in the atmosphere) noticed immediately after thunderstorms, seemed to confirm this hypothesis.

By the early seventeenth century references to nitre and sulphur in the air, and to their role in the presumed production of thunderstorms and earthquakes, are by no means uncommon. The theory we have outlined has become accessible to prose writer and poet in various scientific works. One of the earliest poetic uses of it occurs in a much-discussed passage of Milton's *Paradise Lost* where Satan the fallen angel is followed in his wild flight through space:

> Fluttring his pennons vain plumb down he drops
> Ten thousand fadom deep, and to this hour
> Down had been falling, had not by ill chance
> The strong rebuff of some tumultuous cloud
> *Instinct with Fire and Nitre* hurried him
> As many miles aloft.[17]

"Instinct with Fire and Nitre"; one need only, as the history of chemistry and poetic license both allow, substitute the word "sulphur" for "fire," in order to see clearly an early poetic application of our theory.[18] A similar reference is to be found in John Dryden's free translations from Ovid's *Metamorphoses*, specifically the verses *Of the Pythagorean Philosophy* where we read:

> If thunder was the Voice of Angry Jove
> Of clouds *with Nitre pregnant* burst above:
> Of these, and things beyond the common reach
> He spoke, and charmed his audience with his speech.[19]

So much for lightning. For a clear poetic allusion to the gunpowder theory of earthquakes, we look in vain in Milton, but such can readily be recognized in the following lines of John Philips' bucolic poem *Cydre*, published in 1708:

> . . . Hence 'gan relax
> The Ground's Contexture, hence Tartarean Dregs,
> *Sulphur, and nitrous Spume,* enkindling fierce,
> Bellow'd within their darksome Caves, by far
> More dismal than the loud disploding Roar
> of Brazen Enginry.[20]

❖

Early in the seventeenth century it became widely known that the addition of salt greatly lowered the temperature of water or of an ice-water mixture. That picturesque inventor and *chymist,* Cornelis Drebbel, had demonstrated the astonishing art of artificial freezing in London as early as 1620, and an echo of his performance before the King is to be found in Francis Bacon's *De Augmentis.*[21] Saltpeter or nitre, which Drebbel apparently employed, was found to be particularly effective; and this "frigorific" property was soon taken to be a marked characteristic of this salt, not unconnected with its peculiarly cool taste and its use as a febrifuge. Soon this property of nitre was invoked to explain the formation of snow and hail in the upper atmosphere, since the speculations about the cause of lightning convinced men that it resided there;[22] and because saltpeter was already known to possess the same fertilizing power as animal manures and rotting vegetation, it was no great step to assume that falling snow carried down with it some of this enrichment. Thus could be simply explained a tenacious belief of the countryman that spring snow greatly increases the fertility of the soil.[23]

With this came also to be associated still another property: since saltpeter (nitre) was found most useful in curing meat and preventing its decomposition, and indeed was thought by Paracelsus to exert a life-preserving function in the body,[24] it seemed obvious that wintry air, being full of nitre (as the production of snow seemed to demonstrate), was particularly healthful and salubrious.[25] Even rain, by the same token, exerted its fructifying effect in part through the nitre that it brought down with it.

The earliest literary reference to this nitre-fertility aspect of our theory that I have encountered is in a sermon preached by Robert South at Oxford on December 10, 1661, entitled *False Foundations Removed, and True Ones Laid for Such Wise Builders as Design to Build for Eternity*:

A person in the state of nature, or unregeneracy, cannot, by the sole strength of his most improved performances, acquire an habit of true grace or holiness. But, as in the rain, it is not the bare water that fructifies, but a *secret spirit of nitre* descending with it, and joined to it, that has this virtue, and produces this effect; so in the duties of a mere natural man, there is sometimes an hidden, divine influence, that keeps pace with those actions, and, together with each performance, imprints a holy disposition upon the soul.[26]

The nitre theme is frequently echoed in the nature poetry of the eighteenth century. Thus in 1712 Sir Richard Blackmore wrote that cold weather

> Th' exhausted Air with *vital Nitre* fills
> Infecyion stops, and Deaths in Embryo kills.

While through the same agency the snow fertilizes the soil:

> Thus are the winter Frosts to Nature kind
> Frosts, which reduce excessive Heats, and bind
> Prolific ferments in resistless Chains,
> Whence Parent Earth her fruitfulness maintains.[27]

And from John Philips again:

> . . . nothing profits more
> Than frequent Snows: O, may'st thou often see
> Thy furrows whiten'd by the woolly Rain,
> Nutricious! *Secret Nitre* lurks within
> The porous Wet, quick'ning the languid Glebe.[28]

This theme is several times embroidered upon the fabric of James Thomson's well-known poem *Winter* (1726); and the debt of Thomson to Philips is nowhere more clearly detectable than in the line in which he refers to "Whate'er the Wintry frost Nitrous prepar'd," or in the following descriptive passage:

> Clear frost succeeds; and thro' the blue serene,
> For sight too fine, *th'ethereal nitre* flies;
> Killing infectious damps, and the spent air
> Storing afresh with elemental life.[29]

So often do figures of the like kind occur in the poets—in John Gay's *Trivia*, in Somerville's *The Chase* and in William Cowper's *The Task*, to cite but a few examples—that McKillop, in his excellent study of Thomson's poetry, does not exaggerate when he says that it was common practice to call cold air or snow "nitrous," and that this word became "a stock adjective in descriptions of eighteenth-century poets."[30]

✿

Let us proceed to examine the scientific authority for these meteorological theories we have seen reflected in the poets of the seventeenth and eighteenth centuries. The earliest reference I have found is still somewhat vague but makes more explicit than did Agricola the obvious similarity between gunpowder explosions and the phenomena of thunderstorms. In his *Pirotechnia* (1540), Vannoccio Biringuccio—the well-known pioneer in metallurgical chemistry—wrote that the noise of gunpowder "is born almost as thunder and lightning are generated for the same reason in the middle region of the air, from thick burning vapors."[31] Somewhat more definitely, Jerome Cardan in his *De Subtilitate* (1550) explained earthquakes in chemical rather than physical terms. They resulted, he said, from the ignition in the interstices of the earth of sulphur, nitre, and bitumen.[32] He does not clearly suggest that the substances react or "ferment" together, but seems merely to list those inflammable materials beneath the earth's surface which, when ignited separately or together, produce earthquakes and volcanic eruptions. These substances differ markedly in their effectiveness and power: nitre, the "master-ingredient" of gunpowder, is the most powerful. Milton's reference to the "combustible and fuel'd entrails" of Mount Etna may be an echo of Cardan's views, though there is no reference to nitre in this passage.[33]

Yet we should recall what Milton wrote in the famous passage, where we learn how Satan taught men to invent gunpowder out of the "materials dark and crude, or spiritous and fierie spume" within the earth:

> Forthwith from Councel to the work they flew,
> None arguing stood, innumerable hands
> Were ready, in a moment up they turn'd
> Wide the Celestial soile, and saw beneath
> Th' originals of Nature in their crude
> Conception; *Sulphurous and Nitrous Foame*
> They found, they mingl'd, and with suttle Art,
> Concocted and adusted they reduc'd
> To blackest grain, and into store conveyd.[34]

This suggests that only the advantage of ellipsis kept Milton from being more specific about the precise nature of the "combustible and fuel'd entrails" mentioned in the earlier passage. Two of Milton's scientific contemporaries, Bernard Varen or Varenius (d.c. 1660) and Athanasius Kircher (1601–80), explicitly attributed earthquakes to the explosion of a mixture of sulphureous and nitrous vapors in the earth;[35] and so, as we shall see, did Sir Thomas Browne.

The first precise application of the gunpowder theory to lightning and thunder I have so far encountered is due to Blaise de Vigenère (1522–96). In his posthumous *Traicté du Feu et du Sel*, published in 1618, there is the following straightforward statement:

> Le salpetre est approprié à l'air, pource qu'il est comme une moyenne disposition de nature entre l'eau de la mer, & le feu ou soulphre dont il participe entant qu'il est si inflammable; & est salsugineux d'autre-part, se resolvant à l'humide, & dans l'eau comme font les sels, desquels il a l'amertume & acuité. Et tout ainsi que l'air enclos & retenu dans des nuées se rompt & esclate en

une impetuosité de tonnerre; de mesme fait le salpetre: le soufre est ce qui cause les esclairs.[36]

The earliest systematic exponent of the role of nitre in meteorological phenomena would seem to have been Daniel Sennert (1572–1637), a nearly forgotten physician and natural philosopher, noted in his day for his brave effort to reconcile Aristotelian physical theories with Lucretian and Heronic atomism.[37] Sennert expounds his theory in his *Epitome naturalis scientiae*, first published in Wittenberg, where he taught for many years, in 1618. In this work he tries to reconcile the gunpowder theory of lightning with the views of Aristotle. Lightning (*fulgur*) and thunderbolts (*fulmina*) differ, he says, little from each other; they are both made of the hot and dry exhalation, a *spiritus igneus & tenuis, nitrosulphureus*; they differ only in the arrangement and quantity of the particles of sulphur and of nitre. A thunderbolt is merely intense lightning; *fulgure magis cōpacta*.[38]

Pierre Gassendi, the well-known Neo-Epicurean philosopher and scientist, gave this meteorological theory extended treatment in the section on physics of his *Syntagma philosophicum*, first published after his death in the collected works of 1658. Gassendi discussed at greater length than Sennert the role of nitrous particles in the production of lightning and thunderbolts;[39] and in addition gave the first detailed description of the supposed freezing action of the nitre of the air. Cold, he says, is produced by "frigorific" corpuscles of a nitrous nature, which in winter produce snow and in summer, hailstones.[40] The Lucretian influence is manifest.

The earliest complete statement in English of the gunpowder theory appears to be that found in Sir Thomas Browne's *Pseudodoxia Epidemica*, where it is used to explain both lightning and earthquakes. He writes as follows of the startling effects of gunpowder:

> But the immediate cause of the Report is the vehement commotion of the air upon the sudden and violent eruption of the Powder. . . . Now with what violence it forceth upon the air, may easily be conceived, if we admit what *Cardan* affirmeth, that the Powder fired doth occupy an hundred times a greater space than its own bulk. . . . And this is the reason not only of this fulminating report of Guns, but may resolve the cause of those terrible cracks, and affrighting noises of Heaven; that is, the nitrous and sulphureous exhalations set on fire in the Clouds; whereupon requiring a larger place, they force out their way, not only with the breaking of the cloud, but the laceration of the air about it. When if the matter be spiritous, and the cloud compact, the noise is great and terrible; If the cloud be thin, and the Materials weak, the eruption is languid, ending in coruscations and flashes without noise, although but at the distance of two miles. . . .
>
> From the like cause may also proceed subterraneous Thunders and Earthquakes, when sulphureous and nitrous veins being fired, upon rarefaction do force their way through bodies that resist them. Where if the kindled matter be plentiful, and the Mine close and firm about it, subversion of Hills and

Towns doth sometimes follow: If scanty, weak, and the Earth hollow or porous, there only ensueth some faint concussion or tremulous and quaking Motion. Surely, a main reason why the Ancients were so imperfect in the doctrine of Meteors, was their ignorance of Gunpowder and Fire-works, which best discover the causes of many thereof.[41]

Edgar Hill Duncan has recently called attention to a passage in a popular compendium or textbook of physics which he suggests may have been the source of Milton's views on thunderbolts.[42] This is Jan Comenius's *Physicae ad Lumen Divium reformatae Synopsis*, published in Amsterdam in 1643; this work appeared in an English translation in 1651 with the title: *Naturall Philosophie Reformed by Divine Light: or, A Synopsis of Physicks*. In the eighth chapter we read:

The world is the Alembick of nature, the air the cap of this Alembick: the sun is the fire: the earth, the water, minerals, plants, etc. are the things which being softened with this fire, exhale vapours upward perpetually. So there ascend, Salt, sulphury, nitrous, etc. vapours, which being wrapped up in clouds, put forth various effects, for example, *when sulphury exhalations are mixed with nitrous,* (the first of a most hot nature, the second most cold) they endure one another so long, as till the sulphur takes fire. But as soon as that is done, presently there follows the same effect as in gun-powder, (whose composition is the same of Sulphur and Nitre) a fight, a rapture, a noise, a violent casting forth of the matter. For thence it is that a viscous flaming matter is cast forth, which presently inflames whatsoever it touches that is apt to flame, and smiting into the earth, it turns to a stone, and being taken out after a time is called a *thunder-bolt*.[43]

Milton, it is evident, could have known either Browne's or Comenius's statement of this gunpowder theory. Since the *Pseudodoxia* first appeared in 1646 (with a second edition in 1650), Sir Thomas, in turn, might have been familiar with the Latin original of Comenius's treatise, but he could not have used the English version of 1651, from which Professor Duncan quotes. It is clear that Browne knew Biringuccio[44] and that he might have been influenced by Cardan, for his reference to him in the lines cited above is to a passage in the *De Subtilitate*; Cardan was, indeed, a favorite authority of Brown's, "of singular use unto a prudent reader," as he put it;[45] but, as we have seen, neither Biringuccio nor Cardan had developed the theory fully. Sennert, a fellow physician whom Browne quotes at least once in the *Pseudodoxia* in another connection, is a more likely source than Comenius. In fact Browne's comparison of the "cloud compact" with the thin cloud composed of weak materials, is a palpable adaptation of the passage of Sennert cited above.

Later in the seventeenth century, and early in the eighteenth, when all these opinions and theories—including the "frigorific" and nutrifying properties of the nitre in the air—were systematically brought together by William Clarke in his *Natural History of Nitre* (1670)[46] and popularized by men like George Cheyne,[47] it is Daniel Sennert and Pierre Gassendi who are credited with introducing these ideas into England. Both Sennert and Gassendi, though their works were not translated, were widely read and appreciated in Britain.[48]

According to McKillop,[49] Cheyne's account of this theory, with all its variations, is repeated verbatim in Miller's *Gardeners and Florists Dictionary* (1724) and in Chambers's *Cyclopedia* (1728). It was from the latter sources, by all evidence, that John Philips drew, and it was probably from Philips that James Thomson, the greatest of the eighteenth-century nature poets, derived his sulphur-nitre theory of earthquakes, and his quasi-scientific description of the winter's cold:

> What are thou, Frost? and whence are thy keen stores
> Deriv'd, thou secret all-invading Power,
> Whom even th'illusive fluid cannot fly?
> Is not thy potent energy, unseen,
> Myriads of little salts, or hook'd, shap'd
> Like double wedges, and diffused immense
> Thro' water, earth and ether?[50]

☼

It should now, I think, be sufficiently apparent, that John Mayow's much-discussed theory of a nitro-aerial spirit, by means of which he interpreted—or perhaps first imagined—his series of revealing experiments, had a long and complex background. Not only had the chemists, from the time of Paracelsus, been meditating about the mysterious properties of air (and from the time of Sendivogius had attributed its support of respiration, and life itself, to a nitrous substance in the air); but the natural philosophers, with their gunpowder theory of lightning and earthquakes, help us to comprehend Mayow's peculiar explanation of muscular contraction, and more important still, his general sulphur-nitre theory of chemical change.

The theory of a nitre in the air, and its use to explain various meteorological phenomena, though much criticized, received considerable scientific support in Mayow's time and for some time after. It is seriously mentioned by such illustrious figures as John Locke, John Wallis, Robert Hooke, and even the great Isaac Newton; it passed to America in the *Compendium Physicae* of Charles Morton and the *Christian Philosopher* of Cotton Mather; and such works as J. Martin Clare's *Motion of Fluids* kept it alive during the eighteenth century. Yet probably no single text can have served to maintain its popularity in the face of growing scientific disapproval so much as the *Opticks* of Newton, that bible of eighteenth-century experimental science; and for this reason, if for no other, we may draw from it our concluding quotation:

> . . . sulphureous Steams abound in the Bowels of the Earth and ferment with Minerals, and sometimes take Fire with a sudden Coruscation and Explosion; and if pent up in subterraneous Caverns, burst the Caverns with a great shaking of the Earth, as in springing of a Mine. And then the Vapour generated

by the Explosion, expiring through the Pores of the Earth, feels hot and suffocates, and makes Tempests and Hurricanes, and sometimes causes the Land to slide, or the sea to boil, and carries up the Water thereof in Drops, which by their weight fall down again in Spouts. Also some sulphureous Steams, at all times when the Earth is dry, ascending into the Air, ferment there with nitrous Acids, and sometimes taking fire cause Lightning and Thunder, and fiery Meteors. For the Air abounds with acid Vapours fit to promote Fermentations, as appears by the rusting of Iron and Copper in it, the kindling of Fire by blowing, and the beating of the Heart by means of Respiration.[51]

The rise, popularity, and eventual eclipse of the sulphur-nitre theory of meteorological phenomena deserves a concluding word. In our habit of over-simplifying the story of the rise of science in the seventeenth century, we often speak as if modern science—that is, a science of some demonstrable validity or at least some cumulative power—grew mushroomlike from the rotting vegetation of scholasticized classical learning. But in this paper I have called attention to a scientific theory, or at least a plausible conceptual scheme, which held the stage during an interregnum between the collapse of classical natural philosophy and the emergence of modern scientific explanations.

A moment's reflection will show that this is by no means an uncommon pattern. In physiology and medical theory we see similar examples in the temporary vogue of the iatrochemists and their theories, and of their successors the iatrophysicists, who sought to apply, often in the crudest fashion, the New Mechanics to an understanding of the living body. Such a description might equally well apply to the Becher-Stahl phlogiston theory in chemistry, and even in seventeenth-century physics to the great bridging system of René Descartes. This may well prove on further examination to be one of the marked characteristics of seventeenth-century science.

Notes

1. *Tractatus Quinque Medico-Physici* (Oxford, 1674). A translation of these four tracts is given in *Medico-Physical Works, being a translation of Tractatus Quinque Medico-Physici by John Mayow, LL.D., M.D. (1674)* (Edinburg and Chicago, 1908), Alembic Club Reprints, no. 17.
2. T. S. Patterson, "John Mayow in Contemporary Setting: A Contribution to the History of Respiration and Combustion," *Isis* 15 (1931), 47–96, 504–46.
3. Henry Guerlac, "John Mayow and the Aerial Nitre," Studies in the Chemistry of John Mayow—I, *Actes du Septième Congrès International d'Histoire des Sciences* (Jerusalem, August 1953), 332–49.
4. "Spiritus nitro-aerus, & sulphur perpetuas inimicitias invicem exercent. Et utique ab eorum mutua congredientem lucta, varioque vicissim succumbentium statu, rerum mutationes quaecunque oriri videntur," (*Tractatus Quinque*, p. 49). The translation is from *Medico-Physical Works of John Mayow*, Alembic Club Reprints, no. 17.

5. Patterson, "John Mayow," 76–77.
6. *Meteorologica*, 369^a–369^b and 370^b–371^b; see also the pseudo-Aristotle *De Mundo*, 394^a, 17–19 and 395^a, 11–24.
7. *Historia Naturalis*, bks. 2, 42, and 43. I have used Rackham's edition and translation in the Loeb Classical Library; the account is confused, and Pliny admits the possibility that lightning can be "struck out by the impact of the clouds, as by two stones." Its origin, however, is from the dry and smoky exhalation from the earth.
8. Seneca deals extensively with meteorological phenomena and earthquakes in his *Quaestiones Naturales*; most of bk. 2 deals with thunder and lightning. Lucretius tells of lightning and thunderbolts in the *De Rerum Natura*, bk. 4, lines 91–434.
9. For Greek and Roman views on earthquakes see Aristotle, *Meteorologica*, 365^{a}14–369^{a}9, where the pre-Socratic views are summarized, and the pseudo-Aristotelian *De Mundo*, 395^b, 26–36; Pliny, *Historia Naturalis*, bk. 2, chaps. 81–86; Seneca, *Quaestiones Naturales*, bk. 6; and Lucretius, *De Rerum Natura*, bk. 6, lines 533 ff.
10. The views of Aristotle and Pliny on thunder, lightning, and earthquakes were understandably predominant in learned circles during the sixteenth century, as one can readily see by consulting such typical and popular works as Francesco de Vieri's *Trattato delle metheore* (Florence, 1573) and Antoine Mizauld's *Le Mirouer de l'air, par bon ordre et breves sentences donnant a chascun veue, et avecques causes cognoissance tres facile presque de toutes choses faictes et engendrees en l'Air: comme sont pluyes, gresles, tonnoirres, fouldres, esclairs, neiges, orages, ventz et autres* (Paris, 1548), a French translation of his *Meterologia* of 1547.
11. Andrew D. White, *A History of the Warfare of Science with Theology in Christendom*, 2 vols. (London and New York, 1898), I, chap. 11, "From 'Prince of the Power of the Air' to Meteorology."
12. Printed in Marcellin Berthelot, *La chimie au moyen âge*, 3 vols. (Paris, 1893), I, 109.
13. Roger Bacon, *Letter Concerning the Marvelous Power of Art and of Nature*, trans. T. L. Davis (Easton, Pa., 1923), 48.
14. As for example by Paracelsus, who says that gunpowder "breaks through walls like a thunder-bolt" and "is with good reason called terrestrial lightning." A. E. Waite, *Hermetic and Alchemical Writings of Paracelsus*, 2 vols. (London, 1894), I, 263.

    Weapons using gunpowder remained largely experimental until after the middle of the fifteenth century. Not until the work of Jean and Gaspard Bureau for Charles VII of France did a professional artillery arm take its place in a modern army. For the early history of gunpowder see Lieut. Col. H. W. Hime, *Gunpowder, Ammunition, Their Origin and Progress* (London, 1904), *passim*; and S. J. von Romocki, *Geschichte der Explosivstoffe*, 2 vols. (Berlin, 1895), I, 83–132. On the rise of artillery see Sir Charles Oman, *A History of the Art of War in the Sixteenth Century* (New York: E. P. Dutton & Co., 1937), 28–29.
15. Georgius Agricola, *De Re Metallica*, trans. Herbert Clark Hoover and Lou Henry Hoover, reprint (New York: Dover publications, 1950), 11.
16. The phrase is Sir Thomas Browne's; see his *Pseudodoxia Epidemica*, bk. 2, chap. 5, in *The Works of Sir Thomas Browne*, ed. Charles Sayle, 3 vols. (London, 1904), I, 276.
17. *Paradise Lost*, II, lines 933–38. Patrick Hume had little grasp of the meaning of these verses; he speaks of nitre as a substance like salt, "the more unknown, the better suiting our description here." See his *Annotations on Milton's Paradise Lost* (London, 1695), 15. The italics in this and subsequent quotations are mine.
18. On the other hand the lightning image in *Paradise Lost* (X, lines 1072–75), appears to echo the theory of Seneca and Lucretius, not the sulphur-nitre theory, as the words put in the mouth of Adam seem to testify:

> Or by collision of two bodies grinde
> The Aire attrite to Fire, as late the Clouds
> Justling or pusht with Winds rude in their shock
> Tine the slant Lightning. . . .

19. *The Poetical Works of Dryden*, ed. George R. Noyes (Cambridge, Mass.: Harvard University Press, 1950), 880.
20. *Poems of John Philips*, ed. M. G. Lloyd Thomas (Oxford, 1927), 50.

21. On Drebbel see Gierit Tierie, *Cornelis Drebbel, 1572–1633* (Amsterdam, 1932). For references to artificial freezing see *The Philosophical Works of Francis Bacon*, ed. J. M. Robertson (London, 1905), 508, 510; and Sir Thomas Browne, *Pseudodoxia*, bk. 2, chap. 1.

22. Guy de la Brosse in 1628 wrote as follows of this phenomenon: "Aussi n'est-ce pas par vne simple qualité froide que le vent de Nord & Nordest gele, c'est par la vertu de leur cause materielle, le Nitre. . . . le nitre mis en l'eau, la rafraichit tressoudainement, & de sorte qu'vn autre vaisseau posé dedans contenant quelque liqueur, est aussi tost rafraischy que s'il etoit dedās vne fontaine bien froide. . . ." *De la nature des plantes* (Paris, 1628), 239–40.

    This idea is the basis of Mayow's later theory of the solid state. He writes: "Probably the rigidity of frozen water is also caused by nitro-aerial particles which, from being fixed like pegs between the aqueous particles, arrest their fluid movement and press them together." And again: "That nitro-aërial and igneous particles in a state of rest produce rigidity and cold may be inferred from the case of nitre itself in which the nitro-aërial particles become extremely cold, and when mixed with vinous liquids almost freeze them. . . ." See Mayow, *Medico-Physical Works*, 49.

23. American farmers can still be heard to say that "snow is the poor man's fertilizer." Among the English proverbs collected by George Herbert in his *Jacula Prudentum* (1639) are "A snow year, a rich year," and "Under water, famine; under snow, bread." See *Works of George Herbert*, based on 1853 ed. of William Pickering, 2 vols. (London, 1859), I, 322, 341. A French proverb has it that "neige qui tombe engraise." And Antoine Mizauld wrote in 1548 that "neige fondue rend la terre tres fertile, & semences en elle iectees par saison, bien nourries." *Le Mirouer de l'air* (Paris, 1548), 33.

24. "I have said of salt that it is the natural balsam of the living body. That is, so long as the body lives, so long the aforesaid salt is its balsam against putridity. By this balsam the whole body of man, as well as that of other creatures, is kept conserved. . . . If salt can preserve the dead body or corpse, much more will it preserve the live flesh." *Hermetic and Alchemical Writings of Paracelsus*, I, 259.

25. Needless to say this interested the physicians. The Danish physician, Thomas Bartholinus, published a book entitled *De nivis usu medico* (Copenhagen, 1661), in which, to use the picturesque summary by William Derham, "he shews of what great use Snow is in fructifying the Earth, preserving from the Plague, curing Fevers, Colicks, Head-Aches, Sore Eyes, Plurisies, (for which Use he saith his Country-Women of Denmark keep Snow-water gathered in *March*,) also in prolonging Life, (of which he instanceth in the *Alpine* Inhabitants, that live to great Age,) and preserving dead Bodies." W. Derham, *Physico-Theology*, 3d ed. (London, 1714), 23–24. Derham, by the way, was no friend of the nitre theory of atmospheric phenomena.

26. Robert South, *Sermons Preached upon Several Occasions, A New Edition*, 4 vols. (Oxford, 1842), II, 144.

27. Sir Richard Blackmore, *Creation*, 2d ed. (London, 1712), 69. Blackmore's references to nitre are discussed by Alan Dugald McKillop, *The Background of Thomson's Seasons* (Minneapolis: University of Minnesota Press, 1942), 61.

28. *Poems of John Philips*, 73.

29. See McKillop, *Thomson's Seasons*, 60. The significance of Phillips's and Thomson's lines, and their relation to speculations about nitre in the air, was pointed out in the eighteenth century by R. Watson, *Chemical Essays*, 4 vols. (1–3, 5) (Cambridge, 1781), II, 78. By Watson's time the scientific foundation of the nitre theory had long since been exploded, as the author is at pains to demonstrate.

30. McKillop, *Thomson's Seasons*, 60. My attention was called by my Cornell colleague, Professor W. R. Keast, to the Dryden lines cited above and to the allusions in Somerville and Cowper.

31. Vannoccio Biringuccio, *The Pirotechnia*, trans. Cyril Stanley Smith and Martha Teach Gnudi (New York: American Institute of Mining and Metallurgical Engineers, 1943), 411.

32. *Hieronymi Cardani Mediolanensis De Subtilitate, Libri XXI* (Nuremburg, 1550).

Cardan's text reads: "Cum enim haec accenduntur nex exitum inveniunt, ut in cuniculis machinisne, terram movent & quatiunt. Pessime quidem Halinitro, mediocriter bitumine, tenuiter sulphure." The comparison to mines (*cuniculi*) and to cannon (*machinae*) is quite explicit.

These views are also set forth by Guy de la Brosse in his *De la nature des plantes*, 326–27, where we read: "Le Nitre enfermé dedans le Canon rarefié par le feu qu'il conçoit, en est vn excellent exemple, & les mines que l'on fait joüer par la pouldre à Canon, des experiences irreprochables, les trembleterres ne se font d'autre sorte que par la resolution & subtiliation des matieres resolutiues, telles que les Sels, principalement le Nitre & l'Armoniac."

33. *Paradise Lost*, I, lines 230–37. F. D. Adams, in the *Birth and Development of the Geological Sciences* (Baltimore: Williams & Wilkins Co., 1938), 30, interprets these Miltonic lines, I believe incorrectly, as derived from Cardan's theory. They suggest equally well (better, since there is no reference to nitre) the views of Georgius Agricola expressed in bk. 2 of his *De ortu et causis subterraneorum* (1558). According to Agricola, the vapors and steams of the earth, excited by subterranean heat, supplement the action of entrapped wind in producing earthquakes. With Milton, the combustible vapors are ignited by the friction of the winds struggling to escape. See Hume's *Annotations*, 15.

34. *Paradise Lost*, VI, lines 507–15.

35. Varenius, whose *Geographia Generalis* (1650) was immensely popular, and was later edited by Isaac Newton in 1672 for his students at Cambridge, seems to have been one of the earliest to apply the sulphur-nitre theory to earthquakes. See F. D. Adams, *Birth and Development*, 409–10, where the passage is given from the English version. For the views of Athanasius Kircher, see his *Mundus Subterraneus* (Amsterdam, 1664), Liber Quartus, Sect. II, Cap. X, pp. 220–23.

36. *Traicté du Feu et du Sel, Excellent et rare Opuscule du Sieur Blaise Vigenere Bourbonnois, trouvé parmy ses papiers apres son decés* (Paris, 1622), 65–66.

37. There is no general study of Sennert, who is worthy of investigation, but see the *Allgemeine Deutsche Biographie* and Kurd Lasswitz, *Geschichte der Atomistik*, 2 vols. (Hamburg and Leipzig, 1904), I, 436–54.

38. I have used a later edition, *Danielis Sennerti vratislaviensis Epitome Naturalis Scientiae* (Paris, 1633), esp. pp. 133–37.

39. See Cap. IV, *De fulgure & Tonitru*, and Cap. V, *De Fulmine & Prestere* of *Sectio Physicae III* of the *Syntagma. Opera Omnia*, ed. H. L. Habert de Montmor and F. Henri, 6 vols. (Lyons, 1658), II, 85–98. The nitre theory is not to be found in the similarly entitled *Syntagma Philosophiae Epicuri* (Lyons, 1649).

40. For Gassendi's nitre theory of snow and hail see his *Opera Omnia*, (Lyons, 1658). The crucial text reads: "Verum rursus est dari Superiorem, quae sit semper frigida (aut certe minus calida, quam sit aliquando inferior) non sane ob distantiam Ignis, qui nullus incumbat, neque absolute ob radiorum reflexorum debilitatem; sed omnino ob semina, seu corpuscula frigoris (cuiusmodi nitrosa sunt) quae illuc vsque abigantur; pauciora quidem per hyemem, vnde illeic solum generatur nix; & copiosiora per aestatem, vnde generatur illeic etiam grando." *Opera Omnia*, II, *Sectio Physicae*, III, 70.

41. *Pseudodoxia Epidemica*, bk. 2, chap. 5, in *Works of Sir Thomas Browne*, I, 272–73. The second paragraph was added in the second edition of the *Pseudodoxia*, published in 1650. It is possible that Varenius's *Geographia Generalis*, which appeared in this year, may have inspired this addition.

42. Edgar Hill Duncan, "Satan-Lucifer: Lightning and Thunderbolt," *Philological Quarterly* 30 (1951), 441–43.

43. Cited by Duncan, "Satan-Lucifer," 442–43.

44. A reference to Biringuccio occurs in the *Pseudodoxia*, bk. 2, chap. 5, p. 274, in the discussion of gunpowder immediately following the passages just quoted. Browne's reference is to a passage in bk. 5 of the *Pirotechnia*, which follows the lines cited above. Since Sir Richard Eden's English translation was limited to the first three books of the *Pirotechnia*, and since no Latin translation existed, Browne must have used the Italian original or the French translation, more likely the former. For

Browne's knowledge of Italian, see Robert Ralston Cawley, "Sir Thomas Browne and his Reading," *PMLA* 48 (1933), 426, n. 3. Except for geography, this study contributed little to our understanding of Browne's scientific sources.

45. *Pseudodoxia,* bk. 1, chap. 8, p. 176. Simon Wilkin's vague reference in his edition of the *Pseudodoxia* (London, 1852), I, 178, n. 1, to the similar opinion of Browne's "great contemporary Dr. Wallis" is vague and misleading. The reference can only be to John Wallis's letter of 1687 in *Philosophical Transactions,* vol. 19, pp. 653–58, "Concerning the Generation of Hail, and of Thunder and Lightning, and the Effect thereof." This is too late to concern us here, although it influenced the eighteenth-century writers.

46. William Clarke, *The Natural History of Nitre: Or a Philosophical Discourse of The Nature, Generation, Place, and Artificial Extraction of Nitre, with its Vertues and Uses* (London, 1670), 19–31. The author, who attempted to draw together all that was believed to be known about nitre, relied heavily upon Sennert, but mentioned also Digby and Sendivogius.

47. George Cheyne's *Philosophical Principles of Religion: Natural and Revealed* (London, 1715), was particularly influential.

48. For references by William Harvey to Sennert, "a man of learning and a close observer of nature, etc.," see *On Generation,* trans. Robert Willis (London, 1847), 356–57, 365. Sennert was frequently alluded to by Robert Boyle in his writings on chemistry and the nature of matter. Gassendi is credited by R. Watson with having put abroad the "false philosophy" which first gave rise to the idea that nitre in the air gives snow its presumed fertilizing power; see *Chemical Essays,* II, 77.

49. McKillop, *Thomson's Seasons,* 60–61.

50. *The Seasons,* Quarto ed. (London, 1730), *Winter* p. 220, lines 671–77.

51. *Opticks,* 2d ed. (London, 1718), Query 31, pp. 354–55. For the theory in Locke see *The Philosophical Works of John Locke,* ed. J. A. St. John (London, 1906), 653–58. Newton was doubtless familiar with the nitre theory in the work of Hooke, Lower, and John Mayow. But it is interesting that the book of Sendivogius was also well known to him; extracts from the *Novum Lumen Chymicum* are to be found among the alchemical notes of Newton in the Keynes Collection in the library of King's College, Cambridge.

For Hooke's version of the theory, see *The Posthumous Works of Robert Hooke,* ed. Richard Waller (London, 1705), 169, 424–25. For Wallis's theory see above, n. 45. It is Wallis and Hooke who are the chief sources of Cotton Mather's summary of these ideas. See *Christian Philosopher* (London, 1721), 60–63. For Morton's *Compendium Physicae,* see *Publications of the Colonial Society of Massachusetts* 33 (1940), 42, 106–7.

17
The Continental Reputation of Stephen Hales

✸

The contribution of the eighteenth-century English clergyman Stephen Hales to plant physiology, and his role as a pioneer in animal experimentation, are well known.[1] But his work in chemistry, as embodied in a chapter of his *Vegetable Staticks* (1727), has been less fully appreciated, even though his influence on the British pneumatic chemists—Black, Cavendish, and Priestley— is often spoken of in general terms.[2] Yet Hales's European reputation in the eighteenth century, such as it was, sprang in good part from this long chapter, deceptively called "Analysis of Air," in which he showed conclusively, if confusedly, that air could take part in chemical processes and was a constituent of many common substances. Though advanced from ancient times as pure speculation, this participation of air was quite generally opposed in Hales's day on the Continent on the authority of such influential figures as Boerhaave and Stahl.[3]

A recognition of this role of air, or more exactly of a particular one of the gases that compose it, in combustion, was, of course, the salient feature of the chemical revolution brought about by Lavoisier. The first link in Lavoisier's chain of researches was his discovery in the autumn of 1772 that when combustible substances are burned or metals are calcined they increase in weight by combining with a portion of the air.

To this initial discovery, as I hope to show, Lavoisier was led, in part at least, by his encounter with Hales's book in French translation. And what he found in Hales seems to have led him to familiarize himself during the fall and winter of 1772–73 with the writings of Black, Priestley, and Cavendish. The works of all three were promptly translated into French, probably under Lavoisier's auspices, and were summarized at length in his first important book, the *Opuscules physiques et chimiques* (1774). An appreciation of this unexpected importance of Hales's chapter has suggested the need for a closer examination of his chemistry and of his reputation on the Continent.

✸

Paper presented at the 6th International Congress of the History of Science, Amsterdam, August 1950.
From *Archives Internationales d'Histoire des Sciences,* no. 15 (1951), 393–404.

Hales's preoccupation with chemistry dated from his Cambridge years, when it was only one of many scientific interests.[4] He was familiar with the writings of Boyle and Mayow, the founders of the British school, both of whom were interested in combustion and the properties of air. But it was Newton, who incidentally was still at Trinity College when Hales was at Benet College (Corpus Christi), who shaped his chemical philosophy. Hales took his departure from those passages in the Queries of the *Opticks*, where Newton speculates about the fixation of air in bodies and tries to explain this phenomenon in terms of corpuscular attractions and repulsions. The Newtonian example, moreover, contributed to foster the hope, which Hales shared with contemporaries like Peter Shaw and John Freind, that chemistry could be made an exact science modeled upon physics. Hence his emphasis, or overemphasis, upon the quantitative aspect of his experiments, and his use of the phrase "chymio-statical experiments" in the work to be discussed, for it was intended to imply a link between the two sciences.[5]

In the early stages of his work on the movement of plant fluids in his bleeding vines, Hales noticed air bubbles rising through the sap in his attached manometers. This brought him to a series of experiments "to prove that Air is freely inspired by Vegetables,"[6] and finally to a systematic attempt to demonstrate the presence of this elastic fluid in a wide variety of different substances and to show its substantial identity with common air. He subjected hog's blood, amber, oyster shells, beeswax, Indian wheat, tobacco, gallstones, urinary calculi, and many other substances to destructive distillation, to the action of mineral acids, or to ordinary fermentation and decay. He scrupulously determined the volume of gas produced in each case in apparatus of his own contrivance. Since he prepared without knowing it a variety of gases and was unaware of their differing solubility in the water over which he collected them, it is small wonder that his results are hard to interpret and his numbers, in the main, quite valueless.[7]

Yet the chapter does not, on this account, deserve the near neglect that has been accorded it by historians of science. Its historical importance, as we shall see, was considerable, and we can point to some definite accomplishments. In the first place, as we have seen, the role of air, or as we would say, of the so-called permanent gases, in the composition of substances, was established in a general way. In the second place, Hales made a significant contribution to the methods and apparatus for handling gases. In the third place, he made observations concerning the absorption of air during combustion that went beyond all earlier observations of this effect and seem to have exerted a directive influence on Lavoisier's thinking.

Hales is quite generally credited with inventing a forerunner of the pneumatic trough, later perfected by Cavendish and Priestley, as a method of collecting gases over water. Actually, his most important experiments were performed with equipment of quite a different kind; and the "pneumatic trough," or musket-barrel equipment, used in only a few experiments, was

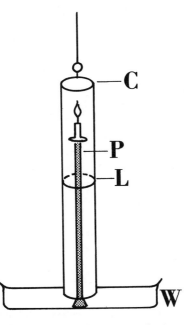

thought of by him as a device for *washing* gases during distillation, not primarily as a method of collecting or storing them. This is made quite clear in the account of Experiment 77.

Historically the most significant piece of equipment was his *pedestal apparatus* for measuring the volume of air given off, or absorbed, in chemical processes within a closed system. A pedestal (P) is placed in a basin of water (W); on this pedestal is placed the candle, living animal, or chemical substance to be studied. A glass cylinder is inverted over the pedestal and suspended so that its mouth is three or four inches under water. Air is withdrawn by a siphon or bellows (not shown) until the water rises to the indicated level (L). The change in the water level in the course of the experiment revealed—subject to solubility and other errors of which he was only dimly aware—the volume of air released or absorbed. Hales employed two sizes of this apparatus: one with a capacity of 2024 cubic inches, in which the pedestal could support a full-grown rat or a small cat; and a small one of 594 cubic inches. It seems to have been this apparatus which Lavoisier meant when, in his early notes, he referred to "Hales's apparatus." Some of Lavoisier's first combustion and calcination experiments were carried out with a slightly modified form of this equipment.

Among Hales's most important observations were those on the absorption of air during combustion, especially in the cases of burning phosphorus and sulphur. Here the effect is a large one and solubility effects are not pronounced. In one of the first experiments, Hales recorded that two grains of phosphorus, ignited by means of a burning glass in the large pedestal apparatus, absorbed

28 cubic inches of air.[8] In a later experiment he similarly ignited matches made of linen rags dipped in melted brimstone (i.e., sulphur) and observed that as much as one-tenth of the volume of air in the vessel was absorbed.[9] A similar absorption of air was observed when equal quantities of iron filings and sulphur, or of antimony and sulphur, were "let fall on a hot Iron on the pedestal."[10]

In the course of these experiments, Hales noticed some interesting effects. The air absorbed by the brimstone seemed, as far as he could tell, to be permanently combined and did not recover its elastic state while he was able to observe it. Moreover, as he expected, a fresh match of sulphur burned a much shorter time in this "infected" air, but also absorbed much air.[11] In particular, by carrying out his experiments in pedestal equipment of the two different sizes, he observed that more air was absorbed in the larger vessel, but that "most air in proportion to the bulk of the vessel" was absorbed in the smaller vessel. This observation is cited by Lavoisier and aroused his curiosity, though he ignored Hales's elaborate "Newtonian" explanation of it.[12]

❁

Very little has been written in any connected fashion about the Continental influences of Hales's work or the extent to which it was appreciated. It has been pointed out that by 1756 both his chief works had been translated into French, German, Dutch, and Italian,[13] and that in 1753 he was chosen *associé étranger* of the French *Académie Royale des Sciences*, to replace Sir Hans Sloane, President of the Royal Society, who died in that year. Almost simultaneously Hales was made a member of the Academy of Sciences of Bologna.[14] How the way was paved for these Academic honors remains to be considered.

Buffon, the eminent French naturalist, was the chief architect of Hales's reputation in France. In 1735 he brought out a French translation of the *Vegetable Staticks* and added an important preface praising the experimental method so well exemplified by the book. This translation was widely known, as Daniel Mornet has shown by his sampling of the contents of eighteenth-century French libraries.[15,16]

A twentieth-century biographer of Buffon, Louis Roule, has emphasized that the Hales episode was centrally important in the development of Buffon's scientific career.[17] It was Buffon's early interest in sylviculture and reforestation which led him to open the English version of the *Vegetable Staticks*. It is Hales the plant physiologist, whom he calls "ce savant observateur qui nous a tant appris de choses sur la végétation," who was, so to speak, at Buffon's elbow as he wrote his early papers on sylviculture and the properties of wood published in the *Mémoires* of the Académie des Sciences between 1737 and 1742.[18] But the extraordinary fecundity of Hales's experimental approach aroused Buffon's interest in this method of studying nature. Buffon's preface is a short but eloquent eulogy of the experimental method; yet there are only a few indications that he was led to follow in Hales's path as an experimenter. Such hints as we possess suggest that it was the "Analysis of Air" which ended by capturing his imagination. He published no results himself, but the English Catholic priest, Needham—famous for his speculations on the problem of generation—mentions that in 1748 he undertook with Buffon experiments similar to those of Hales: "une suite très longue et très variée d'expériences et d'observations sur la composition et la décomposition des substances animales et végétales."[19] An echo of Hales's influence is also to be found in Buffon's chief passages of chemical speculation: in the introduction to *L'Histoire des minéraux.* Buffon here suggests that fire, like air, may exist in a "fixed and concrete" form in nearly all bodies, and that perhaps some successful experimenter may emulate "le docteur Hales," who was able to disengage the air fixed in all bodies and determine its quantity, and that this man may yet find a way to separate the fluid of fire from all substances where it is found in fixed form![20]

Considerable publicity for Hales's opinions came also from the publication of the *Collection Académique.* This series of reprints and translations of the scarce, early publications of the first academies had been begun by a physician of Auxerre, Jean Berryat, and was carried to completion by a number of persons in Buffon's immediate circle. The fourth volume, published in 1755, was a French translation of the *Saggi* of the Accademia del Cimento. It was prepared not from the Italian original but from the Latin version of Pieter van Musschenbroek, and embodied the extensive notes in which the early Florentine results were compared by the Dutch editor with later investigations, his own included. Musschenbroek, who died in the same year as Hales, greatly admired the English experimentalists and did much, as is well known, to make their work popular. It is not surprising, therefore, to find Robert Boyle many times quoted in the notes to the *Saggi,* but somewhat unexpected to discover so many references to the *Vegetable Staticks* of Hales. Hales is quoted almost exclusively from the "Analysis of Air," and on a variety of subjects: on the thermal expansion of air; on the combustion of amber; on the air produced by burning coal or distilling oil; and, last but not least, on the important question of the absorption of air by burning sulphur.[21]

It is not my intention to exaggerate the popularity of these ideas of Hales. It is certainly true, as Alfred Maury long ago observed, that more than a quarter of a century elapsed before the full significance of his discoveries was grasped.[22] The chemical dogmas of the day were in opposition to his principal conclusion: that air was commonly to be found intimately mixed or chemically combined in familiar solid or liquid substances. Yet, as we have seen, the discoveries did not vanish wholly from sight. There are, it is true, few references to Hales in the chemical literature of France in the generation before Lavoisier. But there is one important exception provided by Lavoisier's mentor and patron, Macquer. In his influential *Dictionnaire de chymie,* first published anonymously in 1766, Macquer quotes both Boyle and Hales to prove that air can be chemically combined in animal and vegetable substances.[23]

Let us now review the evidence which suggests that the writings of Hales, kept alive in the tenuous fashion just described, exerted an influence of a very specific sort on the first important researches of Lavoisier.

Just what impelled Lavoisier to enter upon the long series of researches that led him to fame has provided a mystery that is still not totally dispelled. It is well known that, without apparent preparation or prolonged interest in the problem of combustion, Lavoisier, on November 1, 1772, entrusted a famous sealed note to the Secretary of the Académie des Sciences. This note recorded, in order to ensure priority according to a not uncommon method, that he had discovered the crucially important fact that sulphur gained weight on burning with the absorption of what he called "a prodigious amount of air." For a long time this was supposed to be the only known beginning of his important series of researches. But in 1932, the late A. N. Meldrum published two memoranda by Lavoisier which prove that as early as September 1772—almost two months before the sealed note—the French chemist had undertaken to discover whether phosphorus absorbed air in burning. Just what led him to ask this significant question is a mystery both to Meldrum and to the author of one of the best general accounts of Lavoisier's work, Douglas McKie.[24] Meldrum suggested that Lavoisier's interest in the problem was aroused by a paper published a short time before in the Abbé Rozier's new journal, *Observations sur la physique*. This paper, by the Italian investigator G. F. Cigna,[25] is itself in the Hales tradition and takes up the problem of animal respiration where Veratti had left it.[26] It is entitled "Dissertation de M. Cigna, sur les causes de l'extinction de la lumière d'une bougie, et de la mort des animaux renfermés dans un espace plein d'air." Hales's experiments on respiration and combustion form the real point of departure for this work and he is repeatedly cited. Careful attention is given by Cigna to Hales's discovery that phosphorus and sulphur "weaken the elasticity" of air.[27] Precisely after the manner of Hales, Cigna interprets this phenomenon as due to the production of vapors which alter the mechanical properties of the air, rather than as an actual decrease in the quantity of air remaining. Cigna's paper may very well have directed Lavoisier's attention to Hales's work; but since Cigna makes no mention of the apparatus used by Hales for these experiments, and since Lavoisier shows himself quite familiar with it, there is every likelihood that the French chemist turned directly to Hales's own work. This meant almost certainly the Buffon translation, since Lavoisier did not read much, if any, English.

Douglas McKie has called the observation on the burning of phosphorus "Lavoisier's first important step in the study of combustion," which in a certain sense it was. Moreover he believes that the experiments which Lavoisier had been conducting on the destruction of diamonds during the previous months "do not include any observation about the function of air."[28] In a paper which I read before the History of Science Society in New York a little time ago, I presented evidence to show that, contrary to McKie's opinion, the diamond investigation may have led directly to the classic experiments with phosphorus and sulphur, and very probably led Lavoisier to make the acquaintance of Hales's book. I can only summarize the evidence here, but I must do so in order to explain Lavoisier's debt to Hales and to conclude the discussion of the latter's Continental reputation.

Earlier in this crucial year 1772 Lavoisier's interest was aroused in the controversy over the cause of the total disappearance of the diamond in the heat of an assay furnace. In April, he had helped demonstrate that this destruction was impossible if air were totally excluded. He became convinced that the process was a combustion, and that like all such processes it probably involved, as he says, the production of "acid vapors, as with phosphorus or sulphur, or certain emanations . . . which can be collected in suitable apparatus."[29] On August 8 he outlined a program of work which is of the greatest interest, for though it involves mainly a discussion of the uses to which the burning glass can be put, it is also concerned in part with discovering what happens to the diamond when it "evaporates." Though published by Dumas in 1865 in the *Œuvres complètes*, this document has not been accorded the importance it may well deserve. In the concluding section of this memorandum Lavoisier, besides outlining experiments to be performed on the diamond with the burning glass, made the significant proposal that it was desirable *to study the air contained in various other bodies.* This is followed by the still more significant suggestion that one should try to use Hales's apparatus—"*l'appareil de M. Hales*"—in conjunction with the burning glass in order to measure the quantity of air produced or absorbed in every operation. This reference to Hales, and the internal evidence of the concluding paragraphs showing that he had read and digested the ideas and the techniques of Hales's "Analysis of Air," is coupled with a remark which makes it probable that he was not yet acquainted with Black's classic work on the constitution of the carbonates, already nearly twenty years old, and by the same token with the work of Cavendish and of Priestley, which was only just beginning to appear.[30] Hales, then, seems to have afforded Lavoisier's first link with the British pneumatic chemists. Lavoisier forthwith made these British works familiar in France by seeing that as many as possible were translated within the year. It seems probable that the speculations, and the methods, of Hales's "Analysis of Air" suggested to Lavoisier a mode of attack and a vague explanation of the disappearance of the diamond. They provided not only the clues about the behavior of phosphorus and sulphur, but the method by which the problem was attacked. Lavoisier did in fact make use of a simple modification of Hales's pedestal apparatus, for his study of the "evaporation" of the diamond by means of the burning glass and a closed system, and apparently also for the earliest combustion experiments. In the *Opuscules chimiques* we find the apparatus already employed in calcining lead and tin under conditions in which the amount of air absorbed could be measured. The apparatus is illustrated by Lavoisier in Fig. 8 of the *Opuscules*.

Because of Buffon's translation, and such references as those of Buffon, Musschenbroek, and Macquer, Hales's work, unlike that of Black, which was virtually unknown in France, had enjoyed an early *succès d'estime* and had by no means disappeared from sight. It was available to Lavoisier, who drew inspiration from it at this critical moment of his career and who gratefully refers to it at one point as "un fond inépuisable de méditations."

❈

[This paper was my first excursion into Lavoisier studies. In later writings I have changed my opinion on certain matters (e.g., the importance of Cigna and of the diamond experiments). But my main thesis—Hales's influence on the young Lavoisier—has held up and even been elaborated by my students, my colleagues, and me. Today I could cite much more evidence of the interest in Hales's work on the part of Continental scientists, notably in his work on ventilators, on the cause of earthquakes, etc.]

Notes

1. Accounts of Hales's physiological work have been given by Julius Sachs, plant physiologist and leading historian of botany in the nineteenth century, in his *Geschichte der Botanik (1530–1860)* (Munich, 1875), and by Sir Francis Darwin in his collection of essays called *Rustic Sounds* (London, 1917). See also R. T. Gunther, *Early Science in Cambridge* (Oxford: The University Press, 1937); Emile Guyenot, *L'évolution de la pensée scientifique: sciences de la vie* (Paris, 1941); and Jocelyn Thorpe in *Notes and Records of the Royal Society* 3 (April 1940). For biographical information the account in the *D.N.B.* has been superseded by the standard biography: A. E. Clark-Kennedy, *Stephen Hales, D.D., F.R.S.: An Eighteenth Century Biography* (Cambridge, 1929).
2. Joseph Black refers directly to Hales in his classic paper on *magnesia alba*. See *Essays and Observations* (Edinburgh, 1770–71), III, 176, 210.
3. Hélène Metzger, *Les doctrines chimiques en France* (Paris, 1923), 179–80; and her *Newton, Stahl, Boerhaave et la doctrine chimique* (Paris: F. Alcan, 1930), 118. See also Lavoisier, *Oeuvres*, publiées par les soins de Son Excellence le Ministre de l'Instruction Publique et des Cultes, 6 vols. (Paris, 1862–93), I, 464. [The late J. R. Partington called my attention to my error in here linking Boerhaave's name to Stahl's. Boerhaave came to value Hales's discoveries concerning air. This matter is put in proper perspective by Milton Kerker, "Herman Boerhaave and the Development of Pneumatic Chemistry," *Isis* 46 (1955), 36–49.]
4. Clark-Kennedy, *Stephen Hales*, 4–21. The principal source for the Cambridge period is worth consulting directly; it is the *Family Memoirs of the Rev. William Stukeley, M.D.*, ed. Rev. W. C. Lukis, 3 vols. (London, published by the Surtees Society, 1882). Hales himself recalls an experiment performed "20 years since" with several others "at the elaboratory in Trinity College Cambridge." See *Vegetable Staticks* (London, 1727), 195.
5. The best source for Hales's chemical ideas is his own writings. I have consulted both the first (1727) and the second (1731) English editions. In the second edition, the *Vegetable Staticks* makes up the first volume of the *Statical Essays*, 2 vols. (London, 1731). All citations are from the first edition: *Vegetable Staticks: Or, An Account of some Statical Experiments on the Sap in Vegetables: Being an Essay towards a Natural History of Vegetation. Also, a Specimen of An Attempt to*

Analyze the Air, By a great Variety of Chymio-Statical Experiments; Which were read at Several Meetings before the Royal Society. By Steph. Hales, B.D. F.R.S. Rector of Teddington, Middlesex, London. Printed for W. and J. Innys, at the West End of St. Paul's; and T. Woodward, over-against St. Dunstan's Church in Fleetstreet. M, DCC, XXVII.

For Hales's chemical theories and his indebtedness to Freind and to the Queries of Newton's *Opticks*, see *Vegetable Staticks*, "The Preface," p. v, and "The Analysis of Air," pp. 165–66, 207–8 ff.

6. *Vegetable Staticks*, 155. These experiments are given in chap. 5, pp. 148–55.

7. Clark-Kennedy, *Stephen Hales*, 97, pays little attention to the chemistry and says "it is not worth while to enter into these experiments in detail, or to attempt to interpret them along modern lines," and J. R. Partington says of these experiments that they constitute "a good example of the poor results obtained when the qualitative chemical characteristics of the substances investigated are neglected," which is a good description of many of them. See his *Short History of Chemistry*, 2d ed. (London: Macmillan & Co., 1948), 91.

8. *Vegetable Staticks*, Experiment 54, p. 169. The apparatus in question is described with care in the introductory section of the "Analysis of Air," pp. 157–64. The pedestal apparatus is shown in fig. 35, p. 206, of the same work.

9. *Ibid.*, Experiment 103, pp. 226–27.

10. *Ibid.*, Experiment 104, pp. 227–28.

11. *Ibid.*, 227–28.

12. Lavoisier, *Oeuvres*, I, 457.

13. The *Vegetable Staticks* was translated into Dutch by P. Le Clercq (Amsterdam, 1734); into French by Buffon (Paris, 1735); into German with a preface by Von Wolff (Halle, 1748); and into Italian by Ardinghelli (Naples, 1756). Clark-Kennedy, *Stephen Hales*, 221, refers only to the combined translation of the *Vegetable Staticks* and the *Haemastaticks* into Dutch in 1750. I am indebted to Dr. A. Schierbeek for information concerning the Le Clercq translation of 1734. The *Haemastaticks* was translated into French by De Sauvages (Geneva, 1744).

14. Clark-Kennedy, *Stephen Hales*, 221. Hales was only infrequently cited in the publications of these academies. Aside from the papers of Buffon (see below, n. 18), the earliest references in the *Mémoires de l'Académie des Sciences* are in a paper by the Abbé Nollet on the freezing of water (*Mémoires* for 1743, pp. 52–53) and one by Duhamel du Monceau where Buffon's translation of Hales is cited extensively concerning the movement of sap in trees (*Mémoires* for 1744, pp. 6–8). There is one reference of little significance in 1748 and one in 1756. Grandjean de Fouchy's eulogy of Hales appeared in 1764 and seems to have made use, for biographical material, of the account of Peter Collinson in the *Gentleman's Magazine* 34 (1764), 273–78. It devotes considerable space to Hales's "Analysis of Air."

As early as 1746 Hales is mentioned in the *Commentarii* of the Bolognese Academy of Sciences; in an article of Vincenzo Menghini, "Deferrearum particularum sede in sanguine," and more significantly by Giuseppi (Josephus) Veratti (1707–93), who several times cites *clarissimus Hales* in a paper entitled "De avium quarundum et ranarum in aere interclusarum interitu." While of no great scientific consequence, this paper helped keep alive an interest in the old problem of animal respiration and suffocation, and is later cited by Cigna. See *De Bononiensi Scientarum et Artium Instituto atque Academia Commentarii*, vol. 2, pt. 1, pp. 244–66, 267–78.

15. *La Statique des Végétaux et l'Analyse de l'Air, Expériences Nouvelles lues à la Société de Londres. Par M. Hales D. D. et Membre de cette Société. Ouvrage traduit de l'Anglais par M. de Buffon, de l'Académie Royale des Sciences* (Paris, 1735).

For the importance of this translation and Buffon's preface, see Daniel Mornet, *La pensée française au XVIIIe siècle* (Paris, 1926), 8; and his *Les sciences de la nature au XVIIIe siècle* (Paris, 1911), 108–9.

16. In Mornet's survey of 500 library catalogues, Buffon's translation of Hales appears 28 times. For comparison Geoffroy's popular *Histoire des insectes* appeared 29 times, and the Abbé Pluche's fantastically successful *Spectacle de la nature*, 206 times.

17. Louis Roule, *Buffon et la description de la nature* (Paris, 1924), 25–29.

18. See Buffon's "Expériences sur la force des bois," his "Conservation et le rétablissement des forêts," and his "Sur la culture et l'exploitation des forêts," in *Oeuvres*, ed. Pierre Flourens, 12 vols. (Paris, 1853–55), XII, *passim*. See also the papers of Duhamel with Buffon, *Oeuvres*, XII, 109–39.
19. *Introduction aux observations sur la physique*, vol. 5 (September 1772), 421.
20. Buffon, *Oeuvres*, IX, 33.
21. *Collection Académique*, 13 vols. (Paris and Dijon, 1755–59), IV (1755), 33, 34, 36, and *passim*.
22. A. Maury, *L'Ancienne Académie des Sciences* (Paris, 1864), 117.
23. *Dictionnaire de chimie* (1766), articles "Air" and "Principes."
24. Douglas McKie, *Antoine Lavoisier: The Father of Modern Chemistry* (London: V. Gollancz, 1935).
25. Giovanni Francesco Cigna (1734–90), the nephew of the electrical experimenter G. B. Beccaria, attained the chair of anatomy at the University of Turin in 1770. He was one of the founders of the Academy of Sciences of Turin.
26. See above, n. 14.
27. *Introduction aux observations sur la physique*, vol. 2 (May 1772), 97.
28. McKie, *Antoine Lavoisier*, 112–13.
29. Lavoisier, "Sur la destruction du diamant par le feu. Deuxième Mémoire," *Oeuvres*, II, 67.
30. Lavoisier, *Oeuvres*, III, 266.

18

Joseph Black

A founder of modern quantitative chemistry and discoverer of latent and specific heats, Joseph Black, although born in France, was by blood a pure Scot. His father, John Black, was a native of Belfast; his mother, Margaret Gordon, was the daughter of an Aberdeen man who, like John Black, had settled in Bordeaux as a factor, or commission merchant, in the wine trade.

John Black and Margaret Gordon were married at Bordeaux in 1716, apparently in the Catholic faith. Joseph, fourth of their twelve children, born on April 16, 1728, was first educated by his mother. At the age of twelve he was sent to Belfast, where he learned the rudiments of Latin and Greek in a private school. About 1744 he crossed the North Channel to attend, as did so many Ulster Scots, the University of Glasgow. Here he followed the standard curriculum until, pressed by his father to choose a profession, he elected medicine. At this point he began the study of anatomy and attended the lectures in chemistry recently inaugurated by William Cullen. These lectures were the decisive influence on Black's career; chemistry captivated him, and for three years he served as Cullen's assistant. So began a close friendship that lasted until Cullen's death.

In 1752 Black left Glasgow for the more prestigious University of Edinburgh, which boasted on its medical faculty the great anatomist Alexander Monro *primus*, the physiologist Robert Whytt, and Charles Alston, a botanist and chemist who lectured on materia medica. Black gained less, he said, from their lectures than from the bedside clinical instruction provided by the university's Royal Infirmary. Alston's lectures pleased him most, although he found him deficient in chemical knowledge, a matter of concern, for, as he wrote to Cullen, "no branch should be more cultivated in a medical college."[1]

In 1754 Black received the M.D. with his now historic dissertation *De humore acido a cibis orto et magnesia alba*. The next year, before the Philosophical Society of Edinburgh, he described the chemical experiments, considerably expanded, that had formed the second half of his dissertation. This classic paper—the chief basis of Black's scientific renown and his only major publication—appeared in 1756 in the Society's *Essays and Observations* under

From *Dictionary of Scientific Biography*, vol. II (1970), 173–83. Copyright © 1970 by the American Council of Learned Societies.

the title "Experiments Upon Magnesia Alba, Quicklime, and Some Other Alcaline Substances." Here Black demonstrated that an aeriform fluid that he called "fixed air" (carbon dioxide gas) was a quantitative constituent of such alkaline substances as *magnesia alba*, lime, potash, and soda.

The same year, 1756, brought Cullen to Edinburgh as professor of chemistry, and saw Black—at the age of twenty-eight Cullen's outstanding student—replace him in Glasgow. Here Black spent the next ten years. Although this period is sparsely documented, we know that he soon emerged as a gifted and effective teacher. His course in chemistry, launched in 1757–58, proved so popular that many students, some with no particular relish for the subject, pressed to attend. Alongside his teaching, Black carried on an active and demanding medical practice; and since Glasgow, unlike Edinburgh, was administered by its faculty, he was constantly pressed upon by multifarious college duties. Yet it was at Glasgow that he developed his ideas about latent and specific heats—the second of his major scientific achievements—and carried out experiments, alone or with his students, to confirm his theories. These important discoveries he could never be induced to publish.

In 1766 Black received the call to Edinburgh, William Cullen having relinquished the chair of chemistry to succeed Robert Whytt as Professor of the Institutes of Medicine. Black took over Cullen's chair at Edinburgh, where he was destined to remain. Although his duties were less onerous than at Glasgow—he limited his medical practice to the care of a few close friends like David Hume—his period of scientific creativity was at an end. Two short papers on insignificant subjects were his only publications. The teaching of chemistry now became his central concern; here, as at Glasgow, he became an idol to the medical students and to many others as well. Each year, from October to May, he delivered a series of more than a hundred lectures, and sometimes offered a course during the summer months.

Black's pedagogic achievement at least equals that of his great French contemporary, G.-F. Rouelle. Although he had no student of the stature of Lavoisier, there were many of great ability. His audience was surprisingly cosmopolitan; although French students were rare, men came from Germany, Switzerland, Scandinavia, and from as far away as Russia and America, attracted by the reputation of the Scottish medical schools and of Black himself. Lorenz Crell, known as editor of early chemical journals, was one of his German students. To Edinburgh from the American colonies came such men as James McClurg, later a successful Richmond physician, and the still more famous Benjamin Rush. Black's British students were no less gifted. At Glasgow there were John Robison, who was to bring out in 1803 his master's *Lectures on Chemistry*, and William Irvine, Black's collaborator in the work on specific heat. His Edinburgh students included Thomas Charles Hope, who succeeded him in 1797; Daniel Rutherford, the discoverer of nitrogen; and John McLean, who emigrated to America in 1795, where he became Princeton's first professor of chemistry. Among the last to hear Black lecture were Thomas Young, the versatile physician, physicist, and linguist; the elegant and prolific Henry Brougham; and Thomas Thomson, chemist and pioneer historian of chemistry.

There are several contemporary descriptions of the appearance, personality, and lecturing skills of this great teacher. On the platform he was an immaculate figure; his voice was low but so clear that he was heard without difficulty by an audience of several hundred. His style was simple, his tone conversational, far different from Rouelle's flamboyance. He spoke extemporaneously from the scantiest of notes; yet his lectures, of which numerous manuscript versions by his students have survived, were models of order and precision: the facts and experiments led a listener by imperceptible degrees to the theories and principles by which he explained them. Vivid accounts of his own discoveries and demonstration experiments conducted with unvarying success were the highlights of his performance. He kept abreast of the progress of chemistry: through the years the outline of the lectures remained the same, but new material was added as chemistry, that "opening science," as he called it, steadily advanced; and Black told his students of new discoveries and theories, and of the men who had made them.

Black was a typical valetudinarian; never robust, he suffered all his life from chronic ill health, perhaps pulmonary in origin. With only limited reserves of energy, he nevertheless managed by careful diet and moderate exercise—hours of walking were part of his regimen—to husband his strength. In his prime, as the portrait by David Martin depicts him, he was a handsome man; and even in old age his appearance was impressive. Henry Cockburn, who saw Black in his last years, gives the following description:

> He was a striking and beautiful person; tall, very thin, and cadaverously pale; his hair carefully powdered, though there was little of it except what was collected in a long thin queue; his eyes dark, clear, and large, like deep pools of pure water. He wore black speckless clothes, silk stockings, silver buckles, and either a slim green silk umbrella, or a genteel brown cane. The general frame and air were feeble and slender. The wildest boy respected Black. No lad could be irreverent towards a man so pale, so gentle, so elegant, and so illustrious. So he glided, like a spirit, through our rather mischievous sportiveness, unharmed.[2]

And so we see him, on one of his increasingly rare strolls, pictured by the sharp eye of the caricaturist John Kay: slim, slightly stooped, an intent and pensive figure.

Black never married, but he was no recluse. Calm, self-possessed, gentle, and a trifle diffident, he nevertheless enjoyed conviviality. Until at last his health failed him, he frequented, besides the Philosophical Society and the Royal Society of Edinburgh, which replaced it in 1783, those informal clubs for which Edinburgh was famous: the Select, the Poker, and the Oyster. The Oyster, a weekly dining club, was his favorite; indeed, with his two closest friends, Adam Smith and the geologist James Hutton, he had founded it. Other members were his cousin Adam Ferguson, William Cullen, Dugald Stewart,

Joseph Black, by David Martin (1737–1798)

John Playfair, and James Hall: in a word, the scientific luminaries of that remarkable Scottish Enlightenment. Since the rising industrialists of the region were often at table—John Roebuck, Lord Dundonald, for example, and visitors like Henry Cort—the Oyster might be compared with the famous Lunar Society of Birmingham, for discussion often turned on the role of science in technological progress.

One after another Black's friends passed from the scene: Cullen in 1790 and William Robertson, principal of the University of Edinburgh, in 1793. Adam Smith was the first of the triumvirate to die, but increasing infirmity afflicted the two surviving members. Hutton, wasted by years and illness, fell gravely ill in the winter of 1796–97. Black, too, found his feeble strength waning. He gave his last full course of lectures in 1795–96; but, aware of his debility, he chose Thomas Charles Hope as his assistant and eventual successor. The next year his health worsened, and Hope in effect took over. For a time Black's health improved slightly, and he lingered on two more years. The manner of his death was so peaceful, in a way so characteristic of his methodical and undramatic life, that it has been several times recounted. Curiously, the early authorities on Black's life are mistaken about the date of his death, variously given as November 26 (Adam Ferguson, John Robison, and Lord Brougham) and November 10 (Thomson, also quoted by his modern biographer, Sir William Ramsay). But a letter from Robison to James Watt settles the point: Black died on December 6, 1799, in his seventy-second year.[3]

By scrupulous frugality, Black had quietly amassed a substantial competence, something in excess of £20,000. In his will, this sum was divided among his numerous heirs according to an ingenious, and of course mathematical, plan. Black had a certain reputation for parsimony—it is said that he weighed on a balance the guineas his students paid to attend his course—and his biographers have felt obliged to set the record straight by citing instances of his generosity: the loans he made to friends, the poor patients he treated without charge, and even the spaciousness of his house and the plenty of his table, "at which he never improperly declined any company."[4] He was, at the very least, as methodical in his financial affairs as he was in his science, his teaching, and all other aspects of his life.

✸

Black's investigation of alkaline substances had a medical origin. The presumed efficacy of limewater in dissolving urinary calculi ("the stone") was supported by the researches of two Edinburgh professors, Robert Whytt and Charles Alston. It interested Cullen as well, and Black came to Edinburgh as a medical student with the intention of exploring the subject for his doctoral dissertation.

But at this moment Whytt and Alston were at loggerheads: they disagreed as to the best source, whether cockleshells or limestone, for preparing the quicklime. And they differed as to what occurs when mild limestone is burned to produce quicklime. Whytt accepted the common view that lime becomes caustic by absorbing a fiery matter during calcination, and thought he had proved it by showing that quicklime newly taken from the fire was the most powerful dissolvent of the stone. Alston, in an important experiment on the solubility of quicklime, showed that this was not the case, and that the causticity must be the property of the lime itself. Both men were aware that on exposure to the air quicklime gradually becomes mild, and that a crust appears on the surface of limewater. For Whytt, this resulted from the escape of fiery matter; but Alston, noting that the crust was heavier than the lime in solution, hinted that foreign matter, perhaps the air or something contained in it, produced the crust. Yet he was more disposed to believe that the insoluble precipitate formed when the quicklime combined with impurities in the water. Black, although he had criticized Alston as a chemist, was soon to profit from his findings.

Preoccupied at first with his medical studies, Black did not come to grips with his chosen problem until late in 1753. When he did so, he found it expedient to avoid any conflict between two of his professors; instead of investigating limewater, he would examine other absorbent earths to discover, if possible, a more powerful lithotriptic agent. He chose a white powder, *magnesia alba*, recently in vogue as a mild purgative. Its preparation and general properties had been described by the German chemist Friedrich Hoffmann; although it resembled the calcareous earths, *magnesia alba* was clearly distinguishable from them.

Black prepared this substance (basic magnesium carbonate) by reacting Epsom salts (magnesium sulfate) with pearl ashes (potassium carbonate). He treated the purified product with various acids, noting that the salts produced differed from the corresponding ones formed with lime. The *magnesia alba*, he observed, effervesced strongly with the acids, much like chalk or limestone.

Could a product similar to quicklime be formed by calcining *magnesia alba*? Would its solutions have the causticity and solvent power of limewater? Black's effort to test this possibility was the turning point of his research. When he strongly heated *magnesia alba*, the product proved to have unexpected properties. To be sure, like quicklime, this *magnesia usta* did not effervesce with acids. But since it was not sensibly caustic or readily soluble in water, it could hardly produce a substitute for limewater.

The properties of this substance now commanded Black's entire attention, notably the marked decrease in weight that resulted when *magnesia alba* changes into *magnesia usta*. What was lost? Using the balance more systematically than any chemist had done before him, he performed a series of quantitative experiments with all the accuracy he could command. Heating three ounces of *magnesia alba* in a retort, he determined that the whitish liquid that distilled over accounted for only a fraction of the weight lost. Tentatively he concluded that the major part must be due to expelled air. Whence came this air? Probably, he thought, from the pearl ashes used in making the *magnesia alba*; for Stephen Hales, he well knew, had shown long before that fixed alkali "certainly abounds in air."[5] If so, upon reconverting *magnesia usta* to the original powder, by combining it with fixed alkali, the original weight should be regained. This he proved to his satisfaction, recovering all but ten grains.

Magnesia usta, he soon found, formed with acids the same salts as *magnesia alba*, although it dissolved without effervescence. Only the presence or absence of air distinguished the two substances: *magnesia alba* loses its air on combining with acids, whereas the *magnesia usta* had evidently lost its air through strong heating before combining with acids.

Could the same process—the loss of combined air—also explain the transformation of lime into quicklime? Tentative experiments suggested something of the sort, but not until the work on magnesia was completed, late in 1753, did he examine this question. When he precipitated quicklime by adding common alkali, the white powder that settled out had all the properties of chalk, and it effervesced with acids. Early in 1754, Black wrote William Cullen that he had observed interesting things about the air produced when chalk was treated with acid: it had a pronounced but not disagreeable odor; it extinguished a candle placed nearby; and "a piece of burning paper, immersed in it, was put out as effectively as if it had been dipped in water."[6] This was an observation clearly worth pursuing. Nevertheless, he could no longer postpone the writing of his Latin dissertation and his preparation for his doctoral examination.

The dissertation is in two parts: the first, dealing with gastric acidity, was clearly added to give medical respectability to the work; Black was never happy about it, and hoped it would pass "without much notice."[7] The second set forth the experiments on *magnesia alba* and the tentative conclusions he drew from them. Nothing of significance was said about other alkaline substances or about what he was to call "fixed air."

The "Experiments," on the other hand, is a longer and more elaborate work. Like his dissertation, it is divided into two parts. In Part I, he recounts the experiments on *magnesia alba*; little or nothing is added, but now his theory is presented without equivocation: this substance he now describes as "a compound of a peculiar earth and fixed air."[8]

In Part II, Black describes experiments that enabled him to generalize the theory and to support his explanation of causticity. When lime is calcined, air is given off in abundance, and the caustic properties of the resulting quicklime do not derive from some fiery matter but from the lime itself. He showed by experiment, in effect confirming what Alston had already done, that all of a given amount of quicklime, not merely a part of it, is capable of solution, if enough water is used. Thus the mysterious property of causticity is associated with a definite chemical entity having a definite solubility. As in the case of magnesia, he showed that calcareous earth combines with the same quantity of acid whether it is in the form of chalk (combined with air) or of quicklime. Again quicklime, made from a measured weight of chalk, when saturated with a fixed alkali can be converted into a fine powder nearly equal in weight to the original chalk. The quicklime was evidently saturated with air obtained from the alkali.

Black's theory also explained the production of strong or caustic alkalies (e.g., caustic potash) prepared by boiling quicklime with a solution of a mild alkali. What must occur is not, as chemists thought, that the acrimony of the potash is derived from the lime, but that fixed air is transferred from the mild alkali to the quicklime, thereby uncovering the inherent causticity of the alkali. Careful experiments confirmed this new extension of his doctrine.

A conclusive test of his theory of inherent causticity was Black's demonstration that both quicklime and *magnesia usta* could be produced by the "wet way," without the use of fire. He argued that if caustic alkali is caustic when not combined with "fixed air," it should separate magnesia from combination with acid and deposit it as *magnesia usta*. This he easily demonstrated. He performed a similar experiment with chalk.

An important collateral investigation stemmed from Black's experiments on the solubility of quicklime. Although he established to his satisfaction that it could be almost completely dissolved, he was puzzled not to find a larger residue of insoluble matter, for the air dissolved in water ought to combine with the quicklime to form a small amount of insoluble earth (carbonate). Perhaps the air had been driven off when the water was saturated with quicklime. The rough experiment to test this was performed after he had presented his major results to the Philosophical Society. In the receiver of an air pump, Black placed a small vessel containing four ounces of limewater; alongside it he put an identical vessel containing the same quantity of pure water. When the receiver was exhausted, the same amount of air appeared to bubble from both vessels. Clearly, the limewater contained dissolved air, but not of the kind that combined so readily with quicklime. In his "Experiments" he wrote: "Quicklime therefore does not attract air when in its most ordinary form, but is capable of being joined to one particular species only."[9] This he proposed to call "fixed air," preferring to use a name already familiar "in philosophy," rather than invent a new one. The nature and properties of this substance, he wrote, "will probably be the subject of my further inquiry."[10]

Black had shown that a particular kind of air, different from common air, can be a quantitative constituent of ordinary substances and must enter, as Lavoisier put it later, into their "definition." But he was not destined to make the investigation of such elastic fluids his "future inquiry." This was to be mainly the work of his British disciples—MacBride, Cavendish, Priestley, and Rutherford—and he published nothing further on the subject. Nevertheless, from his *Lectures* and other bits of evidence we learn that his discoveries did not end abruptly with the publication of his "Experiments." He knew that "fixed air" did not support combustion, that it had a density greater than common air, and that its behavior with alkaline substances resembled that of a weak acid. By experiments with birds and small animals, he soon demonstrated that this air would not support life. Using the limewater test, he showed that air expired in respiration consisted mainly of "fixed air"; and likewise that the elastic fluid given off in alcoholic fermentation, like that produced in burning charcoal, was identical with the "fixed air" yielded by mild alkalies when they effervesce with acids.

Black's doctrines did not have the prompt success on the Continent that they enjoyed in Britain. His influence on the early stages of the chemical revolution in France was far less than scholars have imagined; indeed, at first it was negligible. Before 1773 French chemists were unfamiliar with his "Experiments," which had appeared in English in an obscure publication. What they knew of his work derived largely from the arguments advanced against him by the German chemist J. F. Meyer, whose rival theory of *acidum pingue* was for a time widely credited. Black's case would surely have been strengthened had he published the simple experiment, performed at Glasgow in 1757 or 1758, of directly impregnating a solution of caustic alkali with the "fixed air" expelled from chalk or limestone, and so obtaining a product both effervescent and mild.

❧

Black's discoveries concerning heat, the major achievement of his Glasgow period, were originally stimulated by William Cullen. In 1754 Cullen noted a striking phenomenon—the intense cold produced when highly volatile subtances like ether evaporate—and he promptly wrote Black about his experiment. At about this time, Black set down in a notebook a curious observation made by Fahrenheit: water can be cooled below the freezing point without congealing; yet if shaken, it suddenly freezes and the thermometer rises abruptly to 32° on Fahrenheit's scale. This, Black speculated, might be due to "heat unnecessary to ice."

Fahrenheit's observation was recorded in Boerhaave's *Elementa chemiae* (1732), a famous work that Cullen, and later Black, recommended to their students in the English version of Peter Shaw. This observation was hard to reconcile with the prevailing view that when water is brought near the freezing point, withdrawal of a small increment of heat must bring prompt solidification. But Fahrenheit's experiment showed that solidification (or liquefaction) required the transfer of substantial quantities of heat; of heat lying concealed and not directly detectable by the thermometer; of heat, to use Black's term, that was *latent*. Upon reflection, Black saw this notion to be quite consistent with commonly observed facts of nature. Snow, for example, requires a considerable time to melt after the surrounding temperature has risen well above the freezing point. A gradual absorption of heat must therefore be taking place, although the temperature of the snow remains unaltered.

Black became convinced of the reality of this latent heat through thoughtful reading and meditation on the familiar phenomena of change of state. He presented his doctrine in his Glasgow lectures, perhaps as early as 1757–58, before he had performed any experiments of his own. Nor did his doctrine arise from any firmly held theory as to what heat might be.

Not until 1760 did Black carry out his earliest experiments on heat. The first fact to be ascertained was the reliability of the thermometer as a measuring tool. Would a thermometric fluid, having received equal increments of heat, show equal increments of expansion? Ingenious and simple experiments on mixing amounts of hot and cold water (*Lectures*, I, 56–58) convinced him that the scale of expansion of mercury, over that limited range, was indeed a reliable scale of "the various heats, or temperatures of heat."

Crucial to Black's experiments was his recognition of the distinction between *quantity of heat* and *temperature*, between what we sometimes describe as the *extensive* and *intensive* measures of heat. Although not the first to note this distinction, he was the earliest to sense its fundamental importance and to make systematic use of it. In his lectures, which always opened (after certain preliminaries) with a careful discussion of heat, he would tell his students:

> Heat may be considered, either in respect of its quantity, or of its intensity. Thus two lbs. of water, equally heated, must contain double the quantity that one of them does, though the thermometer applied to them separately, or together, stands at precisely the same point, because it requires double the time to heat two lbs. as it does to heat one.[11]

Temperature, of course, is read directly from the thermometer. But how to measure the *quantity* of heat? Black's answer is implied in the above quotation: the time required to warm or cool a body to a given temperature is related to the amount of heat transferred. This elusive quantity required a *dynamic* measurement: the heat gained or lost should be proportional to the temperature and the time of heat flow "taken conjointly."[12] Here, as so often in his career, the influence of Newton is quite apparent. In a famous paper of 1701, Newton had used a dynamic method to estimate temperatures beyond the reach of his linseed-oil thermometer. Black in his lectures gave a clear account of Newton's experiments and how his law of cooling (as we now call it) was used to estimate relative temperatures above that of melting tin. Black's own experiments made use of the law of cooling, and it is not hard to imagine that Newton's dynamic method was the key to Black's.

Characteristically, Black was not satisfied to demonstrate qualitatively that there is such a thing as the latent heat of fusion: he proposed to measure it. The method occurred to him in the summer of 1761. First cooling a given mass of water to about 33°F., he would determine the time necessary to raise its temperature one degree, and compare this with the time required to melt the same amount of ice. Conversely, he would compare the time necessary to lower the temperature of a mass of water with the time necessary to freeze it completely. Assuming that both systems received heat from, or gave up heat to, the surrounding air at the same rate, as much heat should be given off in freezing a given amount of water as in melting the same amount of ice. Obliged to wait until winter, Black carried out the experiment in December 1761 in a large hall adjoining his college rooms. The following April he described his results to his Glasgow colleagues and friends.

Black soon saw that latent heat must play a part in the vaporization of water as well as in the melting of ice. The analogy was so persuasive that as early as 1761—before testing his conjecture by experiment—he presented this version of his doctrine to his students. The success of the freezing experiments soon led him to investigate the latent heat of vaporization. But the method he first employed, a precise analogue of Fahrenheit's observation on super-cooled water, was unsuited to measurement. A better method could be modeled on his freezing experiments. Tin vessels, containing measured amounts of water, would be heated on a red-hot cast-iron plate. The time necessary to heat the water from 50°F. to the boiling point would be compared with the time necessary for the water to boil away. The chief obstacle was to find a source of heat sufficiently unvarying so that the absorption of heat could be safely measured by the time. A "practical distiller" informed Black that when his furnace was in good order, he could tell, to a pint, the quantity of liquor that he would get in an hour. When Black confirmed this by boiling off small quantities of water on his own laboratory furnace, he was ready for the experiments.[13] These he performed late in 1762. From the average of three experiments he calculated that the heat absorbed in vaporization was equal to that which would have raised the same amount of water to 810°, were this actually possible. This gives a figure of 450 calories per gram for the latent heat of vaporization of water, compared to a modern figure of 539.1 calories per gram. More accurate figures were obtained in later experiments, but several years elapsed before Black took up the subject again.

The second, and closely related, discovery made by Black concerning heat was that different substances have different heat capacities. It is commonly assumed that Black discovered *specific heats* before his work on latent heat. This is not the case. To be sure, the clue, once again, was found in Boerhaave's textbook; Fahrenheit had made certain experiments at his request and had obtained the surprising result that when he mixed equal quantities of mercury and water, each at a different temperature, the mercury exerted far less effect in heating or cooling the mixture than did the water. At first Black was puzzled; but he soon realized that mercury, despite its greater density, must have a smaller store of heat than an equal amount of water at the same temperature. If so, the capacities of bodies to store up heat did not vary with their bulk or density, but in a different fashion "for which no general principle or reason can yet be assigned."[14] Now Black could explain a peculiar effect reported twenty years earlier by George Martine, an authority on thermometers. Martine had placed equal volumes of water and mercury in identical vessels before a fire, and observed that the mercury increased in temperature almost twice as fast as the water. Black saw that since less heat was required to bring mercury up to a given temperature, a thermometer placed in it should rise more rapidly. Not until 1760 did Black perceive the significance of this effect, but he did not pursue the subject; he was principally absorbed with the more striking phenomena of changes of state. In 1764—a year that James Watt made memorable in the history of invention—Black returned to the study of heat. His experimental inquiry into specific heats, and his attempts to obtain a more accurate value for the latent heat of vaporization, were stimulated by the activities of Watt.

The year that Black began his Glasgow lectures, James Watt, a young man of nineteen, skilled in making mathematical instruments, was taken under the wing of the university as what we might today call a technician. He was soon called upon by Black to make things he needed for his experiments. Watt, in turn, after repairing the now-famous model of a Newcomen engine and undertaking experiments to improve its performance, turned to Black to explain an effect he could not comprehend. He was astonished at the large amount of cold water required to condense the steam in the engine cylinder, until Black explained his ideas about latent heat.

Watt was many months, and many experiments, away from hitting upon the historic invention of the separate condenser, and Black may be pardoned for believing that this disclosure inspired Watt's radical improvement of the steam engine, a claim advanced even more strongly by John Robison, who spoke of Watt as Dr. Black's most illustrious pupil. This, in a strict sense, Watt never was; and, despite his lifelong attachment to Black, he later insisted that the invention of the separate condenser had not been suggested by his knowledge of the doctrine of latent heat. But he readily credited Black with having clarified the problems he encountered and with teaching him "to reason and experiment in natural philosophy."[15]

On the other hand, Watt's ingenuity and questioning mind, and the practical problems he raised, revived Black's interest in heat. The problems he now investigated with John Robison and William Irvine were closely related to those Watt needed to elucidate. Irvine was set to work determining a more accurate value of the latent heat of steam. Using a common laboratory still as a water calorimeter, Irvine obtained improved values, although these were not high enough to be really accurate. Black, it should be remarked, never made or used the mythical ice calorimeter associated with his name, although it occurred to him in the spring or summer of 1764 that his knowledge of the latent heat of fusion of ice could be used to measure the latent heat of steam. Plans to put this to the test were set aside when Watt, late in 1764, began to obtain values that Black deemed sufficiently precise. Years later, Black gave the French scientists full credit for the independent invention and first use of an ice calorimeter.[16]

The measurements made at this time by Black, Irvine, and Watt on the specific heats of various substances are the earliest of which we have any trace.

Watt seems to have been the first to stress the importance of investigating the subject systematically. He carried out experiments of his own, and Black put Irvine to work on the problem. Using the method of mixtures, Black and Irvine determined the heats communicated to water by a number of different solids. These joint experiments continued until Black left for Edinburgh. After Black's departure, Irvine continued these investigations, but his results were not published in his lifetime.

✿

Joseph Black's view of chemistry, his chemical doctrines, can be derived from his single major paper and from the several versions of his lectures. Chemistry, to him, was a subject with wide practical application to medicine and to the progress of industry. But he insisted, as Cullen had done, that it is a science, albeit an imperfect one, not merely an art: "the study of the effects of heat and mixture, natural or artificial, with a view to the improvements of arts and natural knowledge."[17]

On the question of the "elements" or "principles" of bodies, Black showed a typical caution. He no longer credited the venerable doctrine of the four elements; little could be known about them, for there was no knowledge of the underlying constitution and forces of nature. It was more sensible to group into several classes, as Cullen did, those substances sharing certain distinguishable properties: the salts, earths, inflammable substances, metals, and water. These were not necessarily elementary; of earths there were several sorts that could not be decomposed further; water, Black believed, can on distillation be converted into earth; and there was reason to suspect that salts were not compounds of earth and water, but of an earth and some other unknown substance.

Black invariably devoted the early lectures of his course to the subject of heat, telling his students that Boerhaave, Robert Boyle, and Sir Isaac Newton followed Lord Bacon in believing that heat is caused by motion, and that the French thought heat to be the vibration of an imponderable, elastic fluid. The fluid theory was the one to which he quite definitely leaned, for it seemed to agree best with the phenomena; he could not, for example, readily conceive a motion of particles in dense, solid bodies. But such questions are involved in obscurity. And he told his students: "The way to acquire a just idea of heat is to study the facts."[18]

In discussing problems of combustion, fermentation, and the calcination of metals, Black—until the close of his career—presented a gingerly version of the phlogiston theory, although he generally avoided the term. Air, of course, was required for combustion; but like his contemporaries he invoked the property of elasticity to explain its role. Combustion was caused by the presence of an inflammable principle, for which different substances had a different "elective attraction." Inflammable and combustible substances, including the calcinable metals, were pervaded with this mysterious substance, the nature of which "we are still at a loss to explain." Although little could be said on the subject, heat and light appeared to be the principles of inflammability. There was, however, a fact that was hard to explain and raised a strong objection to this theory. When it was possible to collect the product of combustion, or weigh a calcined metal, this product was heavier, despite the loss of the inflammable principle. Possibly this was a kind of matter that defied the general law of gravitation. Yet speculations of this sort were not Black's cup of tea. These doubts almost certainly prepared him to accept, in the main, Lavoisier's discoveries.

As late as 1785 Black was reluctant to adopt the new "French chemistry," a term he heartily disliked. The geologist Sir James Hall was the earliest of Black's circle to sense the winds of change. A visit to Paris in 1786 convinced him; and on his return to Scotland, in the course of long discussions, he brought Thomas Charles Hope around to his opinions. Early in 1788 Hall read before the Royal Society of Edinburgh a paper entitled "A View of M. Lavoisier's New Theory of Chemistry." Black may well have been present— we cannot be sure—but his intimate friend James Hutton was, and defended the phlogistic hypothesis with a paper of his own. Nevertheless, an Italian visitor to England, an admirer of Mme. Lavoisier, wrote her from London in 1788 that Black and Watt ("in my opinion the two best heads in Great Britain") were on the verge of being convinced by Lavoisier's antiphlogistic theory. Soon thereafter, Black began to mention the new doctrine in his lectures; he wrote to Lavoisier, in a famous letter of October 1790, that he had begun to recommend the new system to his students as simpler, and more in accord with the facts, than the old. Robinson's edition of Black's lectures, based on what Black was telling his students between 1792 and about 1796, amply confirms his statement. Yet it is clear that Black strongly disapproved of the new nomenclature, while recognizing that advances in chemistry made some such reform necessary. He objected to obliterating the work of the earlier chemists completely; he felt that the new terms were "evidently contrived to suit the genius of the French language," and he perceived in the new scheme a clever stratagem of the French chemists to give their doctrines "universal currency and authority."[19]

The most interesting and pervasive of Black's doctrines is the theory of chemical affinity. Here his debt to Cullen, and beyond Cullen to Newton's *Opticks*, is clearly evident. Chemical reactions result from the differential or "elective" attraction of chemical individuals for one another. Simple elective attractions are those produced by heat; "double elective attractions," reactions of double decomposition, are chiefly those that take place in solution. Black saw no reason for avoiding the term "attraction," as did the French chemists who spoke instead of *affinités* or *rapports*. As Newton had insisted, "attraction" should be taken as a descriptive term, not a causal explanation.

Black's earliest use of this concept, and of Geoffroy's well-known table of affinities, appears in his "Experiments." He employs it to show the differential behavior of alkaline substances toward acids and "fixed air." For Black, as for Cullen, this became a centrally important pedagogical device; invariably he devoted several lectures to elective attractions, describing the table, and referring to it elsewhere when speaking of particular reactions. In the lectures of the early years Black set forth these reactions with the diagrams Cullen had invented, adding numbers to indicate the relative force of attraction between substances:

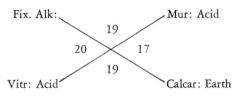

Later, he used a diagram consisting of segmented circles, but without numbers, to indicate the relative attractions.[20] The following example illustrates what takes place when a compound of volatile alkali with any acid (e.g., ammonium chloride) reacts with a mild fixed alkali (e.g., sodium carbonate); here the "fixed air" or "mephitic air" represented by **MA** combines with the volatile alkali, ♉ ; and the fixed alkali, ♑ , joins itself to the acid, ⟩ :

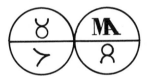

Did Black wish to represent molecules in which the atomic partners are interchanged? Perhaps this occurred to him, but he probably conceived these diagrams primarily as what we call visual aids. Unlike his master, William Cullen, Black did not explicitly link his discussion of elective attractions with the corpuscular or atomic doctrine, about which he had little to say in his lectures.

Robison records that in an early conversation Black "gently and gracefully" checked his disposition to form theories and warned him to reject "even without examination, every hypothetical explanation, as a mere waste of time and ingenuity."[21] Like Newton, whom he so greatly esteemed—at least Newton as he understood him—Black chose not to "deal in conjectures."

In mid-career he told his students, well before the fulfillment of what we call the Chemical Revolution:

> Upon the whole, Chymistry is as yet but an opening science, closely connected with the usefull and ornamental Arts, and worthy the attention of a liberal mind. And it must always become more and more so: for though it is only of late, that it has been looked upon in that light, the great progress already made in Chymical knowledge, gives us a pleasant prospect of rich additions to it.

The Science is now studied on solid and rational grounds. While our knowledge is imperfect, it is apt to run into errour: but Experiment is the thread that will lead us out of the labyrinth.[22]

[Certain points in my attempted reconstruction of Black's experiments on *magnesia alba* have recently been challenged by A. L. Donovan in his stimulating and careful study *Philosophical Chemistry in the Scottish Enlightenment* (Edinburgh, 1976).]

Notes

1. John Thomson, *Account of the Life, Lectures and Writings of William Cullen, M.D.*, 2 vols. (Edinburgh-London, 1859), I, 573.
2. Henry Cockburn, *Memorials of His Times* (Edinburgh, 1856), 48–49.
3. James Patrick Muirhead, *Origin and Progress of the Mechanical Inventions of James Watt*, 3 vols. (London, 1854), II, 261–63.
4. Adam Ferguson, "Minutes of the Life and Character of Joseph Black, M.D.," *Transactions of the Royal Society of Edinburgh* 5 (1805), 116.
5. See Black's *Dissertation medica inauguralis* . . . (Edinburgh, 1754), English translation by A. Crum Brown in *Journal of Chemical Education* 12 (1935), 272; and "Experiments upon Magnesia Alba, Quicklime, and Some Other Alcaline Substances," in *Essays and Observations, Physical and Literary. Read Before a Society in Edinburgh* 2 (1756), 17. Compare his *Lectures on the Elements of Chemistry*, ed. John Robison, 2 vols. (Edinburgh, 1803), II, 63–64.
6. Thomson, *Cullen*, I, 50.
7. Letter to Cullen, June 18, 1754; *ibid.*, 50–51.
8. "Experiments," 25.
9. *Ibid.*, 30–31.
10. *Ibid.*, 31.
11. Alexander Law, "Notes of Doctor Black's Lectures on Chemistry" (University of Edinburgh Library), I, 5.
12. *Ibid.*, 18.
13. *Lectures*, I, 157.
14. *Ibid.*, 79.
15. Muirhead, *Origin and Progress*, 264.
16. *Lectures*, I, 175.
17. Thomas Cochrane, "Notes from Black's Lectures 1767/8" (Andersonian Library, University of Strathclyde, Glasgow), 3. Black saw no reason to change his definition: see *Lectures*, I, 12–13.
18. Law, "Notes," I, 5. Compare *Lectures*, I, 35.
19. *Lectures*, I, 489–93.
20. Henry Guerlac, "Commentary on the Papers of Cyril Stanley Smith and Marie Boas," in Marshall Clagett, ed., *Critical Problems in the History of Science* (Madison, Wis.: University of Wisconsin Press, 1959), 515–19. For a fuller treatment, see M. P. Crosland, "The Use of Diagrams as Chemical 'Equations' in the Lecture Notes of William Cullen and Joseph Black," *Annals of Science* 15 (1959), 75–90. Crosland argues that Black wished to illustrate the course of chemical reactions without implying mechanical explanations and that his symbols were generalized expressions for reactions of a similar type.
21. *Lectures*, I, p. vii.
22. Law, "Notes," III, 88.

Bibliography

I. Original Works

Black's published writings are the following:
Dissertatio medica inauguralis . . . (Edinburgh, 1754). Reprinted in *Thesaurus medicus Edinburgensis novus*, II (Edinburgh-London, 1785). English translation by A. Crum Brown in *Journal of Chemical Education* 12 (1935), 225–28, 268–73, with facsimile of title page and dedication.

"Experiments upon Magnesia Alba, Quicklime, and Some Other Alcaline Substances," in *Essays and Observations, Physical and Literary. Read Before a Society in Edinburgh* 2 (1756), 157–225. Republished, together with Cullen's "Essay on the Cold Produced by Evaporating Fluids" (Edinburgh, 1777; reprinted 1782). Black's famous paper is most readily available as Alembic Club Reprints no. 1 (Edinburgh, 1898). The first French translation was "Expériences sur la magnésie blanche, la chaux vive, & sur d'autres substances alkalines par M. Joseph Black, Docteur en Médecine," in *Observations sur la physique* 1 (1773), 210–20, 261–75. A short summary of Black's work on magnesia alba had been published in the *Journal de médecine, chirurgie, pharmacie* 8 (1758), 254–61.

"On the Supposed Effect of Boiling upon Water in Disposing It to Freeze More Readily," in *Philosophical Transactions of the Royal Society of London* 65 (1775), 124–29.

"Lettre de M. Joseph Black à M. Lavoisier," *Annales de chimie* 8 (1791), 225–29. The English original of Black's letter was printed by Douglas McKie in *Notes and Records of the Royal Society of London* 7 (1950), 9–11.

"An Analysis of the Water of Some Hot Springs in Iceland," *Transactions of the Royal Society of Edinburgh* 3 (1794), 95–126.

Lectures on the Elements of Chemistry, ed. John Robison, 2 vols. (Edinburgh, 1803; American edition, 3 vols. Philadelphia, 1806–7). Robison omitted Black's introductory lecture and much of the next two or three lectures. In a letter of September 16, 1802 to James Black, Joseph's brother, he tells of his difficulties in putting together, from Dr. Black's sparse notes, a coherent text. He even speaks of having to "manufacture" one lecture; obviously, although Robison's edition probably represents Black's opinions, the language may sometimes be Robison's. The text should be compared with the MS. versions. See McKie, *Annals of Science* 16 (1960), 131–34, 161–70.

"Case of Adam Ferguson, Drawn up by Joseph Black, M.D., in May, 1797," *Medico-Chirurgical Transactions* 8 (1816). Cited and summarized by Crowther.

"A Letter From Dr. Black to James Smithson, Esq. Describing a Very Sensible Balance," *Annals of Philosophy*, n.s. vol. 10 (1825), 52–54. Black's letter is dated September 18, 1790.

The earliest pictures of Black are two ink sketches made by Thomas Cochrane in his notes while attending Black's lectures in 1767–68 in what appears to be an anatomical theater. Reproduced by McKie in *Annals of Science* 1 (1936), 110, and in his edition of Cochrane's "Notes 1767/8."

Not much later (c. 1770) is a fine oil by David Martin, the teacher of Sir Henry Raeburn (Collection of the University of Edinburgh), published by Guerlac in *Isis* 48 (1957), and by McKie as frontispiece to his edition of the Cochrane "Notes 1767/8."

The most familiar portrait of Black is by Sir Henry Raeburn; occasionally exhibited at the University of Edinburgh, it is in a private collection in London. It has often been reproduced, sometimes (as in the frontispiece of Robison's edition of Black's *Lectures*) from an inferior engraving.

Roughly contemporaneous with the Raeburn portrait are the sketches of John Kay: one shows Black walking in the country; another places *en face* a birdlike Hutton and a pensive Black. The most interesting is a close view of Black lecturing, spectacles in hand, and before him a spate of scattered notes, a siphon, a burning candle, and a small bird in a cage, all to demonstrate the properties of "fixed air." See *A Series of Portraits and Caricature Etchings by the Late John Kay*, 2 vols. (Edinburgh, 1837), I, pt. 1, 52–57.

There are reproductions of Kay's drawing by Ramsay in his biography of Black and, more attractively, by John Read in *Humour and Humanism in Chemistry* (London: G. Bell, 1947), plates 41, 44–45.

No published census of manuscript versions of Black's lectures exists, but a few may be mentioned, together with their present locations. The earliest is Thomas Cochrane's (Andersonian Library, University of Strathclyde, Glasgow); dating from Black's early teaching at Edinburgh, it is otherwise notable only for the caricatures of Black. Very sketchy, it has recently been published as *Notes from Doctor Black's Lectures on Chemistry 1767/8*, ed. with intro. by Douglas McKie (Cheshire, 1966).

More complete are the closely written set of 120 lectures recorded by James Johnson, 1770 (University of Edinburgh Library), and three volumes bearing the name of Joseph Freyer Rastrick and covering Black's lectures of 1769–70 (History of Science Collections, Cornell University Library). A fine set is the Beaufoy MS. 1771/72 (University of Saint Andrews). Quite different from others is Alexander Law's "Notes of Doctor Black's Lectures on Chemistry" (University of Edinburgh Library), with fifty-seven lectures from June 13 to December 22, 1775, and elaborate notes and appendices. For Black's later period there are the notes of George Cayley, 6 vols., 118 lectures, 1785–86 (York Medical Society) and a similar set, without date or name of the original owner, now at University College, London. McKie believes this set to be contemporary with the Cayley notes. "Some MS Copies of Black's Chemical Lectures," *Annals of Science* 15 (1959), 73.

II. Secondary Literature

A brief, uninspired account of a visit to Black in 1784 (largely devoted to a description of Black's portable furnace) is given by the geologist Barthélemy Faujas de Saint-Fond, in his *Voyage en Angleterre et en Ecosse*, 2 vols. (Paris, 1797), II, 267–72. The earliest biographical sketch is the short, anonymous account, probably by Alexander Tilloch, in *Philosophical Magazine* 10 (1801), 157–58. But the most important source is Adam Ferguson, "Minutes of the Life and Character of Joseph Black, M.D.," *Transactions of the Royal Society of Edinburgh* 5 (1805), 101–17. Ferguson was Black's cousin and close friend. John Robison, "Editor's Preface," in *Lectures*, vol. I, pp. v–lxvi, draws heavily on Ferguson, yet adds much useful information. Thomas Thomson, *History of Chemistry*, 2 vols. (London, 1830–31), vol. I, chap. 9, relies on Ferguson and Robison, but adds personal impressions. Thomson also contributed short accounts of Black to the *Annals of Philosophy* 5 (1815), 321–27, and to the *Edinburgh Encyclopedia*. Lord Brougham—like Thomson, one of Black's last students—devotes an interesting chapter to Black in his *Lives of Philosophers of the Time of George III* (London-Glasgow, 1855), 1–24. George Wilson's brief note, in *Proceedings of the Royal Society of Edinburgh* 2 (1849), 238, corrects the date of Black's death as given by Ferguson and Robison, citing newspaper accounts and Muirhead.

John Playfair's "Biographical Account of the Late Dr. James Hutton," in *Transactions of the Royal Society of Edinburgh*, vol. 5, pp. 39–99 of the "History of the Society," has references to Black. James Patrick Muirhead, *Origin and Progress of the Mechanical Inventions of James Watt*, 3 vols. (London, 1854), published numerous letters from Watt to Black, and one from Black to Watt. Also valuable is John Thomson's *Account of the Life, Lectures and Writings of William Cullen, M.D.*, 2 vols. (Edinburgh-London, 1859), with several early letters of Black to Cullen.

For Glasgow University in Black's time, see Henry G. Graham, *Social Life of Scotland in the Eighteenth Century*, 2 vols. (London, 1899), vol. II, chaps. 12–13; W. Innes Addison, *Roll of Graduates of the University of Glasgow* (Glasgow, 1898), and *Matriculation Album of the University of Glasgow* (Glasgow, 1913). Worth consulting is William Robert Scott, *Adam Smith as Student and Professor* (Glasgow: Jackson, Son & Co., 1937). Letters of Thomas Reid in *The Works of Thomas Reid, D.D.*, ed. Sir William Hamilton, 7th ed., 2 vols. (Edinburgh, 1872), I, 39–50, describe Black's Glasgow lectures. For the chair of chemistry at Glasgow, see Andrew Kent, ed., *An Eighteenth Century Lectureship in Chemistry* (Glasgow: Jackson, Son & Co., 1950). For Edinburgh, consult Alexander Bower, *History of the University of Edinburgh*, 2 vols.

(Edinburgh, 1817); and Sir Alexander Grant, *Story of the University of Edinburgh*, 2 vols. (London, 1884).

Agnes Clarke's article on Black in the *Dictionary of National Biography* is disappointing and sometimes inaccurate, but gives the correct date for Black's death. The most scientifically eminent of Black's modern biographers is Sir William Ramsay. He first discussed Black's work in his *Gases of the Atmosphere* (London, 1896), 527–31, and again in his *Joseph Black M.D.: A Discourse* (Glasgow, 1904). His *Life and Letters of Joseph Black, M.D.* (London, 1918) is the only full-length biography. Published posthumously, it is valuable chiefly for the use made of letters and papers of Black, including an autobiographical sketch, which have been otherwise inaccessible to scholars. Ramsay's book is unsatisfactory; when a scholarly biography is written, and one is badly needed, use will surely be made of Henry Riddell's "The Great Chemist, Joseph Black, His Belfast Friends and Family Connections," in *Proceedings of the Belfast Natural History and Philosophical Society* 3 (1919/20), 49–88.

Black is, of course, discussed in the familiar histories or studies of early chemists. Most can be ignored; an exception is Max Speter's "Black," in G. Bugge, *Das Buch der grossen Chemiker*, 2 vols. (Weinheim, 1929), I, 240–52. J. R. Partington, *History of Chemistry*, 4 vols. (London: Macmillan & Co., 1962), III, 131–43, appraises Black's proficiency as a chemist and gives detailed citations of the literature. John Read has a readable, if not wholly reliable, account of Black in his *Humour and Humanism in Chemistry*, chap. 8, and there is a brisk chapter in Kent's *Eighteenth Century Lectureship*, 78–98. A longer and more informative account of Black, with new insights and some inaccuracies, is J. G. Crowther, *Scientists of the Industrial Revolution* (London: Cresset Press, 1962), 9–92. Archibald and Nan L. Clow, *The Chemical Revolution* (London: Batchworth Press, 1952), has a misleading title; it deals with the applications of chemistry to industry in the eighteenth and early nineteenth centuries, and is valuable for many passing references to Black's involvement in such matters.

Douglas McKie's paper on the Cochrane "Notes," in *Annals of Science* 1 (1936), 101–10, has been superseded by his edition of that MS. But see his "Some MS Copies of Black's Chemical Lectures," *Annals of Science* 15 (1959), 65–73; *ibid.* 16 (1960), 1–9; 18 (1962), 87–97; 21 (1965), 209–55; 23 (1967), 1–33. E. W. J. Neave, "Joseph Black's Lectures on the Elements of Chemistry," *Isis* 25 (1936), 372–90, merely outlines the contents of Robison's edition of Black's *Lectures*.

Black's influence on the progress of scientific medicine and biology is treated by Heinrich Buess, "Joseph Black und die Anfänge chemischer Experimentalforschung in Biologie und Medizin," *Gesnerus* 13 (1956), 165–89. Henry Guerlac, "Joseph Black and Fixed Air," *Isis* 48 (1957), 124–51, 433–56, attempts to clarify the chronology of Black's early life and to reconstruct the steps in his chemical investigations. Guerlac's *Lavoisier—The Crucial Year* (Ithaca, N.Y.: Cornell University Press, 1962), 8–35, 68–71, sees Stephen Hales, rather than Joseph Black, as the chief British influence on Lavoisier before 1773. M. P. Crosland has studied Black's teaching symbols in "The Use of Diagrams as Chemical 'Equations' in the Lecture Notes of William Cullen and Joseph Black," *Annals of Science* 15 (1959), 75–90. Twenty-six recently discovered letters by or concerning Black, including twenty-one written to his brother Alexander, have been published by Douglas McKie and David Kennedy, "Some Letters of Joseph Black and Others," *Annals of Science* 16 (1960), 129–70. Included is the important letter by John Robison on the problems encountered in publishing Black's *Lectures*.

For Black's work on heat, consult Ernst Mach, *Die Principien der Wärmelehre* (Leipzig, 1896), 153–81; Douglas McKie and Niels H. de V. Heathcote, *The Discovery of Specific and Latent Heats* (London: E. Arnold & Co., 1935), 1–53; and Martin K. Barnett, "The Development of the Concept of Heat from the Fire Principle of Heraclitus Through the Caloric Theory of Joseph Black," *Scientific Monthly* 42 (1946), 165–72, 247–57.

19

Joseph Priestley's First Papers on Gases and Their Reception in France

✿

In March of 1772—before we find any evidence that Lavoisier had turned his attention to the chemistry of gases—Joseph Priestley presented to the Royal Society of London, at a succession of meetings, the chief results that were to be embodied in his first important publication on gases, his *Observations on Different Kinds of Air*. It is commonly stated that this long paper—in which Priestley described his improved pneumatic trough, and the use of mercury in place of water for handling soluble gases, announced his discovery of new kinds of air (e.g., nitric oxide and hydrochloric acid), discussed the results Cavendish had obtained on "inflammable air" (hydrogen), and showed that nitrous air could be used to measure the "goodness of air"—was first printed later in the same year in the *Philosophical Transactions*.[1] The actual time of its appearance, if it can be ascertained more precisely, is a matter of some importance, especially if we wish to assess the early influence of Priestley upon Lavoisier. For this was a crucial year in Lavoisier's scientific career: it was in the autumn of 1772 that he carried out his dramatic experiments on the combustion of phosphorus and sulphur;[2] and it was on February 20, 1773, that he set down in a newly begun laboratory register his famous memorandum outlining the research program he planned to pursue "on the elastic fluid that is set free from substances . . . and also on the air absorbed in the combustion of a great many substances." Here, for the first time, he listed the names of those who had been pioneers in the study of fixed air: "Messrs. Hales, Black, Magbride [Macbride], Jacquin, Cranz, Prisley [Priestley], and de Smeth. . . ."[3] At first glance we would take this as evidence that Lavoisier was already familiar with Priestley's *Observations on Different Kinds of Air*, and that this paper may well have been one of the earliest influences upon him. The matter, however, is somewhat more complex than appears at first sight.

Priestley's epoch-making paper, as the reader can readily verify, appeared in Volume 62 of the *Philosophical Transactions*, which bears on the title page the name of the Royal Society's printer, Lockyer Davis in Holbourne, and the date 1772. But an examination of this volume of the *Transactions* makes it

From *Journal of the History of Medicine and Allied Sciences* 12 (1957), 1–12. Copyright 1957 by Journal of the History of Medicine and Allied Sciences, Inc.

clear that it did not actually appear in the year given, but early in 1773. Proof of this is contained in the preliminary material which includes a resolution passed at a meeting of the Royal Society's council on January 28, 1773, referring to a proposed change in the publication schedule of the journal.[4] For reasons probably connected with the new schedule, Volume 62—which could hardly have appeared before February, and probably not until March 1773—was predated to 1772. Priestley himself, in his *Memoirs*, clearly places the publication of the paper in the year 1773.[5]

This innocent printer's legerdemain explains why historians of chemistry have assumed that Priestley's paper appeared in the *Transactions* in the same year in which it was delivered, although there is much evidence that Priestley included in it results that he obtained during the summer and autumn of 1772. Anyone would be forgiven—since the time of the supposed appearance of the *Transactions* in 1772 is never specified—who assumed that Lavoisier might have learned of it, not only before he wrote his memorandum of February 20, 1773, but perhaps earlier, i.e., by the time he began his long series of investigations, on combustion and on the air combined with various substances, in the autumn of 1772.

Oddly enough, Priestley's paper *was* published in a different form before the end of 1772, though rather too late and in too small an edition, I am certain, to have influenced Lavoisier at this time. Hoefer—alone of the sources I have consulted—makes mention of what he describes as a reprint of the *Observations*: "réimprimées à part; Lond., 1772."[6] Priestley himself makes no reference to it in his *Memoirs*, and neither do any of Priestley's biographers. It is not listed in the chemical bibliographies of Ferguson, Bolton, or Duveen (1949); nor is it included in the list of his publications appended to the sketches of Priestley in Poggendorff or the *Dictionary of National Biography* (by Sir Philip Hartog). Dr. John F. Fulton, who made a preliminary survey of printed works of Priestley in twenty-nine libraries, does not record it.[7] I find no copy listed in the catalogues of the British Museum or of the Library of Congress; there is none in the Edgar Fahs Smith Memorial Collection at the University of Pennsylvania nor in the Library of the American Philosophical Society. It is, however, clearly described in the catalogue of the *Bibliothèque Nationale* and in that of the Surgeon General's Library. A copy from Benjamin Franklin's library is in the collection of the Historical Society of Pennsylvania. There is a well-preserved copy in the Evans-Rosenwald Collection, recently given to the Institute for Advanced Study (Princeton), where I first became aware of its existence and was able to examine it.

A careful perusal revealed that the pamphlet is identical in substance with the paper in the *Transactions* and convinced me that it is the separate to which Hoefer refers. It is a slim quarto, bearing the date 1772, printed by W. Bowyer and J. Nichols. The pagination does not correspond with the paper in the *Transactions*: the separate has an unnumbered title page and numbered pages running from 3 to 120. Internal evidence shows that it could not have appeared until after October 29, 1772, for a postscript to a letter of William Hey to Priestley (p. 118) bears this date. Indeed, it could hardly have appeared before the end of November. A letter of Priestley to Dr. Price, dated November 11, 1772, includes the following paragraph:

> I sent my papers for the Royal Society by a private hand on Tuesday last, (this day sennight), to be delivered by Mr. Johnson to Dr. Franklin, and by him to be sent to Dr. Maty. I hope they have arrived, though I have this day received a letter from Dr. Maty, desiring me to send them as soon as I can get them ready.[8]

The tenor of Priestley's correspondence with Franklin and Price, and Priestley's exclusive scientific preoccupation in these months with his experiments on air, leave no doubt that he is referring to the *Observations*.

Now there is a possibility, despite the date of the title page, that this separate did not actually appear until January or February 1773,[9] but I do not think this is the case. It was clearly intended to appear before the next volume of the *Transactions* and is certainly an early instance of Priestley's impatience at the delays of publication through the Royal Society, which led him, as Hartog has argued, to bring out his subsequent researches in book form.[10] On the title page of the "pre-print," as we may call it, is the statement that it was "printed for" Volume 62 of the *Philosophical Transactions*, 1773; and the mention of this year supports our inference that the volume of the *Transactions* in question, despite the date that it bears, did not appear in 1772. Yet there is evidence that Volume 62 had been at least partly set up in type before the preprint appeared, for the latter contains the famous engraved folding plate, with the indication that it was to be inserted opposite page 252 of the *Transactions*, where it does in fact appear.

The publishers of the preprint were William Bowyer the younger (1699–1777), "the learned printer," and John Nichols, his partner since 1766.[11] Bowyer was a close friend of Davis's, the younger man who was bookseller to the Royal Society and at least nominally their printer. It is possible that Davis, preoccupied with the task of bringing out the *Transactions*, appealed to Bowyer, or sent Priestley to Bowyer, so that Priestley's results might appear without delay. But I think it more likely that the actual printing of the *Transactions* had been entrusted to Bowyer, who could strike off separate copies, merely changing the title page and the pagination, once the paper had been set up for the Royal Society.

This rare preprint is in any case the earliest form in which Priestley's classic paper appeared. By a convenient cancellation of errors, the date assigned by historians of science to the first appearance of Priestley's work is probably correct; but it is clear that the *Observations* did not, as is usually stated or implied, first see the light of day in the *Philosophical Transactions*.

Lavoisier and the French chemists are unlikely to have seen a copy of this preprint; its rarity today suggests that the edition was extremely small, as would have been logical, since the paper was soon to secure wider circulation in the *Transactions*. Although we know that news of Priestley's reports to the Royal Society during the course of March 1772 reached France in the same month,[12] it was not until the appearance of Volume 62 of the *Transactions* in the spring of the following year that Lavoisier and his fellow chemists could know at first hand what Priestley had accomplished.[13] There is good evidence to support this conclusion.

The interest of Frenchmen in Stephen Hales's "fixed air"[14] had been greatly stimulated by the publication in the Abbé Rozier's *Observations sur la physique* (issue of February 1773) of a summary of Jacquin's attack on Meyer's *acidum pingue* theory of causticity, with Jacquin's defense of the views of Joseph Black;[15] and by the appearance in the same journal (issue of March 1773) of a translation of the first part of Black's *Experiments on Magnesia Alba*.[16] There is reason to believe that Lavoisier may have had a hand in these translations; if so, he may have read them before they were actually printed.

The subject of fixed air was first discussed at meetings of the Academy of Sciences in April and May 1773. On April 24, in what appears to be the earliest session devoted to this subject, the gifted young chemist Jean Bucquet (1746–80) reported on experiments he had performed in the laboratory of the Duc de la Rochefoucauld d'Enville on the properties of Black's fixed air, in order to ascertain (1) if fixed air is the same as ordinary atmospheric air and (2) whether it is the same regardless of the source from which it is obtained.[17] This paper—the earliest original contribution on fixed air to be presented publicly in France—is prefaced by a review of the literature. Bucquet refers to Van Helmont, Boyle, Black, Macbride,[18] and Jacquin, but—significantly —makes no mention of Cavendish or Priestley. Yet on June 12 a report praising Bucquet's work, prepared jointly by N. Desmarest (1725–1815) and Lavoisier, begins its concluding paragraph with this sentence: "Quoique ces expériences aient beaucoup de rapport avec celles publiées avant M. Bucquet et *surtout celles de M. Priestley*, elles n'en sont pas moins précieuses pour la physique."[19]

There was every reason why, by mid-June 1773, Priestley's work should have been known to the initiated: in the April issue of Rozier's journal, which apparently was not distributed until early May, there appeared not only the concluding sections of the translation of Black's *Experiments*, but the first part of a translation of Priestley's *Observations*; the rest of the Priestley translation was printed in the May issue, which, by the same token, probably did not appear until early June.[20]

As we know, Lavoisier lost little time in familiarizing himself with Priestley's work, for a long summary of it appears in the historical section of the *Opuscules physiques et chimiques* (1774), on which he was at work in the summer and early fall of 1773.[21] It was the knowledge of Priestley's work that brought this book into somewhat premature existence. Indeed, just as Bucquet's paper—and the series of articles on fixed air appearing in Rozier's journal—may have impelled Lavoisier to present to the Academy the results he had been quietly accumulating since the previous September (it was on May 5, 1773, that he empowered the Secretary of the Academy to open his famous sealed note describing his earliest experiments),[22] so it seems that knowledge of Priestley's work led Lavoisier's colleagues in the Academy to urge him to bring his results out in book form as soon as possible. Fourcroy's testimony is valuable on this point:

> Jaloux de ne publier que des faits nouveaux [Lavoisier] . . . ne se pressa point d'abord de les faire connoître à mesure qu'il les découvroit . . . & c'est vraisemblablement forcé & en quelque sorte malgré lui, qu'il présenta à l'académie vers la fin de 1773 l'ouvrage annoncé [the *Opuscules*] qui contient le germe ou la première idée de toutes ses découvertes postérieures. Celui du docteur Priestley venoit de paroître en Angleterre; la vaste étendue de ses expériences, l'ensemble que le physicien anglois embrassoit, sembloit faire craindre aux amis de Lavoisier, qu'il ne fût prévenu dans beaucoup de points par Priestley, & qu'il perdît ainsi le fruit de ses travaux; ils le pressèrent de publier ses recherches, & Lavoisier céda à leurs instances, sur-tout à celles de Trudaine-de-Montigny, amateur éclairé, avec lequel il avoit répété une partie de ses expériences.[23]

And Lavoisier himself later wrote:

> Il y avait, à cette époque, une correspondance habituelle entre les savants de France et ceux d'Angleterre; il régnait, entre les deux nations, une sorte de rivalité, qui donnait de l'importance aux expériences nouvelles, et qui portait quelquefois les écrivains de l'une ou de l'autre nation à les contester à leur véritable auteur.[24]

Indeed a trace of this rivalry is to be found in a note appended by Lavoisier to his discussion of Priestley's work in the *Opuscules*, where he writes that "il y avait déjà du temps que je m'occupais du même objet, et j'avais annoncé dans un dépôt fait à l'Académie des sciences le 1er novembre 1772, qu'il se dégageait une enorme quantité d'air des réductions métalliques."[25]

One question, nevertheless, remains unanswered. If I have correctly described the sequence of events in this crucial period of Lavoisier's early career, and if he did not know Priestley's *Observations* at first hand until the late spring of 1773, how can we account for the inclusion of Priestley's misspelled name in the list of workers on fixed air he gave in the memorandum of February 20, 1773? I believe the answer is simple. Priestley, at this date, could only have been cited for a much less important work, his *Directions for Impregnating Water with Fixed Air*, which had been published in London in the summer of 1772[26] and which was almost immediately translated into French.[27] Though there was little in this pamphlet on artificial soda water to have directed Lavoisier's interests into new channels, and only a general reference to the fact that Priestley was at that time studying and discovering other sorts of air besides the "fixed air" encountered in the famous brewery, our carbon dioxide, it was natural to include his name in a list of those who had written on this mysterious elastic fluid. Because of the presumed medical value of "fixed air," which Macbride's work had seemingly established, Priestley's pamphlet was widely praised. The subject was considered at a meeting of the Royal College of Physicians, before whom Priestley demonstrated his method, and the discovery was recommended by that body to the Lords of the Admiralty "as likely to be of use in the sea scurvy."[28] It played a part, though certainly not the principal part, as some have claimed, in the award to Priestley of the Copley Medal in the autumn of 1773.[29]

As early as February 1773 Lavoisier was clearly not citing Priestley for his *Observations*, as is usually supposed.[30] It does not seem to have been stressed that in his memorandum Lavoisier shows himself not only undecided as to whether the air that is fixed in various substances is the same as common air, as Hales believed, or whether it is "another substance, to which they [Joseph Black and his followers] have given the name of 'fixed air,'" but also quite unaware that Henry Cavendish and Joseph Priestley had already shown there were other elastic substances differing both from ordinary air and from Black's "fixed air." Had he known of the work of Cavendish (published in the *Philosophical Transactions* in 1767), or of Priestley's recently published separate, with the significant title of *Observations on Different Kinds of Air*, Lavoisier's view of the problem would have been quite different. The chemical opinions he summarizes make it quite likely that at the time he wrote the memorandum Lavoisier knew of Priestley's work on gases only through the *Directions for Impregnating Water with Fixed Air* and that he had not seen the classic paper of Cavendish, whom he does not mention.[31]

Finally, this little analysis can, I think, be used to settle once and for all a disputed point concerning Lavoisier's much-discussed February memorandum. As all historians of chemistry are doubtless aware, the original bears the date 1772; but a generation ago A. N. Meldrum quite persuasively argued that Lavoisier was guilty of a familiar clerical slip, so common at the beginning of a new year, and had written 1772 when he meant 1773.[32] Though Meldrum's emendation was widely accepted—for it conformed to what we know of the chronology of Lavoisier's work, which the date Lavoisier actually wrote did not— doubt has recently been cast upon it.[33] But Lavoisier's reference to Priestley is conclusive. As early as February 1772, Priestley had published *nothing* concerning fixed air, and his name would never have been included in such a roster as Lavoisier gives. Meldrum's conclusion would seem to be confirmed.

Notes

1. See, for example, J. R. Partington, *Short History of Chemistry*, 2d ed. (London: Macmillan & Co., 1948), 113; Sir Philip J. Hartog, "The Newer Views of Priestley and Lavoisier," *Annals of Science* 5 (1941), 25; Douglas McKie, *Antoine Lavoisier, the Father of Modern Chemistry* (London: V. Gollancz, 1935), 173; Ferdinand Hoefer, *Histoire de la chimie*, 2d ed., 2 vols. (Paris, 1866–69), II, 477 and n. 2; H. Kopp, *Geschichte der Chemie*, 4 vols. (Braunschweig, 1843–47), pt. 1, p. 243.

2. A. N. Meldrum, "Lavoisier's Three Notes on Combustion: 1722," *Archeion* 14 (1932), 15–30.

3. Printed in full in M. Berthelot, *La Révolution Chimique—Lavoisier* (Paris, 1890), 46–49. An English translation is given in Meldrum, *The Eighteenth Century Revolution in Science—the First Phase* (Calcutta: Longmans, Green & Co., 1930), 8–10. Meldrum's translation, which I give here, is reproduced by McKie, *Antoine Lavoisier*, 120–23.

4. Here we read, *Phil. Trans.*, vol. 62, p. iv:

> At a COUNCIL, January 28, 1773
>
> Resolved, That after Volume LXII. the *Philosophical Transactions* be published twice in a year; the first publication to be of the months of November and December of the preceding year, and January and February of the current year, as soon as may be after February, under the name of the "first part" of the volume: and the second publication to be of the remaining months unto the recess of the Society, as soon as may be after the recess, under the name of the "second part" of the volume.

5. Memoirs of the Rev. Dr. Joseph Priestley (London, 1809), 56–57. Here he writes: My first publication on the subject of air, was in 1772. It was a small pamphlet, on the method of impregnating water with fixed air; which being immediately translated into French, excited a great degree of attention to the subject, and this was much increased by the publication of my first paper of experiments, in a large article of the 'Philosophical Transactions,' *the year following*, for which I received the gold medal of the society. [Italics mine.]

6. Hoefer, *Histoire de la chimie*, II, 477, n. 2.

7. John F. Fulton and Charlotte H. Peters, *Works of Joseph Priestley, 1733–1804: Preliminary Short Title List* (New Haven, Laboratory of Physiology, Yale University School of Medicine, 1937), 20 pp., lithoprinted. Dr. Fulton has recently instituted, at my request, a search in the extensive Priestley materials in the Yale Historical Medical Library, and informs me that no copy of the Bowyer-Nichols separate is to be found there. There is, however, a copy in the Rare Book Room of the Yale University Library.

8. J. T. Rutt, *Life and Correspondence of Joseph Priestley*, 2 vols. (London, 1832), I, 185–86.

9. A copy of Priestley's *Observations on Different Kinds of Air* was sent by Franklin to Benjamin Rush. See Franklin's letter of February 14, 1773, in the Library of the University of Pennsylvania, cited by Denis I. Duveen and Herbert S. Klickstein, "Benjamin Franklin (1706–90) and Antoine Lavoisier (1743–94). Part I," *Annals of Science* 11 (1955), 111 and n. 54. This was almost certainly the "preprint" which we know Franklin to have received, for a copy from his library is now in the rare-book collection of the Historical Society of Pennsylvania, as the librarian of the Society has been kind enough to inform me.

10. Hartog, "Newer Views of Priestley and Lavoisier," 26.

11. For the Bowyers, father and son, and for Lockyer Davis, see the excellent articles in the *Dictionary of National Biography*.

12. See the letter of Jean Hyacinthe de Magellan or Magalhaens (1722–90) to the French chemist, P. J. Macquer (1718–84), dated "London, 20 Mars, 1772" (Bib. Nat., Fr. 12306, vol. 2, letter 18). In this unpublished letter Magellan, a most important link between British and French scientists in this period, writes that

they have continued at the Royal Society to read Priestley's "excellent Mémoire sur l'air fixe & le méphitique. . . ." This implies that Magellan had written Macquer concerning the earlier reports, but the other letters have not turned up. Priestley's reports were read on March 5, 12, 19, and 26, 1772.

13. So Fourcroy: "Mais bientôt tout changea de face à cet égard dès la publication de la première distribution de Priestley, publiée en 1772 à Londres, & qui ne fut connue que plus d'un an après à Paris." See the article "Chimie" in *Encyclopédie méthodique —chimie, pharmacie et métallurgie* (Paris, An IV), III, 412.

    In his *Opuscules* of 1774 (chap. 6) Lavoisier describes his experiments proving that air is absorbed during the calcination of metals (lead, tin, iron). These experiments were carried out late in March 1773, as his entries in the laboratory *registres* indicate. Berthelot, *La Révolution Chimique*, 235–37. It is therefore significant to find Lavoisier writing in a note appended to this chapter: "Je n'avais point connaissance des expériences de M. Priestley, lorsque je me suis occupé de celles rapportées dans ce chapitre." *Oeuvres de Lavoisier*, 6 vols. (Paris, 1862–93), I, 621.

14. For this subject see my "Continental Reputation of Stephen Hales," *Archives Internationales d'Histoire des Sciences*, no. 15 (1951), 393–404. The term "fixed air," used broadly by Hales, was applied restrictively by Joseph Black to the gas we call carbon dioxide.

15. *Observations sur la physique, sur l'histoire naturelle et sur les arts*, quarto reprint ed. (Paris, 1784), I, 123–34.

16. *Ibid.*, 210–20.

17. "Extrait d'un mémoire de M. Bucquet, etc." in *Oeuvres de Lavoisier*, I, 548–50.

    In Roux's *Journal de Médecine* there appeared in the issue for May 1773 a short article by Hilaire-Marie Rouelle (who had succeeded his more famous brother, the teacher of Lavoisier, as "démonstrateur de chimie au Jardin du Roi"). This appeared just at the time that Lavoisier was presenting his own discoveries to the Academy of Sciences; Lavoisier reproduced this article without change in his *Opuscules*, chap. 18 of the "Précis historique"; see *Oeuvres de Lavoisier*, I, 538–47. Rouelle shows himself familiar with the work of Hales, MacBride, Black, and Jacquin and with Priestley's *Directions*. He starts, in fact, from this work of Priestley—and that of Lane, reported by Priestley—in an effort to produce artificial ferruginous waters with the aid of dissolved fixed air. But just as his paper was completed he learned of the publication of Priestley's *Observations*, and he adds (*Journal de Médecine*, 547) the following note:

    > Je viens d'apprendre qu'il paraît, depuis peu une dissertation en anglais de M. Priestley, dans laquelle on trouve une très-belle suite d'expériences sur l'air fixe, l'air inflammable et l'air méphitique ou de putréfaction. J'ai regret de ne l'avoir pas connue plus tôt: la manière dont sont faites les expériences que nous avons déjà de lui est un garant sur de l'usage excellent qu'on peut faire de tout ce qui vient de sa main.

    The French translation of Priestley's *Observations* had not yet appeared, and Rouelle is reporting, perhaps by hearsay, what he had learned of the contents of the paper in the *Philosophical Transactions*.

18. David MacBride (1726–78), a physician of Dublin, was Joseph Black's earliest disciple. His *Experimental Essays on Medical and Philosophical Subjects* (Dublin, 1764) was published in a French translation in 1766 and aroused great interest. The translator was Vincent Abbadie. A copy of this book was in Lavoisier's personal library.

19. "Rapport sur un mémoire de Bucquet sur l'air fixe," in *Oeuvres de Lavoisier*, IV, 155–58. The italics in the quotation are mine. Compare *Procès-Verbaux de l'Académie des Sciences* 92 (1773), fol. 133.

20. *Observations sur la physique* (edition cited), I, 261–75 ("Suite des expériences de M. le Docteur Black"); 292–325 ("Observations et expériences sur différentes espèces d'air; par M. Joseph Priestley"); 394–426 ("Suite des observations et expériences . . . par M. Joseph Priestley"). A review of the April number of Rozier's *Journal* appeared in the June issue of the *Journal Encyclopédique* (1773, vol. 4, pt. 2, pp.

200 ff.) with a three-page summary of Priestley's paper, beginning: "Les physiciens liront avec plaisir des observations & des expériences du docteur Priestley sur différentes espèces d'air, & dont nous connoissons les trois principales."

21. *Oeuvres de Lavoisier*, I, 512–35. In his appended note Lavoisier writes: "Ces expériences de M. Priestley ont été publiées en anglais à la fin de l'année 1771." This, of course, is a slip. In the text, indeed in the sentence to which the note refers, he speaks, quite correctly, of the "suite nombreuse d'expériences communiquée l'année dernière à la Société royale de Londres par M. Priestley." Since Lavoisier wrote this in 1773, his dating—not always reliable—is here correct.

22. *Procès-Verbaux de l'Académie des Sciences* 92 (1773), fol. 116v, where the secretary writes: "J'ai ouvert en présence de l'Académie le dépôt . . . fait par M. Lavoisier le 2 9bre 1772 et j'ai paraphé son écrit pour lui conserver la date."

It is of some interest that in the next entry (May 8, 1773) the secretary records: "J'ai présenté à l'Académie l'exemplaire du mois d'Avril du Journal de M. L. Rozier." This issue, as we have seen above, contained the second part of the translation of Black's paper and the first part of Priestley's.

23. *Encyclopédie méthodique—chimie, pharmacie et métallurgie*, III, 415. Trudaine de Montigny (1733–77), a close friend of Lavoisier, had been made an honorary member of the Academy of Sciences in 1764; in this capacity he was vice-president of the Academy in 1772 and president in 1773. On this man and his friendship with Lavoisier see Ralph E. Oesper's "Priestley, Lavoisier, and Trudaine de Montigny," *Journal of Chemical Education* 13 (1936), 403–12. For Trudaine's participation see Berthelot, *Révolution Chimique*, 245. Trudaine was on the committee of the Academy which in September 1773 saw a number of Lavoisier's experiments repeated. See "Verification des expériences de M. Lavoisier, etc." (Bib. Nat. Fr. nouv. acq. 5, 153).

24. *Oeuvres de Lavoisier*, II, 103. The editor (Dumas) guesses that this undated paper may have been written in 1792.

25. *Oeuvres de Lavoisier*, I, 512, n. 1. This is the date of the sealed note; it was deposited with the Academy the following day. See above, n. 22.

26. See above, n. 5. In a letter (London, May 4, 1772), Benjamin Franklin wrote to Priestley: "I think with you that there cannot be the least Occasion for my explaining your Method of impregnating water with fix'd air to Messrs. Banks and Solander, as they were present and I suppose are as well acquainted with it as myself: however, I shall readily do it, if they think it necessary. I am glad you intend to improve and publish the process." *The Writings of Benjamin Franklin*, ed. A. H. Smyth, 10 vols. (New York: Macmillan Co., 1905–7), V, 394. This book had therefore not been printed when Franklin wrote, but the French translation was reviewed in Rozier's journal in the August issue: *Introduction aux observations sur la physique* (August 1772), 323–31. Obviously the work had been published in the interval. Priestley's dedication, in fact, bears the date: "Leeds, 4th of June, 1772." A facsimile of the *Directions*, with an historical introduction by John J. Riley, was printed in 1945 by the American Bottlers of Carbonated Beverages, Washington, D.C.

27. Priestley's *Directions* had been sent to France through the agency of Magellan, and the translation was due to the efforts of Magellan's correspondent, Lavoisier's friend, Trudaine de Montigny; see *Introduction aux observations* (August 1772), p. 323, n. 1. Maurice Daumas has seen a letter of July 14, 1772, in which Trudaine asks Lavoisier to repeat Priestley's experiments on fixed air. See Daumas, *Lavoisier, théoricien et expérimentateur* (Paris: Presses Universitaires de France, 1955), 28. This must refer to the *Manière d'imprégner l'eau d'air fixe*, and not, as Daumas seems to imply, to the more important work reported by Priestley to the Royal Society in March, which had not yet been published. But see above, n. 12.

28. *Memoirs of the Rev. Dr. Joseph Priestley*, 57. See also T. E. Thorpe, *Joseph Priestley* (London and New York, 1906), 77–80.

29. See above, n. 5. But compare the portion of Sir John Pringle's address to the Royal Society on the occasion of the award of the Copley Medal (November 1773) given in Thorpe, *Joseph Priestley*, 80, and in J. M. Stillman, *Story of Early Chemistry* (New York, 1924), 481–82. Professor John Winthrop of Harvard, to whom Benjamin Franklin had sent a copy of the *Directions*, deemed it a "very important

discovery." See I. Minis Hays, ed., *Calendar of the Papers of Benjamin Franklin* (Philadelphia, 1908), I, 145. Franklin was one of those instrumental in obtaining the Copley award for Priestley. See Franklin's letter of recommendation mentioned by Duveen and Klickstein, "Benjamin Franklin," 111 and n. 52.

30. See, for example, Hartog, "Newer Views," 29.

31. Cavendish's paper, *Phil. Trans.* 56 (1766), 141–84, was cryptically mentioned in France, soon after its appearance, in a summary of the contents of this volume of the *Transactions* that appeared in the *Journal Encyclopédique* (1767), vol. 7, pt. 3, pp. 90–103. Here we read: "Cet article est composé de trois mémoires contenant diverses expériences faites sur l'air factice, par M. Henri Cavendish, Membre de la Société Roy. Ces expériences nous ont paru très-ingénieuses; mais nous les rendrions mal sans le secours des planches." This early reference to the paper seems to have been without influence.

32. Meldrum, *Eighteenth Century Revolution*, 7–13. See above, n. 3.

33. By Maurice Daumas in his review of Douglas McKie's *Antoine Lavoisier* in *Archives Internationales d'Histoire des Sciences*, no. 22 (1953), 125; and more recently in his *Lavoisier, théoricien et expérimentateur*, 27.

20
Lavoisier and His Biographers

✦

It is my purpose in this essay to review, in as brief a compass as the complexity of the subject permits, the progress that has been made in Lavoisier studies. This I shall do by examining as closely as my abilities allow certain of the most recent biographies of the so-called founder of modern chemistry. I shall emphasize the use that has been made of the available printed sources, both primary and secondary, in the hope that this paper may serve as a bibliographical introduction to a subject that still awaits thorough and critical treatment despite the number of books that have appeared dealing with it.

✦

Although a few short biographical sketches of Lavoisier appeared in the first half of the nineteenth century—by Cuvier, Thomas Thomson, Hoefer, and others—and although his accomplishments were summarized with varying degrees of sympathy and accuracy in the histories of chemistry by Thomson, Hoefer, and Kopp, we may date the beginning of serious Lavoisier scholarship with the vivid eulogy which the eminent chemist J. B. Dumas accorded his great predecessor in his lectures delivered in 1836 at the Collège de France (*Leçons sur la philosophie chimique*). This is not because Dumas's contribution was a work of serious scholarship—it can best be classed as an impassioned act of piety—but because in the course of these lectures he made the suggestion that the French government should sponsor the publication of a collected edition of the works of Lavoisier, and promised that if this suggestion were acted upon, he himself would undertake the editorial work. That some such effort was needed to rescue Lavoisier from neglect was quite evident. Copies of his published books were no longer easy to obtain, and many of his most important scientific writings were scattered through the *Mémoires* of the Academy of Sciences. After interminable delays, both the official approval and a subsidy were granted, and the first four volumes appeared between 1862 and 1868 under the editorship of Dumas. They consisted almost entirely of reprints of Lavoisier's printed scientific works; letters and private papers were not included.

From *Isis* 45 (1954), 51–62. Copyright 1954 by the History of Science Society, Inc.

A full ten years elapsed after the publication of the fourth volume of the *Oeuvres de Lavoisier* before the appearance of a serious study of Lavoisier based on these materials and on other, still unpublished, sources. In 1888 Edouard Grimaux brought out his famous biography, *Lavoisier, 1743–1794, d'après sa correspondance, ses manuscrits, ses papiers de famille et d'autres documents inédits*.[1] In most respects this was an outstanding achievement; even today the book is indispensable, and is likely to remain so until the letters and other papers which Grimaux used (and others of which he was ignorant) have been collected and published. Despite the reliance that of necessity has been placed upon it, Grimaux's book is not everywhere dependable even on purely biographical matters. In matters of detail he was sometimes careless, at times almost slovenly. More important still, he was always conscious of the favor that the heirs of Lavoisier had done him by putting their papers at his disposal, and he was scrupulously careful to avoid offending their family pride and their extremely conservative political opinions. Because he glossed over evidence unfavorable to Lavoisier, and even suppressed documents which might have given offense to the Chazelles family, a somewhat idealized and stereotyped portrait emerged, impregnably defended by an impressive display of scholarly apparatus and by the subsequent inaccessibility of many of the sources he used. Excessive caution—together with a strong measure of French patriotism—seems to account for Grimaux's behavior, for he did not share the political and social views of the Chazelles family, as at least one recent writer has implied. In 1939, when this reviewer was allowed, through the kindness of the late Mlle. de Chazelles, to study the Lavoisier papers at the Château de la Canière, there was placed at his disposal a mass of Lavoisier documents only recently returned to Aigueperse by Grimaux's heirs. On this occasion, Mlle. de Chazelles explained that the Dreyfus Affair had been the parting of the ways for the two families. Grimaux had been staunchly pro-Dreyfus, as a document in my possession—one of the petitions circulated among scientists and men of letters in favor of *revision*—amply confirms. M. de Chazelles and his family were, and indeed long remained, anti-Dreyfusard. These deep political differences led to an estrangement which was never mended, and the Lavoisier documents were for a time forgotten. This episode shows that there was perhaps good reason for Grimaux's caution, however much as historians we must deplore it.

Most of the later biographical studies of Lavoisier have done little more than make available in other forms and in other languages (chiefly English) the substance of Grimaux's investigations; not only his facts, but in many cases his interpretations as well.[2] Surprisingly little use was made until relatively recently of the rich material (much of it dealing with Lavoisier's various governmental and quasi-official occupations) published by Grimaux, in 1892 and 1893, in the two concluding volumes of the *Oeuvres*. Yet, since Grimaux did not come close to exhausting this material—most of which he must have known while preparing his biography—this constitutes a most important supplement to Grimaux's book.

One of the first efforts to draw upon the Grimaux mine for the benefit of the English-speaking public, Mary Louise Foster's *Life of Lavoisier* (Northampton, Mass., 1926), is a slight work which can be passed over without comment. Much more successful is J. A. Cochrane's *Lavoisier* (London, 1931), a well-written book which, though relying heavily upon Grimaux, displays much independence of judgment. It is unfortunate that the chapters on Lavoisier's accomplishments in chemistry are no longer up-to-date, for this is in some respects the best-balanced and most readable of Lavoisier biographies in English.[3]

Between 1890 and the Second World War much progress was made in our knowledge of Lavoisier's chemical accomplishments, an aspect that Grimaux treated very cursorily. The point of departure for all later studies was Marcellin Berthelot's *Révolution Chimique* (Paris, 1890), based on a persual of Lavoisier's unpublished laboratory *registres*. Lavoisier's work in chemistry was studied in Germany by Kahlbaum, Max Speter, and others, before and after the First World War.[4] In France Emile Meyerson published in 1922 his penetrating paper entitled *La résistance à la théorie de Lavoisier*,[5] and his brilliant disciple, Hélène Metzger, followed her illuminating monographs on seventeenth- and eighteenth-century chemistry[6] with her brief but important *Philosophie de la matière chez Lavoisier* (Paris, 1935). But by far the most striking contributions to Lavoisier scholarship were made by Andrew Norman Meldrum in his *Eighteenth Century Revolution in Science-The First Phase* (Calcutta, 1930) and in his papers that appeared in *Archeion* and in *Isis* during the early thirties. Meldrum did not live to carry his studies beyond the early elaboration of the oxygen theory, but by a careful study of published and unpublished sources he corrected Berthelot on a number of points and clarified the early steps by which Lavoisier was led to his great discoveries on combustion.[7]

In 1935 much of this new material was brought together by Douglas McKie, then Lecturer and now Reader in the History of Science at the University of London, in his useful study of Lavoisier's scientific work: *Antoine Lavoisier, the Father of Modern Chemistry* (London, 1935). Since then *Annals of Science*, of which McKie is an editor, has published a steady stream of valuable papers relating to our subject.[8] By the outbreak of the Second World War the Lavoisier material had been enriched by the study of M. Lenglen, *Lavoisier Agronome* (Paris, 1936); by J. H. White's *History of the Phlogiston Theory* (London, 1932), and by valuable biographies of certain of Lavoisier's contemporaries.[9]

Despite all this activity, Lavoisier scholarship has labored under severe handicaps. There is as yet no adequate bibliographical guide to Lavoisier's own works or to studies devoted to him.[10] The great edition of the *Oeuvres de Lavoisier* (1862–1893), which has conveniently republished Lavoisier's chief writings, is by no means exhaustive even for his scientific work.[11] Important unpublished material, other than letters, still exists in the archives of the *Académie des Sciences* and other repositories, public and private.[12] The thirteen large *registres de laboratoire* of Lavoisier remain unpublished.[13] Not until 1948 were steps finally taken that should lead in the near future to the publication of a *Correspondance de Lavoisier*.[14] Yet even with these handicaps it was to be expected that Lavoisier's later biographers would have made competent use of the widely scattered published materials, and that a serious effort would have been made to familiarize themselves with the archival sources. To what extent this has been the case should shortly emerge.

※

McKie's first book on Lavoisier was, as we have seen, not a biography, but an extremely useful survey of Lavoisier's work in chemistry. In 1941 two books appeared which sought to place Lavoisier in the framework of his time and to treat his varied accomplishments as economist, administrator, and reformer, as well as his scientific work. Both were directed mainly towards a general audience; neither is wholly successful, and each displays considerable ignorance of the available sources. S. J. French's *Torch and Crucible: The Life and Death of Antoine Lavoisier* (Princeton, 1941) makes use of some new material,[15] but is the work of an enthusiastic amateur, who, though aware of the work of Meldrum and McKie, is hardly at home in the French materials. His documentation is casual, and he is often inaccurate. He seems not to have consulted the materials in the *Oeuvres de Lavoisier*, and his unfamiliarity with this basic corpus is indicated by his citation of it in his list of sources: the dates are incorrect, and he believes that Grimaux was the editor of the early volumes and J. B. Dumas of the later ones, whereas the reverse is true.

Maurice Daumas's *Lavoisier* (Paris, 1941) marked the debut under the strained conditions of the German occupation of one of the most promising Lavoisier scholars. This vividly written book is primarily a study of Lavoisier's character as revealed not only in his scientific work but in his other manifold activities as economist, reformer, and public servant. As a literary production, it is skillfully written, sometimes almost too much so, for it comes perilously close to the style of a successful *roman historique*. Daumas makes use of a number of useful collateral published sources, and quotes from several unpublished letters by Lavoisier's contemporaries. His book is documented,[16] though not impeccably so; but its weakness is the total unawareness it displays of English and American studies of Lavoisier's work in chemistry; for Daumas relied entirely on Berthelot, and was at the time he wrote apparently ignorant of the work of Meldrum, Cochrane, McKie, and others. So he was blissfully unsuspecting that Lavoisier's famous opening entry of his laboratory *registres*, a veritable program of future research, was incorrectly dated February 1772 instead of 1773; the result is hopeless confusion in Daumas's account of the early phases of Lavoisier's work on combustion. Since he was also not aware of Lavoisier's habit of rewriting early memoirs in the light of subsequent knowledge, before final publication, he credits him with having known in 1774 that it was *"la portion salubre"* of the air which combines with metals on calcination. Yet the memoir Daumas quotes was revised before second publication in 1778 in the light of Priestley's work and Lavoisier's own later experiments. There was still less excuse for asserting, as

Daumas did (p. 106), that the results recorded in the famous *pli cacheté* of November 1, 1772, show that Lavoisier had already caught a glimpse of (*a entrevu*) Priestley's *"air très pure"* before he heard of it from the English chemist. The whole section dealing with the relations of Priestley and Lavoisier would have profited in other respects had he been cognizant of the papers of Oesper and Hartog and the books of Cochrane and McKie. These were the oversights of a *débutant*; as a number of his recent papers testify—and they are excellent— Daumas would not now fall victim to these errors.[17]

❁

Douglas McKie in his *Antoine Lavoisier: Scientist, Economist, Social Reformer* (New York: Schumann, 1952) offers us an ambitious study evidently intended for the general reader, in which the events of Lavoisier's life, and his activities as scientific statesman, economist, and liberal reformer, are treated as fully as his contributions to chemistry.[18] Nowhere else is so much information about this extraordinary man to be found between a single pair of covers. The chapters on chemistry are solid and valuable, drawn in good part from his earlier book, the stock of which was destroyed during the late war. There are interesting facts about the tax collecting *Ferme Générale*;[19] important chapters on Lavoisier's estate at Fréchines and the agricultural experiments he conducted; and a long account of the great chemist's share in the events leading up to the Revolution, of his struggle to preserve the Academy of Sciences from its political enemies, and of the final tragic events of Lavoisier's arrest, trial, and execution.

Yet a reviewer is confronted with a serious dilemma. Should McKie's book be treated as a popular introduction to Lavoisier, or should we, in full appreciation of the author's reputation and knowledge, and of the diligent research which went into this work, examine it closely as a scholarly production? McKie does not write with particular brilliance, and a general reader might well be repelled by the lack of selectivity, the frequency and excessive length of direct citation, and the rather dogged adherence to a rigid chronological scheme. Nor can it be said that the personality of Lavoisier emerges as clearly as the quality of his scientific and administrative accomplishments. Little is made of Lavoisier's only close friendship, that with Trudaine de Montigny, or of those blemishes of character which influenced his relations with other men and even left a mark on his scientific work. McKie may be justified in his refusal to recount once again at length the controversies with Priestley over oxygen or with Cavendish, Priestley, and Watt over the discovery of the composition of water. Yet to ignore completely these important episodes, over which so much ink has been spilt, and so omit all evidence of Lavoisier's ambivalent attitude in the matter of according scientific credit, is to paint a false picture of this extraordinary man. McKie's judgment on these matters would have been valuable to have.

Yet despite the author's serious, factual tone, his close familiarity with the central corpus of Lavoisier printed sources, and the wealth of information he supplies, this work—which Lavoisier students will refer to for some time to come—has a number of defects as a scholarly work, and not infrequent errors. Let us first discuss his sources—no easy task, since the book is almost completely undocumented—and see what use he makes of them and what he has perhaps overlooked.

McKie is, of course, deeply indebted to Grimaux, to whom he makes a generous, if only general, acknowledgment. He refers in his short bibliography to the biographies by French and Daumas; he has used the papers on eighteenth-century chemistry and on Lavoisier by Partington, Sir Philip Hartog, Duveen, and others; he calls attention to Lenglen's *Lavoisier Agronome* and to the disappointing book by R. Dujarric de la Rivière, *Lavoisier Economiste* (1949).[20] Except for these, he seems to rely, as far as French materials go, largely upon Berthelot and Grimaux, and upon the rich material in the later volumes of the *Oeuvres de Lavoisier*. Yet it is impossible to discover, at any particular point, what he owes to any of the authors he is using, for—perhaps at the urging of his publisher—there are almost no footnotes. The book is, moreover, nearly devoid of critical judgments and comparisons; the author's interpretation, in the main, follows Grimaux. Never does he warn the reader against inaccuracies in the secondary works he recommends; for example, he fails to point out Daumas's serious blunders or show where in other respects he differs from him. His only direct reference to any of these books is unfortunate; overgenerously, he credits Professor French (p. 419) with having detected from the letters of Du Pont de Nemours, written from the prison of La Force, Du Pont's burgeoning love affair with Mme. la veuve Lavoisier. This was pointed out long before French.[21]

Professor McKie shows no evidence of having returned to certain of the printed sources used by earlier scholars. He makes no mention of the early *notices* by Lalande and Fourcroy,[22] and makes no use of the remarkable memoirs of E. M. Delahante, our only eyewitness account of the arrest and imprisonment of the Farmers General.[23] And when we find him accepting as probably authentic the apocryphal phrase attributed to Coffinhal at the drumhead trial ("La République n'a pas besoin de savants"), we can only conclude that he has not studied the important paper in which J. Guillaume long ago demolished this myth.[24] As Denis Duveen has pointed out, McKie's statements about Lavoisier's religious views suggest that he is equally unfamiliar with Guillaume's stimulating paper "Lavoisier anti-clérical et révolutionnaire."[25] Such indifference to important and indeed indispensable published sources on the Revolution explains, in part, such minor inaccuracies as that the Société de 1789 (he calls it the '89 Club) was founded in 1789 (p. 297), when the correct date is April 26, 1790;[26] why he can refer to Du Pont de Nemours as a physician (p. 418), and can ascribe the editorship of the *Moniteur* to Brissot de Warville (p. 297).

A few questions should be raised even about the chapters on chemistry, in the main so reliable. To say of the aerial nitre or nitro-aerial spirit of Hooke and Mayow that "we now know" it to be oxygen is surely an act of disinterment. However much T. S. Patterson may have overstated his case, he made it abundantly clear in his well-known article on Mayow that this hypothetical substance can hardly be identified with the precisely defined gaseous constituent of the atmosphere we call oxygen.[27]

In speaking of Lavoisier's chief teacher of chemistry, the elder Rouelle, McKie leaves the usual impression that Lavoisier must have heard him at the Jardin du Roi. Yet in 1763–64, when Lavoisier probably attended the lectures, and even before, Rouelle was lecturing in his house or apothecary's shop in the rue Jacob.[28] But he does not seem to have retired from his official chair at the Jardin until 1768,[29] so it is difficult to ascertain in which of the two places Lavoisier attended these famous lectures.

Though McKie has something to say of the influence which Stephen Hales exerted upon Lavoisier in the crucial year 1772, he lays greater stress upon the influence of Joseph Black, as all previous biographers have done.[30] This reviewer has already suggested elsewhere that it was through Hales that Lavoisier was first introduced to British pneumatic chemistry, and that Hales, whose *Vegetable Statics* was accessible in Buffon's French translation, was probably the only such writer with whose works Lavoisier was directly familiar in 1772.[31] McKie even asserts that the famous experiment on the presumed transmutation of water into earth "was clearly inspired by Black's researches."[32] Now there is no mention of Black in this famous paper of 1770, whereas Lavoisier refers at some length to the "expériences très-ingénieuses rapportées dans la Statique des végétaux," and even recalls that certain of them were repeated in public "un grand nombre de fois" in the lectures of Rouelle.[33] So far as I can ascertain Lavoisier's earliest reference to Black as a chemist is in the famous entry of February 20, 1773.[34] Black's work did not appear in French dress until later in 1773 when, perhaps at Lavoisier's instigation, a number of translations and discussions of pneumatic research appeared in Rozier's Journal.

These minor criticisms are relatively unimportant in view of the serious defect of the book: its total lack of footnotes and critical documentation. Even direct quotations are never clearly identified, and one must be content with the consolation that many are probably to be found by riffling through Grimaux or the more than 45,000 pages of the *Oeuvres*. It is difficult, and in some cases impossible, under these circumstances to control the author's statements and sources, to check his translations, and to determine his precise debt to other scholars.

I should like to show by a casual sampling that this work is just as much in need of control and confirmation as any other. McKie quotes at some length from Lavoisier's memoir of 1788 on the convocation of the Estates General (pp. 293–94). At one point Lavoisier says that the Estates General have the right "de faire des règlements généraux en matière de législation, de police, de commerce, aussi bien qu'en matière d'imposition."[35] McKie translated the word *police*, quite gratuitously, as "their own internal government," meaning that of the Estates. But the context shows clearly that it refers to the executive administration of the realm. In the concluding sentence the phrase "de faire constater et punir les infractions" is rendered by the words "to pursue and punish those who break" the law. *Constater* means, of course, to determine or establish; and this is a rather important distinction when we remember the emphasis placed in the *cahiers* on the humane reformation of the criminal law.

The author summarizes at length the famous *cahier de doléances* (pp. 295–97) drafted by Lavoisier for the nobility of Blois. Whereas a careful reading of the original document[36] makes it clear that it was the work of a committee, to whom were submitted a great number of observations and memoranda, McKie gives Lavoisier, the secretary, sole credit for everything that appears in it.

McKie does not even do justice to his own recent investigations. Although he seems not to have consulted the Lavoisier materials at Aigueperse and Clermont, he gives (pp. 412–17) the results of his inspection at the *Archives Nationales* of a packet of seventeen letters which had been taken from Lavoisier by the Revolutionary Government shortly before his arrest. This includes letters, all but one of them unpublished, from such scientific contemporaries as Joseph Black, Josiah Wedgwood, Robert Kerr (who translated the *Elements* into English), Priestley, and Spallanzani. There is also a supposedly compromising letter written to Madame Lavoisier in September 1792 by an unknown friend. The reader is not made privy to the secret that McKie printed these letters with a commentary in the *Notes and Records of the Royal Society of London* (vol. 7, no. 1, December 1949, pp. 1–41). A careful perusal of these published letters, immensely interesting in other aspects, has not persuaded this reviewer that they can have had much importance in the accusations against Lavoisier, an opinion which McKie advances somewhat less cautiously in the book than in his article.

This reviewer has been particularly disturbed by McKie's use of a most important letter of Lavoisier to Benjamin Franklin (pp. 308–9). The source is not given, and the date is inaccurate.[37] Written to accompany complimentary copies of Lavoisier's *Traité Elémentaire de Chimie* sent to Franklin and to the American Philosophical Society,[38] it begins with an admirable account of what Lavoisier thought of his own accomplishments and of the reception accorded to the antiphlogistic chemistry in Europe. Here he says that these changes have brought about a "revolution" in chemistry (a phrase he used prophetically[39] as early as 1773). This part of the letter is omitted by Dr. McKie. The second part begins: "Après vous avoir entretenu de ce qui se passe dans la chimie, ce serait bien le cas de vous parler de notre révolution *politique*.[40] Nous la regardons comme faite et comme faite sans retour." McKie translates this in part: "It would be well to give you news of our Revolution; we look upon it as over, and well and irrevocably completed." Ignoring the liberties taken here, it is worth noting that the omission of the adjective *politique* (the result of severing this passage from what came before) obscures the parallel Lavoisier sought to establish between the two revolutions; yet this symmetry is what gives the letter—surely one of the most striking in the history of science—its special flavor and interest.

The rest of the passage is even more poorly translated. For example, Lavoisier makes the significant remark—highly important for his own political opinions—that it is *fâcheux* ("unfortunate" is a weak rendering) to have been obliged to arm the people and "impolitique de placer la force entre les mains de ceux qui doivent obéir et qu'il est à craindre que l'établissement de la nouvelle constitution ne trouve des obstacles de la part de ceux mêmes en faveur de qui elle a été faite." This is translated: ". . . that it is inexpedient to put power into the hands of those who ought to obey, into the hands of the very people from whom it is to be feared that the setting up of a new constitution will meet with obstruction on account of what it has established." The meaning of its original has not been conveyed; yet it is only by chance that a reader might discover how much this translation diverges from the French text.

⚙

Now that an effort is at last being made in France to collect and publish the letters and other personal papers of Lavoisier, we can only hope that a moratorium will be declared on biographies of this amazing man until a more or less definitive study can be made. And we can only hope that this will be undertaken by someone who, if a Frenchman, will have a familiarity with the work that has been done in Britain and elsewhere, and if by a foreign scholar, by one who is a thorough master, not only of the history of eighteenth-century chemistry, but of the history of the Enlightenment and of the French Revolution.

It is self-defeating for both author and publisher to bring out plausible books, based on great labor and thought, and with many of the attributes of serious scholarship, yet which do not bring to bear explicitly the critical judgment and experience of the writer upon the works he has used; which are denuded of precise documentation; which use secondary works without precise reference; which publish and loosely translate texts which cannot be controlled or even readily located. Yet this, as I have tried to show, appears to have been the practice—lamentable though it is—of some of Lavoisier's recent biographers.

Notes

1. Second ed. (Paris, 1896), with a few corrections and a single important addition: a letter of Lavoisier to the Convention, discovered by J. Guillaume in the Archives Nationales; 3d ed. (1899), a reprint.
2. Among the earliest examples are Marcellin Berthelot's *Notice historique sur Lavoisier* (Paris, 1899) and the paper "Antoine-Laurent Lavoisier," by Sir Edward Thorpe, reprinted in his *Essays in Historical Chemistry* (London, 1894). Thorpe, however, was extremely critical of Lavoisier's scientific ethics and of Berthelot's interpretation of Lavoisier. See his "Priestley, Cavendish, Lavoisier and La Révolution Chimique," delivered as the Presidential Address to the Chemical Section of the British Association at Leeds in 1890. The debate that ensued is reviewed in Daumas's "Les polémiques au sujet des priorités de Lavoisier," *Revue d'histoire des sciences* 3 (1950), 133–55.

3. The reliance on Grimaux by other recent biographers in French and English will be discussed in this paper. A. Mieli's *Lavoisier* (Rome, 1926) is a pleasant essay but adds nothing. Two recent studies on Lavoisier have appeared in Cairo: 'Abd al-hamid Yunos and 'Abd al-aziz Amin, *Lavoisier* (1944), and 'Abd al-hamid Ahmad, *Lavoisier et son activité scientifique* (1945). I have seen neither of them.

I. G. Dorfman's *Lavoisier* (Moscow: The Academy of Sciences of the U.S.S.R., 1948) is a documented study in Russian that relies upon the obvious printed sources, such as Grimaux's *Lavoisier* and the *Oeuvres*, and cites the work of Andrew Meldrum, Douglas McKie (1935), and S. J. French (1941). I am not able to study it closely, but the short notice by R. Portal in the *Archives Internationales d'Histoire des Sciences*, no. 14 (1951), indicates that the author tries to establish Lavoisier's debt to Lomonosov, on what evidence I am uncertain.

4. See especially Kahlbaum's *Die Einführung der Lavoisier'schen Theorie im besonderen in Deutschland* (Leipzig, 1897); and Max Speter, *Lavoisier und seine Vorläufer*, in Ahrens and Herz, *Sammlung chemischer Vorträge*, vol. 15 (Stuttgart, 1910), 108 f. In 1926 Speter revived Volhard's attack on Lavoisier's originality: *Zeitschrift für angewandte Chemie* 39 (1926), 578–82; he was refuted by A. N. Meldrum in *Archeion* 14, no. 1 (1932), 15–30.

5. Published as an appendix to his *De l'explication dans les sciences* (Paris, 1921).

6. *Les doctrines chimiques en France du début du XVIIe à la fin du XVIIIe siècle* (Paris, 1923); and her *Newton, Stahl, Boerhaave et la doctrine chimique* (Paris: F. Alcan, 1930).

7. See his "Lavoisier's Three Notes on Combustion," *Archeion* 14 (1932), 15–30; "Lavoisier's Work on the Nature of Water and the Supposed Transmutation of Water into Earth (1768–1773)," *Archeion* 14 (1932), 246–47; and "Lavoisier's Early Work in Science," *Isis* 19 (1933), 330–63, and *Isis* 20 (1934), 396–425.

8. See especially J. R. Partington and Douglas McKie, "Historical Studies on the Phlogiston Theory," *Annals of Science* 2 (1937), 361–404; vol. 3 (1938), 1–58, 337–71; vol. 4 (1939), 113–49. Also Sir Philip Hartog, "The Newer Views of Priestley and Lavoisier," *Annals of Science* 5 (1941), 1–56.

9. Leslie J. M. Coleby, *The Chemical Studies of P. J. Macquer* (London: G. Allen & Unwin, 1938); Georges Bouchard, *Guyton-Morveau, chimiste et conventionnel (1737–1816)* (Paris: Perrin, 1938); J. Salwyn Schapiro, *Condorcet and the Rise of Liberalism* (New York: Harcourt, Brace, and Co., 1934); and similar works.

10. An incomplete chronological list of Lavoisier's publications is given by Grimaux (1896 ed., pp. 336–58), who also supplied, pp. 358–64, a short summary of those who had written on Lavoisier. This can now be supplemented by the *Catalogue of Printed Works by and Memorabilia of Antoine Laurent Lavoisier* (New York, The Grolier Club, n.d., but 1952), a catalogue of an exhibition of books and documents from the library of Denis I. Duveen exhibited at the Grolier Club in February and March 1952.

    A beginning has already been made by Duveen in his "Antoine Lavoisier's Traité Elémentaire de Chimie, A Bibliographical Note," *Isis* 41 (1950), 168–69. Maurice Daumas provides the background in his "L'Elaboration du Traité de Chimie de Lavoisier," *Archives Internationales d'Histoire des Sciences*, no. 10 (1950), 570–90, a valuable article. On the instruments of Lavoisier, M. Truchot's old article in the *Annales de Chimie et de Physique* (1879) is supplemented by Daumas's paper in *Chymia* 3 (1950), 45–62. [After this essay appeared there was published by Denis I. Duveen and Herbert S. Klickstein, *A Bibliography of the Works of Antoine Laurent Lavoisier 1743–1794* (London: E. Weil and Wm. Dawson & Sons, 1955). The Duveen collection, on which this bibliography was based, is now in the Cornell University Library.]

11. As one example, the *Oeuvres* gives only the *revised* versions of such crucial papers as the memoir on calcination on April 14, 1774, and that describing the experiments on oxygen, delivered April 26, 1775. Both were printed, much as they were delivered, in the Abbé Rozier's *Observations sur la physique*. As finally published in the *Mémoires* of the Academy (in both cases three years later) these papers were substantially revised. It is these that are published in the *Oeuvres*.

12. Such as the remarkable material studied by Daumas in the paper cited above, n. 10.

13. The only ready access to their contents is provided by the extracts and summaries given in Berthelot's *Révolution Chimique* (Paris, 1890), 209–310. Not only are these incomplete, but Berthelot was unable to consult the second volume, which he believed was lost. This later turned up in the Bibliothèque de Perpignan, and Berthelot was able to give an abstract of its contents in his "Sur les Registres de laboratoire de Lavoisier," *Comptes rendus de l'Académie des Sciences* 135 (1902), 549–57. René Fric has recently announced that he has prepared a manuscript of the *registres* for publication. [In 1976 the *registres* still remain unpublished.]

14. On this project see Jean Pelseneer, "Pour l'édition de la correspondance de Lavoisier," *Archives Internationales d'Histoire des Sciences*, vol. 1, no. 2 (1948), 259–60; and René Fric, *Catalogue préliminaire de la correspondance de Lavoisier* (no place or date, but published by the Union Internationale d'Histoire des Sciences, which supported this project). The meetings of the committee on publication have been regularly reported in the *Archives*. [Three volumes, or fascicles, of *Correspondance* were published under the editorship of René Fric between 1955 and 1964. For comments on this work see my *Antoine-Laurent Lavoisier, Chemist and Revolutionary* (New York: Charles Scribner's Sons, 1975), 143.]

15. Such as Ralph E. Oesper's excellent article "Priestley, Lavoisier, and Trudaine de Montigny," *Journal of Chemical Education* 13 (1936), 403–12; and B. G. Du Pont, *Life of Eleuthère Irénée du Pont from Contemporary Correspondence*, 2 vols. (Newark, Del., 1923–27), which prints many important Du Pont family letters, a number of them containing references to Lavoisier.

16. Among unpublished documents Daumas makes use of the Prospectus of G.-F. Rouelle's chemistry lectures for 1759 and one of the numerous manuscript versions of Rouelle's course of lectures. He calls attention to the articles of Charles Henry and of Grimaux on Rouelle's lectures; *Revue scientifique* (2ᵉ semestre, 1884), 97 ff., 184–85, but seems to have missed Henry's second article, *Revue scientifique* (1ᵉ semestre, 1885), 801–2, where the author modified his views.

    Daumas has used the *Journal de Paris* and the *Mercure de France*; Lescure's *Correspondance secrète* and various memoirs of the period, including Delahante and Chaptal; as well as studies like Bouchard's *Guyton-Morveau*, Anastasi's *Nicolas Leblanc*, Pigeire's *Chaptal*. He has made use of Lalande's *Notice sur la vie et les ouvrages de Lavoisier* (1795), the first biographical sketch. He does not cite Fourcroy's *Notice* (An IV), or his article "Chimie" in the *Dictionnaire de chimie* (1797) of the *Encyclopédie méthodique*. The annoyance of Lavoisier enthusiasts at Fourcroy's conduct does not justify overlooking these important primary sources. Daumas nowhere gives the precise location or *cotes* of his manuscript material. No English titles of any sort, even on British chemists, are given.

17. See, for example, his "Polémiques au sujet des priorités de Lavoisier." In his review of McKie's recent book, Daumas still appears to believe that the famous opening entry of the *registres* may well have been February 20, 1772, *Archives Internationales des Sciences*, no. 22 (1953), 125. The weight of opinion is strongly against this view. Daumas has recently (June 13, 1953) defended before the Faculté des Lettres an elaborate doctoral dissertation, *Lavoisier théoricien et expérimentateur*, based on a careful study of the laboratory *registres* and other papers.

18. The English edition, brought out simultaneously by Constable & Co., Ltd., London, is superior in appearance. Certain of the illustrations are suppressed in the American edition, as well as a small amount of text. But neither book is documented. In the American edition the illustrations are poorly reproduced, and the interesting *perspectives cavalières* of Paris (from the *Plan de Turgot*, 1734–1739, and thus somewhat earlier than Lavoisier's time) are virtually indecipherable.

19. McKie's account of the dispute over the famous wall of the Farmers General would have profited by his knowing Marcel Raval's *Claude-Nicolas Ledoux, 1756–1806* (Paris, 1945), where an excellent plan of the wall is given, together with sketches of the *barrières* and photographs of those that survived into the mid-nineteenth century and later.

20. There are no references to biographies of Lavoisier's scientific contemporaries like Bouchard's *Guyton-Morveau*, Louis de Launay's *Monge*, or the article of Pierre Lemay

and Ralph Oesper, "Claude Louis Berthollet (1748–1822)," *J. Chem. Educ.* 23 (1946), 158–65, 230–36, to mention some examples.

21. Gilbert Chinard, *Un épilogue de neuf Thermidor. Lettres de Du Pont de Nemours écrites de la prison de la Force, 5 Thermidor-8 Fructidor An II* (Paris, 1929), first published the French originals of the letters. These letters, with others referring to the Lavoisiers, were earlier brought to light in English translation in the *Life of Eleuthère Irénée du Pont* which French used. See n. 15. There is still information on Lavoisier to be drawn from this correspondence. Dujarric de la Rivière is completing a new study based on these and other Du Pont letters.

22. "Notices [*sic*] sur la vie et les ouvrages de Lavoisier, par Jérome Lalande," *Magasin encyclopédique* 5 (1795), 174–88. *Notice sur la vie et les travaux de Lavoisier* [by Fourcroy], *précédée d'un discours sur les funérailles et suivi d'une ode sur l'immortalité de l'âme* (Paris, An IV). These two short accounts by contemporaries of Lavoisier have been unduly neglected by recent biographers.

23. Adrien Delahante, *Une Famille de finance au XVIIIe siècle*, 2 vols. (Paris, 1880). See especially vol. 2.

24. "Un mot légendaire: 'La République n'a pas besoin de savants.'" *Etudes révolutionnaires*, 1st ser. (Paris, 1908), 136–55.

25. James Guillaume, *Etudes révolutionnaires* (Paris, 1908–1909), 354–79. See D. Duveen, "Antoine Laurent Lavoisier and the French Revolution," *J. Chem. Educ.* 31 (1954), 60–65. This helps explain also McKie's failure to mention the assistance which Lavoisier rendered to Talleyrand in preparation of the latter's plan of national education. The letter of Talleyrand to Lavoisier on this subject is discussed at length by Guillaume. Incidentally, Talleyrand refers to this consultation in his *Memoirs*; see Georges Lacour-Gayet, *Talleyrand*, 4 vols. (Paris: Payot, 1928–34), I, 137–38.

26. See the account of Augustin Challamel, *Les clubs contre-révolutionnaires* (Paris, 1885), 390.

27. T. S. Patterson, "John Mayow in Contemporary Setting: A Contribution to the History of Respiration and Combustion," *Isis* 15 (1931), 47–96, 504–46. In this connection, McKie perpetuates (p. 44, fig. 7) a mistranslation found in the Alembic Club version of Mayow's *Tractatus Quinque*, where he labels the flask used by Mayow a "cupping glass." When Mayow at one point actually refers to the medical cupping glass, he calls it a *cucurbitula* (a little gourd), which was the accepted term. But his word for flask is *cucurbita*, a common term in early chemistry for a gourd-shaped or round-bottomed flask.

28. Daumas, *Lavoisier*, 17. To the evidence adduced by Daumas we may add that Rouelle announced his course in chemistry "ou analyse des substances végétales, animales & minérales" as beginning "le lundi 26 novembre 1764, à trois heures après midi, en sa maison, rue Jacob, au coin de la rue des Deux-Anges, fauxbourg S. Germain." See the *Journal de Médecine, Chirurgie et Pharmacie* 21 (November 1764), 478–79. A similar announcement appeared in 1762.

29. Paul-Antoine Cap, *Rouelle* (Paris, 1842), 26. Daumas asserts, however, that Rouelle had already retired form his chair in 1763; yet in his advertisement of 1764 Rouelle still refers to himself as "démonstrateur en chymie au Jardin du Roi."

30. Grimaux explicitly credits an interest in Black's work with having drawn Lavoisier's attention to the chemistry of gases (1892 ed., p. 101). His opinion has been echoed by Mary Louise Foster, *Life of Lavoisier*, 29; and by J. A. Cochrane, *Lavoisier*, 50.

31. "The Continental Reputation of Stephen Hales," *Archives Internationales d'Histoire des Sciences*, no. 15 (1951), 393–404. I had overlooked the parenthetical assertion of Sir Philip Hartog, "Newer Views of Priestley and Lavoisier," 28, that "Lavoisier, like Priestley, goes back to Hales." He gives no evidence, however.

32. McKie goes so far as to discuss Black before Hales in his background chapter, thus placing greater emphasis upon the former.

33. *Oeuvres de Lavoisier*, 6 vols. (Paris, 1862–1893), II (1862), 7.

34. For this text see Berthelot, *Révolution Chimique*, 46–49; the English translation of Meldrum (1930), pp. 8–10, is reproduced by McKie (1935), pp. 120–23. The misspelling by Lavoisier of the names of two of these chemists, MacBride and Priestley, suggests his unfamiliarity at this date with British pneumatic chemistry.

35. *Oeuvres*, VI, 321.

36. *Oeuvres*, VI, 335–63, see esp. pp. 350–51.
37. The correct date is February 2, 1790, not, as McKie says, February 5, the date given by Grimaux, who published (p. 201) only the second half of this letter, and who is evidently McKie's source. A rough draft was discovered by René Fric and published in his "Une lettre inédite de Lavoisier à B. Franklin." Extrait du *Bulletin de l'Académie des Sciences, Lettres et Arts de Clermont-Ferrand* (Clermont-Ferrand, n.d. but 1923). A competent English translation, without indication of source, appeared in Edgar F. Smith's *Old Chemistries* (New York: McGraw-Hill Book Co., 1927). A typescript copy in the possession of the Edgar Fahs Smith Library at the University of Pennsylvania shows that it is based on the Fric article, not on the copy received by Franklin, which seems to have been lost. The French text, with a facsimile of the first page and a good English translation, was published, without reference to Fric's work, by H. S. van Klooster, "Franklin and Lavoisier," *J. Chem. Educ.* 23 (1946), 107–9.
38. The copy sent to the American Philosophical Society has disappeared. But the Society purchased a few years ago from the Bache sale an unbound, uninscribed copy which is believed to have belonged to Franklin and may be one of the copies in question.
39. Berthelot, *Révolution Chimique*, 48. Lavoisier used the same expression ("J'en conclus que la révolution est faite en chimie") in a letter to Chaptal of 1791 quoted by Grimaux (1892 ed., p. 126); but Daumas is incorrect when he wrote recently apropos of this letter that "c'est là une phrase de circonstance et le mot révolution a été suggéré par la grande révolution politique qui était en train de s'accomplir sous ses yeux." "L'Elaboration du Traité de Chimie de Lavoisier," 570.
40. My italics.

21
A Note on Lavoisier's Scientific Education

❖

All our authorities on Lavoisier refer to his years of schooling as a day student (*externe*) at the Collège Mazarin (Collège des Quatre–Nations),[1] and credit the future chemist's first introduction to the exact sciences to one of his teachers at this school, the well-known astronomer, the Abbé de Lacaille. And it is also common knowledge that after leaving Mazarin, Lavoisier combined the study of law with informal scientific instruction under Lacaille, Bernard de Jussieu, Jean Etienne Guettard, and—in chemistry—Guillaume-François Rouelle. But since no official records seem to have survived to tell us precisely when Lavoisier entered the school, how long he remained, or when he left to study law, the accounts of this period of his life are vague and confusing. Yet a particular fact has long been known which, if properly interpreted, can straighten out some chronological difficulties, and help us establish with reasonable precision the moment when Lavoisier began his independent scientific study.

Edouard Grimaux records in his much consulted biography that Lavoisier "obtint en 1760, le second prix de discours français, au concours général, dans la classe de rhétorique."[2] Grimaux does not elaborate, and merely uses this fact to stress that Lavoisier was a diligent and successful student. Recent writers have misread this passage and have assumed that the phrase, "classe de rhétorique," refers to that branch of the competition which included a contest in French oratory. Actually, it tells us something more important: namely, that at the time of this competition Lavoisier was in a particular class or form (the *classe de rhétorique*) at the Collège Mazarin.[3]

Now the sequence of studies in the great secondary schools of Paris during the eighteenth century resembles the one prevailing in a modern French *lycée*. While Mazarin, as we shall see, had a curriculum that was in certain respects unique, a student began, as in the other *collèges*, by entering a sixth form at about the age of twelve and reached the last year of his humanistic studies, called the *classe de rhétorique* (corresponding roughly to the *première* of a modern *lycée*), after five years of study.[4]

The *concours général*, mentioned by Grimaux, was a literary competition among the students of the secondary schools of the University of Paris; it had been established in 1746 and was restricted to students who were in the last

From *Isis* 47 (1956), 211–16. Copyright 1956, by the History of Science Society, Inc.

years of their study of the humanities (i.e., it was not open to students in *philoso-phie*). The compositions and orations were written during the month of June, and the distribution of prizes took place at the end of the school year, in August.[5] We may reasonably conclude, therefore, that when Lavoisier competed for the prize in the summer of 1760 he was completing his year in the *classe de rhétorique*.[6]

With this as a fixed point, we may infer that Lavoisier—if he began, as we know he did, in the lowest class[7]—entered Mazarin as an *externe* in the autumn of 1754, shortly after his eleventh birthday. Now the full course of study at Mazarin occupied *nine* years, three more years of work being required after *rhétorique* for the degree of Bachelor of Arts: after the *classe de rhétorique* came a *classe des mathématiques*, and this was followed by *two* years devoted to philosophy. This interesting arrangement—a year devoted to mathematics and the exact sciences, and two years of philosophy rather than one—was peculiar to Mazarin, and was much admired by those who approved a scientific emphasis in the curriculum. It is clear that if Lavoisier completed the normal course for the baccalaureate of arts, he must have remained at Mazarin until the summer of 1763.

It is never specifically claimed that Lavoisier completed the regular course and received the degree of Bachelor of Arts, yet we are left to assume that he did. The strongest evidence that he did not, and that he left the school at an earlier date, is the fact—exhumed by Grimaux from the records of the old *Faculté de Droit*—that Lavoisier, on September 6, 1763, received the degree of Bachelor of Law.[8] I see no reason to believe that it was common practice for future lawyers to earn both baccalaureates, especially students at Mazarin, where the course was two years longer than elsewhere. And even if we make the greatest allowance for Lavoisier's tireless energy and his extraordinary capacity for work, it is difficult to believe that he attended simultaneously the lectures at Mazarin and at the *Faculté de Droit*, while embarking at the same time on his scientific studies. Still less can we believe that he could have satisfied—or would by the regulations have been allowed to satisfy—the two-year course[9] required of bachelors of law in a matter of a month or so after leaving secondary school. It is much more likely that he left the Collège Mazarin in the summer of 1761 and—following what would have been a sensible procedure for Mazarin students intending to study law—substituted the two-year program leading to the baccalaureate in law for the two years which were to be devoted to philosophy. Indeed, this is what Cuvier, who wrote an early sketch of Lavoisier's life, making use of notes supplied by Mme. Lavoisier, appears to tell us:

> Arrivé à la philosophie, il conçut tant de goût pour les sciences, qu'il résolut de s'y consacrer tout entier. . . . Ainsi le jeune Lavoisier, au sortir du collége, s'occupa aussitôt à approfondir les mathématiques et l'astronomie dans l'observatoire de l'abbé de La Caille, à pratiquer la chimie dans le laboratoire de Rouelle, et à suivre Bernard de Jussieu dans ses herborisations. . . .[10]

Note that Cuvier says that Lavoisier abandoned Mazarin after having *reached* philosophy, not after having completed it. As we have already seen, had he left Mazarin at this point (1761) to study law, we would expect him to have received the degree of Bachelor of Law precisely when he did, viz., in September, 1763.

❀

But what becomes of our belief that Lavoisier studied with the Abbé de Lacaille while a student at Mazarin, and before he was admitted—as Cuvier tells us—to Lacaille's small observatory at the school for advanced instruction in mathematics and astronomy? In the *collèges* of Paris such instruction, if any of importance was given, was offered in the course of *philosophie*, which, it appears, Lavoisier did not pursue. But here the special curriculum of Mazarin comes to our aid. Lacaille did not teach the students of philosophy; he was *régent* (professor) of the *classe des mathématiques*, the form which, as we have said, came between rhetoric and the first year of philosophy.[11] In the normal course of events Lavoisier, having completed the work in rhetoric in the summer of 1760, would have sat under Lacaille in the academic year 1760–61. I feel sure that this was what happened, and that Lavoisier did not leave the school after his year of rhetoric; for not only does Cuvier say that Lavoisier *reached* philosophy, but it would be incredible if Lacaille, a notoriously reserved and forbidding man, had admitted Lavoisier into the intimacy of his observatory or special instruction if the latter had not already demonstrated his promise in the regular course of mathematics and physical science at the school.[12] We must conclude, therefore, that Lavoisier left the Collège Mazarin at the end of the academic year 1760–61, and that—his interest in science having been awakened by Lacaille's instruction—he determined to combine the extracurricular study of science with the study of law, which had doubtless been urged on him by his family.

Lavoisier could not have spent much time in Lacaille's observatory after leaving the school, for the astronomer died in March 1762. Indeed it would seem to have been the death of Lacaille that turned Lavoisier's scientific interests into other channels, even if—as Meldrum contended—Lacaille's influence persisted in the meteorological and barometric observations Lavoisier began at about this time and which continued for many years thereafter.[13]

For the guidance of Lacaille, Lavoisier soon substituted the informal instruction and advice of the geologist Guettard. Though the two men may have been acquainted earlier—Guettard seems to have frequented the Lavoisier house in the rue du Four as a friend of the family—we have no evidence that they were on terms of scientific intimacy until after the death of Lacaille. Our earliest record of Lavoisier's geological apprenticeship is a surviving, but unpublished, letter of Guettard to Lavoisier, dated May 18, 1763,[14] which shows Lavoisier already interested in geological problems and in collecting mineralogical specimens. Grimaux has assigned Lavoisier's earliest geological notes to this year—the

Observations d'histoire naturelle sur les environs de Villers-Cotterets,[15] and the *Observations d'histoire naturelle sur les environs de Lizy, la Ferté-sous-Jouarre and Meaux*[16]—and indeed the manuscript catalogue of Lavoisier's mineralogical collection lists samples, evidently the earliest he collected, dating from 1763.[17]

Meldrum has remarked that Guettard's influence is more clearly reflected in Lavoisier's earliest scientific work, with its central focus on geology, mineralogy, and hydrology, than is that of Lacaille, Jussieu, or even Rouelle.[18] This is certainly accurate; but I would go a step further and suggest that it was owing to Guettard that Lavoisier was introduced to the science of chemistry. We should not forget that mineralogy and metallurgy had contributed greatly to the intensified interest in chemistry in France during Lavoisier's boyhood. Largely under the influence of German and Scandinavian writers, chemistry and mineralogy had come to be treated in close connection with each other; in fact this new emphasis had much to do with the general acceptance of the phlogiston theory during these same years. Guettard's own work shows that he was well aware of the importance of chemistry and that he was by no means ignorant of it.[19] Rouelle, Lavoisier's teacher of chemistry, was one of the most enthusiastic supporters of the new mineralogical emphasis. We know him to have encouraged and assisted the Baron d'Holbach, who was making available the resources of German mineralogy and metallurgy by his articles in the *Encyclopédie* and by his French translations of important German books.[20] Most important of all, Rouelle devoted considerable time in his chemical lectures not only to mineral chemistry but even to broader geological questions.[21] To my knowledge this has never been pointed out, and it seems to have escaped notice that Rouelle's famous lectures had a considerable importance as an introduction to geology—a subject, so far as I know, publicly taught nowhere else in France. It was doubtless because Rouelle could supply this kind of background that Lavoisier, already interested in Guettard's bold program for a mineralogical survey of France and the preparation of a mineralogical map, determined to attend these famous lectures. There is support for this hypothesis in Lavoisier's own printed references to Rouelle. I have always been troubled by the infrequency with which this supposedly influential teacher is quoted by Lavoisier. In checking these passages, I was struck by the fact that Rouelle is as often cited on geological questions as he is on points of chemistry.[22] In one place, Lavoisier links Rouelle with Buffon and Guettard as one of his three great authorities on geological matters.[23]

I think it likely, therefore, that Guettard, noting the enthusiasm of his young law-student friend, urged him to prepare himself for serious mineralogical and geological investigations and to acquire the requisite chemical knowledge by attending Rouelle's lectures and working with Rouelle in the laboratory. It is reasonable to assume that such advice was given early in Lavoisier's period of geological apprenticeship. In rough confirmation of this hypothesis, we may cite the passage where Lavoisier appears to assign his first work in chemistry to the year 1763 or somewhat earlier. In 1773 he wrote: "Depuis plus de dix ans que je m'occupe de physique et de chimie. . . ."[24] I am tempted to conclude that, with geological interests foremost in his mind, Lavoisier attended Rouelle's lectures in the fall and winter of 1762–63, certainly not before, though there is nothing to exclude the following year, or even the possibility that he heard Rouelle more than once.

As I have pointed out elsewhere, we can no longer assume that Lavoisier attended Rouelle's famous lectures in the auditorium of the Jardin du Roi,[25] although Rouelle did not retire from his post of *démonstrateur* at the Royal Botanical Garden until 1768. At least as early as 1762–63 he was offering a course of lectures on chemistry at his house and apothecary shop on the corner of the rue Jacob and the rue des Deux-Anges in the Saint-Germain quarter. Two reasons make me suspect that Lavoisier may have worked with Rouelle at the shop rather than at the Jardin du Roi. The first is the closer proximity of the rue Jacob to the quarter where Lavoisier at this time was presumably attending the lectures on the Roman Institutes and on Canon Law. The second is that if, as Cuvier says, Lavoisier worked with Rouelle in the laboratory,[26] it is more likely that Rouelle had greater freedom for such private laboratory instruction at his shop than at the Jardin.

Notes

1. For this school see Alfred Franklin, *Recherches historiques sur le Collège des Quatre-Nations* (Paris, 1862). Founded by the will of Cardinal Mazarin to provide free education for noble youths from the four provinces united to France by the treaties of Münster (1648) and the Pyrenees (1659), it opened its doors in 1688 in what is now the Palais de l'Institut. In the eighteenth century, the number of scholarship boarding students (*boursiers*) varied between thirty and forty; but like the other *collèges* in Paris, its classes were open to day students without charge. At Mazarin these day students exceeded 1,000, far greater than the number attending any other school. Among its distinguished alumni were the Président Hénault, the physicist d'Alembert, the astronomer J.-S. Bailly, the painter David, and the chemist Cadet de Gassicourt.
2. Edouard Grimaux, *Lavoisier*, 2d ed. (Paris, 1896), 3. Grimaux's source was a note in the *Intermédiaire des chercheurs et curieux*, 19ᵉ année, no. 438 (August 1886), cols. 447–80. The official entry reads:

Antonius Laurentius Lavoisier

Parisinus

e Collegio Mazarinaeo

Concours de 1760

Classe de rhétorique

2ᵉ prix de discours français

3. This is clear from the discussion in the *Intermédiaire*, where the contest subjects for the various forms are briefly indicated. For members of the "classe de rhétorique" the subjects included "discours latin, discours français, vers latin, version grecque, version latine."
4. "Lettres patentes portant règlement pour le collège des Quatre-Nations," in Franklin, *Recherches historiques*, 166–75; see also pp. 109–10.
5. Charles Jourdain, *Histoire de l'Université de Paris*, 2 vols. (Paris, 1888), II, 265–74.

6. Mr. Denis I. Duveen, who kindly read this paper in manuscript, has offered important confirmatory evidence of the point I am trying to make. He informs me that the prize consisted of a calfbound copy of the *Oeuvres de M. de Tourreil, de l'Académie Royale* (Paris: Brunet, 1721), with the arms of the University of Paris in gold on the covers. He further informs me that while in the "classe de rhétorique," Lavoisier was awarded two other prizes. For success in "éloquence française," he received a similarly bound copy of *La Religion chrétienne, prouvée par les faits . . . par M. l'Abbé Houtteville* (Paris: Dupuis, 1722); and for a "5° accessit de version grecque," he was awarded a copy of Frézier's *Relation du voyage de la mer du Sud aux côtes du Chily et du Pérou, Fait pendant les années 1712–1713 et 1714* (Paris: Nyon, 1716)

7. Again Mr. Duveen comes to my aid. He informs me that Lavoisier received at least three prizes from the Collège Mazarin in the form of calfbound volumes with the arms of the school embossed on the covers. As a member of the "classe de sixième," he won a second prize "de thème latin" and was rewarded with a copy of *Les Fables d'Esope, gravées par Sadeler* (Paris: Thiboust, 1743). This volume, like the others mentioned, is in Mr. Duveen's collection. [The collection is now at Cornell University.]

8. Grimaux, *Lavoisier*, 4, n. 1.

9. "Déclaration portant règlement pour les études de droit," in Jourdan, Decrusy, Isambert, eds., *Recueil général des anciennes lois françaises*, 29 vols. (Paris, 1821–1833), XX, 349–53.

10. *Biographie Universelle*, ed. Michaud, art. "Lavoisier."

11. The first professor of mathematics at Mazarin was the distinguished Pierre Varignon, who held the chair from 1688 until 1704. The success of his mathematical lectures had much to do with establishing the reputation of the school. Late in the eighteenth century a writer on education expressed regret that mathematics was commonly studied as part of the work in philosophy, and urged that other schools follow the example of the Collège Mazarin in giving it a special place. Cited by Jourdain, *Histoire de l'Université de Paris*, II, 382–83.

12. A well-known astronomer and member of the Academy of Sciences, chiefly famous for his catalogue of southern stars observed during an expedition to the Cape of Good Hope, Lacaille was the author of elementary textbooks of mathematics, mechanics, astronomy, and optics; in all likelihood these reflect his teaching at the Collège Mazarin and indicate the scope of his lectures. A good account of the man, his teachings, and his personality is given by his pupil, J.-S. Bailly, "Eloge de M. l'Abbé de la Caille," in *Discours et Mémoires par l'auteur de l'Histoire de l'Astronomie*, 2 vols. (Paris, 1790), I, 139–80. See also Franklin, *Recherches historiques*, 98.

13. It would seem to be with reference to this work with Lacaille that Lavoisier's friend De Troncq alludes to him in the earliest surviving piece of correspondence (March 28, 1762) as "mon cher Et aimable mathématicien," and warns him against overwork. This letter, first cited by Grimaux, has been printed in its entirety by M. René Fric in *Oeuvres de Lavoisier: Correspondance*, recueillie et annotée par René Fric, 3 fascicles (Paris: A. Michel, 1955–64), fascicle I, pp. 1–2.

14. Guettard papers. Muséum d'Histoire Naturelle (Paris), MS. 1929, IV. This seems to have escaped the attention of the editor of Lavoisier's *Correspondance*, and I regret that I did not inform M. Fric of its existence. The earliest known letter of Lavoisier to Guettard is dated October 21, 1766. *Correspondance*, 14–16.
   [Professor Rhoda Rappaport has shown that I incorrectly described this letter, which was written by an unidentified correspondent to Lavoisier in 1765, not 1763. See her Cornell doctoral dissertation, "Guettard, Lavoisier, and Monnet: Geologists in the Service of the French Monarchy," Ithaca, 1964, p. 110.]

15. *Oeuvres de Lavoisier*, 6 vols. (Paris, 1862–1893), V, 48–52.

16. *Oeuvres de Lavoisier*, V, 84–90.

17. "Catalogue d'histoire naturelle qui compose le cabinet de M. Lavoisier—1781." Lavoisier manuscripts, Archives de l'Académie des Sciences, dossier 150, fols. 1 *v*, 7 *v*, 11, and 17 *v*.

18. A. N. Meldrum, "Lavoisier's Early Work in Science," *Isis* 19 (1933), 335.

19. For example, Guettard analyzed the shales from widely scattered regions of France

by treating them with mineral acids, and showed that only the coarser specimens (*d'une pâte moins fine*) effervesced, denoting the presence of calcareous matter. See his paper "Sur les ardoisières d'Angers," in *Mém. Acad. Roy. Sci.* for 1758 (1762). A contemporary summary of what Guettard reported is given in *Rapport des ouvrages qui ont été lus dans les assemblées de l'Académie des Sciences, depuis la rentrée de la St. Martin 1757, jusqu'à celle de Paques. Fait à l'Académie des Belles Lettres le 2 May 1758. Par Mr. de Montigny de l'Académie des Sciences.* MS. in possession of the author.

20. Pierre Naville, *Paul Thiry d'Holbach et la philosophie scientifique au XVIIIᵉ siècle* (Paris: Gallimard, 1943), 68, 181–83, 185–86. D'Holbach translated the *Mineralogy* of Wallerius and various writings of Gellert, Henckel, Lehmann, Orschall, and others. Of particular interest is his French version of G. E. Stahl's *Treatise on Sulphur*, which contains an important exposition of the phlogiston theory.

For a list of articles on mineralogy contributed by d'Holbach to the *Encyclopédie*, a number of which refer to Rouelle, see Naville, 407.

21. *Cours de Chymie de Mr. Rouelle rédigé par Mr. Diderot & éclairci par plusieurs Notes*, 9 vols. (Bibliothèque de Bordeaux, MS. 564). Of the nine volumes, five are devoted to the *règne minéral* (Vols. V–IX), and they contain much material of mineralogical and geological interest.

22. In his second memoir, "Sur le Gypse," Lavoisier, referring to the geology of Sweden, Switzerland, and part of Germany, writes, "Tout le monde sait combien ces pays sont riches en mines; ils appartiennent, par conséquent, à la bande schisteuse de M. Guettard, à l'ancienne terre de M. Rouelle." *Oeuvres de Lavoisier*, III, 135. In the same paper Lavoisier lists three samples of gypsum supplied by Rouelle for his experiments (*ibid.*, 132–34). Elsewhere Lavoisier writes: "Quelques naturalistes, et principalement M. Rouelle, ont distingué cet ordre de bancs [strata] sous le nom de *nouvelle terre*; il est à remarquer qu'en général ils occupent toujours la partie basse du globe." *Oeuvres*, V, 227. Farther on we read: "M. Rouelle et quelques naturalistes pensent que l'aliment du feu des volcans consiste dans des charbons de terre embrasés dans le sein de la terre." *Ibid.*, 237. But Rouelle is also cited, as I have shown elsewhere, for having repeated "un grand nombre de fois, aux yeux de tout le public," the experiments of Stephen Hales on air fixed in various substances, and on some less important points of general chemistry. See my "Lavoisier and His Biographers," *Isis* 45 (1954), 59.

23. *Oeuvres de Lavoisier*, V, 226, where he writes: ". . . je vous tracerai l'abrégé de ce que j'ai retenu des leçons, des ouvrages et des conversations des minéralogistes les plus célèbres, MM. de Buffon, Guettard, Rouelle." The "ouvrages" would be Buffon's, the "conversations" certainly with Guettard, and the "leçons" those of Rouelle, the only one of the three who taught formally and publicly. A more explicit reference to Lavoisier's attendance at Rouelle's lectures, again in a mineralogical context, appears in a joint paper with Guettard, "Description de deux mines de charbon de terre"; "L'un de nous se rappelle avoir entendu dire à M. Rouelle l'aîné, dans ses leçons de chimie, que le charbon de terre de Balleroy, en Normandie, présentait le même phénomène, et qu'il donnait également de l'acide par la distillation, au lieu d'alcali volatil." *Oeuvres de Lavoisier*, II, 243.

24. *Oeuvres*, I, 439.

25. *Isis* 45 (1954), 59 and nn. 28, 29.

26. It would seem to have been from personal observation that Lavoisier recorded in his *Traité élémentaire de chimie* that "le célèbre Rouelle avait fait tracer en gros caractères dans le lieu le plus apparent de son laboratoire" the peripatetic motto: *Nihil est in intellectu quod non prius fuerit in sensu.* See *Oeuvres de Lavoisier*, I, 246.

22

A Lost Memoir of Lavoisier

❀

In 1951 the present writer called attention to a long-neglected Lavoisier document, first published in the *Oeuvres de Lavoisier* in 1865, and entitled "Réflexions sur les expériences qu'on peut tenter à l'aide du miroir ardent."[1] Dated August 8, 1772, and therefore written shortly before Lavoisier's famous experiments on the combination of air with sulphur and phosphorus, it is the earliest dated document we have in which Lavoisier makes any significant reference to the problem of combustion. As the title indicates, it is largely devoted to proposed experiments to be carried out with powerful burning glasses. But in it Lavoisier speculates on the mysterious disappearance of the diamond when strongly heated—a question which had engaged him during the previous months —and discusses the possibility that air may play a significant role in chemical combinations. He hints at some discontent with the phlogiston theory; he refers to Stephen Hales's pedestal apparatus which he was subsequently to use in modified form in some of his own experiments; and he includes a passage which makes it highly improbable that at this date, despite frequent assertions to the contrary, Lavoisier was at all familiar with the work of Joseph Black on "fixed air." Had he been, he could scarcely have asserted that no chemist had been able to show that air enters into the "definition"—that is, into the quantitative chemical constitution—of any inorganic substance (*corps minéral*).[2] Altogether this valuable document can tell us much.

At first I did not think to question the curious annotation printed by the editor, J. B. Dumas, in which it is stated that Lavoisier read this paper before the Academy of Sciences on August 19, 1772, under the title "Mémoire sur le feu élémentaire."[3] This seemed likely enough, at first glance, since the introductory paragraphs deal with phlogiston and discuss the similarity of the doctrines of Stahl and of the French chemist E. F. Geoffroy (1672–1731). But after studying the printed document more closely, I began to have my doubts. In the first place, the title which Lavoisier gave to it ("Réflexions sur les expériences qu'on peut tenter, etc.") describes it much more accurately than does the title under which this annotation says it was read. In the second place, its form and style suggest a personal memorandum rather than a formal memoir.[4] And third, it is only by courtesy that the document can be said to discuss "feu élémentaire"; for even if

From *Isis* 50 (1959), 125–29. Copyright 1959 by the History of Science Society, Inc.

we take this term to mean phlogiston, which is perhaps not accurate, only the first few paragraphs are devoted to this subject.

I now believe that two different documents have been confused. I do not think that the "Réflexions" was ever read to the Academy, or was ever intended to be read, but that it was only a private memorandum of research plans, intended for his co-workers on the burning-glass experiments, Macquer, Cadet, and Brisson. Moreover, it is probable that the title "Mémoire sur le feu élémentaire" is that of another work, a lost memoir on the theory of heat, that may be the earliest thing of the kind written by Lavoisier. Nor was *this* memoir ever read to the Academy. Let me give the evidence for these conclusions.

The first point to stress is that the annotation in question has little authority. Inspection of the original manuscript will show that it is not in Lavoisier's hand, unlike the date added at the top of the page. Indeed it would seem to have been written (perhaps early in the nineteenth century, as the handwriting suggests) by someone—not J. B. Dumas—who had undertaken to sort and classify Lavoisier's scientific manuscripts.[5]

It was probably from a reference in the *Procès-verbaux* of the Academy of Sciences that the annotator took the title he said Lavoisier used. On this date of August 19, 1772, there is an entry in which the Secretary, M. Grandjean de Fouchy, has recorded: "M. Lavoisier a dit qu'il avait un *mémoire sur le feu élémentaire*. Il est sorti pour l'aller chercher. Il l'a rapporté et je l'ai paraphé à l'instant."[6] Nothing is said, be it noted, about Lavoisier's *reading* the memoir in question. Quite the contrary; he must have brought it in to be initialed (*paraphé*) by the Secretary, because he did *not* intend to read it. The purpose of this well-established procedure was, of course, to secure priority without having to disclose the contents of a piece of work in progress.

Further light is shed on the question by a reference Lavoisier himself made in print two months later. In the Abbé Rozier's Journal, the *Observations sur la physique*, for October 1772, Lavoisier—as we shall see in more detail below—describes some experiments on mixtures of ice and water. In commenting upon the unexpected temperature effects, he writes:

> Je crois être en état de rendre une raison satisfaisante de ce phénomène: mais comme l'explication que j'en donnerois tient à *un systeme sur les élémens, & qui est déjà paraphé par M. de Fouchy*, je remets à en entretenir l'Académie dans un autre tems.[7]

In all likelihood this is a reference to what Grandjean de Fouchy called Lavoisier's "Mémoire sur le feu élémentaire." Although the brief description of the memoir as "un système sur les élémens" accords only roughly with the title given by M. de Fouchy, I can find no evidence in the *Procés-verbaux* that the Secretary had initialed any other similar paper for Lavoisier at this time, although it was customary to record such occurrences.

The discrepancy in the titles is not as great as appears at first glance. At this moment in his career, Lavoisier meant by *élémens* not the various unresolvable substances familiar to the chemist (as he was later to define the term in the *Traité élémentaire de chimie*) but the four peripatetic principles: Fire, Air, Water, and Earth. This should occasion no surprise, since (largely as a result of the experiments of Stephen Hales, which showed air to be a constituent of many common substances) the old four-element theory was enjoying a revival. This theory was clearly held, with only slight reservations, by Rouelle, Macquer, and Baumé.

We can interpret Lavoisier's reference to a "systeme sur les élémens" to mean that he had in hand a work in which he intended to treat each of the four substances corresponding to Macquer's *principes primitifs*. This plan may well have been an outgrowth of his detailed study of water and of his famous paper on the presumed transformation of water into earth. He seems already, by August 1772, to have begun the study of air. In the manuscript of the "Réflexions" there is a significant passage which Lavoisier crossed out, and which Dumas therefore did not give in the printed text. Here Lavoisier speaks of having a work on air "déjà fort avancé, même en partie redigé." The "Mémoire sur le feu élémentaire" evidently also formed a part of the larger plan, and it was sufficiently complete for Lavoisier to have brought it to Fouchy for initialing at the meeting of August 19, 1772. This could explain why Lavoisier gave this work a title different from that referred to by Fouchy.

I can only conclude that there is not the slightest connection (except an approximate coincidence in dates) between the paper published by Dumas in the *Oeuvres de Lavoisier* under the title "Réflexions sur les expériences, etc." and the "Mémoire sur le feu élémentaire." The latter, written sometime in 1771–72, probably as part of the projected larger work on the four elements, was never published, nor even (as Lavoisier appears to tell us) read to the Academy. Its disappearance is regrettable, for it would seem to have been the first of Lavoisier's studies on the nature of heat, a subject which preoccupied him throughout his career. In it we might expect to find the earliest form of his "caloric" theory,[8] as well as some insight into his attitude towards the phlogistic doctrine on the eve of his first great experiments on combustion. Perhaps he amplified the doubts suggested in the "Réflexions" where he writes: ". . . il faut avouer que, même aujourd'hui, nous ne connaissons pas encore assez bien la nature de ce que nous nommons phlogistique pour pouvoir rien prononcer de très-précis sur sa nature."[9]

Can we make some reasonable conjectures as to the views Lavoisier might have held at this time concerning the *feu élémentaire?* How far, in this lost memoir, had he elaborated the caloric theory which by 1777 he had developed in all its essentials? Some light can perhaps be shed on these questions by the events which seem to have led Lavoisier, in such a precipitate fashion, to leave the meeting of August 19, 1772, in search of a memoir he wished to have promptly initialed.

Attention was drawn in 1935 by Douglas McKie and Niels H. de V. Heathcote, in their scholarly little classic *The Discovery of Specific and Latent Heats*, to the existence of a French summary, printed in 1772, of Joseph Black's work on latent heat.[10] Though not the earliest account of these discoveries—Black himself, it will be remembered, never published his work on heat—it was the first version accessible to the French reader.[11] This anonymous précis appeared in the September 1772 number of Rozier's *Observations*, and the editor describes it as having been communicated from Edinburgh by one of Black's disciples.[12] Black was at this time scarcely known in France, and Rozier's introductory remarks constitute the opening salvo of an effort to bring Black's work—not only on heat, but also on "fixed air"—to the attention of French scientists.[13] In the issue of the following month (October 1772), as an obvious sequel to the Black summary, appeared the paper of Lavoisier from which I have already quoted. It was entitled "Expériences sur le passage de l'eau en glâce." In it Lavoisier reported results similar to Black's and said he had obtained them "dans le courant de Septembre de l'année dernière"[14]—i.e., in 1771.

In this paper Lavoisier makes it clear that the discoveries of Black on heat had been reported to the Academy of Sciences not long before their publication by Rozier. The communication had been made by "Monsieur Desmarets," whom I take to be Lavoisier's colleague in the Academy, the well-known geologist and *inspecteur des manufactures*, Nicolas Desmarest (1725–1815). Desmarest's communication was made at the meeting of August 29, 1772 (*Procès-verbaux*, 1772, T. 91, fol. 304). Lavoisier seems to have learned something of its contents ten days before; for it was probably some reference to Black's findings, so like his own, that caused Lavoisier to leave the meeting of August 19 and return with his own memoir on heat for the Perpetual Secretary to initial.

Lavoisier, at all events, tells us that his memoir contained an explanation of the thermal effects both he and Black had observed in the case of melting ice: e.g., the phenomenon of latent heat. Is it not likely, therefore, that in its main lines this explanation resembled the theory Lavoisier presented later in 1777, and which, as in Black's theory, described heat as a fluid capable of existing in two forms or states: one free (*feu libre*) and capable of influencing the thermometer; the other latent or combined (*feu combiné*)?[15] Lavoisier surely believed in 1772, as we know he did later, that heat—or properly speaking the "matter of fire"—was a very subtle and very elastic fluid enveloping our planet on all sides, fine enough to penetrate more or less easily into all bodies, and tending to reach a state of equilibrium among them. This was, with certain variations in detail, the view most commonly held on the Continent; it was, as Lavoisier later noted, the theory favored by Benjamin Franklin and by that influential early teacher of chemistry, Hermann Boerhaave.[16] But it is likely, from the evidence just presented, that Lavoisier had gone further and had already developed, on the basis of his own experiments, something approaching the theory of the two forms or states of the matter of fire which he was to set forth so clearly in 1777,[17] and which was to be an inseparable part of his oxygen theory of combustion.

Nothing, then, would be a more interesting contribution to Lavoisier studies than to have this lost memoir, if indeed it has managed to survive in some hidden corner, turned up by some fortunate scholar who could then confirm, reject, or modify the conjectural reconstruction given above. We would then know how closely this theory resembled Black's and how far a knowledge of Black's results contributed to Lavoisier's mature theory of caloric.

However that may be, I hope my brief discussion will at least serve to dispel some of the misapprehensions that have clustered around Lavoisier's August memorandum of 1772. We should restore to it the title which Lavoisier gave it, and see it for what it is: the earliest dated document we possess in which Lavoisier records his interest in the work of Stephen Hales and links the study of air in bodies with speculations about combustion and calcination.

۞

[Soon after the appearance of this paper M. René Fric published what is certainly my "lost" document together with later *inédits*. See his article in *Archives Internationales d'Histoire des Science* 12 (1959), actually published in 1960. For the discussions this document aroused see the bibliography in my *Antoine-Laurent Lavoisier: Chemist and Revolutionary* (New York: Charles Scribner's Sons, 1975), 152.]

Notes

1. H. Guerlac, "The Continental Reputation of Stephen Hales," *Archives Internationales d'Histoire des Sciences* 4 (1951), 393–404. For the document see *Oeuvres de Lavoisier*, 6 vols. (Paris, 1862–1893), III, 261–66.
2. *Oeuvres de Lavoisier*, III, 266.
3. *Ibid.*, 261, n. 1. This has been accepted by Meldrum (see below, n. 4); by Douglas McKie, *Antoine Lavoisier: Scientist, Economist, Social Reformer* (New York: H. Schumann, 1952), 100; and by Duveen and Klickstein, *A Bibliography of the Works of Antoine Laurent Lavoisier* (London: E. Weil and Wm. Dawson & Sons, 1954), 388. McKie, however, going beyond the implications of the annotation, asserts that this was an official proposal made by Lavoisier to the Academy requesting the use of the great lens of Tschirnhausen so that he could conduct experiments using a stronger heat than that afforded by ordinary fire. This is not the case. As we learn from Macquer, it was Cadet and Brisson who, somewhat earlier, had asked the Academy for permission to take from its cabinet the Tschirnhausen lens. These experiments which were carried out in common—and included many in which Lavoisier was not especially interested—were begun on August 14, five days *before* this presumed request to the Academy (see "Détail des expériences executées au moyen du grand verre ardent," *Oeuvres de Lavoisier*, III, 284 ff.). For Macquer's account, see "Mémoire lu par M. Macquer, à la Séance publique de l'Académie Royale des Sciences, le 14 Septembre 1772, sur des expériences faites en commun, au foyer des grands verres ardens de Tschirnhausen, par MM. Cadet, Brisson & Lavoisier," *Introduction aux observations sur la physique*, 2 vols. (Paris, 1777) II, 612–16. The *Introduction* is the readily available quarto reprint of the first years of the *Observations*. The original duodecimo version is almost unobtainable. [A set is now in the Cornell University Library.]
4. Meldrum, who refers to it as the *Memoir on Elementary Fire*, and did not perceive its

importance, seems to have had some suspicions, for he calls it "a preliminary note, rather than a memoir." *The Eighteenth Century Revolution in Science—The First Phase* (Calcutta: Longmans, Green & Co., 1930).

5. Through the courtesy of Mme. Pierre Gauja at the archives of the Académie des Sciences, I have been able to study a clear photograph of the original manuscript. The 15 pages of text, written on the right-hand side only, are the work of an amanuensis. There are numerous corrections and additions in Lavoisier's hand, and these have been incorporated into Dumas's printed text. Between the heading and the opening line is the date "Du 8 août 1772," in Lavoisier's writing. A numbered footnote (1) refers to the bottom of the page, where someone has written: "Lu à l'Académie des Sciences, le 19 août 1772, sous le titre de *Mémoire sur le feu élémentaire.*" This is an elegant and legible hand which is neither that of Lavoisier nor of Dumas. I have not identified the writer.

6. *Procès-verbaux, Académie Royale des Sciences* (1772), vol. 91, fol. 292 *v.* My italics.

7. *Introduction aux observations*, II 1772 (1777), 510–11. My italics.

8. The term "caloric" was substituted for the terms "fluide igné" and "matière de la chaleur" in the course of the reform in nomenclature about 1786–1787. The theory antedated the new terminology by at least ten years, though it did not appear in print until 1780. See Lavoisier's remarks in his *Traité élémentaire de chimie* in *Oeuvres de Lavoisier*, I, 19.

9. *Oeuvres de Lavoisier*, III, 262.

10. Douglas McKie and Niels H. de V. Heathcote, *The Discovery of Specific and Latent Heats* (London: E. Arnold & Co., 1935), 45–48.

11. The earliest published account is the anonymous *Enquiry into the general effects of Heat: with observations on the Theories of Mixtures. In Two Parts, etc.* (London: J. Nourse, 1770). For this work see McKie and Heathcote, 50–52.

    In the spring of 1772 the Swedish Academy's *Handlingar* published Johann Carl Wilcke's paper in which he described experiments, like those of Black and Lavoisier, made by mixing ice with water at different temperatures; these led him to a clear statement of the doctrine of latent heat. But Wilcke's famous paper was not generally available until 1776, when A. G. Kästner published his German translation of the *Handlingar* for 1772.

12. "Expériences du Docteur Black, sur la marche de la Chaleur dans certaines circonstances," *Introduction aux observations*, II, 1772 (1777), 428–31.

13. I shall discuss in a forthcoming study the delayed introduction of Black's chemical ideas into France. [Subsequently published in *Lavoisier—The Crucial Year* (Ithaca, N.Y.: Cornell University Press, 1961), see esp. 17–19, 23–24, 68.]

14. "Expérience sur le passage de l'eau en glâce, communiquée à l'Académie des Sciences, par M. Lavoisier," *Introduction aux observations*, II, 1772 (1777), 510–11.

15. Lavoisier, "De la combinaison de la matière du feu avec les fluides évaporables." *Oeuvres de Lavoisier*, II, 212–24. This paper was read to the Academy in 1777 and published in the *Mémoires* for that year in 1780.

16. Lavoisier, "Mémoire sur la combustion en général." *Oeuvres de Lavoisier*, II, 228. This paper of 1777 was published in 1780 in the *Mémoires* of the Academy for 1777.

17. Support for this theory would also have been provided by another discovery Lavoisier made about this time: that of the latent heat of crystallization. On November 25, 1772, he read to the Academy of Sciences a memoir on the crystallization of Glauber's salt (*Procès-verbaux*, vol. 91, fol. 344 *v*, and subsequent entries) in which he reported that heat was evolved during crystallization of the salt. This memoir was published in Rozier's *Observations* for January 1773 (I, 10–13); cf. *Oeuvres*, V, 243–47.

23
Some French Antecedents of the Chemical Revolution

✿

By the phrase "Chemical Revolution" we commonly mean the great reform in chemical theory and nomenclature brought about by Lavoisier and his co-workers between 1772 and 1789.[1] But the term can also be employed in a broader sense: when Archibald and Nan L. Clow entitled their excellent study of the rising chemical industry of eighteenth-century Britain "The Chemical Revolution,"[2] they focused our attention on the chemical phase of the Industrial Revolution and on that swift emergence of chemistry both as science and as technology that is such a marked feature of scientific progress during the second half of the century, not only in Britain but on the European continent. This broader "chemical revolution" set the stage for the great achievement of Lavoisier, by giving chemical studies a new urgency and importance, elaborating new techniques, describing new elements and compounds, and contributing to the pneumatic chemistry of Britain and the analytic chemistry of the Continent.

In this paper I use the term "Chemical Revolution" in the familiar and narrow sense. But I wish to defend the seemingly obvious proposition that, in order to understand Lavoisier's accomplishments, one should know something of the state of chemistry in France when this extraordinary genius came upon the scene; and something, too, of the way in which the Industrial Revolution in France stimulated interest in chemistry, altered the character of the chemist's profession, and presented him with new problems and with new techniques for their solution.

French Chemistry in Lavoisier's Youth

Historians of chemistry who discuss Lavoisier's "chemical heritage" have often contented themselves with listing a few of his illustrious predecessors and contemporaries, mainly the British pneumatic chemists Robert Boyle, Stephen Hales, Joseph Black, Henry Cavendish, and Joseph Priestley.[3] Important as these men certainly were in Lavoisier's later career, i.e., especially after 1772, it is to the scientific environment of France that we must look for the influences which drew Lavoisier to chemical studies, molded his early thinking, and brought him

From *Chymia* 5 (1959), 73–112. © by the Trustees of the University of Pennsylvania.

to the threshold of his great discoveries. Yet on this subject the biographers of Lavoisier are largely silent, chiefly, I suppose, because the French chemists before Lavoisier fall short of the first rank, and because the problems that preoccupied them seem remote from those upon which Lavoisier was to center his attention.

France in 1763, the year in which Lavoisier began his chemical studies, had yet to produce a chemist we can fairly class with England's Robert Boyle; with Joseph Black in Scotland; with Hermann Boerhaave of Leiden; or with such influential German scientists as Georg Ernst Stahl, Friedrich Hoffmann, or A. S. Marggraf. French chemistry, as was true almost everywhere but in England in the early eighteenth century, was still closely tied to medicine and pharmacy. The leading French chemists of Lavoisier's youth, and the majority of his contemporaries as well, were trained as physicians or pharmacists. Formal chemical instruction in lecture and textbook still bore the stamp of a pharmaceutical purpose.

To be sure, in the closing years of the seventeenth century French chemists began to free themselves from a complete subservience to the medical arts. At least two men of the early Academy of Sciences shared the belief that chemistry could earn for itself an honored place as a branch of the New Experimental Philosophy. Guillaume Homberg (1652–1715), a widely traveled and versatile scientist who had studied in Boyle's laboratory, published useful chemical papers on many different subjects. In the next generation, E.-F. Geoffroy (1672–1731), among his other contributions, invented his famous table of chemical affinities, and adopted for this purpose a scheme of quasi-alchemical symbolism widely copied during the eighteenth century.

The impulse provided by these men—and by Louis Lemery (1677–1743), an important theorist—soon faded out, partly, I suppose, because all three men published only short technical memoirs and no general or pedagogical work. The next generation was markedly less distinguished. Until 1742, when G.-F. Rouelle (1703–1770) began his teaching there, the chemical lectures at the Jardin du Roi, the center of chemical instruction in Paris, were at a low ebb and still mainly pharmaceutical in character. The general sterility was reflected in the membership of the Academy of Sciences during Lavoisier's youth:[4] of the three senior Academicians (*pensionnaires*) for chemistry, only one, Jean Hellot (1685–1766), of whom we shall have reason to speak, was a man of real ability; the others were La Condamine, the mathematician and traveler, whose contributions to chemistry are largely illusory; and Louis-Claude Bourdelin (1696–1777), the mediocre Professor of Chemistry at the Jardin du Roi. Of the two chemists of the second rank (*associés ordinaires*) we need mention only Rouelle who, from his position as demonstrator in chemistry at the Jardin, was already stimulating and transforming the study of chemistry in France. The two junior members (*adjoints*) were also unequal in ability. Théodore Baron (1715–68) was a mediocrity, best known for his re-edition (1756) with copious notes of the elder Lemery's *Cours de Chymie*; but his companion, P. J. Macquer (1718–84), was already the outstanding chemist among Lavoisier's seniors.[5] A pupil of Rouelle, he had published an "Elements of Chemistry,"[6] which rivaled Lemery's textbook, and was soon to publish his famous *Dictionnaire de chymie* (1766).[7]

To Rouelle's teaching and to Macquer's expository writing we must attribute in substantial measure the ferment of activity in chemistry already apparent when Lavoisier entered upon his career. Though both Rouelle and Macquer were trained in the medico-pharmaceutical tradition, both sought to treat chemistry as a branch of natural philosophy; and each in his own way was profoundly influenced by, and in turn helped to promote, the new current of technological curiosity in which is clearly evident in France by mid-century; each was deeply interested in the relation of chemistry to the useful arts.

The Industrial Revolution in France

It is England we first think of when we speak of the Industrial Revolution. Yet by 1763, despite the strains of the recently concluded Seven Years' War, France was already entering upon a period of steady and cumulative industrial growth.[8] If she was slow to mechanize industry and slower still to introduce Britain's factory system, if her technical improvement in the extractive industries lagged behind both Britain and Germany, we nevertheless note substantial progress in the introduction of new processes, the improvement of the old, and the manufacture of new products. France's most rapid technological development took place precisely in those areas where chemistry was crucially important. In the period between 1740 and 1760 profound influences were at work that the government set itself to encourage and direct, at first in a strict Colbertist spirit of detailed regulation and later in a somewhat more liberal manner. A marked feature of this movement was the use, to an extent heretofore unparalleled, of the advice and cooperation of men of science. For the first time in French history we hear of men whom we can only describe as professional industrial scientists— like Jean Hellot and Gabriel Jars—as well as academic scientists, among whom Réaumur is the earliest pioneer, who took an active interest in improving industrial processes, while reserving their main efforts to advancing fundamental science.

René-Antoine Ferchault de Réaumur (1683–1757), best known perhaps for his study of thermometry and for his biological writings, foreshadowed in his own work during the 1720s the century's later preoccupation with applied chemical research.[9] He was a chief promoter, if not indeed the instigator, of a program to bring the Academy of Sciences closer to the industrial problems of the day and to interest the Royal Government in its potential utility. To this end he drafted, or at least influenced, a memorial drawn up sometime before 1727 that argued at length for the great potential usefulness of the sciences represented in the Academy.[10] The paragraph dealing with chemistry is particularly revealing:

> Chemistry, whose investigations seem rather frivolous [*assez vaines*] to those
> who do not know its true purpose, could become one of the most useful parts

of the Academy. Let us not boast of the help that medicine could draw from it; let us look at it only from the standpoint of the arts, to which it could be more useful even than mechanics itself. The conversion of iron into steel, the methods of plating or whitening iron to make tin-plate, the conversion of copper into brass, three great industries which the Kingdom lacks, are in the province of chemistry. It is the business of chemistry, also, to investigate the mineral substances used in dyeing and ores and minerals. Glass-works, pottery works, faïenceries, porcelain—industries which all need to be improved—also concern it.[11]

Here, in brief, is a broad research program for applied chemistry, referring especially to areas in which Réaumur himself was actively at work, yet also clearly prophetic of later developments. Let us glance at certain of these areas and record the progress made in them before the time of Lavoisier.

Textiles and Dyeing

France's textile industry—the spinning and weaving of linens, woolens, and silks—had suffered, like the rest of the French economy, from the prolonged wars of Louis XIV and the speculative debacle of the Regency. But recovery was rapid: every year after the 1730s saw the rehabilitation of old establishments, the creation of new ones, and the naturalization of foreign specialties such as the manufacture of English serge, of velvet on the Dutch model, and the spinning and weaving of cotton. This last was greatly stimulated by a decree of 1762, which allowed the countrymen to spin and weave at their pleasure;[12] and still more by the activity of such men as the Manchester weaver John Holker (1719–1786) who, with the active support of the Royal Government, instructed French spinners and weavers, introduced English methods and machinery, and imported English workers.[13] A most successful new departure was the manufacture of light cotton prints—the brightly colored *toiles peintes* or *indiennes*—which it had long been forbidden to produce in France. Soon after an edict had legitimized this industry, the best-known factory was established in 1760 by a Bavarian immigrant, Christophe-Philippe Oberkampf. Here were manufactured, at Jouy-en-Josas in the valley of the Bièvre, those delicate *toiles de Jouy* now sought after by collectors.[14]

This rapid development of the textile industry, as might be expected, led to the proliferation of dye works, the introduction of new dyestuffs (*vert de Saxe, rouge d'Adrianople*, for example), and an effort to improve and standardize the methods of dyeing.

France's luxury fabrics had long owed their fame to the brilliance and fastness of their colors; and the French were noted for the meticulous care and the minute regulations by which they policed the quality and durability of the dyes employed. For long the techniques of dyeing had been craft secrets of a few skilled artisans, but shortly before the middle of the eighteenth century, with government encouragement, scientists began to concern themselves with its mysteries. In 1737 the Controller General of Finance, Philibert Orry (1689–1747), issued a new set of instructions for the regulation of dye works. Like the earlier edicts of Colbert, on which they were modeled, these instructions distinguished the *teinturiers de grand teint* (Great Dyers)—who used fast colors and worked for the luxury market—from the *teinturiers de petit teint* (Little Dyers), who employed cheaper, more brilliant but fugitive colors to dye the cheaper kinds of cloth.[15] Regulations forbade the Great Dyers to use, or even keep in their possession, the materials used by the Little Dyers. Enforcement, however, was attended with difficulties: the earlier regulations had prescribed a test solution, or proof liquor, which inspectors could use on the spot to detect fraud; but the proof liquor, a mixture of alum and tartar, was not wholly satisfactory, nor was it even certain which dyes ought to be classed as fast and which as fugitive. To clarify these questions, Orry called upon Charles de Cisternay Du Fay (1698–1739)—a man best known to historians of science as an electrical experimenter and as Buffon's predecessor as director (*Intendant*) of the Jardin du Roi—to investigate the entire problem.[16] Turning his own house into an experimental dye works, Cisternay Du Fay studied all the common dyestuffs and many others also, to determine, by exposing bits of colored woolens to the action of sun and air for extended periods, which substances gave true and which false colors.[17] He experimented also with the proof liquors, to find some that might be constant and reproducible in their effects on different dyes; his results supplied the technical basis for Orry's instructions of 1737.[18]

In 1740, Cisternay Du Fay was succeeded as "Inspecteur général des teintures," the title Orry had given him, by Jean Hellot.[19] We may call Hellot the first French industrial chemist, for though he was a pupil of the pharmacist Geoffroy, and for a time had pursued science as an avocation, his reputation was made as a scientific adviser to the Royal Government.[20] Although he made useful contributions to mining, metallurgy, and ceramics, his outstanding role was as an expert adviser to the dye industry, where he soon became the acknowledged authority. In 1740–41 he presented to the Academy of Sciences two memoirs on the theory of dyeing; and in 1750 published his *L'Art de la teinture des laines*, the first scientific book on the subject, and long a standard work.[21] In it he described the equipment and organization of dye works; took up the various vegetable substances used in dyeing (sharply distinguishing the two classes of substances we have mentioned), and outlined the procedures appropriate to each coloring matter used. He also briefly summarized his theory of dyeing, according to which true or fast colors were those where the particles of coloring matter can be deposited and firmly held in the pores of the fibers, and false colors those where the dye merely adheres to the surface or is held in pores inadequate to retain it.[22] Nothing better than this physical theory made its appearance until Berthollet advanced his chemical theory of the action of dyes and mordants in 1791.

Hellot's book was supplemented by the appearance of Macquer's *Art de la teinture en soie* (1763). More original than Hellot, with whom he had col-

laborated and whom he succeeded in 1766 as Inspector General, Macquer proposed the use of Prussian blue as a fast blue dye that might compete with indigo
and pastel. In 1768 he announced a successful method of dyeing silk with scarlet,
the cochineal product long used for coloring woolens. With Claude-Louis Berthollet (1748–1822), the distinguished disciple of Lavoisier, who succeeded
Macquer in 1784 as *Inspecteur des teintures*, the tradition of Cisternay Du Fay,
Hellot, and Macquer, linking chemistry to the needs of the textile industry,
reached its culmination. A pioneer in the use of chlorine for bleaching, Berthollet
published in 1791 the first really modern work on the chemistry of dyeing: his
two-volume *Eléments de l'art de la teinture*.[23]

The Basic Chemical Industries

Industries closely allied to chemistry, and making use of basic chemicals, grew
with great rapidity during the period we are considering. By 1760, for example,
Marseilles boasted twelve sugar refineries, seven candle works, and thirty-eight
soap factories. Like the manufacture of glass, these industries were deliberately
restricted during the early part of the century in order to conserve the dwindling
supplies of wood and charcoal for the iron industry.[24] Gradually, however, the
shortage was relieved by a technological change that freed the fuel-consuming
industries to a marked extent from their dependence upon the forests. This was
the widespread substitution of coal for wood in household heating and in the
manufacture of glass and pottery, and its increasing use by soap boilers, sugar
refiners, candlers, and the like. After about 1730 the government no longer opposed the creation of new glassworks. The great factory of Saint-Gobain grew
and flourished, but lost its monopoly for the production of plate glass. In the
1750s and 1760s several important new establishments were founded, among
them the great crystal works of Baccarat.[25]

The voracious demand of these industries for alkalis and mineral acids led
in turn to serious attempts at producing these substances in large supply, without
recourse to importation. Potash, obtained from wood ashes, was always inadequate in amount for the needs of the textile trades and the makers of soap and
glass. Soda, more abundant, was produced in primitive fashion along the seacoast
by burning seaweed, but yet in insufficient quantity, so that France remained
tributary for this substance to Italy and especially to Spain, where the best
quality was produced in the neighborhood of Alicante. In 1757 a certain Jean de
Fontanes, encouraged by the botanist Bernard de Jussieu and by Hellot, attempted to grow seaweed in artificial plantations, but without notable success.[26]
The real solution, it was soon recognized, was to hit upon a method of converting
sea salt into soda. The earliest success in this direction was obtained, more than a
decade before the famous Nicolas Leblanc, by a Benedictine abbé, Père Malherbe,
on whose method Macquer reported favorably in 1778.[27]

More success was obtained in the production of mineral acids precisely in
the decade of the sixties on which we are focusing our attention. The French had

long known how to make sulphuric acid by distilling vitriols, and by the cheaper, and therefore more practical, method of burning sulphur in moist air under a glass bell jar; this *per campanum* process was described in the well-known textbooks of N. Lefèvre (1660) and the elder Lemery (1675). But it was an Englishman, Joshua Ward, who had learned of the *per campanum* method during a stay in France, who first employed it on an industrial scale at Twickenham in 1736. A decade later, the cost of manufacture was still further reduced when Roebuck and Garbett substituted lead chambers for glass globes in their plants at Birmingham and Prestonpans. With this major technical advance, Britain captured the European market for sulphuric acid.[28] It was John Holker, once more, who introduced the new methods and helped free his adopted country from dependence upon Britain. With government encouragement, Holker founded near Rouen about 1767 a prosperous lead-chamber plant, and Holker's vitriol works provided a model for others in the region. Charles Ballot does not exaggerate when he takes the founding of Holker's factory to mark the birth of French industrial chemistry.[29] Yet precisely how sulphuric acid was formed by burning sulphur in moist air was left for Lavoisier to explain a decade or more later.

Pottery and Fine Ceramics

The eighteenth century, with its porcelain figurines and Dresden shepherdesses, was a great age of ceramic art in France as elsewhere in western Europe. Already in the previous century, inspired by Italian majolica and the faïence of Delft, the French potters of Nevers, Rouen, and Moustiers had perfected their art. Hampered for a time by the economic restrictions and the fuel shortages of the early years of the eighteenth century, they soon prospered once more but had to face new rivals at Marseilles, Strasbourg, and Sceaux. Beautiful though they were and are, their masterpieces were technically little in advance of those by Bernard Palissy: rather heavy pieces of enameled pottery.

To duplicate the hardness, whiteness, and translucency of delicate oriental porcelain was the central aim of European ceramists during the latter part of the seventeenth century and the greater part of the eighteenth.[30] The chemists of Augustus II of Saxony were, as we all know, the first to succeed. At Meissen in 1715, after prolonged experimentation, they produced the first successful imitation of Chinese porcelain. From Saxony, through the dishonesty of fugitive workmen, the secret spread to other centers in Germany. The King of Prussia commissioned the distinguished chemist J. H. Pott (1692–1777) to attempt to duplicate the porcelain of Meissen. In his *Lithogeognosia* (Potsdam, 1746) Pott described his improved furnaces and the results of prolonged experiments on the effect of heat upon mineral substances. Though the secret of making porcelain was not disclosed, this work became the *vade mecum* of the ceramist and the most popular treatment of the fundamentals of the subject.[31]

The French were not as fortunate as the Germans. Gradually they succeeded in producing a ware known as "soft paste" that resembled true porcelain in appearance, but scratched too easily and could not withstand high temperatures. This *pâte tendre* was being manufactured early in the eighteenth century at Saint Cloud, at Chantilly, and at the factory at Vincennes which, after its acquisition by Louis XV and its transfer to Sèvres in 1756, became the Royal Porcelain Works. Here, with full government support, there was established the first industrial research laboratory of France. Hellot served as technical adviser from 1751 to 1766, concentrating, it would seem, on problems of pigments. Macquer, brought in as his assistant, assumed full responsibility from 1766 until his death; his chief aim, fulfilled after ten years of work and a signal stroke of good fortune, was the production of true porcelain.

Réaumur had been the first in France to investigate the problem scientifically.[32] He showed that porcelain was a substance intermediate between pottery and glass, produced from a mixture of two substances, one fusible and vitrifiable (petuntse), the other earthy and refractory (kaolin). Having received a small amount of these materials from an official of the Chinese Missions, he baked them into a little cake of genuine porcelain. But evidently despairing of finding such materials in France, he attacked the problem as it were in reverse and attempted to transform glass vessels into something resembling porcelain, by heating them imbedded in a mixture of fine white sand (*sablon*) and powdered gypsum. The result, called "Réaumur's porcelain," while striking in appearance was defective in color and physical properties and little more than a curiosity.[33]

The real solution was to find an adequate source of true kaolin in France. In 1750 the Academician Jean-Etienne Guettard (1715–87), already distinguished as a botanist and soon to emerge as one of France's pioneer geologists, had his attention directed to the problem by his patron Louis, duc d'Orléans (1703–52).[34] From specimens in his mineralogical collection, he discovered the existence near Alençon of a source both of kaolin and of suitable feldspar (petuntse). In a laboratory set up for him at Bagnolet by his patron, Guettard succeeded in approximating oriental porcelain, though because the kaolin was gray and inferior, the results were disappointing.[35]

Having learned of Guettard's discovery, a soldier-scientist, the Comte de Lauraguais (1733–1824), set up a laboratory in 1758 in his château of Lassay, taking into his employ a certain Le Guay, who had worked with Guettard at Bagnolet, and two chemists, Jean Darcet (1725–1801) and Augustin Roux (1726–76).[36] In 1766 he was able to present to the Academy of Sciences samples of the best porcelain so far made in France; it duplicated the grain, firmness, and infusibility of the products of China and of Meissen, but totally lacked their whiteness and luster.

Meanwhile Macquer and his assistants were hard at work in the laboratory at Sèvres; experiments "multipliées presque à l'infini" were carried out on promising samples supplied by travelers and by the engineers of the *Ponts et chaussées*.[37] Macquer improved Pott's ceramic furnaces and tested the characteristics of more than eight hundred samples of clay, but he found none that yielded a porcelain of the much desired whiteness and beauty. At last, in 1765, he had his stroke of good fortune. A chemist of Bordeaux, M. Villaris, forwarded to him, through the good offices of the Archbishop of Bordeaux, samples of a white unctuous earth found at St. Yrieix near Limoges. Macquer delightedly recognized this as the object of his prolonged search: it was kaolin of the finest quality, and preliminary experiments proved that it could yield excellent porcelain. Macquer had little difficulty persuading the Minister and Secretary of State, Bertin, to send him on a reconnaissance mission. He found the pure kaolin in great abundance, and discovered that the quarries of St. Yrieix also produced the indispensable pure white feldspar, or petuntse. The techniques of large-scale porcelain manufacture were worked out with the assistance of two skilled artisans of the plant, Millot and Bailli. In 1769 the production of the first true native porcelain was begun at Sèvres.

The problem of porcelain manufacture, which preoccupied his elders during the first years of his chemical career, was not without influence on Lavoisier's early work. In the spring of 1768 Lavoisier heard Darcet, the associate of the comte de Lauraguais, read to the Academy of Sciences the results of experiments, carried out with Lauraguais' improved furnace, on the effects of strongly heating a wide variety of different mineral substances. It was on this occasion that Darcet startled his audience of academicians by reporting that the diamond, believed to be not only the hardest but also the most refractory of substances, could be totally destroyed by heat.[38] The subject was explored in the course of 1771–72 by a number of chemists—by Darcet and the younger Rouelle, by A. Roux, and by Macquer, who brought to bear the techniques developed in the long studies of porcelain. Early in 1772 Lavoisier collaborated in similar experiments with Macquer and Cadet, to confirm the fact of the destruction of the diamond and to determine its cause.[39] It was these experiments that first lured Lavoisier away from the strictly mineralogical and hydrometric interests that had occupied him, and brought him for the first time face to face with the problem of combustion.

Mining and Metallurgy

Of greatest significance for the progress of chemistry was the determined effort of energetic individuals and of the Royal Government to bring French mining and metallurgy to a higher level of production and technical mastery.[40] In these industries the Germans and Swedes had long been the acknowledged leaders; and after the middle of the eighteenth century France's great rivals, the English, were completely transforming their iron industry by the introduction of steam power and the use of coke for smelting.

France was well endowed with mineral wealth. Iron, of course, had been mined and smelted in nearly every region of France from medieval times. And Louis XIV's conquests to the north and east had added new mineral-rich regions, notably Franche-Comté, Alsace, and Lorraine. France had few precious metals and no tin; but there were copper mines in Lorraine, at Giromagny near Belfort, and in the Lyon region; and lead was mined in lower Brittany, Alsace, Forez, and especially in Dauphiné, where the Blumensteins, a family of German immigrants, had been accorded a monopoly in 1728.[41] Dauphiné was also rich in products ancillary to the building trade—limestone and gypsum, for example—and was second to none in the number and prodigious variety of its mineral waters.

Iron was centrally important to the French economy, and the demand was steadily increasing: for nails, tools, and agricultural implements; for the springs and tires of carriages; for decorative work and cutlery; for the guns and cannon of the Royal army, and the anchors and fittings of the navy. The naturalizing of a tin-plate industry after 1730 still further increased the demand for iron.

Yet the iron industry was widely dispersed in small units to be found almost everywhere.[42] In outlying regions such as the Pyrenees and Corsica the old Catalan forge was still in use even at the end of the century, but elsewhere blast furnaces and ironworks had moved to the watercourses in wooded areas in search of power to work the bellows and the tilt hammers and for a supply of fuel. This still increased the problem of dispersion and the difficulties of transportation. Considerable study was devoted to the proper shape and utilization of blast furnaces by practical men, technologists, and scientists. Buffon, in the forge he built near Montbard, systematically explored this problem, and it was later studied with great care by the Chevalier de Grignon.

In a deplorable state at the beginning of the eighteenth century, and at its lowest ebb in the period from 1724 to 1730, the iron industry gradually overcame a number of severe handicaps: inadequate transportation, lack of available capital for investment, reluctance of many owners to change their traditional habits, and, perhaps most serious of all, the shortage of fuel.[43] So serious was the depletion of the forests that in 1723 a decree forbade the opening of new blast furnaces. The gradual exploitation of the coal mines and the substitution of coal for wood in many of the fuel-consuming industries and in ironworks themselves brought a gradual improvement, though the problem was never fully solved in the eighteenth century.[44]

For raw steel and steel products France relied largely on importation from Sweden, Germany, and England. Steel plants were indeed in existence, but the quality of their product was poor and the amount they produced inadequate to satisfy the market of the cutlers and other workers in fine steel.[45] The steel made from native iron was wholly inferior; and a true steel industry, to all intents and purposes, did not exist. The problem of steel manufacture was a central preoccupation from early in the century to its close.

It was Réaumur, once more, who published, in the second decade of the century, the first book devoted exclusively to the metallurgy of iron and steel. In his *L'Art de convertir le fer forgé en acier* (1722), composed of memoirs read to the Academy of Sciences between 1720 and 1722, Réaumur sought to elucidate from his own painstaking experiments and such theoretical concepts as lay at hand the mysteries of the production of steel, an empirical art long practiced but little understood.[46] He correctly described steel as largely dependent for its properties, not on the particular ore from which it was made, but on the addition of outside materials. He came close to the true answer—which is, of course, that steel depends for its properties upon the proportion of carbon—when he asserted that steel has more "sulphur" and "salt" than wrought iron, and that the best source of sulphurous matter was pure charcoal or the *"suie de cheminée."* He successfully prepared samples of good steel by melting scrap with carbonaceous substances.

The restrictions on the use of wood and charcoal, the increasing use of coal for household heating and for soap boiling, candlemaking, and the like, led to the swift expansion of coal mining. The search for high-grade coal occupied technologists and entrepreneurs during the early decades of the century, and with remarkable results. The rich Saint-Etienne basin, from Firminy to Rive de Gier, was rapidly developed.[47] Mines were opened, or more actively exploited, at Fresnes-sous-Condé and Valenciennes in the north. After an intensive exploration of this region came the discovery of the rich coal deposits of Anzin (1734), which has grown into the greatest coal-mining center of modern France. It was here, in the 1730s and 1740s, that Newcomen steam engines in France were first set to work for industrial purposes.[48]

The most striking feature of this rapid growth in mining and metallurgical operations was not so much the creation of Royal metallurgical factories,[49] nor the multiplication of often ephemeral joint-stock companies, nor even the role of the government as patron and moneylender,[50] but the steady concentration of ownership. Many of the mines and forges belonged to great nobles or ecclesiastical foundations, though especially in the eastern and northern provinces, members of the *noblesse de robe* and bourgeois outnumbered the noble proprietors.[51] Here and there the actual owners exploited the mines themselves or directed the work of a forge; but frequently the right of exploitation was farmed out to an ironmaster (*maître de forges*) who had available capital to invest or the professional skill required. Steadily during the century a small number of enterprising proprietors or *maîtres de forges*, especially foreigners or men from the provinces bordering on Germany, increased their holdings and concentrated the mineral wealth and the ironworks in their own hands, either by uniting many small, scattered establishments under a common management, or less commonly by forming a single large industrial aggregate.[52] This concentration of industry was chiefly noticeable in Lorraine and Alsace. At Hayange, near the Moselle at Thionville, the energetic Wendel family put together, by the gradual acquisition of neighboring properties, one of the great industrial empires of France.[53] The Baron Jean de Dietrich (1719–95) is another outstanding example. Of a family that originated in Lorraine but had settled in Strasbourg, the Baron Jean, who had made a fortune in banking, invested his profits in blast furnaces and iron foundries. By 1785 he had six plants in operation, most of which he had expanded, acquired, or established between 1768 and 1771, as well as large holdings of forest land and mines to sustain his operations.[54]

In the next generation, a nephew of Jean de Dietrich, Frédéric de Dietrich

(1748–93), a successful industrialist and the first mayor of Strasbourg during the Revolution, became a member of the Academy of Sciences,[55] published an important account of the mines of France,[56] translated a work of Scheele,[57] and as a convert to Lavoisier's doctrines was one of the founders of the *Annales de chimie*, the journal launched in 1789 to propagate the new doctrines.

These men and their associates were progressive, imaginative, and often scientifically inclined. At Hayange, Charles de Wendel employed coal instead of wood for auxiliary heating, and in 1769 encouraged a pioneer attempt to use coke for smelting. In 1779 his son, F. I. de Wendel, an artillery officer, took over the organization and direction of the Naval foundry on the Ile d'Indret where William Wilkinson, the brother of John Wilkinson, the greatest of English ironmasters, was introducing English methods for founding cannon. And it was the younger Wendel who became the guiding genius in the establishment of the great ironworks of Le Creusot in 1781. Here, in France's largest metallurgical factory, the iron industry was at last fully mechanized along English lines and coke was for the first time successfully employed in France.[58]

Besides the men of science who tried their hand as ironmasters, often with indifferent practical success—among them Réaumur and Buffon[59]—were more humble *maîtres de forges* who had some scientific qualifications. Etienne-Jean Bouchu (1714–73), a Burgundian ironmaster who had studied physics and chemistry, was a corresponding member of the Academy of Sciences, and the author of the article "Forges" for Diderot's *Encyclopédie*. With the Marquis de Courtivron, a retired soldier and member of the Academy, Bouchu published an *"Art des forges et des fourneaux à fer"* (1762), one of the great series of works on French industry published by the Academy.[60] Appended to it was Réaumur's *"Nouvel art d'adoucir le fer fondu"* and a translation of Swedenborg's important treatise on the metallurgy of iron. This book became, as Bertrand Gille has written, "le *vade-mecum* de tout maître de forges dans la dernière partie du XVIIIe siècle."[61] In 1767, Courtivron and Bouchu published their *Observations sur l'art du charbonnier*. And in 1775, Nicolas Rigoley, ironmaster and friend of Buffon, brought out his *"Art du charbonnier, suivi d'observations sur les moyens d'améliorer les fers aigres et de leur ôter leur fragilité."* Antoine de Gensanne, a corresponding member of the Academy and a mining expert at Plancher-les-Mines, published in 1770 one of the earliest French works on the reduction of iron ores by means of coke.[62]

Of especial interest is Guillot Duhamel (1730–1816), an engineer trained at the newly created *Ecole des Ponts et Chaussées*, who accompanied Jars on his early metallurgical missions, and who in 1765 founded a plant at Ruffec for the manufacture of steel by the cementation process.[63] Made a corresponding member of the Academy of Sciences in 1775, he was appointed by Necker *Inspecteur général des mines* in 1781, and two years later chosen to be one of the two professors of the newly founded *Ecole des Mines*.[64]

Perhaps the outstanding example of this type of person was the Chevalier de Grignon (1723–84). A *maître de forges* in Champagne, an authority on the design of blast furnaces, he was made a corresponding member of the Academy of Sciences in 1768 and assigned to Duhamel du Monceau. A friend and adviser of Buffon in his work on iron and steel, Grignon put his son in charge of Buffon's forge near Montbard from 1774 to 1777. In 1778 Grignon was sent by the government on an inspection tour through Brittany, Burgundy, the Lyonnais, Forez, and Dauphiné.[65] In 1780 he carried out at the request of the government, at Buffon's forge and at the recently established steel works at Néronville of which he was the director, large-scale experiments to determine whether French iron was suitable for the production of steel by cementation.[66] Cyril Stanley Smith has recently called attention to the importance of Grignon's observations on the polycrystalline structure of iron, and his pioneer work on the crystallization of metals from a melt.[67] His work on the chemistry of iron and steel, though handicapped by the state of chemical knowledge in his day, was probably the most important carried out in France from the time of Réaumur until the publication in 1788 of the joint memoir of Berthollet, Monge, and Vandermonde—all disciples of the New Chemistry—on *Les différents états de fer*, the classic paper that definitely established the role of carbon in the production of steel from iron.[68]

The Role of the Government

We have already referred in passing to the part played by the Royal government in seeking to modernize various industries—textiles, ceramics, heavy chemicals—by the founding of Royal factories, the encouragement of private initiative, and by drawing into its service some of the ablest scientific talent. From 1750 onward the same concern is evident in the case of the mineral resources and the metallurgical industries of France. Indeed from 1750 through the 1760s a special effort was made by government officials to inventory France's resources and to compare the methods of extraction and processing with those used by her competitors.

From 1750 to 1755, two of Rouelle's students, Venel and Bayen, were charged by the government to make an exhaustive analysis of the mineral waters of France, a subject neglected after the work of Duclos in 1675, and now of renewed interest since the German physician, F. Hoffmann (1660–1743), had founded the subject as a serious branch of analytical chemistry. This important mission was brought to a close when Bayen joined the French forces in Minorca as "pharmacien en chef" of the army.[69]

Soon after having founded the *Ecole des Ponts et Chaussées* in 1747, the enterprising intendant of finance, Daniel Trudaine,[70] conceived the idea of establishing a school of mines.[71] To prepare competent experts, and at the same time collect as much information as possible about the state of mining and metallurgy, Trudaine selected two young graduates of *Ponts et Chaussées*, Guillot Duhamel, of whom I have spoken, and Gabriel Jars (1732–69), a native of Lyon, where his father and older brother were working the pyrites deposit of nearby Saint-Bel and Chesy. Between 1754 and 1756 these two young men were sent to visit the principal mines of France: in the Forez, the Pyrenees, and the Vosges. In 1756, Duhamel went to England, where he doubtless learned about the cementation method of steel manufacture he was later to exploit. In the same year Jars was dispatched on a tour of inspection of the mines and forges of Saxony, Bohemia, Austria, the Tyrol, Styria, and Carinthia.[72]

At the close of the Seven Years' War, concerned over the evidence of British industrial superiority and her rapid progress in metallurgy, the French government resolved to send to England a technician capable of understanding what he saw and rendering reliable reports. Jars had demonstrated his ability; in June 1764, he received instructions to visit the coal mines of Newcastle and the Scottish border and "to ascertain the various uses made of the different types of coal, of their prices at the pit-head; if it were true that large coal was employed in blast furnaces to smelt iron and also copper ore; if it were necessary to treat it for this purpose and reduce it to the material called 'coucke' in England." He was to visit other mines, especially the tin mines, to investigate the manufacture of sulphuric acid, and "to obtain data on the method of making English steel by cementation and [determine] if it is true that Swedish iron is preferred." And finally, so read his instructions, "Le Sieur Jars, will, above all, ascertain the reason why industry is pushed much further in England than it is in France, and whether this difference, as there is every reason to suppose, is due to the fact that the English are not hindered by regulations and inspections and that they have few means of gaining wealth other than by trading and manufacturing."[73]

Jars was in England between July 1764 and September 1765. Six long and informative reports were brought back to Trudaine. Jars concluded, from his observations in England, that only Swedish iron was suitable for conversion into steel, and reported that the only process used was cementation, in which iron bars were heated for five days and five nights in contact with charcoal dust. He remarked on the widespread use of machinery and of steam engines, and on the extent to which division of labor was employed in complicated industries. His observations at Birmingham convinced him that free competition and *laissez faire* largely accounted for this city's thriving prosperity.[74] But most important of all were his observations on the use of coke for smelting.

Upon his return to France, Jars was promptly sent on another mission, this time to the mines and ironworks of Liège and Holland; to Hanover, the Hartz Mountains, and Saxony; and finally to Sweden and Norway. From these missions came sixteen detailed reports.[75] In 1768 Jars was dispatched throughout France to disseminate techniques that he had learned in his travels. In the summer of that year he traveled through Burgundy, Franche-Comté, Alsace, Lorraine, and Champagne. At the colliery of Montcenis, later to supply the great forges of Le Creusot, he taught the method of transforming coal into coke, and sent a recommendation to the minister, Bertin, stressing that the location at Le Creusot was especially well suited for the installation of coke blast furnaces. In January of 1769 he was at Charles de Wendel's great plant at Hayange, where he carried out a successful demonstration of coke smelting, the first in France. Jars died seven months later, but the following year there appeared from his pen in the October 1770 number of the *Journal d'agriculture* a pioneer paper on the use of coke. From this time on, the problem of coke smelting was seriously explored by a number of chemists and industrialists.[76]

It was just at this time that the government took under its wing another important project, Guettard's proposal for a mineralogical atlas of France, on which he planned to indicate by appropriate symbols, resembling those of the alchemists, the occurrence of different minerals, rock formations, and fossils. As early as 1746, Guettard had presented to the Academy of Sciences a memoir on the subject, accompanied by a preliminary map.[77] An indefatigable traveler, he had already amassed materials and information, and as early as 1763 he had interested the young Lavoisier in this project.[78] But it was not until 1767 that their efforts received government encouragement. In this year the minister and Secretary of State, Bertin, gave his official approval, obtained a modest subsidy for the travelers, and instructed them to concentrate first on a survey of the Vosges, Alsace, and Lorraine, a region rich in mines and mineral resources.[79] This mission, which lasted from June to late October 1767, has been often described by students of Lavoisier's work.[80] It was in May of the following year that the Academy of Sciences held an election to fill the vacancy caused by the death of the chemist Baron. The votes were about equally divided between Lavoisier and Jars, with Lavoisier slightly ahead.[81] But the decision rested with the King's minister, the Comte de St. Florentin,[82] who determined in favor of Jars, because of his age, and because "il a été employé dans plusieurs circonstances par ordre de sa Majesté pour les objets intéressants à Son Service et même à l'Etat." For Lavoisier, "aussi un sujet très distingué de Sa Majesté," a second and temporary post of *adjoint* was to be created.[83] This well-known episode is comprehensible in the light of Jars's record of devotion to the public interest, and one wonders whether Lavoisier would have fared as well had he not also demonstrated by his work with Guettard his usefulness to the State.

The Technical Literature

Réaumur early took a special interest in a project of the Academy of Sciences, which had languished until he devoted his attention to it, and which, although he speaks of it in the memorandum from which I have quoted, came to fruition only after mid-century. This was the proposal for a great "Dictionnaire des Arts" that would describe in detail and illustrate with engraved plates the chief processes and secrets of French industry.[84] At Réaumur's death in 1757 nothing had been published, but among his literary remains were notes and papers related to the project and a large number of engraved plates. These were left to the Academy, which deposited them in the Jardin du Roi, under

Buffon's care, until someone could be found to continue the work.[85] This turned out to be Buffon's friend and collaborator, Duhamel du Monceau (1700–1782) who, beginning in 1761, brought out a series of many volumes under the general title of the *Description des Arts et Métiers*.[86] Among the first volumes to appear were, quite appropriately, several works of Réaumur. In the series, also, was the *Art des forges* of Courtivron and Bouchu, Morand's great work on coal, and Macquer's *Art de la teinture en soie*. But it was Duhamel himself who wrote the greater number of books devoted to the chemical and metallurgical arts: on starch making, on charcoal burning, on the metallurgy of copper, on wire drawing, sugar refining, soap making, etc.[87]

The slow incubation of this academic project, when Réaumur was in charge of it, allowed another and more energetic group, the editors of the great *Encyclopédie*, to enter the field with explosive effectiveness. It was one of the express purposes of Diderot and his collaborators to arouse interest in the facts of technology and to make more widely available the knowledge that had long been kept secret by the artisans themselves. By the time Réaumur died, this Baconian project was well advanced; the first seven volumes of the *Encyclopédie* had been published; and work on the sumptuous volumes of plates was progressing rapidly, though the first of these did not appear until 1762.[88]

Far more effectively than the *Description des arts*, the *Encyclopédie* focused attention upon the importance of fundamental science for the improvement of the arts and crafts. The editors made much of the importance of chemistry and the chemical arts, emphasizing the close kinship of chemistry with the world of industry. The author of the basic article "Chymie," Venel, stressed this point, and was at pains to list the practical arts that depend upon a knowledge of chemistry or in which chemical knowledge and experience play a part. In his summary of the history of chemistry down to his own time, he shows little enthusiasm for the medical and pharmaceutical tradition. By contrast he praises "trois chimistes célèbres qui ne doivent rien à Paracelse," and who "illustrent une branche de la chimie des plus étendues & des plus utiles, je veux dire la métallurgie." These men are Georgius Agricola, Lazarus Ercker, and Modestin Fachsius. Venel has a long paragraph in praise of Bernard Palissy (c. 1510–1589), stressing the practical nature of Palissy's work, and he treats at greatest length the early German chemists, especially Glauber, Becher, and Stahl.[89]

If Venel has little to say of the German chemists of his century, this was not true of the later volumes of the *Encyclopédie* to which Venel ceased to contribute. In fact, the *Encyclopédie* was one of the most important vehicles for introducing a knowledge of German analytical and metallurgical chemistry into France. A key role in this very significant process was played by the Baron d'Holbach, a man best known as the author of the atheistical *Système de la nature*. D'Holbach contributed some of the most important articles dealing with specific chemical and especially mineralogical topics, and drew much of his material from Swedish and German writers.[90] Moreover, between 1752 and 1766, d'Holbach devoted his literary energy—later to be turned into more controversial channels—to translating into French books of Kunckel,[91] Wallerius,[92] Gellert,[93] Lehmann,[94] Orschall,[95] and Henckel,[96] authors whom he cited in his *Encyclopédie* articles. In 1764 he published a translation of selected memoirs on chemistry and metallurgy from the *Acta* of the Academy of Upsala and the *Handlingar* of the Royal Academy of Stockholm. Here were printed in French for the first time the important memoirs of George Brandt on arsenic and cobalt, of A. F. Cronstedt on gypsum and on nickel, and various papers of Wallerius, H. T. Scheffer, and others.[97]

Behind this important activity we discern the influence of G.-F. Rouelle, the teacher of Venel and probably of d'Holbach, and also—it should not be forgotten—of Diderot himself. D'Holbach acknowledges the assistance that he obtained from Rouelle, both in his translations[98] and in his articles for the *Encyclopédie*. He was also aided by Darcet and Roux, two young disciples of Rouelle who—like Venel and perhaps the master himself—were assiduous visitors at d'Holbach's house in the rue Saint-Roch.[99]

We can scarcely exaggerate the importance that this movement of translation had for French chemistry in this period.[100] Just as the study of mineral waters—the subject of Lavoisier's first early works—had already been greatly stimulated in France by knowledge of the work of Friedrich Hoffmann,[101] so the serious study of chemical petrology by French chemists dates from 1753 when d'Arclais de Montamy published his translation of the *Lithogeognosia* of J. H. Pott.[102] To the importance of this work for ceramic investigations I have already referred. A similar influence was exerted in another department by the appearance in French, in 1755, of Johann Andreas Cramer's classic work on the chemical theory and practice of assaying, his *Elementa artis docimasticae*,[103] not supplanted in France until the appearance in 1772 of B.G. Sage's *Elémens de minéralogie docimastique*.

But of all the German chemists whom the French came to know in this period, none deserves our admiration more—for the dignity of his professional life, the care of his investigations, the clarity of his descriptions, and the range of his accomplishments and influence—than Andreas Sigismund Marggraf (1709–82).[104] Admired and quoted by Lavoisier and his contemporaries, he was the first German chemist—indeed, if we except Stephen Hales, who was primarily a physiologist, and the physician, Hermann Boerhaave, the first foreign chemist—to be made a Foreign Associate of the Royal Academy of Sciences (1777). This was a fitting honor, for better than anyone else Marggraf marks the birth of the analytical school in Europe and the emergence of chemistry from its thralldom to pharmacy and medicine. Trained by his father, a Berlin pharmacist, and by Caspar Neumann (1683–1737), the discoverer of camphor, Marggraf afterward studied at Frankfort, Strasbourg, Halle, and—significantly for his interest in mineral chemistry—the Freiberg Mining School. The best of Marggraf's work appeared as papers published during the 1740s and 1750s in the *Memoirs* of the Berlin Academy, with which he had been associated since 1738, and of which he became Director in 1760. His most fundamental accomplishment was to emphasize analysis by the "wet way," and to free chemists from a nearly total dependence upon fusions and distillations.[105] He made important studies of platinum and zinc; he discovered beet sugar; he investigated the action of vegetable acids on metals. In several instances he undertook problems of a sort that later interested Lavoisier. In a memoir on water analysis, which Lavoisier was to quote, Marggraf observed the earthy turbidity produced by prolonged agitation of samples of clear distilled water. He anticipated Cronstedt (and also Lavoisier in his maiden paper) by showing that gypsum is composed of lime and sulphuric acid. He studied phosphorus, prepared it in a new way that for a time was standard, and described phosphoric acid. Here he noted (though he may not have been the first to do so) that phosphorus increased markedly in weight when burned. Toward the end of the decade of translations we have described, Marggraf's most important early papers were published by Formey, under the title *Opuscules chimiques* (2 vols., Paris, 1762). The twenty-seven translations this book contained—fifteen from Latin and twelve from German—were carefully reviewed by Marggraf himself.[106]

The Decade of Translations

The most significant by-product of the Decade of Translations, as we may call it, is the rise to prominence of the phlogiston theory. Because the writings of Becher and of Stahl, the authors of this theory, all date from the late seventeenth or the very early eighteenth century, it is generally assumed that the phlogiston idea had been long accepted, indeed had already passed its prime, when Lavoisier came upon the scene. This does not seem to be true. Its real popularity and authority dates from this influx of German chemical writings.

An early attempt to bring these Stahlian views to the attention of chemists in France was made by the author of an inferior book entitled *Nouveau Cours de Chymie suivant les principes de Newton et de Stahl* (1723).[107] Fourcroy also mentions a certain chemist named Grosse, about whom little seems to be known.[108] But the fact is that the textbooks most popular in France in the early and middle eighteenth century, Boerhaave's and Lemery's, are silent on the subject.[109] Indeed the new Stahlian devotees complained, for this and other reasons, of the excessive vogue of Boerhaave, and even denied him the right to be called a chemist.[110]

Once again, it is Rouelle and his circle who, not content with bringing in metallurgical and other chemical writings written in the phlogistic dialect, were responsible for popularizing the writings of Stahl. It is indeed for this service, as much as for the effectiveness of his teaching, that Rouelle was canonized by the Encyclopedists and by his disciples as the "founder of chemistry in France."[111] Through the articles of Venel and d'Holbach, disciples of Rouelle, the *Encyclopédie* propagandized the new chemical doctrines, even

TRAITÉ
DU SOUFRE,
O U
REMARQUES SUR LA DISPUTE
Qui s'eſt élevée entre les Chymiſtes,
au ſujet du Soufre, tant commun,
combuſtible ou volatil, que fixe, &c.

Traduit de l'Allemand de STAHL.

À PARIS,
CHEZ PIERRE-FRANÇOIS DIDOT, LE JEUNE,
Quai des Auguſtins, à Saint-Auguſtin.

M. D. CC. LXVI.
Avec Approbation, & Privilege du Roi.

Title page of the d'Holbach translation
from the German of Stahl's *Treatise on Sulfur*

though, as Daumas has recently pointed out, the article "Feu," written by d'Alembert, was a "condemnation of phlogiston."[112]

In 1756, Baron's re-edition of Lemery's *Cours de chymie* devoted many of the notes to adding a phlogistic commentary to Lemery's text.[113] A year later, Demachy brought out a French translation of a major phlogistic textbook, Johann Juncker's *Conspectus chymiae theoretico-practicae*, giving it the title *Eléments de chimie suivant les principes de Becher et de Stahl*. In 1766 d'Holbach made his last, but perhaps his most important, contribution to this movement, when he brought out anonymously an early work of Stahl, translating the long German title as the *Traité du soufre*.[114] In many respects this proved the simplest, best, and most readable exposition of the new doctrines.

I need not here review the features of the phlogiston theory. The concept of a sulphureous, inflammable principle long antedated Stahl and Becher. Many of the early French chemists of the eighteenth century (Geoffroy and Réaumur, for example) had theories that resembled it in some respects; even the name had been coined before. But what Stahl did was to develop his ideas into a well-rounded and ambitious theory, by which he explained not only combustion but also the calcination and reduction of metals. The recognition that calcinations were really slow combustions was a most important feature of the famous theory.

That the phlogiston theory should have invaded France with the influx of these mineralogical and metallurgical writings should occasion no surprise. It was in explaining metallurgical processes that it was particularly useful and commonly employed, as for example by J. T. Eller (1689–1760) in various papers published by the Berlin Academy, and by Gellert in his *Chimie métallurgique*. Indeed it seemed to have originated with Becher and also with Stahl in an attempt to interpret the manifold observations of the practical metallurgist. Stahl himself makes this quite clear:

In my youth [Stahl writes in the *Treatise on Sulphur*] I often asked what could be the use and necessity of smelting by putting powdered charcoal at the bottom of the furnace. Nobody could give me any other reason except that the metal, and especially lead, could bury itself in this charcoal, and so be protected against the action of the bellows which would calcine or dissipate it. . . . Nevertheless it is evident that this does not answer the question. I accordingly examined the operation of a metallurgical furnace and how it was used. In assaying some pure litharge, I noticed each time a little charcoal fell into the crucible, I always obtained a bit of lead.

Certain founders gave me as a reason that the lead was cooled in the charcoal dust, which meant the same thing, that the lead hid itself there and was thereby protected from the great heat of the fire. So I do not think that up to the present time foundry-men ever surmised that in the operation of founding with charcoal there was something which became corporeally united with the metal.[115]

This "something" Stahl identified as phlogiston. When added to a metallic calx it gives the perfect, shiny, ductile metal. When withdrawn from metals "fusion turns them into a glass which is reduced to powder, and this changes into a dry powder which fire cannot attack."

Conclusion

I have tried to indicate in broad lines the historically significant "revolution" in chemistry that took place in France about the middle of the eighteenth century. In various branches of industry, the importance of which had been foreseen by Réaumur long before—in textiles, ceramics, mining, and metallurgy —chemistry in this period was called upon to play a central role. Chemists who had been trained as pharmacists and physicians were drawn into this new orbit and confronted with new problems of practical import in which private groups as well as the government took an intense interest. This new turn of affairs is manifested in the broader range of problems treated in the lectures of that first great teacher of chemistry, G.-F. Rouelle. It was Rouelle by his lectures and Macquer by his writings who trained the great generation of chemists of which Lavoisier became the acknowledged leader. And it was disciples of Rouelle who made available to France, in a flood of important translations, the most important German and Scandinavian works of analytical and applied chemistry. In a broad sense this invigoration of chemical science set the stage for the great theoretical transformation brought about by Lavoisier and his associates, when, after 1772, they amalgamated the current of English pneumatic chemistry (really a branch of natural philosophy) with the analytical chemistry of the Continent, to produce a new synthesis of chemical theory.

I have been concerned to sketch in a background, not to suggest a direct causal connection between the developments in mining and metallurgy, for example, and Lavoisier's Chemical Revolution. The specific lines of stimulus and influence upon the Chemical Revolution are far less clear to me than they are to a recent Soviet writer who tries to show how the metallurgical revival directly led to Lavoisier's great achievement. In a recent work entitled *Sketches on the History of Science and Technology of the Period of the French Bourgeois Revolution, 1789–1794*, the author applies the familiar arguments of economic materialism to prove her point that the Chemical Revolution was a direct result of the development of capitalistic forms of production in France.[116] Stimulated by the demand for metals, by the needs of heavy industry for materiel of war, and by the use of coal in the iron and steel industry, capital— seeking profitable investment—was directed more and more to mining and metallurgy. French chemists and physicists, so her argument runs, were impelled to inquire into the chemical and metallurgical industries of neighboring countries, and the government sent men to Germany, Bohemia, and England to gather data on the extractive and metallurgical industries. Oversimplified though this is, I do not question the main outlines. But I cannot follow her when she tries to find in these economic forces—to the neglect of what she terms "ideational factors," whose existence she admits—the causative factors in Lavoisier's work. For example, she would have us believe that there is some connection between the development of new blast furnaces and the growing necessity of understanding the role of air in combustion (hence an interest in pneumatic chemistry). More definitely still she thinks that the overthrow of the phlogiston theory was forced upon French chemists by its failure to account for metallurgical phenomena. "The key to Lavoisier's stubborn struggle against phlogiston," she writes, "lies to a significant extent in the indubitable fact that Stahl's theory, which arose from the demands of metallurgical practice, ceased to satisfy that same practice, metallurgical technology in the new conditions of its development."[117] Lavoisier was therefore faced with the task of unifying in a synthetic theory the achievements of French metallurgy, on the one hand, and the progress of pneumatic physics (largely English), on the other.

It is significant that the pioneers of Lavoisier's Chemical Revolution were deeply interested in mineralogical, geological, and technological problems. But what evidence is there that specific metallurgical problems led them to inquire into British pneumatic chemistry and to abandon the phlogistic hypothesis?

Lavoisier, to be sure, had begun his career in science as a mineralogist and geologist. The central interest of his early years had been Guettard's project for a mineralogical map; and his earliest chemical work—the paper on gypsum, the study of mineral waters, the famous paper on the supposed conversion of water into earth—all bear the stamp of this central interest. Yet the most important influence of the metallurgical and mineralogical revival upon Lavoisier was that it brought him into chemistry in the first place; elsewhere I have argued that it was to learn what he could of mineralogy and geology that Lavoisier, perhaps at Guettard's suggestion, was impelled to attend the lectures of Rouelle, and so received his first knowledge of chemistry.[118] But there is not a shred of evidence that Lavoisier's investigation of the role of air in combustion—the experiments on the burning of sulphur and phosphorus—had anything to do with the operation of blast furnaces. Nor did his interest in the calcination of metals derive from any firsthand interest in the industrial problems of smelting ores or processing metals. In Lavoisier's case the explanations must be sought on the "ideational" level.

It was the appearance of Guyton de Morveau's *Digressions académiques* early in 1772 that attracted Lavoisier's attention to the problem of calcination and combustion. In one of the essays in this neglected work, Guyton demonstrates the generality of the phenomenon that all metals increase in weight on calcination, a fact that struck Lavoisier as incompatible with the phlogiston theory. Guyton, to be sure, was interested in metallurgical problems; between 1768 and 1771 he had been engaged in an analysis of the coal deposits of Montcenis; and early in 1771 he carried out a successful experiment on smelting iron with coke.[119] But his most active metallurgical work seems to have come after, rather than before, the experiments on calcination reported in the

Digressions; and the origin of his interest in the calcination problem has been well established; it was strictly "ideational."[120] Moreover, neither his interest in metallurgy nor the results of his experiments caused him, at this time, to abandon or even to question seriously, the phlogistic hypothesis.[121]

Like Lavoisier, the two French chemists who were among the first to interest themselves in the study of gases, J. B. M. Bucquet (1746–1780) and Pierre Bayen (1725–98), were pupils of Rouelle with strong mineralogical interests. About 1770 Bucquet, in his early twenties, launched the first course devoted to chemical mineralogy, the meat of which was published as a textbook: *L'Introduction à l'étude des corps naturels, tirés du règne minéral* (1771). In the years 1771–73 Bucquet, in company with the Duc de la Rochefoucauld d'Enville, in whose laboratory the experiments were carried out, began the study of fixed air to determine whether it was different from atmospheric air.[122] But there is not the slightest hint that Bucquet's interest in "fixed air" had any connection with mining, metallurgy, or mineralogy. Instead it was probably stimulated by the presumed *medical* value of fixed air advocated in a book by David Macbride, and by the appearance in the summer of 1772 of Priestley's little book on making artificial soda water.

Bayen's position was somewhat different. Alone of the workers I have mentioned, he was led by a direct path from work of a mineralogical—though hardly a metallurgical—nature to the studies that brought him to question the phlogiston theory. Like Lavoisier, he had focused especially upon the study of mineral waters. In June 1774—too late for his results to have other than historical interest—he read a paper before the Academy of Sciences on the analysis of a sample of iron ore, and included in it mention of experiments on "fixed air" that he had carried on at about the time that Lavoisier and Bucquet were independently investigating the same subject. More important still, his famous studies on mercury (1774)—in which he showed that its calx could be reduced without the use of reducing substances presumably containing phlogiston—were a by-product of his work on mineral waters; in these studies he had used the oxides and salts of mercury as reagents.[123] The behavior of mercury confirmed him in his belief that the phlogiston theory was false, for it showed that all metals did not contain this hypothetical substance.[124]

To neglect the industrial and craft background in a science like chemistry, and to ignore the role of such great historical and social forces as we have described, is to impoverish our understanding of such intellectual events as the Chemical Revolution. But on the other hand, to ignore the emergence of chemistry as a self-contained discipline, with its own theoretical problems, its own methods of thought and inner logic, is to err in the opposite direction. If there is any aspect in the birth of a science that demands our respectful attention, it is its emancipation from narrowly conceived questions of utility and from a reliance upon the requirements of the useful art that nurtured its infancy. Chemistry, in the second half of the eighteenth century, freed itself from a dependence upon pharmacy and medicine and the circle of problems they posed; but it did not do so to become totally the creature of productive industry. As Auguste Comte long ago observed, it is perhaps the outstanding feature of Lavoisier's Chemical Revolution that chemistry emerged as an autonomous discipline, a body of theoretical knowledge at last worthy of standing beside such older sciences as astronomy and physics.

My interpretation of the role played by this social and industrial revolution in chemistry can now be summarized:

1. The study of chemistry in mid-eighteenth-century France was immensely stimulated not only by the growth of mining and metallurgy, but by the demands of all the simultaneously burgeoning chemical industries.

2. Chemists who had been trained as pharmacists and physicians found a new and steadily expanding use for their skills in the solution of industrial problems. Some of these men, like Hellot, Jars, and Macquer, were the earliest industrial scientists in France.

3. The Decade of Translations not only enriched technological skill in France, by making available the work of German and Scandinavian scientists; it brought in new concepts and new analytical techniques for chemistry itself.

4. In chemical theory, the most important consequence was the widespread adoption of the phlogiston theory in the period 1750–1760. Doubts concerning its validity came later. And these doubts were the result of prolonged consideration and experimentation by men who, among the first, thought of chemistry as a branch of theoretical science.

Notes

1. The phrase was first used by Lavoisier himself in his famous memorandum of February 20, 1773, which has been frequently printed, and again in a letter to Benjamin Franklin dated February 1790. It has been given wide currency in the titles of two of the most important studies of Lavoisier's chemistry: Marcellin Berthelot's *La Révolution Chimique—Lavoisier* (Paris, 1890); and A. N. Meldrum's *The Eighteenth Century Revolution in Science—The First Phase* (Calcutta: Longmans, Green & Co., 1930).
2. Archibald Clow and Nan L. Clow, *The Chemical Revolution: A Contribution to Social Technology* (London: Batchworth Press, 1952).
3. See, for example, the chapter entitled "Lavoisier's Chemical Heritage" in a recent and very detailed biography, Douglas McKie's *Antoine Lavoisier: Scientist, Economist, Social Reformer* (New York: H. Schuman, 1952).
4. For the composition of the Académie des Sciences at this time see the *Almanach Royal* for the years 1763 to 1770 and the *Index biographique des membres at correspondants de l'Académie des Sciences de 1666 à 1939* (Paris, 1939) and later editions. There is useful information in Jean-Paul Contant, *L'Enseignement de la chimie au Jardin Royal des Plantes de Paris* (Cahors: A. Coueslant, 1952).
5. For lesser worthies such as Bourdelin and Malouin there is some information to be found in F. Hoefer's *Histoire de la chimie*, 2d ed., rev. and aug., 2 vols. (Paris, 1866–69). On the elder Rouelle one can still consult Paul-Antoine Cap, *Rouelle* (Paris, 1842). But see Paul Dorveaux, "Apothicaires membres de l'Académie

Royale des Sciences. IX. Guillaume-François Rouelle," *Revue d'histoire de la pharmacie*, no. 84 (1933), 169–86. There is an amusing account of Rouelle in Eric John Holmyard, *Makers of Chemistry* (Oxford, 1941), 189–96; and a brief popular survey by Douglas McKie, "Guillaume-François Rouelle (1703–70)," *Endeavour* 12 (1953), 130–33. My student Rhoda Rappaport is making a careful investigation of Rouelle's manuscript lectures. A new study of Macquer is badly needed; Leslie J. M. Coleby's *The Chemical Studies of P. J. Macquer* (London: G. Allen & Unwin, 1938), though useful, is quite inadequate and often misleading. The Macquer correspondence in the *Bibliothèque Nationale* deserves careful study. [Professor Rappaport's studies on Rouelle have been published in *Chymia* 5 (1959), 73–112 and 6 (1960), 68–101.]

6. Macquer's *Elémens de Chymie Théorique* was first published in 1749 (Ferguson's reference to an edition of 1741 is probably a mistake) and was reprinted in 1751 and 1753. His *Elémens de Chymie Pratique* was published in 1751. The two works appeared together in 1756 in revised form, and there is an English translation by Andrew Reid of this double work that appeared later in German, Dutch, and Russian versions. See John Ferguson, *Bibliotheca Chemica*, 2 vols. (Glasgow, 1906), I, 376–77.

7. The *Dictionnaire* first appeared anonymously in two vols., 8vo. This was promptly translated into German (1768), Danish (1771), and English (1771). Greatly expanded second editions appeared in 1777 and 1778, with Macquer's name on the title page, and a second English edition was brought out by Keir in 1777; the second edition was also translated into German and Italian.

8. The literature on the Industrial Revolution in France is far less rich than that available for England. Of great value for general economic history are the various works of Henri Sée, especially his *L'Evolution commerciale et industrielle de la France sous l'Ancien Régime* (Paris, 1925). The most important work dealing directly with the transformation of industry is Charles Ballot, *L'Introduction du machinisme dans l'industrie française* (Paris, 1923). Still of importance, in view of the state of the field, are Germain Martin, *La grande industrie en France sous le règne de Louis XV* (Paris, 1900); and Alfred Des Cilleuls, *Histoire et régime de la grande industrie aux XVIIe et XVIIIe siècles* (Paris, 1900). There is useful information in W. O. Henderson, *Britain and Industrial Europe, 1750–1870* (Liverpool: Liverpool University Press, 1954); and in Shelby T. McCloy, *French Inventions of the Eighteenth Century* (University of Kentucky Press, 1952). See also A. Wolf, *A History of Science, Technology, and Philosophy in the Eighteenth Century* (New York: Macmillan Co., 1939). Though it deals with a somewhat later period than concerns us here, Douglas Dakin's *Turgot and the Ancien Régime in France* (London: Methuen & Co., 1939) gives some idea of industrial problems as seen from the standpoint of the Royal Government.

9. See especially Jean Torlais, *Réaumur* (Paris: Desclée de Brouwer, 1936). The reader should also consult *Réaumur's Memoirs on Steel and Iron, translated by Anneliese Grünhaldt Sisco from the original printed in 1722. With an introduction and notes by Cyril Stanley Smith* (Chicago: University of Chicago Press, 1956). Dr. Smith's introduction has much of value.

10. "Réflexions sur l'utilité dont l'Académie des sciences pourroit être au Royaume, si le Royaume luy donnoit les Secours dont elle a besoin," in Ernest Maindron, *L'Académie des Sciences* (Paris, 1888), 103–10.

11. *Ibid.*, 104–5.

12. Ballot, *Machinisme*, 41–42; and Charles Schmidt, "Les débuts de l'industrie cotonnière en France, 1760–1806," *Revue d'histoire économique et sociale* 6 (1913), 261–95.

13. Ballot, *Machinisme*, 43–45. On Holker, see the study of André Rémond, *John Holker, manufacturier et grand fonctionnaire en France au XVIIIe siècle, 1719–1786* (Paris: M. Riviére, 1946). See also Henderson, *Britain and Industrial Europe*, 10–24.

14. Henri Clouzot, *La Manufacture de Jouy et la toile imprimée au XVIIIe siècle* (Paris, 1926); and Alfred Labouchère, *Oberkampf (1738–1815)* (Paris, 1866), see esp. pp. 13–46.

15. This was an ancient distinction, reflected in the recognition of separate guilds. The two methods of dyeing were also referred to as *teinture en grand et bon teint* and *teinture en petit ou faux teint*. My discussion is largely based on Jean Hellot's account in *The Art of Dyeing Wool, Silk, and Cotton, translated from the French of M. Hellot, M. Macquer, and M. Le Pileur d'Apligny* (London, 1789). I have also consulted the long article "Teintures" in the *Encyclopédie*, and the papers by H. Wescher, "Dyeing in France before Colbert," "Great Masters of Dyeing in 18th Century France," and "The French Dyeing Industry and Its Reorganization by Colbert" in *Ciba Review* 18 (1939), 618–46.

16. See Fontenelle's "Eloge de Du Fay," *Oeuvres*, 8 vols. (Paris, 1790–1792), VII, 522–38; and Hellot, *Art of Dyeing*, 12–13. According to Fontenelle, Du Fay's appointment was by an *arrêt de conseil* of February 12, 1731. On December 13, 1731, and again on August 14, 1732, Du Fay reported on his experiments to the *Conseil de Commerce*. The following year the Council adopted provisional regulations that Du Fay suggested; his revised proposals were taken up by the Council at a meeting of January 10, 1737. See *Conseil de Commerce et Bureau du Commerce 1700–1791. Inventaire des Procès-verbaux, par Pierre Bonnassieux* (Paris, 1900), 194b, 202a, 213b, 251a.

17. Du Fay exposed his bits of fabric to the sun and air for twelve days, during which time the true colors were little affected, whereas the false colors were almost entirely faded. But since the intensity of sunlight varied, and the actual exposure was not equal in every case, Du Fay resolved the problem by an early use of a colorimetric standard:

> He pitched upon one of the worst colours, that is to say, the colours upon which the sun had had the most sensible effect in the space of 12 days. This colour served him as a standard through the whole course of his experiments; for whenever he exposed his patterns, he also exposed a pattern of this piece of stuff at the same time, no longer attending to the number of days, but to the colours of his standard, which he left out, 'til it became as much faded, as those which had been exposed twelve summer days. [Hellot, *Art of Dyeing*, 14.]

18. See above, n. 16. Through Du Fay's efforts, indigo (which because of fear of dangerous competition with the native woad could be used in very limited quantities) was freed from all restrictions by Orry's instructions. These also prescribed the use of Du Fay's proof liquors, chosen to produce the same effect in a few minutes as exposure to sun and air for twelve or fifteen days. Hellot was not fully satisfied with Du Fay's results, for he found that some colors that were true when exposed to air and sun were yet affected by the trial liquors. Action of air and sun, he felt, was still the best criterion. Hellot, *Art of Dyeing*, 16–17.

19. "Eloge de M. Hellot," Grandjean de Fouchy, *Hist. Acad. Roy. Sci.*, 1766 (1769), 167–79. See also Hoefer, *Histoire de la chimie*, II, 375–77. With Hellot, there seems to have begun the practice of attaching scientists to the *Bureau de Commerce* with the right to attend sessions. Among these scientists were the chemists Macquer and Berthollet; the physicist J. B. LeRoy; the naturalists Buffon and Daubenton; and other well-known personages. See *Inventaire Analytique*, pp. xxvii–xxviii.

20. Hellot had passed some time in England, where he had made the acquaintance of members of the Royal Society of London. Upon his return, to eke out a depleted fortune, he became editor of the *Gazette de France*, holding this position from 1718 to 1732. He continued his chemical studies with enough success to permit his election to the Academy as *adjoint chimiste* in 1735.

21. The official dyer's manual of the period, *Le Teinturier Parfait* (1716), was little more than a French rendering of Giovanni Rosetti's *Plichto de l'arte tentori*, published in 1540. Hellot called the *Teinturier* "a monstrous collection of imperfect and worthless recipes." Wescher, in *Ciba Review*, 629.

22. Du Fay's theory of the mechanism of dyeing had been described in *Hist. Acad. Roy. Sci.*, 1737 (1740), 58–62. For Hellot's papers see *Mém. Acad. Roy. Sci.*, 1740 (1742), 126–48, and 1741 (1744), 38–71. The theory is briefly summarized in Coleby's *Macquer*, 88–89, and by A. Wolf, *History*, 513. It appears, however, that

Du Fay more clearly appreciated the role of chemical specificity in the action of dyes and mordants than did Hellot and that the theories were largely independent.
23. For Macquer's work on dyeing see Coleby, *Macquer*, 52–59, 84–96. On Berthollet see Pierre Lemay and Ralph E. Oesper, "Claude Louis Berthollet," *Journal of Chemical Education* 23 (1946), 158–65, 230–36. For the work of all these men, including the writings of Lepileur d'Apligny (1776) and Chaptal (1807) on the dyeing of cotton fibers, see Wescher in *Ciba Review*, 626–41.
24. Martin, *La grande industrie*, 151. Paul Baud, *L'Industrie chimique en France, étude historique et géographique* (Paris: Masson, 1932), should be consulted for information on such industries as tanning, soapmaking, etc. The bibliographies are especially useful. For the rapid growth of the glass industry between 1728 and 1789 see Warren C. Scoville, "State Policy and the French Glass Industry, 1640–1789," *Quarterly Journal of Economics* 56 (1941), 430–55.
25. Martin, *La grande industrie*, 108, 149; Ballot, *Machinisme*, 552. In 1760 the Academy of Sciences offered a prize for the best proposals to perfect the French glass industry. Scoville, "State Policy," 444.
26. Paul Baud, "Les débuts de l'industrie chimique en France," *Annales de l'Université de Paris*, 7ᵉ année (1932), 223–41.
27. See the excellent article by Charles C. Gillispie, "The Discovery of the Leblanc Process," *Isis* 48 (1957), 152–70.
28. Clow and Clow, *Chemical Revolution*, 131–37.
29. Ballot, *Machinisme*, 542–43. Cf. Baud, "Les débuts," 225–26.
30. For this section I have made use of Joseph Marryat, *A History of Pottery and Porcelain*, 2d ed., rev. and aug. (London, 1857); Jeanne Giacomotti, *La céramique*, in the series *Les arts décoratifs depuis l'Antiquité jusqu'au XIXᵉ siècle*, 3 vols. (Paris: R. Ducher, 1935); and, among eighteenth-century works, *L'art de la porcelaine*, par le Comte de Milly (Paris, 1771)—one of the volumes of the *Description des Arts et Métiers*—and the article "Porcelaine" in Macquer's *Dictionnaire de chymie*.
31. The first Prussian manufacture of hard procelain did not derive from Pott's basic investigations, but from the spread of trade secrets from the Rhineland. A factory was established in Berlin by one Wilhelm Caspar Wegeli in 1750; the Royal Porcelain Factory was founded in 1763. For the details see G. Kolbe, *Geschichte der Königlichen Porcellanmanufactur zu Berlin* (Berlin, 1863), 135 ff.
32. Torlais, *Réaumur*, 93–105.
33. The "porcelaine de Réaumur" is treated under a completely separate heading in Macquer's *Dictionnaire*.
34. For Guettard, see the valuable éloge by Condorcet, *Oeuvres*, ed. A. Condorcet O'Connor and F. Arago, 12 vols. (Paris, 1847–49), III, 22–40, that Sir Archibald Geikie has used in his *Founders of Geology*, 2d ed. (London, 1905), chap. 4; and consult E. Lamy, *Les cabinets d'histoire naturelle en France au XVIIIᵉ siècle et le Cabinet du Roi (1635–1793)* (Paris, n.d. [1931?]), 8; and Victor Champier and G. Roger Sandoz, *Le Palais-Royal d'après des documents inédits*, 2 vols. (Paris, 1900), I, 397, 410.
35. Torlais, *Réaumur*, 98–105, has a good account of the work of Guettard. He cites M. de Brébisson, *Histoire du kaolin d'Alençon* (Paris, 1912), which I have not seen. See also Macquer, *Dictionnaire de chymie*, 2d ed., 4 vols. (Paris, 1778), III, 223–24; and Marryat, *History of Pottery*, 308.
36. On Louis Félicité de Brancas, comte de Lauraguais (1733–1824), see, besides the sketches in Didot-Hoefer and Michaud, Maurice d'Ocagne, *Hommes et choses de sciences, propos familiers*, 3d series (Paris, 1935), 144 f. For Darcet consult Michel J. J. Dizé, *Précis historique sur la vie et les travaux de Jean d'Arcet* (Paris, An X); and for Roux, the anonymous "Eloge de M. Roux." Extrait du *Journal de Médecine* (Janvier 1777).
37. Macquer, *Dictionnaire de chymie*, 221–23. According to Giacomotti, the *cahiers d'expériences* of Hellot and Macquer at Sèvres are extant. *La céramique*, III, 28–29.
38. *Procès-verbaux de l'Académie des Sciences*, vol. 87, fol. 72 *v.*, 95.
39. "Résultat de quelques expériences faites sur le Diamant, Par MM. Macquer, Cadet & Lavoisier lu à la séance publique de l'Académie Royale des Sciences le 29 avril

1772," *Introduction aux observations sur la physique*, vol. 2 (ed. 1777, for 1772), 108–11.

40. In the early 1740s the French government set up a new *régime* for the exploitation of its mineral resources. Up to this time (as a result of an *arrêt de conseil of 1698*) the mining of coal had been freed from all restrictions. As far as other mines were concerned, the exploitation was under the "grand-maître et intendant des mines," a post held by members of the Condé family. The *grand-maître* was able to concede to any individual or company of his choice the right to exploit mines. At the death of the incumbent, the Duc de Bourbon, his functions were taken over by the office of the *Contrôleur général des finances*. See Des Cilleuls, *Histoire et régime de la grande industrie*, 59–60; Martin, *La grande industrie*, 117; and Marcel Rouff, *Les mines de charbon en France au XVIIIᵉ siècle* (Paris, 1922), 460–61.

41. Martin, *La grande industrie*, 111, 159. For French mineral resources see the relevant articles in the *Encyclopédie* ("Mines," "Fer," "Cuivre," etc.) and Buffon's *Histoire Naturelle des Minéraux. Oeuvres*, ed. Pierre Flourens, 12 vols. (Paris, 1853–55); see X and XI. Buffon makes considerable use of the contemporary accounts by Hellot, Monnet, Guettard, de Gensanne, Jars, Morand, and others.

42. The small size of most of the industrial units emerges clearly from the inventory given by Hubert and Georges Bourgin, *L'Industrie sidérurgique en France au début de la Révolution* (Paris, 1920).

43. For the difficulties of the iron industry, see the fine account by Bertrand Gille, *Les origines de la grande industrie métallurgique en France* (Paris, n.d. [1947]).

44. Martin, *La grande industrie*, 151. On the problem of deforestation, see Gille, *La grande industrie métallurgique*, 67–84; and Rouff, *Les mines de charbon*, chap. 3. This problem led to the pioneer studies on forestry and forest conservation by Réaumur (1721), and by Buffon and Duhamel du Monceau (1737–42).

45. See Martin, *La grande industrie*, 110, 153, 172. In 1751 Diderot in the *Encyclopédie* mentions that steel was made in Champagne, Nivernais, Franche-Comté, Limousin, and Périgord, but lamented that the quality was poor. He cited a one-time cutler who supplied him with much of the information on steel as saying: "Il est étonnant qu'en France on ne soit pas encore parvenu à faire du bon acier, quoique le royaume soit le plus riche en fer & en habiles ouvriers." *Encyclopédie*, I (1751), art. "Acier," pp. 107a–107b.

46. Torlais, *Réaumur*, 57–74. See also *Réaumur's Memoirs on Steel and Iron*, trans. Sisco, pp. xxii–xxxiii.

47. Martin, *La grande industrie*, 171. The first important authority on coal and coal mining in France is Jean-François-Clément Morand (1726–1784), a physician who had entered the Academy of Sciences in 1759 as *adjoint-anatomiste*. He was a pioneer in population studies, but was best known for his work on coal. In 1770 he prepared for the government a memoir on the advantages of coal for domestic heating (See Rouff, *Les mines de charbon*, 27–28), and published between 1768 and 1777 his technical classic: *L'Art d'exploiter les mines de charbon de terre*. Buffon relied heavily on Morand for his treatment of coal in his *Histoire Naturelle des Minéraux. Oeuvres*, X, 213–76. For Morand, see Condorcet, *Eloge de M. Morand, Oeuvres*, III, 161–68.

48. Ballot, *Machinisme*, 385. Rouff, *Les mines de charbon, passim*. The earliest Newcomen engine was set up in 1726 at Passy to pump water from the Seine, but the first used in mines was put into operation at Fresnes in 1732. Cf. Henderson, *Britain and Industrial Europe*, 43–44.

49. See, for example, the Royal Arms Factory of Saint-Etienne, given this status in 1769 by letters patent, and the Forges Royales at Cosne, 1755. One should distinguish three classes of Royal involvement in manufacture: (1) The *manufactures du Roi*, owned and operated by the state, as were the tapestry factory of Gobelins and the porcelain works at Sèvres; (2) the *manufactures royales*, which were privately owned, but enjoyed special favors from the government, such as the Royal Plate Glass Factory of Saint-Gobain; and (3) the *manufactures privilégiés*, privately owned but chartered by the state and given limited privileges. For this distinction, see the article by Scoville, "State Policy," 433 and n. 2. The metallurgical plants in

which the government took an interest were mostly in the second and third categories, at least toward the middle of the century. In the 1780s the forge at Ruelle in the Charente, the foundry on the island of Indret, the Forges Royales at Cosne, and the tin-plate factory at Blandecques in the Pas-de-Calais were purchased in the King's name. See H. and G. Bourgin, *L'industrie sidérurgique, passim*.

50. Gille, *La grande industrie métallurgique*, 133–36; and E. Depitre, "Les prêts au commerce et aux manufactures, de 1740 et 1789," *Revue d'histoire économique et sociale* 7 (1914–1919), 196–217. State loans are recorded to the Blumensteins, to the miners of Dauphiné, to Holker for his vitriol plant, and for the textile industry.

51. Ballot, *Machinisme*, 425–26; Gille, *La grande industrie métallurgique*, 160–67.

52. Gille, *La grande industrie métallurgique*, 173–86.

53. Ballot, *Machinisme*; and Gille, *La grande industrie métallurgique*, 186–89. On Hayange see H. and G. Bourgin, *L'industrie sidérurgique*, 276–77; and Henri Grandet, *Monographie d'un établissement métallurgique sis à la fois en France et en Allemagne* (Chartres, 1909).

54. Ballot, *Machinisme*, 426–27.

55. Ballot, 427. He cites Louis Spach, *Frédéric de Dietrich, premier Maire de Strasbourg* (Paris, 1857). See the sketch of Dietrich in Didot-Hoefer; the *Index biographique des membres et correspondants de l'Académie des Sciences*; and G. G. Ramon, *Frédéric de Dietrich, premier Maire de Strasbourg sous la Révolution française* (Nancy, Paris, Strasbourg, 1919).

56. Baron Dietrich, *Description des gîtes de minerais et des bouches à feu de la France*, 3 vols. (Paris, 1786).

57. Karl Wilhelm Scheele, *Traité Chimique de l'Air et du Feu, avec une introduction de Torbern Bergmann (sic), traduit de l'Allemand, par le Baron Dietrich* (Paris, 1781). This is the first French edition; the German version from which this had been translated had appeared in 1777. See D. I. Duveen, *Bibliotheca Alchemica et Chemica* (London: E. Weil, 1949), 533.

58. Gille, *La grande industrie métallurgique*, 194–99; Jean Chevalier, *Le Creusot, berceau de la grande industrie française*, nouvelle édition (Paris: Perspectives, 1946).

59. Germain Martin, *Buffon Maître de Forges* (Paris, 1898). Buffon had his forge on his lands at Buffon, six kilometers from Montbard. Here, at the cost of more than 300 livres, he erected a blast furnace, two forges, a foundry, two ore crushers, two tilt hammers and two "batteries" for making sheet iron. *Oeuvres*, X, 477, note a.

60. On Bouchu, see the sketch in Didot-Hoefer and the *Index biographique des membres et correspondants*. For Courtivron see also Condorcet, *Eloge de M. le marquis de Courtivron*, *Oeuvres*, III, 187–95; and P. Brunet, "Gaspard de Courtivron," *Mémoires de l'Académie de Dijon* (1927–1931), 115–34.

61. Gille, *La grande industrie métallurgique*, 86.

62. Gille, *La grande industrie métallurgique*, 82. In 1773 Gensanne carried out at Hayange an unsuccessful attempt at coke smelting. Henderson, *Britain and Industrial Europe*, 39.

63. Alfred Lacroix, *Figures de Savants*, 4 vols. (Paris: Gauthier-Villars, 1932–38), I, 19–23; Gille, *La grande industrie métallurgique*, 99; Martin, *Buffon*, 172.

64. For the *Ecole des Mines*, see below, n. 71.

65. *Index biographique des membres et correspondants*; Gille, *La grande industrie métallurgique*, 87, 93; and Rouff, *Les mines de charbon*, 43–44.

66. *Correspondance de Buffon de 1729 à 1788*, recueillie et annotée par M. Nadault de Buffon, 2ᵉ ed., revue et corrigée, 2 vols. (Paris, n.d. [1884–1885]), I, 270–71 and notes, 413–15. On the experiments of 1780 see *Oeuvres de Buffon*, X, 504–7 and note a.

67. Cyril Stanley Smith, "The Development of Ideas on the Structure of Metals." This paper was presented at the Institute of the History of Science, held at Madison, Wisconsin, September 1–11, 1957. [Subsequently published in Marshall Clagett, ed., *Critical Problems in the History of Science* (Madison: University of Wisconsin Pres, 1959), 467–98.]

68. Grignon translated T. Bergman's important memoir on iron: "Analyse de Fer, par M. Torb. Bergman, Chevalier de l'Ordre Royal de Vasa, traduite en françois avec

des notes & un Appendice, & suivie de quatre Mémoires sur la Métallurgie par M. Grignon, Chevalier de l'Ordre du Roi, Correspondant de l'Académie Royale des Sciences" (Paris, 1783).

69. *Opuscules chimiques de Pierre Bayen,* 2 vols. (Paris, 1789), p. xlii.

70. Daniel Charles Trudaine (1703–1769) is perhaps the crucial government figure in this industrial development. Trained in the law, he was successively a *maître des requêtes* (1727), intendant of Auvergne (1729–34), and *Intendant des finances* (1734–69). In 1743 Orry, the Controller General, put Trudaine in charge of the roads and bridges of France, and this remained his chief responsibility to the end of his life. His earliest important accomplishment was the creation of the *Ecole des Ponts et Chaussées,* at the head of which he placed the famous engineer J. R. Perronet (1708–94). In 1749 Trudaine entered the Bureau de Commerce and henceforth played an increasing part in stimulating the commercial and industrial development of France. His son, Trudaine de Montigny (1733–77), was associated with him in all his functions from 1757 onward until the elder Trudaine's effective retirement in 1764. A detailed study of both Trudaines is badly needed, but see Ernest Choullier, "Les Trudaines," *Revue de Champagne et de Brie* 14 (1883), 19–25, 131–39; [Trudaine de Montigny,] "Eloge de M. Trudaine," *Hist. Acad. Roy. Sci.* 1769 (1772), 135–50. Both men were *académiciens honoraires* in the Academy of Sciences and were close friends of the more important scientists.

71. When the direction of the mines of France was taken over by the Controller General, Orry, the responsibility was passed on to his subordinate, Trudaine, in 1743–44. In 1764 the direction of the mines was transferred to the Minister and Secretary of State, Henry-Leonard-Jean-Baptiste Bertin (1720–92), who took the first steps towards the creation of a separate school of mines. Despite this interest, nothing was accomplished until 1778, when a chair of "minéralogie et métallurgie docimastique" was established at the Royal Mint for the chemist B.-G. Sage (1740–1824). It was not until 1783 that a School of Mines was officially established. See Louis Aguillon, *L'Ecole des Mines de Paris, Notice historique* (Paris, 1889); and Rouff, *Les mines de charbon,* 480–88.

72. On Jars see Grandjean de Fouchy, "Eloge de M. Jars," *Hist. Acad. Roy. Sci.,* 1769 (1772), 173–79.

73. Gille, *La grande industrie métallurgique,* 80–81; Jean Chevalier, "La Mission de Gabriel Jars dans les Mines et les Usines Britanniques en 1764," *Transactions of the Newcomen Society* 26 (1947–48 and 1948–49), 57–68.

74. Chevalier, "Gabriel Jars," 63.

75. Gabriel Jars, *Voyages métallurgiques, ou recherches et observations sur les mines et forges de fer,* 3 vols. (Lyon, 1774–81).

76. Gille, *La grande industrie métallurgique,* 81–84. See also C. Ballot, "La révolution technique et les débuts de la grande exploitation dans la métallurgie française. L'Introduction de la fonte au coke en France et la fondation du Creusot," *Revue d'histoire des doctrines économiques et sociales* 5 (1912), 29–62.

77. *Mém. Acad. Roy. Sci.,* 1746 (1751), 363–92.

78. See my paper, "A Note on Lavoisier's Scientific Education," *Isis* 47 (1956), 211–16. There is a brief account of Bertin's role by Lavoisier in *Oeuvres de Lavoisier,* 6 vols. (Paris, 1862–93), III, 259; V, 205–6.

79. Bertin, to whom the responsibility for the mines of France had been transferred in 1764 (see above, n. 71), was an enthusiastic patron and amateur of science. He had been made a *membre honoraire* of the Academy of Sciences in 1761, and was chosen vice-president in 1763 and president in 1764, both posts being reserved for honorary members. Bertin had a *cabinet d'histoire naturelle,* which Lavoisier, doubtless because it was largely mineralogical, was asked to classify in 1771. See Edouard Grimaux, *Lavoisier, 1743–1794,* 2d ed. (Paris, 1896), 26. [There is useful material on Bertin in Rhoda Rappaport, "Government Patronage of Science in Eighteenth-Century France," *History of Science* 8 (1969), 119–36.]

80. Grimaux, *Lavoisier,* 11–23. The maps resulting from this early effort were published in A. G. Monnet (1734–1817) under the title *Atlas et Description minéralogiques de la France. Entrepris par ordre du Roi, par MM. Guettard & Monnet. Publiés par M. Monnet, d'après ses nouveaux voyages* (Paris, 1780). For a description of

this work, see Denis I. Duveen and Herbert S. Klickstein, *A Bibliography of the Works of Antoine Laurent Lavoisier* (London: E. Weil and W. Dawson & Sons, 1954), 236–44. Monnet, a pharmacist from Rouen, had attracted the attention of Malesherbes for a treatise on mineral waters (1768) and a *Traité de la vitriolization et de l'alunation, ou l'art de fabriquer l'alun et le vitriol* (Paris, 1769). He was brought to Paris and given (1778–79) the newly created post of Inspector General of Mines. Monnet was to remain an antagonist of the new chemistry of Lavoisier. On Monnet see Didot-Hoefer, and Rouff, *Les mines de charbon, passim.*

81. Grimaux, *Lavoisier*, 27–30. The *Procès-verbaux* (vol. 87, p. 90) merely says that "la pluralité des voix a été pour MM. Lavoisier et Jars." The order of the names supports the story that Grimaux derived from Lalande, "Notice sur la vie et les ouvrages de Lavoisier," *Magasin encyclopédique* 5 (1795), 175, that Lavoisier's supporters were in the majority.

82. Louis-Phelypeaux, Comte de Saint-Florentin (1705–77), who like Bertin was one of the four Secretaries of State in the Conseil d'Etat, had in his department responsibility for the clergy, for matters concerning the Protestants, and for the Maison du Roi, under which fell supervision of the several academies.

83. *Procès-verbaux*, vol. 87, p. 110. Saint-Florentin's letter has been printed by René Fric in the *Oeuvres de Lavoisier. Correspondance*, recueillie et annotée par René Fric, fascicle I (Paris, 1955), 114–16.

84. A useful study is that of Arthur H. Cole and George B. Watts, *The Handcrafts of France as recorded in the Descriptions des Arts et Métiers 1761–88*, publication no. 8 of the Kress Library of Business and Economics (Cambridge, Mass., 1952).

85. In this I have followed Torlais, *Réaumur*, 382, instead of Cole and Watts, *Handcrafts of France*, 10, who make no mention of Buffon's role. But Torlais cites the text of Réaumur's will, and the Royal *ordonnance* of January 2, 1758, handing over to Buffon the natural-history collection and that part of the manuscripts (including the engraved plates) having to do with the *Description des Arts*.

86. Cole and Watts, *Handcrafts of France*, 4, n. 4, point out how difficult it is to give a figure for the number of monographs or of volumes in this series. The Boston Athenaeum has a set of thirty-two volumes, while that in the Baker Library (Kress Library) of the Harvard Graduate School of Business Administration numbers fifty-three volumes. But in these sets, monographs on different subjects are bound together. A conservative estimate of the number of items making up a set is seventy-three, running to almost 13,500 pages of text and over 1800 plates. Some crafts are treated in twenty pages or less, while Morand's great study of coal runs to 1656 pages.

87. See Cole and Watts, *Handcrafts of France*, 26–28, for the volumes, to the number 20, attributed to Duhamel du Monceau.

88. Of special interest, though of unequal value, is the collection of papers published by the *Centre International de Synthèse, Section d'Histoire des Sciences* and entitled *"L'Encyclopédie" et le progrès des sciences et des techniques* (Paris, 1952). See particularly the paper by Bertrand Gille, "L'Encyclopédie, dictionnaire technique," 187–214. On the question of the extent to which Diderot and his collaborators made illicit use of some of Réaumur's plates see, besides Gille's article, the paper of Georges Huard, "Les planches de *l'Encyclopédie* et celles de la *Description des Arts et Métiers* de l'Académie des Sciences," 35–46. Huard here expands the views briefly expressed in *Diderot et l'Encyclopédie* (Paris, 1951), 68–99, a catalogue of the exposition held at the *Bibliothèque Nationale* to celebrate the bicentenary of the *Encyclopédie*.

89. "Chymie," *Encyclopédie*, III (1753), 408–37.

90. Pierre Naville, *Paul Thiry d'Holbach* (Paris: Gallimard, 1943), 64–69. Naville lists in his bibliography (pp. 407–20) a number of articles on chemistry and metallurgy contributed by d'Holbach to the *Encyclopédie*, as well as d'Holbach's important scientific translations.

91. *L'Art de la Verrerie de Neri, Merret et Kunckel, auquel on a ajouté Le Sole Sine Veste d'Orschall; l'Helioscopium videndi sine veste solem Chymicum; le Sol Non Sine Veste; le chapitre XI du Flora Saturnizans de Henckel, Sur la Vitrification des*

Végétaux; Un Mémoire sur la manière de faire le Saffre; Le Secret des vraies Porcelaines de la Chine & de Saxe . . . Traduits de l'Allemand par M. D*** (Paris, 1752). Listed by Duveen, *Bibliotheca,* 427.

This is a collection of pieces dealing with the manufacture and coloring of glass, enameling, the problem of porcelain. The main item is a translation of the *De Arte Vitraria Libri VII* of the Florentine priest, Father Antonio Neri; published in 1612, this is one of the earliest important treatises on glassmaking. An English translation was made by Christopher Merret, which was also brought out in a Latin version in Amsterdam in 1668 by Andreas Frisius. A German version, also with Merret's notes, was published in 1678. In the following year, Johann Kunckel (1630–1703) brought out his *Ars Vitraria Experimentalis, Oder Volkommene Glassmacher-Kunst.* This enlarged German version of Neri with Merret's notes and additions by Kunckel is the work that d'Holbach translated.

92. *Minéralogie, ou description générale des substances du règne minéral, par Mr. Jean Gotshalk Wallerius. Ouvrage traduit de l'Allemand,* 2 vols. (Paris, 1753). J. G. Wallerius (1709–85), the predecessor of Bergman in the chair of chemistry at Upsala, gave in this work, published in Swedish in 1747 (German trans., 1750), the first rational classification of minerals. Contemporaries compared him with Linnaeus, and this book, which included also his *Hydrologie,* a treatise on mineral waters, became the standard introduction to mineralogy in France. He was criticized for his "profondeur obscure" and for having "trop multiplié les espèces et les descriptions." See *Introduction aux observations sur la physique,* vol. 1 (ed. 1777, for December 1771), 400.

93. *Chimie métallurgique, dans laquelle on trouvera la Théorie et la pratique de cet Art . . . par M. C. E. Gellert, conseiller des mines de Saxe, et de l'Académie impériale de Petersbourg,* 2 vols. (Paris, 1758). The first German edition, published in 1751–55 in two volumes, is cited by Henry Carrington Bolton, *A Select Bibliography of Chemistry, 1482–1892* (Washington, D.C., 1893), 473.

94. *Traités de physique, d'histoire naturelle, de minéralogie et de métallurgie, par M. Jean-Gotlob Lehmann* (Paris, 1759). On Lehmann, see Geikie, *Founders of Geology,* 195–97.

95. *Oeuvres métallurgiques de M. Jean-Christian Orschall* (Paris, 1760). Orschall, about whom little seems to be known, flourished in the 1680s. A mining official of Hesse, with alchemical leanings, he was better known for his *Sol sine Veste,* in which he discusses the preparation of ruby glass, than for the works printed here. See Ferguson, *Bibliotheca Chemica,* II, 156–57; and Herman Kopp, *Geschichte der Chemie,* 4 vols. (Braunschweig, 1843–47), III, 254.

96. *Pyritologie, ou Histoire naturelle de la pyrite* (Paris, 1760); *Introduction à la minéralogie* (Paris, 1756). This is a translation of a posthumous work, *Henckelius in Mineralogia Redivivus* (Dresden, 1747). Johann Friedrich Henckel (1679–1744) was trained as a physician, practiced in Freiberg in the Erzegebirge, and became a mining councilor there. The German version of the *Pyritologie* was published in Leipzig in 1725. Ferguson, *Bibliotheca Chemica,* I, 385–86.

The *Pyritologie* covers more ground than the title indicates. As d'Holbach writes, the author passes in review all the substances of the mineral kingdom, and "par la liaison de ces substances il a fait le Traité le plus complet & le plus profond que nous ayons sur toutes les branches de la Minéralogie & de la Métallurgie." In d'Holbach's translation, the *Pyritologie* itself occupies the first of two volumes bound together (pp. 1–403). The second volume is composed of several shorter works, of which the most interesting is the "Idée générale de l'origine des pierres," II, 393–455. Here Henckel describes Boyle's experiments on strongly heating precious stones and his own failure to detect any volatility. But in a long and interesting note (II, 413) d'Holbach first called the attention of French scientists to the experiments of Emperor Francis I (1708–1765) and of Cosimo III, Grand Duke of Tuscany (1642–1723), on the destruction of the diamond by fire. This note of d'Holbach's provided Lavoisier with the background material for his first memoir on the destruction of the diamond by heat. *Oeuvres de Lavoisier,* II, 41–44.

97. *Recueil des mémoires les plus intéressants de chymie et d'histoire naturelle, contenus*

dans les Actes de l'Académie d'Upsal et dans les Mémoires de l'Académie royale des sciences de Stockholm (1720–1760), traduit du Latin et de l'Allemand, 2 vols. (Paris, 1764).

98. In his preface to Wallerius' *Minéralogie,* d'Holbach writes (I, p. viii) : "Je ne puis trop me hâter de reconnoître les secours que MM. Bernard de Jussieu & Rouelle ont bien voulu prêter à ma traduction. Ces deux illustres Académiciens ont permis que je leur fisse lecture de mon manuscript & m'ont communiqué un très-grand nombre d'observations utiles & judicieuses."

99. Naville, *d'Holbach,* 185. On Roux's role in helping to see these works through the press, see the "Eloge de Roux" attributed to Niageon, Diderot's literary executor, which Naville summarizes (p. 431, n. 16). In substance this closely resembles the "Eloge de M. Roux" that appeared in the *Journal de Médecine* for January 1777. Here we read :

> M. Roux a dirigé les éditions Françoises que nous avons des *Oeuvres de Henckel,* des *Traités du soufre, & des sels* du célèbre Staahl, & d'une collection en deux volumes des meilleurs *Mémoires de Chymie de l'Académie d'Upsal.* La traduction de ces ouvrages est dûe au zèle infatigable de M. le Baron d'Olback, & c'est de lui que nous tenons, sans contredit, une grande partie des connoissances de chymie et de physique, & d'histoire naturelle qui rendent aujourd'hui la France au moins la rivale d l'Allemagne, puisque cet excellent citoyen nous a fait connoître tout ce qu'elle a produit de plus parfait dans ces trois genres. ["Eloge de M. Roux," extrait du *Journal de Médecine,* (Janvier 1777), 15.]

100. We may consider Hellot as having launched this movement of translation with his version of Christoph Andreas Schlüter's *Gründlicher Unterricht von Hütte-Werken* (Braunschweig, 1738) the most important book on mining of the early eighteenth century. Hellot's translation appeared in two parts: *Traité des essais des mines & métaux* (Paris, 1750); and *De la fonte des mines, des fonderies, grillages* (Paris, 1753). Duveen, *Bibliotheca,* 534–35, describes the German first edition but does not refer to the French translation. Schlüter is not mentioned by Bolton, Ferguson, or Hoefer. Kopp, *Geschichte der Chemie,* I, 218, mentions Hellot's translation, and it is cited by Fourcroy in the "Essai de bibliothèque chimique" appended to the article "Chimie" in the *Dictionnaire de chimie,* 6 vols. (Paris, An IV), III, 761, of the *Encyclopédie méthodique.*

101. The work of Friedrich Hoffmann (1660–1742) on mineral waters was available in several Latin editions. Bolton, *Select Bibliography,* 537, records a French translation, which I have not seen, of his *Observationum physico-chimicarum selectiorum libri III* (Halle, 1722). This is entitled: *Observations physiques et chymiques, dans lesquelles on trouve beaucoup d'expériences curieuses . . . traduites du Latin de F. H.* (Paris, 1754). It is unknown to Hoefer, Ferguson, and Duveen. [But see J. R. Partington, *History of Chemistry,* 4 vols. (London: Macmillan & Co., 1961–70), II, 692.]

A. A. Parmentier later wrote, "Eloge de Pierre Bayen," *Opuscules chimiques de Pierre Bayen,* ed. P. Malatret, 2 vols. (Paris, 1798), I, pp. xli–xlii: "Avant les travaux d'Hoffmann, l'histoire des eaux minérales n'étoit qu'un tissu de mensonges et d'erreurs; on s'en rapportoit aux impressions qu'elles produisoit sur les organes, pour prononcer sur leur nature."

102. "Avant Pott, les chimistes ne s'occupoient que fort peu, ou plutôt ne s'occupoient point de l'examen des pierres; mais la Lithogéognosie de ce savant et laborieux auteur, ayant paru parmi nous en 1753, y produisit une révolution, dont la partie de la physique qui s'occupe de l'histoire naturelle, devoit retirer les plus grands avantages." Pierre Bayen, "Examen de différentes pierres." *Opuscules,* II, 41.

The German edition appeared in 1746 (Ferguson, II, 221). It was translated by d'Arclais de Montamy as *Lithogéognosie ou Examen Chymique des Pierres et des Terres* (Paris, 1753). Montamy, a friend of d'Holbach, was a pioneer in the efforts to produce porcelain; we have his posthumous *Traité des couleurs pour la peinture en émail & sur la porcelaine* (Paris, 1765).

Pott's collected papers were published by J. F. Demachy in French translation: *Dissertations Chymiques de M. Pott*, 4 vols. (Paris, 1759). Duveen, *Bibliotheca*, 483.

103. *Elémens de Docimastique ou de l'art des essais divisés en deux parties, la première théorique et la seconde pratique, traduit du Latin* (Paris, 1755). Duveen, *Bibliotheca*, 147. The translator is unknown.

104. There is a good account of Marggraf's life and work in John Maxson Stillman, *The Story of Early Chemistry* (New York, 1924), 435–42. But compare also Kopp, *Geschichte*, I, 208–11; and Hoefer, *Histoire de la chimie*, II, 407–21. In his eulogy of Marggraf, Condorcet emphasizes that Marggraf was the first chemist to be honored by appointment as foreign member of the Academy of Sciences. *Oeuvres de Condorcet*, II, 598–610.

105. Marggraf, more even than Hoffmann, deserves the title of the founder of the analytical school of continental chemistry. His methods were brilliantly applied by his pupil Martin Klaproth (1743–1817), a contemporary of Lavoisier, who in 1809 became the first *ordinarius* in chemistry at the newly founded University of Berlin. For Klaproth's work see, Kopp, *Geschichte*, I, 343–49, and *passim*.

106. Hoefer, *Histoire de la chimie*, II, 407, n. 1. This translation is not mentioned by Ferguson, Bolton, or Duveen.

107. It appeared anonymously, but has been generally attributed to the physician Jean-Baptiste Senac (1693–1770). Fourcroy is probably responsible for putting abroad the view that the phlogiston theory was popularized as early as 1723 by Senac's book. It heads the list he gives of the most important chemical writings of the century. *Système des connaissances chimiques*, 6 vols. (Paris, 1801–1802), I, 22–23. And he says, with evident exaggeration, that it produced "la même révolution dans notre *chimie*, que les reflexions sur l'attraction que publia M. Maupertuis dans son discours sur les différentes figures des astres, ont operée dans notre physique." Article "Chimie '"in the *Dictionnaire de chimie*, 6 vols. (Paris, An IV, 1796), III, 302, of the *Encyclopédie méthodique*.

108. "Parmi ces hommes éclairés, sectateurs et promoteurs de l'école de Stahl, on doit ranger spécialement en France les Grosse, les Baron, les Macquer, les deux Rouelle; en Allemagne et en Suède les Pott, les Cronstedt, les Wallerius, les Lehman, les Gellert, les Marggraf, les Neumann; en Angeterre les Freind, les Shaw, les Lewis; en Hollande les Gaubius, etc. etc." *Système des connaissances chimiques*, I, 21. Grosse has been described as a follower of Geoffroy, who "a donné quelqu'extension à la table de M. Geoffroy, sans mieux expliquer les mots rapports & affinités." See *Observations sur la physique*, vol. 1 (ed. 1784, for 1773), 197. He was made *adjoint chimiste* of the Academy of Sciences in 1731 and died in 1744. Voltaire sought information from him during his period of scientific study and referred to him as "ce gnôme de Grosse." For his few published memoirs, see Kopp, *Geschichte*, II, 297; III, 102; IV, 304, 348, 351. The most important were those on ether and tartaric acid.

109. Condorcet was of the opinion that in 1740 the doctrines of Stahl were not yet generally known in France. See *Oeuvres Complètes de Voltaire*, ed. Beaumarchais, Condorcet, et Decroix, 70 vols. (1785–89), XXXI, 304, n. 9.

110. "On a jugé assez unanimement que Boerhaave n'étoit pas Chymiste, qu'il n'avoit allumé d'autre feu que celui de sa lampe et qu'il n'avoit fait aucune opération propre à éclairer la théorie." *Cours de Chymie de Mr Rouelle, rédigé par Mr Diderot*, Bibliothèque de Bordeaux, MSS. 564–65, I, 34. This introduction is generally attributed to Diderot. On this "rivalité posthume" between Stahl and Boerhaave, see Hélène Metzger, *Newton, Stahl, Boerhaave et la doctrine chimique* (Paris: F. Alcan, 1930), 192.

111. *Correspondance littéraire, philosophique et critique par Grimm, Diderot, Raynal, Meister, etc.*, ed. Maurice Tourneux, 16 vols. (Paris, 1877–1882), IX, 106; *Oeuvres complètes de Diderot*, ed. J. Assézat and M. Tourneux, 20 vols. (Paris, 1875–77), VI, 404–10. In his introduction to the lectures of the elder Rouelle, Diderot writes: "Le Staalianisme [sic] a été connu en France par les leçons de M. Rouelle et par les ouvrages de M. Macquer." *Cours de Chymie de Mr Rouelle*, I, 34.

112. Maurice Daumas, "La chimie dans *l'Encyclopédie* et dans *l'Encyclopédie méthodique*,"

in *"L'Encyclopédie"* et le progrès des sciences et des techniques (Paris: Presses Universitaires de France, 1952), 135.

113. *Cours de Chymie contenant la manière de faire les opérations qui sont en usage dans la médecine, par M. Lemery, de l'Académie Royale des Sciences, Docteur en Médecine. Nouvelle édition, revue, corrigée & augmentée d'un grand nombre de Notes, & de plusieurs préparations Chymiques qui sont aujourd'hui d'usage & dont il n'est fait aucune mention dans les Editions de l'Auteur, par M. Baron, Docteur en Médecine, & de l'Académie Royale des Sciences* (Paris, 1756).

114. *Traité du soufre, ou remarques sur la dispute qui s'est élevée entre les Chymistes, au sujet du Soufre, tant commun, combustible ou volatil, que fixe, &c., traduit de l'Allemand de Stahl* (Paris, 1766).

115. *Traité du soufre*, 108–10. This free rendering of Stahl's words is my own. D'Holbach's translation of the treatise on sulphur is not mentioned by Bolton, Ferguson, or Duveen. The Duveen collection has the translation by d'Holbach of Stahl's treatise on salts: *Traité des Sels, dans lequel on démontre qu'ils sont composés d'une terre subtile, intimement combinée avec l'eau, traduit de l'Allemand* (Paris, 1771). Duveen makes no mention of d'Holbach as the translator, nor does Ferguson who also cites it (*Bibliotheca Chemica*, II, 397). Inexplicably, it is also missing from Naville's list of d'Holbach's writings and translations. Yet it is certainly the work of d'Holbach, who ends the unpaginated "Avertissement du traducteur" of the *Traité du soufre* with the remark: "Le Traité des Sels, dont nous venons d'achever la traduction, suivra de près celui-ci; les connoisseurs ne le jugeront ni moins curieux ni moins intéressant que celui que nous publions maintenant."

116. O. Starosel'skaia-Nikitina, *Ocherki po istorii nauki i tekhniki perioda frantzuskoĭ burzhauznoĭ revolursii, 1789–1794* (Moscow-Leningrad, 1946). I am indebted to Dr. David Joravsky for having translated at my request the crucial sections of this very provocative book. On a number of points the author has anticipated the present study.

117. Starosel'skaia-Nikitina, 59.

118. H. Guerlac, "A Note on Lavoisier's Scientific Education," 216.

119. Georges Bouchard, *Guyton-Morveau, chimiste et conventionnel (1737–1816)* (Paris: Perrin, 1938), 58–63, 88–89, 105–11.

120. J. R. Partington and Douglas McKie, "Historical Studies on the Phlogiston Theory. I. The Levity of Phlogiston," *Annals of Science* 2 (1937), 373–89.

121. On this see D. I. Duveen and H. S. Klickstein, "A letter from Guyton de Morveau to Macquart [*sic*] relating to Lavoisier's attack against the *phlogiston* theory (1778); with an account of de Morveau's conversion to Lavoisier's doctrines in 1787," *Osiris* 12 (1956), 342–67. It was pointed out to Messrs. Duveen and Klickstein, independently by Maurice Daumas and myself, that they had erred in identifying the addressee of the letter from Guyton de Morveau. A note of rectification on this minor point has been published in *Isis* 49 (1958), 73–74.

122. Lavoisier, "Rapport sur un mémoire de Bucquet sur l'air fixe," *Oeuvres de Lavoisier*, IV, 155–58.

123. A. A. Parmentier, "Eloge de Pierre Bayen," *Opuscules chimiques de Pierre Bayen*, I, pp. xlix–li.

124. Thomas Thomson long ago expressed the opinion that Bayen harbored doubts about the phlogiston theory before Lavoisier. See his *System of Chemistry*, vol. 1, bk. 1, p. 83. Parmentier (*Eloge*, I. p. li, n. 1) was of the same opinion. It is clear, at all events, that Bayen *published* his doubts before Lavoisier, viz., in 1774.

24
The Origin of Lavoisier's Work on Combustion

☼

This paper summarizes some recent findings concerning an important aspect of Lavoisier's scientific career: the events that led up to his first experiments on combustion. These experiments, as everyone knows, were performed in the autumn of 1772, not long after Lavoisier's twenty-ninth birthday.[1] During September and October of that year he made the important discovery that when phosphorus and sulphur are burned they gain markedly in weight through the addition of a "prodigious" amount of air. Before the end of October he had performed a different but closely related experiment, and showed that when the calx of lead is reduced with charcoal, an abundance of air is given off. Convinced, as he put it, that these results were among the most interesting obtained since the time of Stahl, and anxious to secure priority for them, he summarized them in the famous sealed note which he deposited with the Secretary of the Academy until he could make his work public. So startling, indeed, did the implications of these experiments seem to him that a few months later, on February 20, 1773, he outlined in a personal memorandum a program of study and research which he felt sure would produce—and these are his own words—a revolution in chemistry and physics.[2]

Mystery had enveloped the origin of Lavoisier's historic experiments of the autumn of 1772. The train of thought that led him to perform them has remained wholly obscure. He seems to have embarked on this epoch-making work without evident preparation, yet from the start with the intention of testing the hypothesis that air combines with substances when they are burned. But before 1772 we know that he was preoccupied with entirely different kinds of scientific problems, and there is no evidence to suggest that until that year he had the slightest curiosity about combustion, or the chemistry of phosphorus, or any interest in the possible chemical role of air.[3] What, then, could have focused his attention so abruptly upon the problem of combustion? What led him to suspect the chemical role that air might play in this process? These are the two chief questions we must try to answer.

The attempts of previous scholars to answer these questions have been

From *Archives Internationales d'Histoire des Sciences*, no. 47 (April–June 1959), 113–35. Copyright by the Conseil de Direction des *Archives*.

highly conjectural and unconvincing; and understandably so, since documents from Lavoisier's hand relating to these matters seemed to be almost totally lacking for the period before September 1772. These conjectural reconstructions have started from a single assumption: namely that it was ordinary combustion, and more particularly the combustion of phosphorus, which must have aroused Lavoisier's curiosity. This was, of course, reasonable, and indeed apparently obvious, since it was the burning of phosphorus which Lavoisier first investigated in September and October 1772.

It was suggested by A. N. Meldrum—one of the most distinguished and prudent of Lavoisier scholars—that Lavoisier's interest in combustion was probably excited by the experiments on the thermal destruction of the diamond, carried out in concert with Macquer and Cadet in the spring of 1772.[4] These experiments proved beyond a doubt that while diamonds could be readily destroyed by exposure to intense heat in the open air, they resisted destruction if protected from the air, and were totally unaffected by high temperatures if protected by "intermediaries" like powdered charcoal. In reporting on these experiments, Lavoisier wrote that the diamond did not vanish because it was inherently volatile, but disappeared either through decrepitation into minute particles or because it was destroyed by combustion. It can be shown, however, that Lavoisier was only reporting the theories favored by his co-workers, and that he himself did not favor the idea of a combustion.[5] This explanation of Lavoisier's interest in combustion has nevertheless been frequently advanced, among others by the present writer.

Meldrum would also have it that, having had his attention turned to combustion by the diamond experiments, Lavoisier was specifically influenced by a paper of the Italian scientist G. F. Cigna to investigate the combustion of phosphorus, and in particular to see if it absorbed air on burning.[6] Cigna's paper appeared in the May 1772 issue of the Abbé Rozier's *Observations sur la physique*, and dealt with the extinction of candles and the suffocation of animals when confined in limited volumes of air. According to Meldrum, Cigna's paper described the absorption of air by burning phosphorus or sulphur. If this were true, and if Lavoisier had read and pondered this paper, this might explain how he was led in September 1772 to plan his first experiments with phosphorus; for the note of September 10, which Meldrum was the first to discover and publish, shows that it was his principal aim to see whether phosphorus did in fact absorb air when it burned. Unfortunately for Meldrum's theory—the most likely that has been advanced—there is reason to believe that Lavoisier did not read Cigna's paper, or at least that if he did, it made no particular impression upon him. More important still, a careful reading of the paper shows that Cigna did *not* believe that air is absorbed by burning phosphorus or sulphur. He took pains to emphasize that when these substances are burned, the volume of air decreases not because air is actually taken up, but because the elasticity of the air decreases. According to this rather odd theory, which Stephen Hales had earlier advanced, the vapors given off by burning phosphorus or sulphur corrupt the air and diminish the repulsive forces between its particles, thereby reducing the volume and giving the impression that air is actually absorbed by the vapors.[7]

Be that as it may, these theories are of diminished significance, for I believe it can be shown that Lavoisier came by a quite different path, and one that it is possible to trace with some precision, to his study of combustion and to his *idée maîtresse* about the role of air. It was, I shall try to show, the calcination and reduction of metals, not combustion in the broader sense, which first enlisted Lavoisier's attention. It was in connection with these processes, and with other reactions involving metals, that it first occurred to him that air might be an important chemical participant in certain significant reactions, and not—as was generally believed—a mere physical auxiliary supporting combustion through its weight and elasticity.

What I am suggesting is that the order in which Lavoisier performed his experiments in the autumn of 1772—first the work on phosphorus and sulphur, then the experiments on the reduction of lead oxide—does not reflect his earlier train of thought, or the sequence of his speculations in the summer of that year. At first glance this proposition seems wholly inconsistent with what Lavoisier himself tells us in the sealed note of November 1. Here he clearly says that the experiment on the reduction of lead oxide not only followed, but was prompted by, his successful experiments on phosphorus and sulphur, carried out just before. The discovery, he writes, that air combines during combustion

> m'a fait penser que ce qui s'observait dans la combustion du soufre et du phosphore pouvait bien avoir lieu à l'égard de tous les corps qui acquièrent du poids par la combustion et la calcination; et je me suis persuadé que l'augmentation de poids des chaux métalliques tenait à la même cause. L'expérience a complètement confirmé mes conjectures; j'ai fait la réduction de la litharge dans des vaisseaux fermés, avec l'appareil de Hales, et j'ai observé qu'il se dégageait, au moment du passage de la chaux en métal, une quantité considérable d'air, et que cet air formait un volume mille fois plus grand que la quantité de litharge employée.[8]

This is a very matter-of-fact statement, and we shall need some pretty convincing evidence if we are to disregard it. Yet I believe it is not literally true. While it correctly describes the sequence of events in the months of September and October 1772, there is persuasive evidence from Lavoisier himself that he had not only been speculating about the role of air in metallic and mineral reactions *before* he undertook the investigation of phosphorus, but that as early as August 8 he envisaged experiments on metals of the very sort he eventually performed on oxide of lead.

Surprisingly enough, there has long been available in print a statement which Lavoisier set down later in life and in which he tells us exactly how he first became interested in the possibility that air might be chemically significant. He was struck, he says, by a phenomenon familiar to all writers on metallurgy and assaying: namely, that a marked effervescence occurs when a metallic calx is reduced and transformed into the metal. From this it was natural to conclude that air is given off to produce the effervescence, and he hit upon an apparatus (his modification of Hales's apparatus) to collect and measure it.[9]

To be sure, this passage might be taken to refer only to his realization that air might be combined with *metals* as well as other substances, and read as an amplification of his train of thought *after* he had already discovered that air combined with phosphorus and sulphur. There is, however, contemporary evidence from Lavoisier's own pen which proves beyond question that his first speculations about the role of air dealt with metals, and not with phosphorus, the diamond, or any other combustible substance. On August 8, 1772—more than a month before he began work on phosphorus—Lavoisier set down a series of proposals for the benefit of the co-workers with whom he was about to begin experiments with the Academy's great burning glass to determine the effect of very high temperatures on a variety of mineral substances. This memorandum, entitled "Réflexions sur les expériences qu'on peut tenter à l'aide du miroir ardent," I have discussed in some detail in a recent article.[10] The concluding section of this memorandum is of particular interest. Under the heading "Sur l'air fixe, ou plutôt sur l'air contenu dans les corps," Lavoisier remarks that air appears to enter into the composition of mineral substances, and even metals, and in great abundance. This is the earliest dated reference we have to Lavoisier's interest in air. That he had begun to speculate seriously about the role of air is made clear by his remark that "ces vues suivies et approfondies pourraient conduire à une théorie intéressante qu'on a même déjà ébauchée." In the manuscript of the "Réflexions" Lavoisier had crossed out an even stronger statement in which he spoke of having a memoir on air "déjà fort avancé, même en partie rédigé."[11] But what is especially significant in this August memorandum is that Lavoisier makes no mention of phosphorus or sulphur; instead he refers to the phenomenon he was later to recall as having first aroused his curiosity: the effervescences observed in the case of metals and mineral substances. These, he remarks, can only be the result of a sudden release of this combined air. And he concludes:

> Il serait bien à désirer qu'on pût appliquer au verre ardent l'appareil de M. Hales pour mesurer la quantité d'air produite ou absorbée dans chaque opération, mais on craint que les difficultés que présente ce genre d'expériences ne soient insurmontables au verre ardent.[12]

This is precisely the experimental arrangement he was to use three months later in his famous experiment on lead oxide. But this was only after the cooperative experiments with the Academy's burning glass had been brought to a close for the year; his colleagues ignored the suggestion, perhaps because they, too, felt that the difficulties might really prove insurmountable.

How was Lavoisier brought to speculate about effervescences, especially those observed when metallic oxides are reduced, and how was he led to interpret them as the release of combined air? He was certainly familiar with the more common type of effervescences observed in the laboratory, and he may also have observed the reduction effect with his own eyes. But I think it is more likely that he encountered the latter in the course of his reading, for he speaks of it as "observé par tous ceux qui ont travaillé aux opérations de docimasie";[13] and there is no evidence that, despite his early preoccupation with geology and mineralogy, he had any firsthand experience with the techniques of assaying. Yet his library was probably already rich in metallurgical and mineralogical books, for we know that at a later date he owned in the original or in translation most of the important writings available on these subjects. Despite Lavoisier's implication to the contrary, not many of these works mention this phenomenon of effervescence during metallic reduction. One important author, however, gives an extended and vivid account of the effect as observed during the reduction of lead oxide. This is J. A. Cramer, whose *Elementa artis docimasticæ*, a standard work on assaying, was translated into French in 1755 and was widely read and quoted. Lavoisier owned the second (Latin) edition of Cramer's book.[14] Here he could find a vivid description of the foaming, the bubbling—even the whistling sound—produced when lead oxide is reduced.[15] Moreover he must have known that Cramer's words were repeated almost verbatim by Macquer in the latter's *Elémens de chymie-pratique*, a work which, needless to say, Lavoisier also owned.[16]

Cramer, and Macquer after him, were content to describe this phenomenon of metallic effervescence, so particularly striking when a calx of lead is reduced; they do not suggest that this effect might be explained by the release of air combined with the metal. Their purpose was the practical one of showing why the fire must be kept moderate during the early stages of the reduction to keep this effervescence under control and avoid mechanical loss of the material. Yet many French chemists believed, on the authority of Stephen Hales, that effervescences were due in most, if not all, cases to the release of a "true air." Indeed, as I have argued elsewhere, it was from the *Vegetable Staticks* of Hales—a book well known in France—and not from any knowledge of the later work of Joseph Black, Henry Cavendish, or Joseph Priestley on gases, that Lavoisier first came to suspect the chemical importance of air.[17] It is certainly from Hales, directly or indirectly, that Lavoisier was led to explain these effervescences by the release of air.[18] Yet it is curious that nobody in France until Lavoisier—not even Macquer, who accepted Hales's conclusions about air and wrote in 1766 that effervescences "sont dûs à l'air qui se dégage, ou qui se développe dans presque toutes les dissolutions"—thought to apply Hales's theory in interpreting the phenomenon described by Cramer.

But perhaps it is not so odd, after all. There was another closely related phenomenon encountered in metals which received special attention in 1771–72 in Lavoisier's scientific circle, and which gave new significance to the observation concerning effervescences. This was the important fact that metals always gain in weight when transformed into calxes by being roasted and calcined. We have long known that this augmentation of the calx played a crucial part in Lavoisier's reasoning. He mentions it in the sealed note; and later in the

Opuscules physiques et chimiques, his first book, he wrote that when he began to suspect that air might combine with metals, the addition of air offered an explanation for what took place during calcination, in particular the augmentation effect.[19] The phlogistic hypothesis supplied no satisfactory explanation of this gain in weight, since according to this theory the phlogiston was supposed to *leave* the metal during calcination. Lavoisier could not understand how the weight could be increased unless something were added. This "something" could well be the air which came off in such quantities during the process of reducing the calx.

Now we have always known that the augmentation effect was a powerful argument for Lavoisier in his rejection of the phlogistic hypothesis. But what I am suggesting here is that knowledge of this effect played an important part, as early as the summer of 1772, in persuading Lavoisier of the role of air in calcinations and combustions. It fitted, like a key into a lock, with his ideas about effervescences; they were two aspects of the same problem. But why was Lavoisier the first to be so troubled by the inherent contradiction between the phlogiston theory and the augmentation effect in calcined metals? Why did this not worry other chemists before him?

The answer is quite simple. It has always been taken for granted that by Lavoisier's time the effect was well known and generally accepted. But this is not the case. All through the eighteenth century, serious doubts were expressed about the generality and the importance of the phenomenon. Instances were reported in which calcined metals showed no change in weight or actually seemed to have lost it. For example, Hermann Boerhaave, in his popular book, expressed doubts that the phenomenon was general or really significant; and he suggested it might be an accidental effect due to impurities added to the calx from the fuel or the vessel. His doubts were shared by Macquer, by the latter's disciple, Baumé, and by Théodore Baron, who remarked in a note to his edition of Lemery's *Chemistry* (1756) that the evidence for the augmentation effect was contradictory and that the wisest course was to suspend judgment until new experiments could settle the question.[20] The question was carefully reviewed by R. A. Vogel in 1753, who concluded (as Jacob Spielmann did ten years later) that the augmentation effect was a peculiar property of lead and had not been reliably reported for other metals.[21]

The reality and generality of the phenomenon—and hence its significance—were first clearly demonstrated, a year or so before Lavoisier became interested in the question, by the Dijon chemist, Louis-Bertrand Guyton de Morveau.[22] It seems to me extremely important that Guyton's work brought the augmentation effect to general attention, and proved it to everyone's satisfaction, at precisely the time Lavoisier was wondering about the cause of metallic effervescences. I think I can justify my conviction that Guyton de Morveau was a most important influence upon Lavoisier at this crucial moment in the history of chemistry.

Guyton's work was carried out in 1769 and 1770. After presenting his results before the Dijon Academy late in 1770, he submitted them to the Academy of Sciences in Paris, in the summer of 1771, in the form of a memoir entitled *Dissertation sur le Phlogistique*.[23] Lavoisier was present at a meeting of the Academy (held on February 8, 1772) at which two senior chemists, Malouin and Macquer, read a laudatory report on Guyton's memoir. They remarked that it dealt with one of the most interesting and disputed points of chemistry; and they praised Guyton's experiments as having at last settled the important question of the augmentation effect in calcined metals, a matter "jusqu'à présent trop peu décidé et trop particulier." They were less enthusiastic about the theoretical part of the memoir, where Guyton tried to reconcile the weight increase with that loss of the fire principle which the phlogiston theory required.[24]

Guyton's memoir was printed as the first and longest essay of his book entitled *Digressions Académiques*, which was published late in May 1772.[25] It was favorably reviewed; and it was praised by men of the scientific acumen of Macquer and Condorcet chiefly for the *Dissertation sur le phlogistique*, and above all for the excellence of the experimental work described in that essay.[26] A careful reading of this unduly neglected work bears out the favorable judgment of Guyton's contemporaries.

Guyton's essay begins with a detailed historical review of the augmentation problem, in which the testimony for and against the weight effect is carefully marshaled. After a short methodological section, setting forth the principles he felt it necessary to follow in settling the question, Guyton devotes fifteen pages to the careful experiments he had performed. He had heated weighed samples of each metal in open crucibles over a fire or under the muffle of a cupellation furnace. A number of precautions were carefully observed such as using pure samples, preheating the crucibles to drive out moisture, and stirring the melt during the operation. He attempted to avoid loss by sublimation or spattering, and sought in every case to bring the reaction to completion. He obtained characteristic, and for the time reasonably accurate, values for the gain in weight of copper, iron filings, tin, antimony, bismuth, and zinc. He ignored the well-established case of lead. In one interesting experiment he observed the impossibility of calcining a metal in the absence of air.[27]

Despite some manifest limitations, it is an admirable piece of work, and it is not surprising that his contemporaries felt that finally someone had settled the long-disputed fact of the weight change in calcined metals.[28] The timing of the appearance of Guyton's work, and the publication of the book in the spring of the year that was to be so historic in Lavoisier's career, would suggest that, apart from focussing discussion on the significant question of calcination, the *Digressions Académiques* must have exerted a profound effect on Lavoisier and contributed to his interest in, and solution of, the calcination problem.

Is there any testimony from Lavoisier himself in support of this idea? At first glance the answer would seem to be in the negative. I have found no place in Lavoisier's printed works (and in at least two places the occasion clearly offered itself) where he acknowledges a debt to Guyton for having finally established the reality and generality of the weight effect. In one place only does he refer to Guyton's calcination experiments: in a memoir of 1783 he actually makes use of Guyton's values for the weight increase of different metals, speaks favorably of Guyton's experiments, and praises him for obtaining good results with zinc, a peculiarly difficult metal to calcine quantitatively.[29]

In private, however, Lavoisier was more generous. Early in 1774 he sent Guyton a complimentary copy of his recently published *Opuscules*, and in the accompanying letter remarks that though he cannot accept Guyton's theory of the role of phlogiston, the author of the *Digressions Académiques* deserves praise for his "génie d'observation" and for having carried out the most complete, most interesting, and most exact experiments hitherto made on the calcination of metals.[30]

More significant and revealing, however, is an undated personal memorandum, written on a single sheet of paper which I was fortunate enough to find in the archives of the Academy of Sciences in Paris. Written in Lavoisier's hand, it is entitled *Sur la matière du feu* and is only four short paragraphs in length. The opening sentence comes straight to the point: "Tous les métaux exposés au feu augmentent de poids très sensiblement." The early authors, the note continues, believed that this was due to the addition of the ponderable matter of fire, while Stahl imagined that calcination removes the phlogiston from the substance being calcined. But Stahl and his sectaries, Lavoisier goes on, have fallen into a labyrinth of difficulties when attempting to explain how one can increase the weight of a body by removing part of its substance. And he concludes: "Quoi qu'il en soit de l'explication, le fait n'en est pas moins constant. Tous les métaux augmentent de poids par la calcination. M. de Morvaux [*sic*] la démonstre complettement dans ses *Digressions Académiques*, page 72 jusqu'à 88."[31]

Even if this were set down at some later date, it would be valuable evidence. But it probably was written in the summer or early fall of 1772, perhaps at about the time Lavoisier wrote his August memorandum. It was obviously written *after* Guyton's book had become available in June, yet *before* Lavoisier convinced himself by the important experiments of October that the addition of air caused the increase in weight of combustible and calcinable substances. At all events, it is surely Guyton who is particularly envisaged when Lavoisier speaks of those sectaries of Stahl who have ended in labyrinthine difficulties when trying to reconcile the gain in weight in metals with the loss of phlogiston.

✸

My chief argument up to this point has been that Lavoisier's historic experiments of the autumn of 1772, and the key idea which he hoped to test by performing them—the notion that air combines with metals during calcination—resulted from the convergence upon him of two important influences: the early work of Stephen Hales and the recent book of Guyton de Morveau. That such was the case is strongly, if indirectly, supported by the surprising fact that another

Frenchman, the eminent economist and public servant Anne-Robert Jacques Turgot, was led by his familiarity with Hales's doctrine of fixed air to interpret Guyton's experiments by a theory not essentially different from that of Lavoisier.[32] Turgot, at that time *Intendant* at Limoges, received from his friend, the young Marquis de Condorcet, a letter from Paris dated August 2, 1771. In it, Condorcet remarked that the Academy of Sciences had just received an excellent chemical memoir dealing with the increase in weight of metals on calcination. The author of this memoir, a councilor at the Dijon *parlement*, had performed a number of new experiments and sought to explain the weight increase as due to the loss of phlogiston, a substance which, animated by forces opposite to that of gravity, tends to decrease the weight of any substance in which it is found.[33] Turgot's reply, dated August 16, 1771, is long and detailed.[34] He attacked Guyton's explanation, and proposed instead that air must combine with the metal, taking the place of the phlogiston that is given off, and accounting for the greater weight of the calx. By many experiments, "et en particulier par celles de Stales [*sic*]," it is proved that air enters into the composition of the hardest bodies, contributing to their texture and their hardness.[35] Whether air combines or is given off in different chemical reactions depends upon whether its affinity for the substance with which it is combined is greater or less than for other materials to which it is exposed. Effervescences so frequently observed in chemical reactions are produced by bubbles of air released in this manner from combination. Turgot, it is clear, is definitely thinking of a chemical combination with air, for he points out that the characteristic increase of weight of metals must mean that for each substance this absorption of air has a definite point of saturation "comme toutes les unions chimiques." Then come the most remarkable passages of Turgot's letter:

> Au surplus, cette calcination des métaux devrait être appelée combustion; ce n'est qu'une branche du grand phénomène de la combustion par lequel le phlogistique uni aux principes terreux s'en dégage à un degré de chaleur constant dans chaque corps. . . . Il suit de là que le phénomène de l'augmentation du poids devrait être générale dans la combustion de tous les corps; je voudrais constater cette conséquence par des expériences. . . . Mais le temps me manque et j'avoue que sans nouvelles expériences les inductions tirées de celles qui sont déjà faites me paraissent donner à cette théorie une probabilité fort approchante de la certitude.[36]

Condorcet replied by a letter of September 10, 1771, expressing great interest in Turgot's theory, and remarking that "il faudrait qu'un chimiste suivît votre explication et imaginât des expériences décisives pour ou contre. C'est une des questions les plus importantes qu'on puisse agiter dans cette science."[37]

Needless to say, Turgot never found time to carry out these experiments, and the theory he set forth in the letter to Condorcet was never published in the lifetime of either man. We have no way of knowing whether echoes of Turgot's theory came to Lavoisier; yet it is hard to believe that Condorcet, who was not the most reticent of men, did not take some of his fellow academicians into his confidence, or at least drop some meaningful hints. Indeed there is evidence to suggest that he did so in the case of at least one of Lavoisier's colleagues in the Academy, the chemist B.-G. Sage. In a later communication to Turgot, written in November 1772—that is, soon *after* Lavoisier had performed secretly his first experiments on combustion—Condorcet announces that he and Sage are about to collaborate on an experiment to determine whether fixed air is produced when the calx of lead is reduced with charcoal. This they proposed to do by passing through lime water any air that might be produced during the reduction.[38] Whether this proposed experiment, probably never carried out, was inspired by Turgot's theory and by an indiscreet revelation of Lavoisier, or whether it had been independently thought of by Sage, we cannot determine with certainty. But it is at least evident that others in the Academy, inspired perhaps by the recent discussions of English work on fixed air, were approaching the problem of combustion and calcination in a manner similar to Lavoisier's. There must have been some exchange of views among this group of academicians who were seeing each other twice each week and whose interest had recently been aroused by the newer English discoveries concerning fixed air. There is a remarkable passage in the original text of Lavoisier's sealed note, a passage which for various reasons he eliminated from the version later printed by Dumas, which casts some light on this question. Here Lavoisier wrote: "Comme il est difficile de ne pas laisser entrevoir à ses amis dans la conversation quelque chose qui puisse les mettre sur la voye de la vérité, j'ay cru devoir faire le présent depost entre les mains de M. le Secrétaire de l'Académie."[39] Clearly it was fear of being anticipated not by British rivals but by fellow chemists of Paris, who were already, like Sage, becoming interested in fixed air, which impelled him to deposit the sealed note.

I do not intend, however, by resurrecting Turgot's unpublished theory, though it anticipated by a full year Lavoisier's own speculations about the chemical role of air, to imply any unavowed indebtedness of Lavoisier to Turgot. I am reluctant to believe that Lavoisier was set on the "path of truth" by an indiscreet revelation, or broad hints, from Condorcet, though of course this is possible. What we know of the stages of Lavoisier's thought makes such an indebtedness unlikely and unnecessary to imagine. More probably we have here a remarkable case of parallel thinking and independent creation, where these two extraordinary men, Turgot and Lavoisier, came to the same conclusion when exposed to the same influences. Both men were familiar with Hales's doctrine of "fixed air," though they could know little, until late in 1772, about recent British work on the subject; both noted that effervescences were caused by the release of combined air; and to each of them an application of these facts leapt to mind when they learned of Guyton's experiments and the strange explanation he advanced to account for the augmentation effect. Both men saw that the combination of air with a metal during calcination was a more likely explanation of the weight increase—and one more in accord with the accepted principles of Newtonian physics—than Guyton's phlogistic fancies. If this is correct, then Turgot's letter is important chiefly for making explicit the principal influences that also shaped Lavoisier's thought and determined the course of his experiments.

◌

If, as I think my evidence shows, Lavoisier first came to his combustion studies via his interest in metals and in calcination, rather than by some other path, an important though subsidiary question remains unanswered. How did it happen that he did not carry out the reduction of lead oxide in a closed vessel (an experiment envisaged in its general lines as early as August 8) until *after* he had proved that phosphorus absorbs air on burning and gains in weight?

Here again we can find an answer, if we are willing to see Lavoisier as his associates saw him, not as one of the great historic figures of chemistry, but as an ambitious, talented, and promising young man. The proposals he set down in the August memorandum were, as I have indicated, a set of suggestions intended for his colleagues—Cadet, Brisson, Macquer, and the pharmacist Mitouard—with whom he was to embark in the summer of 1772 on experiments using the Academy's great burning glass. We have a record of these experiments,[40] which were carried out almost daily between August 14 and October 13. By comparing the experiments actually performed with the proposals Lavoisier made, it is clear that he was unable to persuade his co-workers to adopt them to any appreciable extent. It was the interests of others, chief among them Cadet and Brisson, the instigators of the project,[41] and Macquer, who had long been interested in the effect of high temperatures on ceramic materials and refractories,[42] which are reflected in the work actually accomplished during the summer of 1772. If we remember that Lavoisier was the junior member of the group, both in age and academic seniority, and that his name had none of the magic that we associate with it, we can understand why this was true, and why his proposal to study the behavior of metals and mineral substances in a closed vessel with the burning glass was ignored, like some of his other suggestions. It was not until these rather miscellaneous experiments were brought to an end that Lavoisier was free to use for his own purposes the equipment set up in the Jardin de l'Infante. It was then, sometime between October 22 and November 1, that Lavoisier turned the Academy's Tschirnhausen burning lens upon a sample of lead oxide, placed in his modification of Hales's apparatus.[43]

In the interval, however, he investigated the behavior of burning phosphorus, and from the beginning sought to determine whether it absorbed air when it burned. How can we account for this? The answer is, I believe, that this investigation of phosphorus was almost an accident, perhaps a by-product of routine Academic business. Lavoisier was, I believe, attracted to the study of phosphorus by an observation of the pharmacist P. F. Mitouard, who had submitted the results of a study of this substance for the consideration of the Academy of Sciences.[44] I am not the first to make this suggestion. More than thirty years ago a German historian of chemistry, Max Speter, called attention to a report made to the Academy by Lavoisier and Macquer on Mitouard's memoir, a report written by Lavoisier, and read to the Academy on December 16, 1772— that is to say, more than a month and a half after Lavoisier had carried out his own experiments on phosphorus.[45] In this report Lavoisier expressly noted a passage in which Mitouard had pointed to the increased weight of phosphorus after it is converted into the acid, and in which the gain in weight was attributed to water and perhaps to the addition of air as well.[46] Speter believed that this suggestion of Mitouard supplied the clue which led Lavoisier to suspect the role of air in combustion; he argued that Lavoisier knew of Mitouard's results, and perhaps had seen Mitouard's memoir before he began his own work. He even suggested that Lavoisier's draft memoir of October 20 had been designed to secure priority over Mitouard, perhaps even to rob him of due credit. A. N. Meldrum strongly attacked Speter's hypothesis, emphasizing that Mitouard's memoir was not formally submitted to the Academy until December 12 (i.e., after Lavoisier's own phosphorus experiments had been performed) and criticizing Speter for having reflected on Lavoisier's good faith and the honor of the Academy.[47] Since then, the problem of Mitouard's possible influence on Lavoisier has been generally, and I believe unwisely, ignored by historians of chemistry.

I have re-examined this question, so clouded by the controversy between Speter and Meldrum. I do not believe we can ignore completely the somewhat mysterious Mitouard episode; for there are facts which were unknown to Speter or to Meldrum. Space does not permit much more than a summary of my tentative conclusions; the supporting evidence I shall publish in detail elsewhere.

The key fact, I believe, is that Mitouard—who was well known and well regarded by Lavoisier and Macquer, and an active participant[48] in the burning-glass experiments in the summer of 1772—was at just this time a candidate for membership in the Academy of Sciences. It is certain that he submitted his unpublished memoir on phosphorus, together with two papers on the diamond which he had read to the Academy the previous spring, to justify his candidacy. This, of course, was standard practice. What is peculiar in this case is that the vacant place was opened up just prior to the annual autumnal vacation, for it resulted from Lavoisier's promotion on August 29, 1772, from the rank of *adjoint* to that of *associé*. Seven candidates for this vacated post of *adjoint* presented their names to the Academy for consideration at the first regular meeting following the *rentrée publique* of St. Martin's day (November 15).[49] It must have been apparent at the time of Lavoisier's promotion that any submitted memoirs would have to be examined before the reconvening of the Academy in mid-November. This suggests that some of the memoirs—in particular, Mitouard's work on phosphorus—may have been submitted to the *rapporteurs*, probably informally, to be read and studied during the recess of September and October. Now the report written by Lavoisier was finally submitted *only four days* after Mitouard had presented the memoir formally to the Academy on December 12. This strikes me, as it did Speter, as a very short time; and it supports my conjecture that Lavoisier had prepared the report, or at least studied Mitouard's results, during September and October. For various reasons, I believe the report may have been drafted before Lavoisier completed his own experiments demonstrating the combination of air with phosphorus. And I think it highly probable that Lavoisier came to study phosphorus because, late in August or early September—at any event, soon after his promotion—he had agreed to support his friend's candidacy and to be one of the referees of Mitouard's memoir. It should be remembered that our first reference to Lavoisier's interest in phosphorus is the fragmentary note, dated September 10, which Meldrum discovered and published in 1932.[50] It begins with a reference to Mitouard, for the opening sentence reads: "J'ay acheté chez M. Mitouard une once de beau phosphore venant d'Allemagne qu'il m'a laissé à 45 Livres prix de la facture." This means that Mitouard let Lavoisier have this uncommon substance *at cost*, a fact not too surprising in view of Mitouard's aspirations, and the support he counted on receiving from Lavoisier. The note goes on to record that Lavoisier placed a medicine vial containing a small piece of phosphorus close to the fire, and though it burst into flame, the vial was not broken. Emboldened by this success, he undertook, in obviously rather primitive fashion, to see if the phosphorus absorbed air on burning.

There is a significant coincidence of dates. Lavoisier began experiments on phosphorus—and it is quite clear from his description that this was his first introduction to this substance—soon after his promotion on August 29, and just at the time when Mitouard must have made known his candidacy. Mitouard supplied him, at cost, sometime before September 10, with the sample of phosphorus needed for his work. I can only conclude that Mitouard had mentioned to Lavoisier his intention of submitting the memoir on phosphorus to the Academy and that Lavoisier had agreed to be one of the referees. Whether Mitouard handed over his memoir to Lavoisier at this time, we cannot know; but at the very least he could have described his results verbally and mentioned his conjecture (which Lavoisier was later to emphasize in his report) that air might play a part in the greater weight of phosphoric acid. That Lavoisier himself began to experiment on phosphorus is not surprising. It was often the practice in the Academy for referees to repeat certain of the experiments in the submitted memoirs; and in view of Lavoisier's unfamiliarity with phosphorus this was especially advisable in this instance.

But what is significant is that Lavoisier, as the note of September 10 reveals, began by testing Mitouard's conjecture concerning the combination with air. It is possible, as Speter long ago suggested, that Lavoisier may have remembered that Guyton de Morveau had recently shown, more clearly than Mitouard, that phosphorus gains weight when transformed into its acid.[51] Guyton reported in the *Digressions*, a book we know Lavoisier to have studied closely, that he had gently heated twenty-two grains of phosphorus until it burst into flame and lined the interior of the flask with a white substance (the phosphorus pentoxide) that weighed thirty-seven grains. Guyton, of course, attributed the increase in weight to a loss of phlogiston, but wrote that though he tried to prevent the effect of deliquescence, he could not be absolutely certain that water did not play a part in this increase in weight.[52] From Guyton—though not from Mitouard, who had weighed the acid after it had deliquesced—Lavoisier could have learned that phosphorus gains in weight on combustion even before the acid takes up water, and have sought the cause in the addition of air during combustion.

Yet we need not—indeed we cannot, if what I have presented earlier in this paper has any force—believe that Mitouard's suggestion, even if somewhat clarified by Guyton's more careful experiment, first led Lavoisier to speculate about the role of air. In this Speter is clearly wrong, for I believe I have shown it was the phenomenon of metallic effervescence which first brought that possibility to his mind.[53] But it is likely—or so I read the rather confusing record—that Lavoisier was struck by Mitouard's suggestion (that air might combine with phosphorus to increase its weight) and saw that it closely resembled his own theory of what takes place in the calcination of metals. Having been obliged, as we have seen, to defer testing his broad hypothesis by the experiment of studying metals in the apparatus of Hales with the burning glass, he seized the opportunity to confirm his hypothesis with the substance that Mitouard's work brought to his attention.

Notes

1. For Lavoisier's work in the autumn of 1772, see A. N. Meldrum, "Lavoisier's Three Notes on Combustion," *Archeion* 14 (1932), 15–30. A good account, largely based on Meldrum, is given by Douglas McKie, *Antoine Lavoisier, The Father of Modern Chemistry* (Philadelphia: J. B. Lippincott Co., 1935), 110–25.
2. First printed by M. Berthelot in his *La Révolution Chimique—Lavoisier* (Paris, 1890), 46–49. It is now accepted that the date of February 20, 1772, found on the memorandum was a slip of Lavoisier's pen, as Meldrum suspected.
3. See especially Meldrum, "Lavoisier's Early Work in Science, 1763–1771," *Isis* 19 (1933), 330–63; and *Isis* 20 (1934), 396–425.
4. Meldrum, *The Eighteenth Century Revolution in Science—the First Phase* (Calcutta: Longmans, Green & Co., 1930), 12.
5. In the memorandum of August 8, 1772 (to be discussed below), Lavoisier proposed an experiment to decide between the two possibilities he thought most likely: that the diamond disappeared as a vapor or that it decrepitated. See *Oeuvres de Lavoisier*, 6 vols. (Paris, 1862–93), III, 264. The extensive diamond experiments performed a year later (in August 1773) show that even at that late date Lavoisier was not convinced that the diamond burned. *Ibid.*, III, 335–42.
6. Meldrum, *Eighteenth Century Revolution in Science*, 3–4. He later wrote "All these things—the work on the diamond, the observations of Sage, of Cigna—doubtless had an influence on Lavoisier by turning his mind towards the subject of combustion." "Lavoisier's Three Notes on Combustion," 16. The suggestion that Lavoisier may have been influenced by the observation of B.-G. Sage, *Elémens de minéralogie docimastique* (Paris, 1772), 5, that phosphoric acid is heavier than the phosphorus from which it is prepared, has little to support it. In a vague passage, Sage attributed the gain in weight entirely to water drawn from the air. A clearer description of this phenomenon based on careful experiment was published about the same time by Guyton de Morveau (see pp. 387–88). Lavoisier probably did not concern

himself with Sage's views concerning phosphorus until the appearance of the expanded edition of the *Elémens de minéralogie docimastique* in 1777. His criticism of Sage dates from this year, and it is only this second edition that appears in the lists of Lavoisier's books.

7. "Les flammes diminuent le ressort de l'air, non en l'absorbant, mais en exhalant des vapeurs qui diminuent la force répulsive des parties de ce fluide avec lesquelles elles se mêlent. . . ." *Introduction aux observations sur la physique, sur l'histoire naturelle et sur les arts*, 2 vols. (Paris, 1777), II, 97.

8. *Oeuvres de Lavoisier*, II, 103. Lavoisier later refined his terminology, and referred to the experiments as having been carried out with *minium*.

9. "Réflexions sur le phlogistique," *Oeuvres de Lavoisier*, II, 628; this paper is of 1783. In his *Opuscules physiques et chimiques* of 1774 there is a similar remark; see *Oeuvres*, I, 598.

10. "A Lost Memoir of Lavoisier," *Isis* 50 (1959), 125–29. Attention was first called to this August memorandum by Max Speter in his *Lavoisier und seine Vorläufer* (Stuttgart, 1910), 33. See also his chapter on Lavoisier in Günther Bugge, *Das Buch der Grossen Chemiker*, 2 vols. (Berlin, 1929), I, 310–12. Speter noted Lavoisier's doubts about the phlogiston theory and his early reference to the possible role of air. I discussed this August memorandum in my "Continental Reputation of Stephen Hales," *Archives Internationales d'Histoire des Sciences* 15 (1951), 393–404, stressing Lavoisier's reference to Hales.

11. "A Lost Memoir of Lavoisier," 127. The substance of this memoir on air is probably given in the first part of the Lavoisier document, dated August 1772, newly discovered by M. René Fric and published by him in this issue of the *Archives*. In this first section, entitled "Réflexions sur l'air et sur sa combinaison dans les minéraux," the ideas set forth do not differ essentially from those in the memorandum of August 8, but are greatly expanded. See below, n. 53.

12. *Oeuvres de Lavoisier*, III, 266.

13. *Ibid.*, I, 598.

14. *Elementa artis docimasticae duobus tomis comprehensa*, etc. 2 vols. (Leiden, 1744). This book is listed in the inventory of Lavoisier's library prepared at the time of his arrest in 1793. *Bibliothèque de l'Arsenal*, MS. 6496, vol. 10, fols. 136–95.

15. *Elémens de docimastique, ou de l'art des essais*, 4 vols. (Paris, 1755), III, 336–37.

16. *Elémens de chymie-pratique*, 2d ed., 2 vols. (Paris, 1756), I, 297. This was the edition that Lavoisier owned. I am indebted for this reference to my friend Dr. Cyril Stanley Smith of the University of Chicago.

17. This matter is discussed in my "Continental Reputation of Stephen Hales," 393–404, and later in my "Joseph Priestley's First Papers on Gases and Their Reception in France," *Journal of the History of Medicine and Allied Sciences* 12 (1957), 1–12.

18. That effervescences are produced by the release of a "true air" was clearly stated in Venel's article "Effervescence" in the Diderot-d'Alembert *Encyclopédie* (1755), V. 404–5. Venel attributes this identification to Hales; and this attribution is repeated by Jacob Reinbold Spielmann, *Instituts de Chymie . . . Traduits du Latin, sur la seconde Edition, par M. Cadet le jeune*, 2 vols. (Paris, 1770), I, 107. The particularly important passages on effervescence were not included in the first edition of Hales's *Vegetable Staticks*; they form part of an appendix, intended for a second edition of that work, that was added to the *Haemastaticks*. This was included by Buffon at the end of his French translation of the *Vegetable Staticks*.

19. *Oeuvres de Lavoisier*, I, 598.

20. N. Lemery, *Cours de Chymie . . . nouvelle édition, revue, corrigée & augmentée d'un grand nombre de notes . . . par M. Baron, Docteur en Médecine, & de l'Académie Royale des Sciences* (Paris, 1756), 113, n. (a).

21. *Rudolphi Augustini Vogel Opuscula Medica Selecta* (Göttingen, 1768), 53–68. See also Spielmann, *Instituts de Chymie*, II, 119–20.

22. There is no really adequate study of Guyton de Morveau's contributions to chemistry. But see A. B. Granville, "An Account of the Life and Writings of Baron Guyton de Morveau," *Journal of Science and the Arts* 3 (1817), 242–96; and Georges Bouchard's useful but rather disorganized *Guyton-Morveau, chimiste et conventionnel (1737–1816)* (Paris: Perrin, 1938).

23. Guyton had sent this work to Macquer, who evidently submitted it to the Academy, for Guyton wrote to Macquer from Dijon, on August 17, 1771, expressing his gratitude and describing his work as "un ouvrage qui n'a eu d'autre titre à votre suffrage que l'indulgence même que vous luy avez promise et qui m'a enhardi à vous l'envoyer." And he continued: "Je juge toute l'étendue des mes obligations envers vous par le choix même des commissaires que l'Académie m'a donné." Macquer Correspondence, Bib. Nat., Fr. 12306, vol. 2, fol. 127 recto and verso.

24. *Procès-verbaux de l'Académie des Sciences*, vol. 91 (1772), fols. 31 verso to 36 recto. This series of manuscript volumes may be consulted in the Archives of the Academy of Sciences, Paris.

25. *Digressions Académiques, ou Essais sur quelques sujets de Physique, de Chymie & d'Histoire naturelle. Par M. Guyton de Morveau, Avocat-Général au Parlement de Dijon*, etc., à Dijon, chez L. N. Frantin, 1762 [*sic*]. Most copies of this rare book bear this wrong date, but some exist with a new title page giving the correct year, 1772. See W. A. Smeaton, "L. B. Guyton de Morveau (1737–1816)," *Ambix* 6 (1957), 18–34. The book was evidently published at the end of May, for Macquer presented a copy to the Academy of Sciences on Guyton's behalf at the meeting of June 3, 1772. *Procès-verbaux*, vol. 91 (1772), fol. 193 verso. On June 5, Buffon wrote Guyton from Montbard to thank him for a complimentary copy. See Jean Pelseneer, "Une lettre inédite de Buffon à Guyton de Morveau à propos du phlogistique," in Léon Bertin et al., *Buffon* (Paris: Muséum Nationale d'Histoire Naturelle, 1952), 133–36.

26. For example, *Journal Encyclopédique* (April 1773), 87; P. J. Macquer, *Dictionnaire de chimie, seconde édition, revue et considérablement augmentée*, 4 vols. (Paris, 1778), I, 346–47; and *Oeuvres de Condorcet*, ed. A. Condorcet O'Connor and F. Arago, 12 vols. (Paris, 1847–1849), II, 38–39.

27. *Digressions Académiques*, 10–75.

28. Guyton himself later claimed credit for this accomplishment. See the article "Air" in the *Dictionnaire de chimie*, 6 vols. (Paris, An IV, 1796) I (1786), 699, of the *Encyclopédie méthodique*. Modern scholars have never given this work its due, but see the passing remark of J. R. Partington and Douglas McKie in "Historical Studies of the Phlogiston Theory. I. The Levity of Phlogiston," *Annals of Science* 2 (1937), 389.

29. "Mémoire sur la précipitation des substances métalliques les unes par les autres," *Oeuvres de Lavoisier*, II, 528–45.

30. *Oeuvres de Lavoisier. Correspondance*, recueillie et annotée par René Fric, 3 fascicles (Paris: A. Michel, 1955–64), fascicle II (1770–75), 404–6.

31. Archives of the Academy of Sciences. Lavoisier papers, dossier 14.

32. Turgot's scientific interests deserve careful study, for they are scarcely touched upon by his biographers. A pupil of the distinguished teacher of chemistry G.-F. Rouelle, Turgot had published in the *Encyclopédie* (VI, 1756, pp. 274–85) the long and interesting article "Expansibilité," which Lavoisier later singled out for praise. For the scientific aspects of Turgot's association with his friend Madame d'Enville (1716–94) and her son, the Duc de la Rochefoucauld d'Enville (1743–1792), see Emile Rousse, *La Roche-Guyon, châtelains, château et bourg* (Paris, 1892).

33. *Correspondance inédite de Condorcet et de Turgot, 1770–1779*, ed. Charles Henry (Paris, 1883), 58.

34. *Ibid.*, 59–63. Reprinted, with other letters of Turgot to Condorcet, by Gustave Schelle, *Oeuvres de Turgot et documents le concernant*, 5 vols. (Paris, 1913–23), III, 542–47.

35. In reprinting this letter, Schelle changed Henry's reading from "Stales" to "Stahl." I have been unable to trace the original of this letter, but the context would suggest that Turgot might have written "S. Hales"; for one could hardly attribute to Stahl (who denied the chemical role of air) experiments on the combination of air. Possibly Turgot was himself at fault, as Rhoda Rappaport has suggested to me, since in Rouelle's teaching Hales's experiments played a part in modifying the doctrines of Stahl, and some of his students may have confused the two. This happened in one of the Poitiers manuscripts of Rouelle's *Cours de Chymie*.

36. *Correspondance inédite*, 62.
37. *Ibid.*, 70.
38. Letter of November 22, 1772; *ibid.*, 108.
39. *Oeuvres de Lavoisier. Correspondance*, II, 389–90. M. Fric has not observed that the original of the sealed note, of which he has given a careful transcription, accompanied by a photograph, differs in its last sentence from the familiar version which is reproduced in the *Oeuvres de Lavoisier*, II, 103, and has been several times reprinted. I shall discuss the significance of this alteration, apparently due to Lavoisier himself, in a forthcoming communication. [Subsequently published as "A Curious Lavoisier Episode," *Chymia* 7 (1961), 103–8.]
40. *Oeuvres de Lavoisier*, III, 284–348.
41. An unpublished account by Lavoisier of the origin of these burning-glass experiments makes it clear that Cadet and Brisson were the originators and organizers of the project. See Lavoisier papers, dossier 72 J, Archives of the Academy of Sciences, Paris.
42. For Macquer's work as a ceramist see Leslie J. M. Coleby, *The Chemical Studies of P. J. Macquer* (London: G. Allen & Unwin, 1938), 96–110. The general background is described in my "French Antecedents of the Chemical Revolution," *Chymia* 5 (1959), 73–112.
43. *Oeuvres de Lavoisier*, I, 599–600.
44. For biographical information on Mitouard, see P. Dorveaux, "Le cervelet de Voltaire et les Mitouart," *Bulletin de la Société d'Histoire de la Pharmacie*, no. 44 (November 1924), 409–21. I do not follow Dorveaux in his rendering (Mitouart or Mitoüart) of this chemist's name but adopt instead the spelling used by the Abbé Rozier, Lavoisier, and by modern writers like Speter and Meldrum.
45. Max Speter, "Kritisches über die Entstehung von Lavoisiers System" *Zeitschrift für angewandte Chemie* 39 (1926), 578–82. Compare *Das Buch der Grossen Chemiker*, I, 313–16.
46. *Observations sur la physique*, III (1774), 421–23; reprinted with minor changes in *Oeuvres de Lavoisier*, IV, 141–43. Mitouard's memoir, which was never published, and the original of Lavoisier's report may be consulted in the Archives of the Academy of Sciences.
47. "Lavoisier's Three Notes on Combustion," 24–27.
48. In the official record of the Academy of Sciences only Cadet, Brisson, Macquer, and Lavoisier were officially designated to carry out experiments with the Academy's burning glass. Lavoisier's unpublished account (see n. 41) says, however, that "MM. Cadet et Brisson . . . inviteront dans cette vue M. Macquer, Lavoisier et Mitouard de concourir à leur objet et ils convinrent entre eux de travailler de concert à ce grand ouvrage." Mitouard was even charged with supervising the construction of the "hangard" to be erected in the Jardin de l'Infante, though the wood and labor were actually paid for by the Government. See Lavoisier papers, dossier 72 J, Archives of the Academy of Sciences. We know that Mitouard actively participated in the experiments and supplied various materials to be exposed to the burning glass, *Oeuvres de Lavoisier*, III, 301 and 320; but he was more than a "passiver Spender von Versuchsmaterial," as Speter calls him. *Das Buch der Grossen Chemiker*, I, 314.
49. For Mitouard's relations with the Academy of Sciences, see P. Dorveaux, "Quelques mots de plus sur Mitouart," *Revue d'histoire de la pharmacie* 19 (1931), 254–60. Additional details about this election are supplied by some manuscript comments of Lalande (Biblioteca Medicea Laurenziana, Florence, MS. Ashburnham 1700, fol. 119). We learn that Mitouard placed fourth out of the seven candidates on the first choice ballot; the winner was Baumé. I am grateful to Mr. Roger Hahn for this reference.
50. "Lavoisier's Three Notes on Combustion," 17–19.
51. Max Speter, "Lavoisieriana," *Chemiker-Zeitung* 55 (1939), nos. 103–4, p. 994.
52. *Digressions Académiques*, 252–53, n. 1.
53. When this study was written, and when it was read at Barcelona, I had not yet heard of Mr. René Fric's discovery of what must surely be Lavoisier's lost "Système sur les élémens." I first learned of its contents when the document was made public on M. Fric's behalf by M. Maurice Daumas at the *Colloque* on the History of Eighteenth-Century Chemistry held in Paris in the week following the Barcelona

Congress. I can only express my pleasure at finding that this important document confirms some of the conjectures I hazarded in my recent *Isis* article, and supports the main contention of the present study, which is that Lavoisier's first speculations on the chemical role of air had their origin in his curiosity about effervescences produced by mineral substances, and owed nothing to any interest on his part in the chemistry of phosphorus.

25
A Curious Lavoisier Episode

❖

Among the documents which M. René Fric has included in the second fascicle of the *Correspondance de Lavoisier* is the famous sealed note (*pli cacheté*), dated November 1, 1772, which the young Lavoisier deposited with the secretary of the Academy of Sciences to assure himself priority for his important first discoveries concerning the combustion of phosphorus and of sulphur and the reduction of lead oxide.[1]

This note is one of the best known, and one of the truly memorable, Lavoisier documents. It has been cited *in extenso* (for it is quite brief) by most of the modern scholars who have treated this period of Lavoisier's career. How startling, therefore, to discover that the text given by M. Fric differs in some significant respects from the familiar version we find in the *Oeuvres de Lavoisier*[2] and in the well-known works of A. N. Meldrum and Douglas McKie.[3] For some reason M. Fric has failed to note this strange fact.

The first discrepancy worth mentioning is of no very great significance. In the version to which we are accustomed, several words are commonly printed in italics. Thus, for example, the words "calcination," "chaux," and "litharge" would seem to have been underscored by Lavoisier himself.[4] No underscored words, however, appear in the original, as the reader can verify by examining the photograph of the manuscript in the *Correspondance*, and as I can further testify from having studied the manuscript itself.

But the final paragraph differs markedly in the two versions. Meldrum, who in his classic paper scrupulously reproduced each of Lavoisier's early notes precisely as he found it, whether in manuscript or in print, gives the last sentence as follows:

> Cette découverte me paroissant une des plus intéressantes qui ait été faitte depuis Staalh [*sic*], j'ai cru devoir m'en assurer la propriété, en faisant le présent dépôt entre les mains du secrétaire de l'Académie, pour demeurer secret jusqu'au moment où je publierai mes expériences.[5]

This is the ending we are accustomed to read. But what Lavoisier actually wrote is quite different; I give his words, from the original *pli cacheté*, with the significant differences set off by square brackets:

From *Chymia* 7 (1916), 103–8. © by the Trustees of the University of Pennsylvania.

Cette découverte me [paroit] une des plus interessantes qui ait ete faitte depuis Sthal [et Comme il est difficile de ne pas laisser entrevoir a Ses amis dans la Conversation quelque chose qui puisse les mettre Sur la voye de la verite] j'ay Cru devoir [faire] le present depost entre les mains de [M. le] Secretaire de lacademie [en attendant que je rende mes experiences publiques].[6]

Apart from minor differences in spelling and punctuation, it is obvious that the sentence has been thoroughly rewritten. The change in style and tense, and the more formal way the original note refers to the Secretary of the Academy, reveal an immediacy and urgency in the original version that are lacking in our more familiar text. But still more striking is the complete omission of the long phrase in which Lavoisier justifies his step in depositing the sealed note. The usual text says nothing about Lavoisier's fear that he might inadvertently disclose to his friends in conversation something that could lead them to the truth he had discovered about combustion and calcination!

A heavy editorial hand has obviously been laid upon the note to produce a later, bowdlerized version which we have long taken as standard. How is this possible? Whom should we hold responsible for these changes?

To answer these questions we must first recall the ultimate source of the document that Meldrum, McKie, Duveen, and others—myself included— have unquestioningly accepted. Meldrum's most careful version differs little (except for some modernization of spelling and punctuation by Dumas, and one of his dubious editorial "improvements") from the text that was printed in the second volume of the *Oeuvres de Lavoisier* in 1862. Dumas's and Meldrum's common source was not a manuscript—the original recently printed by M. Fric obviously escaped the attention of Lavoisier's nineteenth-century editors and of Meldrum—but a printed version. This first appeared in Lavoisier's posthumous *Mémoires de chimie* (1805), where it is introduced towards the end of a paper lengthily entitled: "Détails historiques, sur la cause de l'augmentation de poids qu'acquièrent les substances métalliques, lorsqu'on les chauffe pendant leur exposition à l'air."[7]

In this paper Lavoisier begins with a discussion of the book of Jean Rey, which he is at pains to insist could not have influenced him; he then cites some of the opinions advanced from Rey to Guyton de Morveau to explain the weight increase of calcined metals, and carries the story down to the time when he himself found the true cause of this phenomenon. At this point he introduces, as evidence of the discoveries he had made before November 1772, what purports to be an exact copy of the original sealed note.

Dumas guessed, probably correctly, that Lavoisier's "Détails historiques" was written in 1792—i.e., twenty years after the events described in the note.[8] So far as I know, the manuscript of this paper has not been preserved; at all events it has not been seen or at least not described. But a careful comparison of the texts we now possess—the original note of 1772 and the modified version printed in 1805—and a close examination of the context in which the latter appeared, can perhaps dispel our little mystery.

The finger of suspicion points inexorably at Lavoisier himself as the author of this mild deception, for I think the revisions served a very definite purpose that will emerge if we ponder the sentences by which he introduced his printing of the note. They are as follows:

> J'étois jeune, j'étois nouvellement entré dans la carrière des sciences, j'étois avide de gloire, et je crus devoir prendre quelques précautions pour m'assurer la propriété de ma découverte. Il y avoit, à cette époque, une correspondance habituelle entre les savans de France et ceux d'Angleterre; il régnoit, entre les deux nations, une sorte de rivalité qui donnoit de l'importance aux expériences nouvelles, et qui portoit quelques fois les écrivains de l'une ou de l'autre nation, à les contester à leur véritable auteur; je crus donc devoir déposer, le 1er novembre 1772, l'écrit suivant, cacheté, entre les mains du secrétaire de l'Académie.[9]

Despite his candid admission that the fervor of youth and a desire for scientific glory had played their part, Lavoisier is anxious to have us believe that his motive in depositing the sealed note was, in a sense, primarily a patriotic one. It was, he says, a rivalry with English scientists (like Joseph Priestley, we are to assume) which led him to protect his discovery. Yet when he reread it, he must have seen that the original note, in its final sentence, did little to support this contention; for it was his colleagues of the Academy, whom he regularly saw twice each week, and perhaps other French scientists as well, to whom he had clearly pointed as potential rivals. Only they, and not the English scientists he had yet to meet or correspond with, could have been classed among those friends to whom he might unavoidably drop some hint in conversation which would disclose the "path of truth."

Perhaps Lavoisier felt, when he set himself to copy the original note, that the motive he had so candidly avowed twenty years before might appear niggardly and unworthy of a man who had attained such eminence in science as he now enjoyed. Possibly, if the memoir was actually written in 1792, he was responding to the patriotic mood of a moment when France was again at war, or on the eve of war, with Britain. But another interpretation suggests itself. We should remember that after 1789 Lavoisier was the acknowledged leader of a triumphing band of disciples who were announcing a Revolution in Chemistry, and that the New Chemistry they collectively advocated was already being described as a New *French* Chemistry. It would hardly do to recall that it was French scientists whom he had feared, years before, as possible rivals. Far better to point to the English.

Yet even this explanation does not fully accord with the evidence. Lavoisier seems to have had a quite different and more personal reason. The concluding paragraph of the "Détails historiques" is quite revealing on this point. By 1772, he insisted, he had formulated "tout l'ensemble du système que j'ai publié depuis sur la combustion." In 1777, he had greatly extended his theory; and he had developed it in detail during subsequent years without assistance or much sympathy from those French scientists who later became his paladins of the

New Chemistry. Their conversion, he tells us, had been slow. Berthollet was still writing in the phlogistic idiom in 1785; Fourcroy did not teach the new system until the winter of 1786–87; it was adopted by Guyton de Morveau still later. And Lavoisier concluded:

> Cette théorie n'est donc pas, comme je l'entends dire, la théorie des chimistes françois: elle est *la mienne*, et c'est une propriété que je réclame auprès de mes contemporains et de la postérité. D'autres, sans doute, y ont ajouté de nouveaux degrés de perfection, mais on ne pourra pas me contester, j'espère, toute la théorie de l'oxidation et de la combustion; l'analyse et la décomposition de l'air par les métaux et les corps combustibles; la théorie de l'acidification . . . la théorie de la respiration, à laquelle Seguin a concouru avec moi: ce recueil présentera toutes les pièces sur lesquelles je me fonde, avec leur date; le lecteur jugera.[10]

To claim the central theory of the New Chemistry as his own; to insist that it should be thought of as *Lavoisier's* Chemistry, not as a French Chemistry: this is the chief burden and purpose of his "Détails historiques." Lavoisier began, we have seen, by disposing of Jean Rey as a possible influence upon him; he showed that later French chemists like Lemery the elder and Moïse Charas had theories of calcination quite different from his own and totally incorrect; then he devoted a long paragraph to Guyton de Morveau's "efforts infructueux" to explain in the *Digressions Académiques* how the increase in weight of calcined bodies could be reconciled with the phlogiston theory.[11] And he ended, after inserting the *pli cacheté* as evidence of his own accomplishment in 1772, by recalling how slow his colleagues in France had been to accept his revolutionary theories.

It was this desire to gain recognition over his French colleagues in the New Chemistry which impelled Lavoisier, I feel sure, to alter the text of the sealed note. For alter it he certainly did. Because the "Détails historiques" focused on a presumed anticipation of his theory of calcination (by Jean Rey) and on other theories of the oxidation of metals, Lavoisier put in italics those words in the sealed note ("calcination," "chaux," "litharge") which emphasized this aspect of his discoveries. But more important was the suppression of the evidence in the note that it was his French contemporaries whom he had viewed in 1772 as his potential rivals.

I need hardly urge that the historical account Lavoisier gives of this period of his early career is misleading and inaccurate. In 1772 he had but scant knowledge of the progress that English pneumatic chemists were making. As I show elsewhere, he had not read Black or Cavendish; and he could know of Joseph Priestley's exciting new discoveries on gases only vaguely and by hearsay, for Priestley's first important paper was not printed until the very end of that year, and was not available to French chemists for several months.[12] The regular correspondence between the chemists of France and England, if such there ever was,[13] can hardly have begun as early in 1772. Still less at that time could there have been such a mutual sense of rivalry as Lavoisier describes, least of all from the British side, for Lavoisier was all but unknown abroad. He has clearly telescoped events and projected back, without much regard for factual accuracy, his later tense relations with Joseph Priestley (over the discovery of oxygen) and the controversy (involving the names of Cavendish and Watt) about the discovery of the constitution of water.

But if recent investigations cannot support Lavoisier's version of these events they do confirm and illuminate the unwilling testimony of the original sealed note. By the fall of 1772, a number of Frenchmen had become interested in the calcination problem and in the chemistry of "fixed air." In August 1771, a year before Lavoisier turned his attention to these matters, the economist Turgot had advanced in a private letter a theory of calcination and combustion almost identical with Lavoisier's. By November of 1772, two of Lavoisier's colleagues in the Academy, Condorcet and Sage, had envisaged an experiment (somewhat similar to the one Lavoisier had just secretly performed) to discover whether "fixed air" is released when lead oxide is reduced with charcoal. Outside the Academy, but in close contact with its members, several Parisian chemists—Bucquet, Bayen, and Rouelle the younger—had begun to speculate about the properties of "fixed air" and had probably begun to experiment. The Abbé Rozier, with the encouragement of Lavoisier's friend Trudaine de Montigny, was already collecting for publication in his journal the principal foreign studies on pneumatic chemistry. There is no doubt that in the autumn of 1772 this was a topic of absorbing interest among Lavoisier's colleagues,[14] and that he was keenly aware of the fact.

We must then discount—if we are to understand the origins of the Chemical Revolution in France—the historical picture that Lavoisier took pains to give us and which he sought to support by his altered version of the *pli cacheté*. We can better appreciate the mood with which he entered upon his historic series of experiments, and disentangle more readily the influences upon him, if we return to the words—all the words—he actually set down on November 1, 1772. To M. Fric we must be grateful for making this important document readily available to us in its original form.

Notes

1. *Oeuvres de Lavoisier. Correspondance*, recueillie et annotée par René Fric, 3 fascicles, (Paris: A. Michel, 1955–64), fascicle II (1770–75), 389–90.
2. *Oeuvres de Lavoisier*, 6 vols. (Paris, 1862–93), II, 103.
3. A. N. Meldrum, "Lavoisier's Three Notes on Combustion: 1772," *Archeion* 14 (1932), 23. An English translation of this note was given by Meldrum in *The Eighteenth Century Revolution in Science—The First Phase* (Calcutta: Longmans, Green & Co., 1930), 3. Slightly differing English translations of the text in the *Oeuvres de Lavoisier* are given by Douglas McKie in his *Antoine Lavoisier, The*

Father of Modern Chemistry (Philadelphia: J. B. Lippincott, 1936), 117–18; and in his later book, *Antoine Lavoisier: Scientist, Economist, Social Reformer* (New York: H. Schuman, 1952), 101–2.

4. Dumas's version in the *Oeuvres* and Meldrum's in his "Lavoisier's Three Notes" both reproduce the italicized words.

5. Meldrum, "Lavoisier's Three Notes," 23.

6. *Oeuvres de Lavoisier. Correspondance*, II, 389–90.

7. *Mémoires de chimie*, 2 vols. in 1, n.d. (1805), II, 78–87. This memoir is reprinted in the *Oeuvres de Lavoisier*, II, 99–104, with modernization of spelling and punctuation and an occasional editorial liberty. [It is now generally agreed that the *Mémoires* were printed c. 1803.]

8. See the prefatory remarks by the editor of the *Mémoires*, Lavoisier's widow, who says that her late husband began in 1792 the project of preparing a collection of his memoirs. The memoir was certainly written, from internal evidence, after 1787, and probably in the year 1792, as a sort of preface to certain of the key papers.

9. *Mémoires de chimie*, II, 84, and *Oeuvres de Lavoisier*, II, 102–3.

10. *Mémoires de chimie*, II, 87, and *Oeuvres de Lavoisier*, II, 104.

11. It is interesting that Lavoisier says nothing of the fact that Guyton contributed to his solution of the problem by his series of careful experiments on the increase in weight of calcined metals. I have tried to show elsewhere that Guyton's work was an important influence on Lavoisier in 1772. See my "Origin of Lavoisier's Work on Combustion," *Archives Internationales d'Histoire des Sciences* 12 (1959), 113–35.

12. See my paper "Joseph Priestley's First Papers on Gases and Their Reception in France," *Journal of the History of Medicine and Allied Sciences* 12 (1957), 1–12.

13. Unless indeed Lavoisier is referring to the singlehanded effort of J. H. Magellan to keep both sides of the Channel informed of progress in science, especially chemistry. It was Magellan who, early in 1771, began to inform French chemists of the British investigations concerning gases. His letters played an important part in arousing French interest in this subject, especially in 1772.

14. See my "Joseph Priestley's First Papers on Gases," 8–9, and my "Origin of Lavoisier's Work on Combustion," 126–29.

26
Laplace's Collaboration with Lavoisier

✾

The collaboration between Lavoisier and Laplace is generally known only through their joint "Mémoire sur la chaleur" of 1783, but it began six years earlier. The partnership was instigated by Lavoisier; and Laplace—six years younger than Lavoisier, and his junior in Academic rank—clearly saw in this association an opportunity for professional recognition and material advancement, matters never far from his mind.

The first experiments they performed together were reported by Laplace at the Easter *séance publique* of the Academy of Sciences in the spring of 1777.[1] These were never published, but we can nevertheless reconstruct them from allusions made by Lavoisier in later papers, and from an unpublished letter to Trudaine de Montigny.[2] They dealt with the vaporization of water, ether, and alcohol in the receiver of an air pump. These fluids, the experiments showed, could be converted into vapors that, like air, existed in an "état d'élasticité permanent." Two factors were involved: the temperature and (a factor that Lavoisier had not previously stressed) the atmospheric pressure.

From Lavoisier's point of view, these experiments served to confirm a theory of the nature of elastic fluids that he had begun to formulate some eleven years earlier, which he sketched out and developed in a series of unpublished drafts dating from 1772, 1773, and 1775, and which he had barely hinted at in his *Opuscules physiques et chimiques* of 1774.[3]

This "covering theory," which was to provide the conceptual framework for much of his later work, affirmed that all fluids were capable, when combined with a sufficient quantity of the tenuous "matter of fire," of assuming a permanent elastic state with all the mechanical properties of air. Air itself was simply the combination of a certain fluid (*la base de l'air*) with a large amount of the matter of fire. In the spring of 1773 he had taken a further important step: he perceived how this theory could illuminate his famous experiments on combustion performed the previous autumn.

Those experiments had shown that when sulphur and phosphorus burned they combined with a large amount of air and in so doing gained in weight. But these experiments were still not sufficient to warrant his advancing a new

---

From XIIe Congrès International d'Histoire des Sciences, Paris, 1968 (Paris: Albert Blanchard, 1971).

theory of combustion, for they did not explain the striking feature of such processes: *the production of heat and light.* Yet this was what the phlogistic hypothesis seemed quite readily to account for; combustibles were supposed to be rich in phlogiston—recently identified by several French chemists with a kind of material fire—and so the spectacle of light and heat during combustion was simply a manifestation of the phlogiston given off during these processes.

For some time Lavoisier had harbored doubts about the phlogistic hypothesis; but even after the discovery that Priestley's "dephlogisticated air" (oxygen) was the effective agent in combustion, he did not admit these doubts in public until late in 1777, nor actually publish them until 1780.[4] This delay should have puzzled students of Lavoisier's work.

What Lavoisier strongly suspected and was obliged to prove, before he could offer an alternative to the phlogiston theory, was that *the heat and light came not from the combustible, but from the "dephlogisticated air."* The first experiments with Laplace seemed to confirm his belief that when vapors are formed there is a combination with the "matter of fire," and this in turn supported his theory that all aeriform fluids were combinations of some "base" with the fire matter. If so, then the heat and light in combustions must come from the "dephlogisticated air" (or *air pur,* as he then called it), not from the combustible. The so-called *base de l'air* has a greater affinity for a combustible than for the matter of fire, which accordingly is liberated.

Thus encouraged by these first experiments with Laplace, Lavoisier was emboldened to present in November 1777 the substance of his "Mémoire sur la combustion en général." This, we know, was his earliest public announcement of his theory of combustion and his first—if somewhat discreet—assault on the phlogiston theory. In this important memoir Lavoisier cited the experiments with Laplace as substantive evidence for his ideas.

Until 1781 there is little trace of further active collaboration between the two men, although they seem to have planned to carry forward the vaporization experiments using a more sophisticated technique. Both men were preoccupied with other pressing matters. But when the collaboration was resumed in the summer of that year Lavoisier and Laplace worked closely together until the late spring of 1784 on various problems directly or indirectly concerned with heat. They measured the thermal expansion of glass and various metals, using an ingenious optical pyrometer which Laplace may have helped design; they devised a combustion apparatus with a double nozzle to explore the product formed when "inflammable air" (hydrogen gas) is burned with "dephlogisticated air." And, on the occasion of Volta's visit to Paris, they investigated the electrification of fluids when they are vaporized. Finally, in 1782–83 they undertook the experiments on the measurement of heat that are reported in the classic "Mémoire sur la chaleur." Most of these experiments were carried out using the ice calorimeter devised by Laplace and built for Lavoisier by one of his instrument makers. They were to supply Lavoisier with still more conclusive arguments in support of his theory of combustion.

The sudden renewed interest of Lavoisier and Laplace in the subject of heat can readily be traced to a series of publications that appeared from 1779 to 1781 in France and abroad. Most important of all was the appearance of Adair Crawford's *Experiments and Observations on Animal Heat* (London, 1779).[5] Their knowledge of the contents of this book was doubtless indirect, derived from a précis published by Jean Hyacinthe de Magellan in Rozier's Journal for May and June 1781.[6]

What especially struck Lavoisier and Laplace—besides Crawford's theory of animal heat, so much resembling Lavoisier's own ideas about respiration— was the discussion of the important concept of *specific heats* and the description of the method of mixtures used by Irvine and Crawford for measuring them. In contrast to the phenomena of the *latent heats* of fusion and vaporization, which Lavoisier had understood in a general way for nearly a decade, the notion of specific heats was new and challenging. It was only after reading Magellan's *Essai* that Lavoisier could understand that different bodies at the same temperature could contain different amounts of concealed heat, or *feu fixé*.

The experiments set forth in the "Mémoire sur la chaleur" were conceived to be repetitions of, or improvements upon, Crawford's investigations. This emerges not only from the frequent references to Crawford in the joint memoir and in Lavoisier's later writings, but from an interesting exchange of letters with the Austrian scientist Schwediaur in the early months of 1783.[7] The first experiments Lavoisier and Laplace performed were, in fact, measurements of specific heats by the method of mixtures. But they soon perceived the limitations of that method; to supplant it by a better and more widely applicable one Laplace —and the conception is his, not Lavoisier's—hit upon the idea of measuring the heats evolved in various processes by weighing the water produced from melted ice. At least by the early months of 1783 the ice calorimeter was in active service for determining specific heats, measuring the heat evolved in various chemical reactions and that produced by a guinea pig confined for several hours in the calorimeter.

This new and powerful method, and the results obtained by means of it, were reported to the Academy in June 1783; significantly the presentation was made by Laplace.[8] And before the end of August this famous memoir was published as a pamphlet (a year was to elapse before its appearance in the Academy's *Mémoires*).[9] This exceptional procedure attests the great importance the two authors attached to these researches.

During the summer of 1783, when in any case the prevailing temperature was unsuited for calorimetric experiments, the two men were mainly pre-occupied with the dramatic new problem of the composition of water. Stimulated by the news of Cavendish's experiment in England, Lavoisier and Laplace confirmed his findings before an audience that included Charles Blagden (who had brought the news) and several members of the Academy of Sciences. Once convinced that the water produced in their double-nozzle combustion apparatus equaled in weight the weights of the gases burned, Laplace made the important suggestion to Lavoisier that the hydrogen given off when iron is treated with dilute acid derives from the decomposition of water, not from the metal. This provided an effective argument against those latter-day phlogistonists who identified as pure phlogiston the "inflammable air" released, as they supposed, from metals. In the month of September, Lavoisier and Laplace carried out a series of experiments intended to verify Laplace's conjecture.

At least twice Laplace had expressed his restlessness at spending so much time away from his beloved mathematics,[10] so it is interesting to note the energy and enthusiasm he displayed in the final phase of the collaboration. In December 1783 Laplace wrote Lavoisier a letter urging that the calorimetric experiments be actively resumed and extended.[11] It was important, he urged, to perform careful experiments on the combustion of carbon; to repeat as often as possible the experiments on animal heat, and to determine with precision the specific heats of metals and their calxes. These experiments were performed, pretty much as Laplace had outlined them, in the winter and spring of 1784. The last trace of this collaboration is to be found in Lavoisier's laboratory *registres* for May 10, 1784.

For Lavoisier, the experiments set forth in the "Mémoire sur la chaleur" and the later ones of 1784 further confirmed his theory of heat and vaporization. Just as the earlier experiments of the spring of 1777 had encouraged Lavoisier to make his theories public in his "Mémoire sur la combustion en général," so these later calorimetric investigations paved the way for Lavoisier's definitive attack on phlogiston in what is perhaps the most famous of his memoirs, the "Réflexions sur le phlogistique."

The "Réflexions" was written a year or so after the "Mémoire sur la chaleur" and not long after the collaboration with Laplace had been amicably terminated. It was read to the Academy in the summer of 1785 and printed the next year in the Academy's *Mémoires*. In spite of the praise lavished on the "Réflexions" by historians of chemistry, the main thrust of Lavoisier's argument has rarely been noted. The paper is not merely a brilliant dialectical performance. It is a closely reasoned refutation, supported by the new experimental evidence, of those followers of Stahl—notably Baumé and Macquer—who had attempted to modify the phlogistic hypothesis to accommodate such inescapable facts as the gain in weight and the absorption of air during combustion. To refute these men Lavoisier drew effectively upon the recent experiments with Laplace on heat. For example, Macquer had argued that when *air pur* (oxygen) combines with a combustible, phlogiston, or the matter of heat and light, is given off. But Lavoisier wrote: "Cependant les expériences de M. Crawford, celles de M. Wilke, celles de M. de Laplace et les miennes, prouvent le contraire."[12]

In the succeeding section of his "Réflexions" Lavoisier, having disposed of these antagonists, prefaces an exposition of his own theory of calcination and combustion with a discussion of his ideas about heat and his theory of vaporization. The close relationship between the "Réflexions sur le phlogistique"—his major attack on the older chemistry—and the "Mémoire sur la chaleur" is plainly evident, although it has escaped general notice. We need only note the space devoted in the "Réflexions" to the doctrine of specific heats, the references to Adair Crawford, and Lavoisier's explicit mention of the earlier memoir and the method of measuring heat "imaginé par M. de Laplace."

Space does not permit a detailed analysis in this paper of the "Mémoire sur la chaleur," or even of that remarkably prophetic sketch of a physics of heat (Article III) that can only have been the work of Laplace. Nor can I discuss those later memoirs of Lavoisier on various problems of chemistry in which the influence of the close and prolonged association with Laplace is so clearly manifested. A new Lavoisier emerged after that famous collaboration with Laplace: a Lavoisier at once more cautious in his theoretical pronouncements and —if I may put it that way—more Newtonian; willing to conceive chemical problems on a deeper level than that *zone intermédiaire* to which Mme. Metzger would have confined him; willing in his later papers, such as his remarkable "Considérations sur la dissolution des métaux dans les acides," to use the language of molecular attractive forces.

But here my object has been quite modest. I have tried to trace the main steps in the collaboration between these two giants of science, and above all to show what role this collaboration played in, and what the experiments on heat contributed to, the full and confident elaboration of Lavoisier's theory of combustion and his major assaults on the citadel of the older chemistry.

[Subsequently expanded and published as "Chemistry as a Branch of Physics: Laplace's Collaboration with Lavoisier," *Historical Studies in the Physical Sciences* 7 (1976), 193–276.]

Notes

1. *Procès-verbaux de l'Académie Royale des Sciences* (1777), fol. 256. On this same occasion Lavoisier first reported his discoveries on animal respiration.
2. Fonds Chabrol. I wish to thank the Comte Guy de Chabrol for allowing me to transcribe this important letter from Lavoisier to Trudaine.
3. The early phases of Lavoisier's theory of vaporization are being investigated by my student, J. B. Gough. See, for Lavoisier's speculations of the spring of 1766, Mr. Gough's article in the *British Journal for the History of Science* 4 (1968), 52–57. Several of the drafts in which Lavoisier elaborated this theory have been published by M. René Fric, "Contribution à l'étude de l'évolution des idées de Lavoisier sur la nature de l'air et sur la calcination des métaux," *Archives Internationales d'Histoire des Sciences* 12 (1959), 137–68.
4. *Oeuvres de Lavoisier*, 6 vols. (Paris, 1862–93), II, 225–33. The memoir was given a second reading in December 1779. *Procès-verbaux* (1779), fol. 306. It was published in *Mém. Acad. Roy. Sci.* 1777 (1780), 592–600.
5. Among the works that appeared in this period were Torbern Bergman's *Opuscula physica et chemica*, 6 vols. (Upsala, 1779–1790), of which the first volume, containing an account of Wilcke's experiments on latent heat, was published in French translation by Guyton de Morveau in 1780. In 1781 appeared the Baron de Dietrich's

translation of Scheele's book on air and fire, with a significant introduction mentioning Lavoisier's theories and the work of Wilcke and Joseph Black.

6. *Observations sur la physique*, vol. 17 (1781), 369–86, 411–22. Magellan's summary account, entitled *Essai sur la nouvelle théorie du feu élémentaire, et de la chaleur des corps*, was published as a pamphlet in London in 1780.

7. *Correspondance de Lavoisier*, ed. René Fric, 3 fascicles (Paris: A. Michel, 1955–64), fascicle III, 734–35. M. Fric has published here what is apparently an undated draft of Lavoisier's letter to Schwediaur. The latter's reply, dated March 14, 1783, speaks of having received Lavoisier's letter of February 18.

8. *Procès-verbaux* (1783), fol. 144, and II, fol. 283.

9. The date of publication can be roughly estimated from the letter of Laplace transmitting to Lagrange two complimentary copies of the joint memoir. See *Oeuvres de Lagrange*, publiées par les soins de M. J.-A. Serret, 14 vols. (Paris, 1867–92), XIV (*Correspondance de Lagrange avec Condorcet, Laplace, Euler et divers savants, publiée et annotée par L. Lalanne*), 123–24.

10. Besides the letter mentioned in the previous note there is the interesting letter of Laplace to Lavoisier (March 7, 1782) first published by Denis I. Duveen and Roger Hahn in "Deux lettres de Laplace à Lavoisier," *Revue d'Histoire des Sciences et de leurs Applications* 11 (1958), 337–42; reprinted in *Correspondance de Lavoisier*, III, 712–14.

11. *Correspondance de Lavoisier*, III, 757–58. This letter was likewise first published by Duveen and Hahn, "Deux lettres de Laplace."

12. *Oeuvres de Lavoisier*, II, 636.

27

The Chemical Revolution:
A Word from Monsieur Fourcroy

✸

This note was inspired by a query from a distinguished confrere in the history of science, whose specialty is Newtonian science, not the history of chemistry. He asked what I could tell him, besides what he already knew (which was considerable), about the history of the phrase "chemical revolution." He asked on what occasions Lavoisier himself used that or a similar phrase, when the expression or its equivalent first appeared *in print*, and when the phrase "chemical revolution" came into widespread use to describe, as my correspondent put it, "the new chemistry established primarily through the labors of Lavoisier."

As my correspondent was well aware, the pat phrase "the chemical revolution" did not receive its present-day currency until after the publication in 1890 of Marcellin Berthelot's indispensable *La Révolution Chimique—Lavoisier*. He also knew of a similar, if less precise, locution in the title of Andrew Norman Meldrum's *The Eighteenth Century Revolution in Science—The First Phase* (1930), a study of Lavoisier's chemical investigations from the autumn of 1772 to what Meldrum called "the effective discovery of oxygen."[1]

Berthelot, who was the first to publish it, and Meldrum, both discuss the now famous programmatic memorandum Lavoisier set down on February 20, 1773 (but which he misdated as February 20, 1772).[2] In this remarkable document, after summarizing some of the discoveries made concerning air fixed in solid bodies, Lavoisier wrote: "L'importance de l'objet m'a engagé à reprendre tout ce travail, qui m'a paru fait pour occasionner *une révolution en physique et en chimie*." This would seem to be the earliest invocation of the idea, if not precisely of the phrase, used by Berthelot.[3] It should be emphasized, however, that Lavoisier's use of the word *"révolution"* was a private expression of his elation and optimism; so far as I can recall, he never used the phrase in any of his published papers: not even in his "Réflexions sur le phlogistique," where one might have expected to find it, in the *Opuscules physiques et chimiques* (1774), in the opening essay of the collaborative *Nomenclature chimique* (1787), or in his *Traité élémentaire de chimie* (1789). But on at least two occasions the phrase appears in his private correspondence. In a famous letter of February 2, 1790, to

From *Ambix; The Journal of the Society for the Study of Alchemy and Early Chemistry* 23 (1976), 1–4.

Benjamin Franklin, Lavoisier wrote, after describing the recent work he and his disciples had accomplished in chemistry: "Voilà donc une révolution qui s'est faite depuis votre départ d'Europe dans une partie importante des connaissances humaines [i.e., chemistry]. Je tiendrai cette révolution pour bien avancée et même pour complètement faite si vous vous rangez parmi nous."[4] In a similar vein he wrote to his disciple Chaptal in 1791: "Toute la jeunesse adopte la nouvelle théorie et j'en conclus que la révolution est faite en chimie."[5]

But who first spread abroad the word "révolution" *in print* with reference to the new developments in chemistry? The answer would seem to be one of Lavoisier's well-known disciples, Antoine François de Fourcroy (1755–1809). Oddly enough, Fourcroy used the phrase well before his conversion to Lavoisier's New Chemistry. It first occurs in his *Leçons élémentaires d'histoire naturelle* (1782), where, speaking with cautious diffidence of Lavoisier's "grand nombre de belles expériences," he writes that the proper course is to wait until further experiments convince us that all the phenomena of chemistry can be explained "par la doctrine de gas" without invoking phlogiston, more especially as Macquer is "très-convaincu de la grande révolution que les nouvelles découvertes doivent occasionner dans la chimie."[6]

This passage appears unchanged in the second edition now entitled *Elémens d'histoire naturelle et de chimie*. Much of this work, as William Smeaton has shown, had been written before July 1784, although the preliminary discourse reveals that before the book was published in 1786 Fourcroy had become a convert to Lavoisier's doctrine.[7] In the third edition (1789), rewritten to accord with the new theories and the reformed nomenclature, the passage in question was altered here and there to reflect the author's change of view.[8] The mention of Macquer, with its reference to a "grande révolution," is unaltered; but Fourcroy added a new paragraph in which he remarked that since Macquer's death (1784) chemistry has been enriched by so many new discoveries "que la théorie moderne acquiert de jour en jour de nouvelles forces" and that those who still defend the phlogistic doctrine "ne sont pas parfaitement au courant de la science, ou qu'il leur manque quelque chose dans l'art des expériences."[9] The great popularity of Fourcroy's *Elémens*, an influential vehicle for disseminating the New Chemistry, unquestionably played its part in adding the phrase "revolution in chemistry" to the lexicon of the history of science. But Fourcroy's repeated use of the phrase in later writings surely enhanced his influence in this respect.

Fourcroy was not alone in speaking of a "revolution in chemistry" before Lavoisier put the capstone on the New Chemistry with his *Traité élémentaire de chimie* in 1789. A year earlier appeared the French translation of Richard Kirwan's *Essay on Phlogiston and the Constitution of Acids*. The translation is anonymous, but has been generally credited (on Grimaux's authority) to Madame Lavoisier. Her preface, in which Lavoisier may have had a hand, explains why footnotes by Lavoisier and his disciples have been added for a point-by-point refutation of the Irish chemist, remarking that without these notes "ce travail n'auroit pas suffisamment avancé *la révolution qui se prépare en Chimie. . . .*"[10]

Nevertheless, the earliest to use the word "revolution" in print, when speaking of the dramatic changes in chemistry, would seem to have been Fourcroy's teacher, J.-B.-M. Bucquet, one of the earliest in France to experiment on "fixed air" and for a time a collaborator of Lavoisier. In a neglected book on gases he published in 1778, Bucquet remarked that few discoveries besides those made concerning gases have "opéré une révolution plus grande dans la chimie, & qui ait plus contribué aux progrès de cette belle science." Yet, after all, it is Fourcroy who most effectively canonized the expression "the revolution in chemistry" or its equivalent.[11]

There are several later occurrences of the phrase in Fourcroy's writings. On February 6, 1789, he read to the Société Royale de Médecine a long review (*approbation* or *compte rendu*) of Lavoisier's most famous book, his *Traité élémentaire de chimie*. Signed by Fourcroy and J. de Horne, it was written and presented by the former, printed for the first time as an appendix to the second issue of the first edition of Lavoisier's *Traité*, and published in subsequent editions. Here Fourcroy wrote: "Les physiciens et tout les hommes qui s'adonnent à l'étude de la philosophie naturelle savent que c'est aux expériences de M. Lavoisier qu'est due la révolution que la chimie a éprouvé depuis quelques années. . . ."[12] The key word appears again in Fourcroy's *Notice sur la vie et les travaux de Lavoisier*, read to the Lycée des Arts on August 2, 1796, and which, with the exception of Lalande's, is the earliest biographical sketch of Lavoisier. Here Fourcroy spoke of "le plan de la grande révolution qu'il [Lavoisier] méditoit depuis plusieurs années."[13] To understand what Fourcroy meant in his *compte rendu* and in his *Notice* by "a revolution in chemistry" we may focus on the sentence of the latter text where he expresses his opinion that the period of greatest and most significant change occurred between the years 1780 and 1788:

> Ces huit années ont été marquées par les découvertes les plus importantes, par les travaux les plus suivis; c'est dans leur cours que se sont opérés les changements les plus singuliers et les plus heureux dans la marche et les fondements de la science chimique.[14]

As for Lavoisier, the leader of the revolution in chemistry, Fourcroy wrote of him:

> En un mot, il a été un de ces philosophes, un de ces génies originaux et rares, qui impriment aux connoissances humaines un caractère différent de celui qu'elles avoient avant eux, et qui leur communiquent un mouvement, une direction que rien n'annonçoit qu'elles dussent prendre.[15]

It should not surprise us that Fourcroy referred again to the revolution in chemistry in the Dictionary of Chemistry of Panckoucke's *Encyclopédie méthodique*. Here, in the article "Chimie," a veritable history of chemistry in eighteenth-century Europe, Fourcroy's vision is broader, and the "révolution" is seen as a more gradual and extended process to which other nations, and not France alone, made major contributions. The time scale of the "révolution" is extended: he speaks of "la grande révolution que la chimie a éprouvée depuis 1756, & surtout de 1766 & 1772 jusqu'en 1788, où son sort a été en quelque sorte fixé. . . ."[16] In this long, detailed article he accords ample credit to the British, and to his countryman Venel. Indeed he dates the revival of activity, the beginning of the revolution, from the decade of Venel's papers and the time of the publication of Joseph Black's famous memoir in the *Essays and Observations* of the Philosophical Society of Edinburgh.

A number of writers after Fourcroy applied the word "revolution" to the New Chemistry; the word, if not the precise phrase, was current well before Berthelot immortalized the phrase "La Révolution Chimique" in the title of his book. I need only cite a few that come readily to mind. For example, Thomas Thomson in his two-volume *History of Chemistry* (1830–1831) referred in the second volume (of 1831) to Lavoisier as a man "destined to produce a complete revolution [in chemistry]." Not long after, J.-B. Dumas, in his *Leçons de philosophie chimique* (n.d., but probably 1836), wrote (p. 130) that Lavoisier's use of the balance and its "application à l'étude des phénomènes naturels devait révolutionner la chimie et pouvait seule la révolutionner. . . ." Finally, Ferdinand Hoefer began the chapter on Lavoisier in his *History of Chemistry* with the words: "La révolution opérée dans la science par Lavoisier coïncide—singularité du destin!—avec une autre révolution, bien plus grande encore, opérée dans le monde politique et social."[17]

Notes

1. The title *The Chemical Revolution* was used in a quite different sense by Archibald and Nan Clow in 1952 in a book devoted to the chemical technology of the eighteenth and early nineteenth centuries. Of the irrelevance of this usage to his questions, my correspondent was well aware.
2. Berthelot in his *La Révolution Chimique—Lavoisier* (Paris, 1890), gave the date as February 20, 1772, just as it appears at the opening of Lavoisier's first *registre*, a notebook with scientific entries, begun early in 1773. Grimaux, without explanation, gives the correct date in his famous biography. Meldrum in his 1930 book, *The Eighteenth Century Revolution in Science*, argued that Lavoisier was guilty of the sort of clerical mistake one often makes in dating letters or checks at the beginning of the new year. Douglas McKie followed Meldrum in this matter, as in so much else, in his *Antoine Lavoisier, the Father of Modern Chemistry* (Philadelphia: J. B. Lippincott, 1935), summarizing Meldrum's rather inconclusive arguments (see p. 123). In his later book, *Antoine Lavoisier: Scientist, Economist, Social Reformer* (New York: H. Schuman, 1952), McKie merely said "Lavoisier wrote a long memorandum . . . on February 20, 1773, misdating it 1772" (p. 103). The French tended to follow Berthelot, favoring the earlier date. In his first book on Lavoisier, *Lavoisier* (Paris: Gallimard, 1941), Maurice Daumas adopted Berthelot's dating; in his review of McKie's 1952 book and in his own *Lavoisier, théoricien et expérimentateur* (Paris: Presses Universitaires de France, 1955), Daumas questioned Meldrum's emendation. It was obvious that if the stages of Lavoisier's work were to be properly understood it was essential that the question be resolved. I discussed the problem briefly at the close of my article, "Joseph Priestley's First Papers on

Gases," *Journal of the History of Medicine and Allied Sciences* 12 (1957), 12. Lavoisier's reference to Joseph Priestley in the memorandum seemed conclusive evidence in support of Meldrum (and Grimaux), since Priestley had published nothing concerning fixed air as early as February 1772. Soon thereafter, in a review of my article, Maurice Daumas graciously accepted the Meldrum position on the basis of my new evidence. See *Archives Internationales d'Histoire des Sciences* 10 (1957), 151–52. Not long afterwards I found what I believe to be the clinching argument: in Lavoisier's list of men who had worked on air fixed in bodies appears the name of a certain De Smeth. His name could only have been included for his inaugural dissertation at Utrecht, *Dissertatio de aere fixo*, which he defended in the autumn of 1772. Nevertheless certain French authors still follow Berthelot. For example Léon Velluz in his *Vie de Lavoisier* (Paris: Plon, 1966) still assigns the programmatic memorandum to February 20, 1772.

3. My italics. It strikes me as trivial whether one speaks of a "chemical revolution" or a "revolution in chemistry" or, as Lavoisier does, of a "revolution in physics and chemistry." I have argued elsewhere (see my article "Lavoisier" in the *Dictionary of Scientific Biography*) that Lavoisier believed that much of his research belonged to experimental physics rather than to chemistry *stricto sensu*. It is worth recalling that his first book was entitled *Opuscules physiques et chimiques*, and that he wrote to Franklin that by the scheme used in his *Traité élémentaire de chimie* "la chimie s'est trouvée beaucoup plus rapprochée qu'elle ne l'était de la physique expérimentale." See below, n. 4.

4. René Fric, "Une lettre inédite de Lavoisier à B. Franklin," *Bulletin historique et scientifique de l'Auvergne*, 2d ser., no. 9 (1924).

5. Edouard Grimaux, *Lavoisier, 1743–94*, 2d ed. (Paris, 1896), 126. The letters of Lavoisier to Franklin and to Chaptal were, of course, known to my correspondent.

6. *Leçons élémentaires d'histoire naturelle et de chimie*, 2 vols. (Paris, 1782), I, 22. That Fourcroy used this phrase before 1789 was called to my attention by my friend W. A. Smeaton, who kindly commented upon an early version of this paper. Neither Smeaton nor I have found any place where Macquer refers to a "révolution" in chemistry, although, as Smeaton pointed out to me, there are passages in the second edition of Macquer's *Dictionnaire de Chymie* where the new discoveries in chemistry are described as capable of "overturning" or "destroying" the long-held phlogiston theory. Smeaton suggests that when he used the word "révolution" Fourcroy was merely paraphrasing Macquer. I cannot follow Smeaton here: Fourcroy seems to have meant something more fundamental and wide-ranging than the overthrow of the phlogiston theory; and this, incidentally, is the view we hold today.

7. W. A. Smeaton, *Fourcroy* (Cambridge: W. Heffer & Sons, 1962), 14–15.

8. For example in the *Leçons* of 1782 (p. 21) Fourcroy writes "Le parti sans doute le plus sage & le seule que l'on doive prendre . . . est d'attendre, etc." In the *Elémens* of 1789 (p. 43) he changed the tense and wrote "le seul que l'on dût prendre . . . étoit d'attendre, etc."

9. *Elémens d'histoire naturelle et de chimie*, 3d ed., 5 vols. (Paris, 1789), I, 44.

10. *Essai sur le phlogistique, et sur la constitution des acides, traduit de l'Anglois de M. Kirwan; avec des notes de MM. Morveau, Lavoisier, de la Place, Monge, Berthollet & de Fourcroy* (Paris, 1788), p. viii. My italics.

11. *Mémoire sur la manière dont les animaux sont affectés par différens fluides aériformes méphitiques, etc.* (Paris, 1778), 2. I am grateful to Professor J. B. Gough for this reference.

12. *Oeuvres de Lavoisier*, 6 vols. (Paris, 1862–1893), I, 417. For the first appearance of this review see Denis I. Duveen and Herbert S. Klickstein, *A Bibliography of the Works of Antoine Laurent Lavoisier* (London: E. Weil and Wm. Dawson, 1954), 161. That Fourcroy was the author is evident from the opening sentence: "La Société nous a chargés, M. de Horne et moi, d'examiner, etc." Also there is Fourcroy's comment in the *Encyclopédie méthodique*: "Chargé alors par la société de médecine de lui rendre compte de cet ouvrage que Lavoisier lui présenta, je traçai dans un rapport lu à cette société, etc." See the article "Chimie" in the *Dictionnaire de chimie*, 6 vols. (Paris, 1786–1815), III, 454, of the *Encyclopédie méthodique*.

13. *Notice sur la vie et les trauvaux de Lavoisier, précédée d'un discours sur les funérailles et suivi d'une ode sur l'immortalité de l'âme* (Paris, Feuille du Cultivateur, An IV, 1796), 36. Fourcroy also wrote in his *Notice*, the following words about Lavoisier's *Traité*: "cet ouvrage commença pour les vrais connoisseurs une révolution dans la science," p. 30.

14. *Ibid.*, 35

15. *Ibid.*, 41.

16. *Encyclopédie méthodique*, the *Dictionnaire de chimie*, III, 715. Earlier in the same work (*ibid.*, 440), as Smeaton pointed out to me, Fourcroy uses the word in the more restricted sense and speaks of the revolution in chemistry that Lavoisier produced in France. The two passages were written at different times; the more extended use of the word on p. 715 is doubtless later. On this page and those immediately preceding, there are references to the year 1797, a year later than the An IV appearing on the title page of vol. III.

17. Ferdinand Hoefer, *Histoire de la chimie*, 2d ed., 2 vols. (Paris, 1866–69), II, 489.

IV
SCIENCE IN FRENCH CULTURE

28

Vauban:
The Impact of Science on War

✺

An almost uninterrupted state of war existed in Europe from the time of Machiavelli to the close of the War of the Spanish Succession. The French invasion of
Italy which had so roused Machiavelli proved only a prelude to two centuries of
bitter international rivalry, of Valois and Bourbon against Hapsburg. For a good
part of this period epidemic civil wars cut across the dynastic struggle, never
quite arresting it, and often fusing with it to produce conflicts of unbridled
bitterness. Toward the end of the seventeenth century, when civil strife had
abated and the chief states of Europe were at last consolidated, the old struggle
was resumed as part of Louis XIV's bid for European supremacy, but with a
difference: for now the newly risen merchant powers, Holland and England,
which had aided France in bringing the Spanish dominion to an end, were
arrayed against her. The Peace of Utrecht (1713) was an English peace. It set
the stage for England's control of the seas, but by the same token it did not
weaken France as much as her continental rivals had fervently desired. It left
France's most important conquests virtually intact; it scarcely altered the instrument of Westphalia which was her charter of security; and above all it left her
army—the first great national army of Europe—weakened but still formidable,
and her prestige as the leading military power of the continent virtually un-
diminished.

The military progress of two hundred years was embodied in that army.
And this progress had been considerable.[1] In the first place armies were larger.
Impressed as we are by the first appearance of mass armies during the wars of the
French Revolution, we are prone to forget the steady increase in size of European
armies that took place during the sixteenth and seventeenth centuries. When
Richelieu, for example, built up France's military establishment to about 100,000
men in 1635, he had a force nearly double that of the later Valois kings; yet this
force was only a quarter as large as that which Louvois raised for Louis XIV.

From Edward Meade Earle, ed., *Makers of Modern Strategy: Military Thought from
Machiavelli to Hitler* (Princeton: Princeton University Press, 1943). Copyright, 1943,
by Princeton University Press.

This expansion of the military establishment was primarily due to the growing importance of the infantry arm, which was only twice as numerous as the cavalry in the army with which Charles VIII invaded Italy, but five times as great by the end of the seventeenth century. The customary explanation for this new importance of infantry is that it resulted from the improvement in firearms; and it is true that the invention of the musket, its evolution into the flintlock, and the invention of the bayonet all led to a pronounced increase in infantry fire power, and hence to an extension of foot soldiery. But this is only part of the story. The steadily mounting importance of siege warfare also had its effects, for here—both as a besieging force and in the defense of permanent fortifications—infantry performed functions impossible to cavalry.

European armies in the seventeenth century were bands of professionals, many of them foreigners, recruited by voluntary or forced enlistment. Except for infrequent recourse to the *arrière-ban*, a feudal relic more often ridiculed than employed, and except for the experiment of a revived militia late in the reign of Louis XIV, there was nothing in France resembling universal service. In still another respect this "national" army seems, at first glance, hardly to have been representative of the nation. Whereas the nobility competed for admission into the elite corps of the cavalry and provided officers for the infantry, and whereas the common infantryman was drawn from the lowest level of society—though not always or preponderantly from the moral dregs, as is sometimes implied—the prosperous peasant freeholder and the members of the bourgeoisie escaped ordinary military service whether by enlistment, which they avoided, or through the revived militia, from which they were exempt.

Did one whole segment of society, then, fail to contribute to the armed strength of the country? By no means. The bourgeoisie made important contributions to French military strength, even though they did not serve in the infantry or the cavalry. Their notable contributions fell into two main categories. First, they were important in the technical services, that is to say in artillery and engineering and in the application of science to warfare; and second, they were prominent in the civilian administration of the army which developed so strikingly during the seventeenth century, and to which many other advances and reforms are attributable. These technical and organizational developments are perhaps the most important aspects of the progress that has been noted above. In both, the French army led the way.

❁

The army which Louis XIV passed on to his successors bore little resemblance to that of the Valois kings. The improvement in organization, discipline, and equipment was due chiefly to the development of the civilian administration at the hands of a succession of great planners—Richelieu, Le Tellier, Louvois, and Vauban—whose careers span the seventeenth century.

Until the seventeenth century army affairs were almost exclusively administered by the military themselves, and there was very little central control. The various infantry companies, which had at first been virtually independent under their respective captains, had, it is true, been coordinated to some extent by uniting them into regiments, each commanded by a *mestre de camp*, subject to the orders of a powerful officer, the *colonel général de l'infanterie*. But the prestige and independence of this high office were such as to weaken, rather than to strengthen, the hold of the crown over the newly regimented infantry. The cavalry, in the sixteenth century, had likewise been only imperfectly subjected to the royal will. By virtue of their prestige and tradition, the cavalry companies resisted incorporation into regiments until the seventeenth century. The elite corps of the gendarmerie, representing the oldest cavalry units, were controlled only by their captains and by a superior officer of the crown, the constable, who was more often than not virtually independent of the royal will. The light cavalry, after the reign of Henry II, was placed under a *colonel général* like that of the infantry. Only the artillery provided something of an exception. Here bourgeois influence was strong, a tradition dating back to the days of the Bureau brothers, and the effective direction was in the hands of a *commissaire général d'artillerie*, usually a man of the middle class. But even here the titular head was the grand master of artillery who, since the beginning of the sixteenth century, was invariably a person of high station. Thus, the army manifested a striking lack of integration. Other than the person of the king, there was no central authority. And except in the artillery there were no important civilian officials.

Richelieu laid the foundations of the civil administration of the army by extending to it his well-known policy of relying upon middle-class agents as the best means of strengthening the power of the crown. He created a number of *intendants d'armée*, who were usually provincial *intendants* selected for special duty in time of war, one to each field army. Responsible to the *intendants* were a number of *commissaires* to see to the payment of troops, the storage of equipment, and other similar matters. Finally it was under Richelieu that the important post of minister of war was to all intents and purposes created. Under two great ministers, Michel le Tellier (1643–68) and his son, the Marquis de Louvois (1668–91), the prestige of this office and the complexity of the civilian administration associated with it increased mightily. Around the person of the minister there grew up a genuine departmentalized government office complete with archives. By 1680 five separate bureaus had been created, each headed by a *chef de bureau* provided with numerous assistants. It was to these bureaus that the *intendants*, the commissioners, even commanding officers, sent their reports and their requests. From them emanated the orders of the minister of war; for only persons of great importance dealt directly with the minister, who had thus become, in all that pertained to important military decisions, the king's confidential adviser.

Judged by modern, or even Napoleonic, standards, the French army of Louis

XIV was by no means symmetrically organized. There were gross defects of all sorts, anomalies of organization and administration, vices of recruitment and officering. But this army was no longer an anarchic collection of separate units, knowing no real master but the captain or colonel who recruited them. If it possessed a clearly defined military hierarchy with clearly defined powers, and if the royal authority could no longer with impunity be evaded by underlings or challenged by rebellious commanders, this was made possible by the painstaking work of the civilian administration during the seventeenth century. The great, semi-independent offices of the crown were abolished or brought to heel. Reforms were effected within the hierarchy of general officers to make powers more clear-cut and to eliminate vagueness of function and incessant rivalry among the numerous marshals and lieutenants general. The principle of seniority was introduced. Unity of command was made possible by creating the temporary and exceptional rank of *maréchal général des armées*, held for the first time by Turenne in 1660. A host of minor reforms were also put through during this creative period, touching such diverse matters as the evil of plurality of office within the army, which was severely checked, venality of office, which proved ineradicable, the introduction of uniform dress and discipline, and improvements in the mode of recruiting, housing, and paying the troops.

Doubtless this sustained effort to systematize and order the structure of the army reflected what was taking place in other spheres. Throughout French political life traditional rights, and confusions sanctified by long usage, were being attacked in the interest of strengthening the central power. This cult of reason and order was not merely an authoritarian expedient, or just an aesthetic ideal imposed by the prevailing classicism. Impatience with senseless disorder, wherever encountered, was one expression, and not the least significant expression, of the mathematical neo-rationalism of Descartes, of the *esprit géométrique* detected and recorded by Pascal. It was the form in which the scientific revolution, with its attendant mechanical philosophy, first manifested itself in France. And it resulted in the adoption of the machine—where each part fulfilled its prescribed function, with no waste motion and no supernumerary cogs—as the primordial analogy, the model not only of man's rational construction, but of God's universe. In this universe the cogs were Gassendi's atoms or Descartes's vortices, while the *primum mobile* was Fontenelle's divine watchmaker. We often speak as though the eighteenth or the nineteenth century discovered the worship of the machine, but this is a half-truth. It was the seventeenth century that discovered the machine, its intricate precision, its revelation—as, for example, in the calculating machines of Pascal and Leibnitz—of mathematical reason in action. The eighteenth century merely gave this notion a Newtonian twist, whereas the nineteenth century worshiped not the machine but power. So in the age of Richelieu and Louis XIV the reformers were guided by the spirit of the age, by the impact of scientific rationalism, in their efforts to modernize both the army and the civilian bureaucracy, and to give to the state and the army some of the qualities of a well-designed machine. Science, however, was exerting other and more direct effects upon military affairs, and to these we must now turn.

⚙

Science and warfare have always been intimately connected. In antiquity this alliance became strikingly evident in the Hellenistic and Roman periods. Archimedes's contribution to the defense of Syracuse immediately springs to mind as the classic illustration. The cultural and economic rebirth of western Europe after the twelfth century shows that this association was not fortuitous, for the revival of the ancient art of war was closely linked with the recovery and development of ancient scientific and technical knowledge.[2] Few of the early European scientists were soldiers, but many of them in this and later centuries served as consulting technicians or even as technical auxiliaries of the army. A number of military surgeons have their place in the annals of medical or anatomical science; while still more numerous were the engineers, literally the masters of the engines, whose combined skill in military architecture, in ancient and modern artillery, and in the use of a wide variety of machines served equally to advance the art of war and to contribute to theoretical science. Leonardo da Vinci, the first great original mind encountered in the history of modern science, was neither the first nor the last of these versatile military engineers, although he is probably the greatest.

Throughout the sixteenth century and most of the seventeenth, before the technical corps of the army had really developed, a number of the greatest scientists of Italy, France, and England turned their attention to problems bearing upon the technical side of warfare. By the year 1600 it was generally realized that the service of outside specialists must be supplemented by some sort of technical training among the officers themselves. All the abortive projects for systematic military education, such as the early plans of Henry IV and of Richelieu, gave some place to elementary scientific training.[3] The great Galileo outlined in a little-known document a rather formidable program of mathematical and physical studies for the future officer. Although organized military education, to say nothing of technical education, had to await the eighteenth century, nearly every officer of any merit by the time of Vauban had some smattering of technical knowledge, or regretted that he had not. The developments of science that brought this about are best described by a brief survey of the changes in military architecture and in artillery.

The art or science of military architecture suffered a violent revolution in the century following the Italian wars of Machiavelli's time. The French artillery—using the first really effective siege cannon—had battered down with ridiculous ease the high-walled medieval fortifications of the Italian towns. The Italians' reply was the invention of a new model *enceinte*—the main enclosure of a fortress—which, improved by a host of later modifications, was to prevail in Europe until the early nineteenth century. It was characterized primarily by its outline or trace: that of a polygon, usually regular, with bastions projecting from each angle, in such a manner as to subject the attacker to an effective cross-fire. As it was perfected by the later Italian engineers this *enceinte* consisted of three main divisions: a thick low rampart, with parapet; a broad ditch; and an outer rampart, the glacis, which sloped down to the level of the surrounding countryside.

Designing these fortresses became a learned art, involving a fair amount of mathematical and architectural knowledge. A number of scientists of the first rank were experts in this new field of applied science. The Italian mathematician Tartaglia and the great Dutch scientist Simon Stevin were as famous in their own day as engineers as they are in ours for their contributions to mathematics and mechanics. Even Galileo taught fortification at Padua.[4]

Francis I of France, aware of the skill of the Italian engineers, took a number of them into his service, using them in his pioneer efforts to fortify his northern and eastern frontiers against the threat of Charles V. This first burst of building activity lasted throughout the reign of Henry II, only to be brought to a halt by the civil wars. When the work was resumed under Henry IV and Sully, the Dutch were beginning to contest the primacy of the Italians in this field, and French engineers like Errard de Bar-le-duc were available to replace the foreigners.[5]

Errard is the titular founder of the French school of fortification, which may be said to date from the publication of his *Fortification réduicte en art* (1594). In the course of the seventeenth century there appeared a number of able engineers, some of them soldiers, others civilian scientists of considerable distinction. Among the men in the latter category can be mentioned Gérard Desargues, the great mathematician; Pierre Petit, a versatile scientist of the second rank; and Jean Richer, astronomer and physicist. In the development of the theory of fortification the great precursor of Vauban, one might almost say his master, was the Comte de Pagan.

Blaise de Pagan (1604–65) was a theorist, not a practical engineer. So far as is known he never actually directed any important construction. In engineering—as in science, where he fancied himself more than the dilettante that he really was—his contributions were made from the armchair. He succeeded, however, in reforming in several important respects the type of fortresses built by the French in the later seventeenth century. Vauban's famous "first system" was in reality nothing but Pagan's style, executed with minor improvements and flexibly adapted to differences in terrain. Pagan's main ideas were embodied in his treatise *Les fortifications du comte de Pagan* (1645); they all sprang from a single primary consideration: the increased effectiveness of cannon, both for offense and in defense. To Pagan the bastions were the supremely important part of the outline, and their position and shape were determined by the help of simple geometrical rules that he formulated, with respect to the outside, rather than the inside, of the *enceinte*.

In the development of artillery there was the same interplay of scientific skill and military needs during the sixteenth and seventeenth centuries. Biringuccio's *De la pirotechnia* (1540), now recognized as one of the classics in the history of applied chemistry, was for long the authoritative handbook of military pyrotechnics, the preparation of gunpowder, and the metallurgy of cannon. The theory of exterior ballistics similarly was worked out by two of the founders of modern dynamics, Niccolo Tartaglia and Galileo. Perhaps it would not be too much to assert that the foundations of modern physics were a by-product of solving the fundamental ballistical problem. Tartaglia was led to his criticisms of Aristotelian dynamics by experiments—among the earliest dynamical experiments ever performed—on the relation between the angle of fire and the range of a projectile. His results, embodying the proof that the angle of maximum range is 45°, brought about the widespread use of the artillerist's square or quadrant. But to Galileo is due the fundamental discovery that the trajectory of a projectile, for the ideal case that neglects such disturbing factors as air resistance, must be parabolic. This was made possible only by his three chief dynamical discoveries, the principle of inertia, the law of freely falling bodies, and the principle of the composition of velocities. Upon these discoveries, worked out as steps in his ballistic investigation, later hands erected the structure of classical physics.

By the end of the seventeenth century the progress of the "New Learning" had become compelling enough to bring about the first experiments in technical military education and the patronage of science by the governments of England and France. The Royal Society of London received its charter at the hands of Charles II in 1662, while four years later, with the encouragement of Colbert, the French Académie Royale des Sciences was born. In both of these organizations, dedicated as they were at their foundation to "useful knowledge," many investigations were undertaken of immediate or potential value to the army and navy. Ballistic investigations, studies on impact phenomena and recoil, researches on improved gunpowder and the properties of saltpeter, the quest for a satisfactory means of determining longitude at sea: these, and many other subjects, preoccupied the members of both Academies. In both countries able navy and army men are found among the diligent members. In France especially the scientists were frequently called upon for their advice in technical matters pertaining to the armed forces. Under Colbert's supervision scientists of the *Académie des Sciences* carried out a detailed coast and geodetic survey as part of Colbert's great program of naval expansion, and what is perhaps more important, they laid the foundations for modern scientific cartography so that in the following century, with the completion of the famous Cassini map of France, an army was for the first time equipped with an accurate map of the country it was charged to defend.

If we ask how these developments are reflected in the military literature of the sixteenth and seventeenth centuries, the answer is simple enough: the volume is, on the average, greater than the quality. Antiquity was still the great teacher in all that concerned the broader aspects of military theory and the secrets of military success. Vegetius and Frontinus were deemed indispensable; and the most popular book of the century, Henri de Rohan's *Parfait Capitaine*, was inspired by Caesar's *Gallic Wars*. Without doubt the most important writings concerned with the art of war fell into two classes: the pioneer works in the field of international law and the pioneer works of military technology.

Machiavelli had been the theorist for the age of unregulated warfare, but his influence was waning by the turn of the seventeenth century. Francis Bacon was perhaps his last illustrious disciple; for it is hard to find until our own day such unabashed advocacy of unrestricted war as can be found in certain of his *Essays*. But by Bacon's time the reaction had set in. Men like Grotius were leading the attack against international anarchy and against a war of unlimited destructiveness. These founding fathers of international law announced that they had found in the law of nature the precepts for a law of nations, and their central principle, as Talleyrand put it once in a strongly worded reminder to Napoleon, was that nations ought to do one another in peace, the most good, in war, the least possible evil.

It is easy to underestimate the influence of these generous theories upon the actual realities of warfare, and to cite Albert Sorel's bleak picture of international morals and conduct in the period of the Old Régime. Actually the axioms of international law exerted an undeniable influence upon the mode and manner of warfare before the close of the seventeenth century.[6] If they did not put an end to political amoralism, they at least hedged in the conduct of war with a host of minor prescriptions and prohibitions which contributed to making eighteenth-century warfare a relatively humane and well-regulated enterprise. These rules were known to contending commanders and were quite generally followed. Such, for example, were the instructions concerning the treatment and exchange of prisoners; the condemnation of certain means of destruction, like the use of poison; the rules for the treatment of noncombatants and for arranging parleys, truces, and safe-conducts; or those concerned with despoiling or levying exactions upon conquered territory and with the mode of terminating sieges. The whole tendency was to protect private persons and private rights in time of war, and hence to mitigate its evils.

In the second class, that of books on military technology, no works had greater influence or enjoyed greater prestige than those of Sébastien le Prestre de Vauban, the great military engineer of the reign of Louis XIV. His authority in the eighteenth century was immense, nor had it appreciably dimmed after the time of Napoleon.[7] And yet Vauban's literary legacy to the eighteenth century was scanty and highly specialized, consisting almost solely of a treatise on siegecraft, a work on the defense of fortresses, and a short work on mines.[8] He published nothing on military architecture and made no systematic contribution to strategy or the art of war in general; yet his influence in all these departments is undeniable. It was exerted subtly and indirectly through the memory of his career and of his example, and by the exertions and writings of a number of his disciples. But by this process many of his contributions and ideas were misunderstood and perverted, and much that he accomplished was for a long time lost to view. Thanks to the work of scholars of the nineteenth and twentieth centuries, who have been able to publish an appreciable portion of Vauban's letters and manuscripts, and to peruse and analyze the rest, we have a clearer understanding of Vauban's career and of his ideas than was possible to his eighteenth-century admirers. He has increased in stature, rather than diminished, in the light of modern studies. We have seen the Vauban legend clarified and documented; we have seen it emended in many important points; but we have not seen it exploded.

⚙

The Vauban legend requires some explanation. Why was a simple engineer, however skillful and devoted to his task, raised so swiftly to the rank of a national idol? Why were his specialized publications on siegecraft and the defense of fortresses sufficient to rank him as one of the most influential military writers?

The answers are not far to seek: these works of Vauban were the authoritative texts in what was to the eighteenth century a most important, if not the supremely important, aspect of warfare. In the late seventeenth century and throughout the eighteenth century, warfare often appears to us as nothing but an interminable succession of sieges. Almost always they were the focal operations of a campaign: when the reduction of an enemy fortress was not the principal objective, as it often was, a siege was the inevitable preliminary to an invasion of enemy territory. Sieges were far more frequent than pitched battles and were begun as readily as battles were avoided. When they did occur, battles were likely to be dictated by the need to bring about, or to ward off, the relief of a besieged fortress. The strategic imagination of all but a few exceptional commanders was walled in by the accepted axioms of a war of siege. In an age that accepted unconditionally this doctrine of the strategic primacy of the siege, Vauban's treatises were deemed indispensable and his name was necessarily a name to conjure with.

Yet only a part of the aura and prestige that surrounded Vauban's name arose from these technical writings. He has appealed to the imagination because of his personal character, his long career as an enlightened servant of the state, his manifold contributions to military progress outside of his chosen specialty, and his liberal and humanitarian interest in the public weal. From the beginning it was Vauban the public servant who aroused the greatest admiration. With his modest origin, his diligence and honesty, his personal courage, and his loyalty to the state, he seemed the reincarnation of some servitor of the Roman Republic. Indeed, Fontenelle, in his famous *éloge*, describes him as a "Roman, whom the century of Louis XIV seems almost to have stolen from the happiest days of the Republic." To Voltaire he was "the finest of citizens." Saint-Simon, not content with dubbing him a Roman, is said to have applied to him, for the first time with its modern meaning, the word *patriote*.[9] In Vauban, respected public servant, organizational genius, enlightened reformer, seemed to be embodied all the traits which had combined, through the efforts of countless lesser persons, to forge the new national state.

Still more felicitously did Vauban's technical knowledge, his skill in applied mathematics, his love of precision and order, and his membership in the Académie des Sciences symbolize the new importance of scientific knowledge for the welfare of the state. Cartesian reason, the role of applied science in society both for war and peace, the *esprit géométrique* of the age: all these were incarnated in the man, visible in the massive outline of the fortresses he designed.

❁

Vauban's career was both too long and too active for anything but a summary account in an essay of this sort. Scarcely any other of Louis XIV's ministers or warriors had as long an active career. He entered the royal service under Mazarin when he was in his early twenties and was still active in the field only a few months before his death at the age of seventy-three. During this half century of ceaseless effort he conducted nearly fifty sieges and drew the plans for well over a hundred fortresses and harbor installations.

He came from the indeterminate fringe between the bourgeoisie and the lower nobility, being the descendant of a prosperous notary of Bazoches in the Morvan who in the mid-sixteenth century had acquired a small neighborhood fief. He was born at Saint Léger in 1633, received his imperfect education—a smattering of history, mathematics, and drawing—in near-by Semur-en-Auxois; and in 1651, at the age of seventeen, enlisted as a cadet with the troops of Condé, then in rebellion against the king. Sharing in Condé's pardon, in 1653 he entered the royal service, where he served with distinction under the Chevalier de Clerville, a man of mediocre talents who was regarded as the leading military engineer of France. Two years later he earned the brevet of *ingénieur ordinaire du roi*, and soon after acquired as a sinecure the captaincy of an infantry company in the regiment of the Maréchal de la Ferté.

During the interval between the cessation of hostilities with Spain in 1659 and Louis XIV's first war of conquest in 1667, Vauban was hard at work repairing and improving the fortifications of the kingdom under the direction of Clerville.

In 1667 Louis XIV attacked the Low Countries. In this brief War of Devolution Vauban so distinguished himself as a master of siegecraft and the other branches of his trade that Louvois noticed his distinct superiority to Clerville and made him the virtual director, as *commissaire général*, of all the engineering

work in his department. The acquisitions of the War of Devolution launched
Vauban on his great building program. Important towns in Hainaut and Flanders
were acquired, the outposts of the great expansion: Bergues, Furnes, Tournai,
and Lille. These and many other important positions were fortified according to
the so-called "first system" of Vauban, which will be discussed below.

This, then, was to be the ceaseless rhythm of Vauban's life in the service of
Louis XIV: constant supervision, repairs, and new construction in time of peace;
in time of war, renewed sieges and further acquisitions; then more feverish construction during the ensuing interval of peace. In the performance of these duties
Vauban was constantly on the move until the year of his death, traveling from
one end of France to the other on horseback or, later in life, in a famous sedan
chair borne by horses. There seem to have been few intervals of leisure. He devoted little time to his wife and to the country estate he acquired in 1675, and
he sedulously avoided the court, making his stays at Paris and Versailles as short
as possible. The greatest number of his days and nights were spent in the inns of
frontier villages and in the execution of his innumerable tasks, far from the
centers of culture and excitement. Such free moments as he was able to snatch in
the course of his engineering work he devoted to his official correspondence and
to other writing. He kept in constant touch with Louvois, whom he peppered
with letters and reports written in a pungent and undoctored prose. As though
this were not enough Vauban interested himself in a host of diverse civil and
military problems only indirectly related to his own specialty. Some of these
subjects he discussed in his correspondence, while he dealt with others in long
memoirs which are included in the twelve manuscript volumes of his *Oisivetés*.

These memoirs treat the most diverse subjects. Some are technical, others
are not. But in nearly all of them he answers to Voltaire's description of him
as "un homme toujours occupé de sujets les uns utiles, les autres peu practicables
et tous singuliers."[10] Besides discussing military and naval problems, or reporting
on inland waterways and the interocean Canal of Languedoc, he writes on the
need for a program of reforestation, the possible methods of improving the state
of the French colonies in America, the evil consequences of the revocation of the
Edict of Nantes, and—in a manner that foreshadowed Napoleon's creation of
the Legion of Honor—the advantages of instituting an aristocracy of merit open
to all classes, in place of the archaic nobility of birth and privilege.

The *Oisivetés* reveal their origin and belie their name. They were written
at odd times, in strange places, and at various dates. They are often little more
than notes and observations collected in the course of his travels over the length
and breadth of France; others are extended treatises. What gives the writings a
certain unity is the humanitarian interest that pervades them all and the scientific spirit which they reveal. The writings and the career of Vauban illustrate the
thesis suggested earlier in this paper that in the seventeenth century scientific
rationalism was the wellspring of reform. Vauban's proposals were based on
firsthand experience and observation. His incessant traveling in the performance

of his professional duties gave him an unparalleled opportunity to know his own country and its needs. His wide curiosity and his alert mind led him to amass facts, with the pertinacity known only to collectors, about the economic and social conditions of the areas where he worked; and his scientific turn of mind led him to throw his observations, where possible, into quantitative form.

These considerations help us to answer the question whether Vauban deserves, in any fundamental sense, the label of scientist, or whether he was merely a soldier and builder with a smattering of mathematics and mechanical knowledge. Was membership in the Académie des Sciences accorded him in 1699 solely to honor a public servant and was Fontenelle thus obliged to devote to him one of his immortal *éloges* of men of science?

Vauban's achievements are in applied science and simple applied mathematics. He was not a distinguished mathematician and physicist like the later French military engineer Lazare Carnot. He made no great theoretical contributions to mechanical engineering, as did Carnot's contemporary Coulomb. He invented no steam chariot like Cugnot. Aside from the design of fortresses, scarcely a matter of pure science, his only contribution to engineering was an empirical study of the proper proportions of retaining walls.[11] Vauban's chief claim to scientific originality is that he sought to extend the quantitative method into fields where, except for his English contemporaries, no one had yet seriously ventured. He is, in fact, one of the founders of systematic meteorology, an honor that he shares with Robert Hooke, and one of the pioneers in the field of statistics, where the only other contenders were John Graunt and Sir William Petty.[12] His statistical habit is evident in his military and engineering reports. Many of these are filled with apparently irrelevant detail about the wealth, population, and resources of various regions of France.

From his harried underlings he exacted the same sort of painstaking survey. In a letter to Hue de Caligny, who was for a time director of fortifications for the northwest frontier from Dunkirk to Ypres, he expressed annoyance at the incomplete information he received in reports about the region. He urged Caligny to supply a map, to describe in detail the waterways, together with the wood supply with the date of cutting, and to provide him with detailed statistical information on population, broken down according to age, sex, profession, and rank. In addition Caligny was to give all the facts he could amass about the economic life of the region.[13] It was by information of this sort, painstakingly acquired as a by-product of his work as an army engineer, that Vauban sought to extend into civilian affairs the same spirit of critical appraisal, the same love of logic, order, and efficiency, which he brought to bear on military problems.

✿

Vauban was one of the most persistent military reformers of the century; his letters and his *Oisivetés* are filled with his proposals. There were few aspects of military life, or of the burning problems of military organization and military technology, where Vauban did not intervene with fertile suggestions or projects for over-all reorganization.[14]

The incorporation of his engineers into a regularly constituted arm of the service, possessed of its own officers and troops and its distinctive uniform, was something for which he struggled, though with little success, throughout his career.[15] His recommendations, however, bore fruit in the following century, as did also his efforts in the matter of scientific education for the technical corps. He enthusiastically praised the earliest artillery schools which were created toward the end of the reign of Louis XIV; and although he never succeeded in creating similar schools for the engineers, he established a system of regular examinations to test the preparation of candidates for the royal brevet, and took some steps to see that they were adequately prepared by special instructors.

Improvement of the artillery arm was a matter in which, as an expert on siegecraft, he was deeply interested. His studies and innovations in this field were numerous. He experimented with sledges for use in transporting heavy cannon; he found fault with the bronze cannon then in use, and tried to persuade the army to emulate the navy in the use of iron; he made numerous, but unsatisfactory, experiments on a new stone-throwing mortar. And finally he invented ricochet fire, first used at the siege of Philipsbourg, where the propelling charge was greatly reduced so that the ball would rebound this way and that after striking the target area, a peril to any man or machine in the near vicinity.

Vauban found space in his correspondence and in the *Oisivetés* to suggest numerous fundamental reforms for the infantry and for the army as a whole. He was one of the most tireless advocates of the flintlock musket and was the inventor of the first satisfactory bayonet. As early as 1669 he wrote to Louvois strongly urging the general use of flintlocks and the abolition of the pike; and shortly thereafter he specifically proposed to substitute for the pike the familiar bayonet with a sleeve or socket that held the blade at the side of the barrel, permitting the piece to be fired with bayonet fixed.

He was preoccupied with the condition and welfare of the men as well as with their equipment. He sought to improve still further the mode of recruiting and paying the troops. To him is due in part the limitation of the practice of quartering soldiers on the civilian population which, after the peace of Aix-la-Chapelle, was supplemented by the creation of *casernes*.[16] These special barracks, many of them designed and built by Vauban, were chiefly used in frontier regions and recently conquered territory.

Vauban made no systematic study of naval construction, and what he knew seems to have been learned from Clerville, who was skilled in this sort of work.[17] His first effort was at Toulon, where he improved the harbor installations; but his masterpiece was the port of Dunkirk. He devoted an interesting study to the naval role of galleys, in which he envisaged extending their use from the Mediterranean to the Atlantic coast, where they could serve as patrol vessels, as a mobile screen for heavier ships close to shore, or for swift harassing descents upon the Orkneys, or even upon the English coast. Closely related to these studies was his advocacy of the *guerre de course*, which he deemed the only feasible strategy after the collapse of the French naval power painstakingly built up by Colbert.

❁

Vauban's most significant contributions to the art of war were made, as was to be expected, within his own specialties: siegecraft and the science of fortification. It was characteristic of Vauban's dislike of unnecessary bloodshed, as much as of the new spirit of moderation in warfare that was beginning to prevail in his day, that his innovations in siegecraft were designed to regularize the taking of fortresses and above all to cut down the losses of the besieging force. Before his perfection of the system of parallels, which he probably did not invent, attacks on well-defended permanent fortifications took place only at a considerable cost to the attackers;[18] trenches and gabions were employed without system, and as often as not the infantry was thrown against a presumed weak point in a manner that left them exposed to murderous fire.

Vauban's system of attack, which was followed with but little variation during the eighteenth century, was a highly formalized and leisurely procedure. The assailants gathered their men and stores at a point beyond the range of the defending fire and concealed by natural or artificial cover. At this point the sappers would begin digging a trench that moved slowly toward the fortress. After this had progressed some distance, a deep trench paralleling the point of future attack was flung out at right angles to the trench of approach. This so-called "first parallel" was filled with men and equipment to constitute a *place d'armes*. From it, the trench of approach was moved forward again, zigzagging as it approached the fortress. After it had progressed the desired distance, the second parallel was constructed, and the trench was moved forward once more, until a third and usually final parallel was constructed only a short distance from the foot of the glacis. The trench was pushed ahead still further, the sappers timing their progress so as to reach the foot of the glacis just as the third parallel was occupied by the troops. The perilous task of advancing up the glacis, exposed to the enemy's raking fire from their covered way, was accomplished with the aid of temporary structures called *cavaliers de tranchées*, which were high earthworks, provided with a parapet, from which the besiegers could fire upon the defenders on the covered way. This outer line of defense could be cleared *par industrie*—that is, by subjecting the defenders to the effects of a ricochet bombardment—or by sending up grenadiers to take the position by assault under cover of a protecting fire from the *cavaliers*. Once the enemy's covered way was taken, siege batteries were erected and an effort was made to breach the main defenses.

The essential feature of Vauban's system of siegecraft, then, was the use he made of temporary fortifications, trenches, and earthworks, to protect the advancing troops. His parallels were first tried out at the siege of Maestricht in 1673, and the *cavaliers de tranchées* at the siege of Luxembourg in 1684. The per-

ATTAQUES REGULIERES

En terrein uni, la Tranchée suppofée ouverte à la portée du Canon.

A} Baftions du front de l'Attaque.
B}
C Demi-Lunes du même front.
D Prolongement des Capitales des Baftions attaquez A. B.
E Prolongement de la Capitale de la Demi-Lune C.
F Piquets bouchonnez de paille ou de mèche allumée fur le Prolongement, pour fervir à la conduite des Tranchées.
G Batteries à Ricochets des deux Faces & du Chemin couvert de la Demi-Lune C.
H Batterie à Ricochets de la Face gauche & du Chemin couvert du Baftion A.
I Batterie à Ricochets de la Face droite & du Chemin couvert du Baftion B.
K Batterie à Ricochets des deux autres Faces de ces deux Baftions A. B. & de leur Chemin couvert.
L Batteries à Ricochets des Faces & Chemins couverts des deux Demi-Lunes collaterales M. N.
O Batteries à Bomber.
P Places fur la feconde Ligne, où l'on pourroit mettre les Batteries, s'il étoit néceffaire de les changer.
Q Cavaliers de Tranchée qui enfilent le Chemin couvert de la Place.
R Demi-Places d'Armes.
S Piquets fur le Prolongement des Faces des Pièces attaquées pour l'établiffement des Batteries à Ricochets.
T Paffages que l'on fait en comblant la Place d'Armes avec des Fafcines, pour mener le Canon & les Mortiers à leurs Batteries.
V Redoutes qui terminent la feconde Place d'Armes.
X Chemin pour la communication des Attaques de la droite à la gauche.
Y Première Parallèle ou Place d'Armes.
Z Deuxième Parallèle où Place d'Armes.
& Troifième Parallèle ou Place d'Armes.

Planche V. pag. 29

Echelle de deux cens Toifes.

Vauban's system of attack by parallels

fected system is described at length in his *Traité des sièges*, written for the Duc de Bourgogne in 1705.

Vauban's work in military architecture has been the subject of considerable dispute, first as to whether the style of his fortresses showed great originality, second as to whether in placing them he was guided by any master plan for the defense of France.

Until very recently even Vauban's most fervent admirers have agreed that he showed little originality as a military architect and added almost nothing to the design of fortresses he inherited from Pagan. Lazare Carnot admired Vauban in the manner characteristic of other eighteenth-century engineers, yet he could find few signs of originality. "The fortification of Vauban reveals to the eye only a succession of works known before his time, whereas to the mind of the good observer it offers sublime results, brilliant combinations, and masterpieces of industry."[19] Allent echoes him: "A better cross section, a simpler outline, outworks that are bigger and better placed: these are the only modifications that he brought to the system then in use."[20] This judgment remained in vogue until very recent times. The most recent serious study, that of Lieutenant Colonel Lazard, has modified in Vauban's favor this somewhat unfavorable opinion.[21]

Lazard made important changes in our interpretation of Vauban's methods of fortification. Whereas earlier writers have had the habit of referring to Vauban's *three systems*, Lazard points out that, strictly speaking, Vauban did not have sharply defined systems; rather, he had periods in which he favored distinctly different designs, all modifications of the bastioned trace discussed above. With this restriction in mind, it is convenient to retain the old classification.

Vauban's *first system*, according to which he built the great majority of his fortified places, consisted in using Pagan's trace almost without modification. The outlines of these forts were, whenever possible, regular polygons: octagonal, quadrangular, even roughly triangular, as at La Kenoque. The bastions were still the key to the defensive system, though they tended to be smaller than those of Vauban's predecessors. Except for improvements of detail and the greater use of detached exterior defenses (such as the *tenailles*, the *demi-lunes*, and other items in Uncle Toby's lexicon), little had altered since the days of Pagan. Since, therefore, most of Vauban's structures were built according to this conservative design, and because this was taken as characteristic of Vauban's work, it is not to be wondered at that later critics could find there little or no originality. The originality, according to Lazard, is evident rather in those other two styles which had little influence on Vauban's successors and which were exemplified in only a few samples of his work.

The *second system*, used for the first time at Belfort and Besançon, was an outgrowth of that previously used. The polygonal structure was retained, but the curtains (the region between the bastions) were lengthened, and the bastions themselves were replaced by a small work or tower at the angles, these being covered by so-called detached bastions constructed in the ditch.

The so-called *third system* is only a modification of the second. It was used for only a single work, the great masterpiece of Vauban at Neuf-Brisach. In this scheme the curtain is modified in shape to permit an increased use of cannon in defense, and the towers, the detached bastions, and the *demi-lunes* are all increased in size.

It is the second system that deserves our attention. Here, although his contemporaries could not see it, Vauban had made an important, even revolutionary, improvement: he had freed himself from reliance upon the main *enceinte* and taken the first steps toward a defense in depth. He had gained a new flexibility in adapting his design to the terrain without imperiling the main line of defense. In all previous cases adaptation had been through projecting crown-works or horn-works that were merely spectacular appendages to the primary *enceinte*; and when these were taken the main line was directly affected. The second system was rejected by Cormontaigne, and later by the staff of the Ecole de Mézières, whose ideas dominated the later eighteenth century, and whose schemes of fortification were based squarely upon Vauban's first system. To them this second system seemed only a crude return to medieval methods. Only late in the eighteenth century do we find a revival of Vauban's second system: the revolt of Montalembert, which the Germans accepted long before the French, consisted chiefly in substituting small detached forts in place of the conventional projecting outworks, in reality part of the main *enceinte*.[22] Montalembert's great revolution, like the later advocacy of fortification in depth, was implicit in Vauban's second system, although whether Montalembert was inspired by it may well be doubted.

The confusion about his ideas that has existed until recently results from the fact that Vauban never wrote a treatise on the art of permanent fortification, never expounded it systematically as he did his theories of the art of attack and of defense. All the books which appeared in his own lifetime and thereafter, purporting to summarize his secrets, were the baldest counterfeits. Only the great work of Bélidor, which treated not of basic design or the problems of military disposition, but only of constructional problems and administrative detail, was directly inspired by Vauban.[23] There are, however, two treatises remaining in manuscript which deal with basic principles of fortification and which were directly inspired by him. One of these was written by Sauveur, the mathematician whom Vauban chose to instruct and to examine the engineer candidates; the other by his secretary, Thomassin. These are the best sources, aside from the works themselves, for learning Vauban's general principles of fortification. It is possible to speak only of general principles, not of a dogmatic system, and these principles are exemplified equally well by all three of the Vauban styles. They are few enough and quite general. First of all, every part of the fort must be as secure as every other, with security provided both through sturdy construction of the exposed points (bastions) and by adequate coverage of the curtains. In general these conditions will be provided for if (1) there is no part of the *enceinte* not flanked by strong points, (2) these strong points are as large as possible, and (3) they are separated by musket range or a little less. These strong points

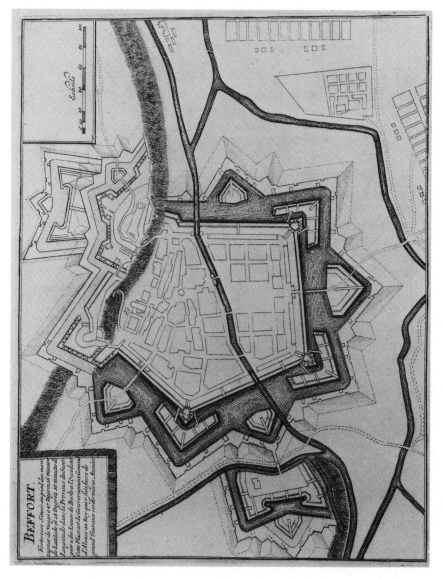

Vauban's second system: The fortress of Belfort

should be so designed that the parts which flank should always confront as directly as possible the parts they are protecting; conversely, the flanking parts should be visible only from the protected parts. A little thought will show that these basic principles are applicable to all of Vauban's schemes. The actual problem of building a permanent fortification consisted in adapting the bastioned trace (or the polygonal trace with detached bastions) to the exigencies of a particular terrain so that none of the basic principles was violated. Clearly this left the engineer a wide range of freedom and an admirable flexibility. It was by this method of work that the second style was developed, for Vauban himself tells us that it was not arrived at as a result of theoretical considerations but was forced on him by the terrain conditions at Belfort.[24]

❀

To what extent was the fortification program of Louis XIV guided by some unifying *strategic* conception; and what is the evidence that this conception, if in truth there was such a thing, was due to the genius of Vauban? These are two of the most important questions, but they are not the easiest to answer.

The earlier biographers of Vauban, with characteristic enthusiasm for their hero, leave us sometimes with the distinct impression that before Vauban France had no system of fortification worthy of the name, and that the ring of fortresses girding the kingdom by the end of his career represented the execution of some cleverly conceived master plan sprung from the mind of the great engineer. To these writers it was just as incredible that anyone but Vauban could have had a hand in organizing this defensive system as it was that this system itself might have been the result of a slow historical growth.

Of late we have drifted perhaps too far in the other direction. Although, as we have seen, Vauban's technical reputation as a military architect has been enhanced by recent studies, there has been a simultaneous tendency on the part of certain writers to reduce him to the level of a great craftsman devoid of strategic imagination. He has been represented as a brilliant technician, executing blindly the tasks dictated by geographical necessity or by the orders of superiors who alone did all the strategic thinking.

Who was there capable of challenging Vauban's authority in the field of his specialty? The answer is, the king himself. Louis XIV, it has been shown, was more than decently proficient in the art of fortification. He had studied it in his youth, and, during the early part of the reign, had profited by the advice and instruction of Turenne, Villeroi, and Condé. Throughout his career he showed a constant interest in the most humble details connected with the art of fortification and on a number of occasions he resolutely opposed insistent recommendations of Vauban. Two important forts, Fort Louis and Mont-Royal, were created on the initiative of the king, and one at least of these was against the express advice of Vauban.[25] To one author, Louis the Diligent was in everything, even in these technical matters, the unquestioned master. Louvois was only an "excellent servant, not to say clerk," while Vauban in his turn "was never anything but the executor of his orders, albeit . . . an excellent one."[26] Another writer describes Vauban as "the chief workman of a great undertaking, the direction of which was never fully entrusted to him."[27] This interpretation is in fact inescapable. Vauban drew or corrected all plans for fortresses that had been decided upon; he submitted technical memoirs and recommendations; he gave his opinion on crucial matters when asked and sometimes when he was not asked. But his presence was not deemed necessary when the decisions were being debated. He was not a policy maker; his was only a consultative voice.

This should not lead us to underestimate his influence upon the royal decisions. Yet even if Vauban had had a master plan for the defense of France, it could only have been imperfectly executed. Many recommendations dear to Vauban's heart were rejected; many of his schemes were shattered by the realities of war and diplomacy. The peace of Ryswick in 1697, for example, marked Louis XIV's first withdrawal from the high watermark of conquest. To Vauban, who was not directly consulted about its terms, this treaty, though not as bad as he feared, was a great deception. Much work had to be done over to make up for the loss of Luxembourg—which he considered one of the strongest places in Europe—and of Brisach, Fribourg, and Nancy.[28]

Did Vauban in reality have a master plan? On this question there is almost complete disagreement. The writers of the last century took it for granted that Vauban had a strategic pattern for his fortresses, although they were not altogether certain in what it consisted. One writer described it as "an assemblage of works sufficiently close to one another so that the intervals between them are not unprotected. Each of these works is strong enough and well provisioned enough to impose upon the enemy the obligation of a siege, yet small enough to demand only a small number of defenders."[29] With this interpretation Gaston Zeller is in categorical disagreement. He points out that Louis XIV and Vauban did not start work with a clean canvas, that neither of these men could have imposed a doctrinaire plan of defense without reference to the work that had gone before; and he indicates that many of the characteristics of the defense system were owing to Francis I, Sully, Richelieu, and Mazarin, to their building programs and their treaties. Just as the actual frontier of the France of Louis XIV was the culmination of a long-sustained national policy, just so the disposition of the fortress towns was "the resultant of a long succession of efforts to adapt the defensive organization of the kingdom to the changing outline of the frontier."[30] In support of Zeller's contention that the fortress system was the work of historical evolution, not the work of a single man, is the evidence from the career of Vauban himself. The greatest number of strongholds that we associate with him were not *places neuves* but older fortresses, some dating back to Errard or his Italian predecessors, which Vauban modernized and strengthened. The fortresses did not in any sense constitute a system as Vauban found them; they were important only as separate units. There was no liaison between them and they were almost always too far apart. Each situation, moreover, had been chosen for its local importance: to guard a bridge, a crossroads, or the confluence of two rivers. Their total value depended not on their relative positions but rather upon their number.[31] Zeller and Lazard both agree that Vauban's general scheme resulted from a process of selection from among these fortresses. He made order out of prevailing chaos by choosing certain forts whose positions made them worth retaining and strengthening, and by suggesting that others be razed. His strategic vision could not work with complete freedom; he was limited—largely for reasons of public economy—to working with what France already possessed. It is easy to discover the principles which guided his process of selection and thus to find the key to his strategic thinking. To Zeller there is nothing remarkable about these principles; the "order" that Vauban effected fell far short of a great strategic conception. But Lazard is much more flattering. He takes the view that Vauban was the first man in history to have an over-all notion of the strategic role of fortresses; that he was not only an engineer but a *stratège*, and one with ideas far in advance of his own day.[32] Only Vauban's own writings can allow the reader to decide between these two interpretations.

It should be remembered that as a result of the War of Devolution against Spain, his first war of conquest, Louis XIV extended his holdings along the northwest frontier deep into Spanish-held Flanders. The new positions—from Furnes near the coast eastward through Bergues and Courtrai to Charleroi—gave France a number of strong points scattered among the Spanish garrisons. Vauban's first great task was to strengthen and refortify these new acquisitions, and this occupied most of his time during the peaceful years from 1668 to 1672. In the spring of 1672, however, Louis launched his war against the Dutch. Vauban took the opportunity to raise for the first time the question of the general organization of the frontier. In a letter to Louvois, dated January 20, 1673, he wrote: "Seriously, my lord, the king should think seriously about rounding out his domain [*songer à faire son pré carré*]. This confusion of friendly and enemy fortresses mixed up pell-mell with one another does not please me at all. You are obliged to maintain three in the place of one."[33]

In 1675, a year which saw him busy consolidating French conquests in Franche Comté and elsewhere, Vauban made more specific suggestions. In September of that year he proposed the sieges of Condé, Bouchain, Valenciennes, and Cambrai. The capture and retention of these places would, he said, assure Louis's conquests and produce the *pré carré* that was so desirable. These towns were accordingly taken: Condé and Bouchain in 1676, Valenciennes and Cambrai in 1677. The Peace of Nimwegen, signed in August of the following year, gave France a frontier approximating the *pré carré*. She gave up some of her Flemish holdings but acquired instead Saint-Omer, Cassel, Aire, Ypres, and a half dozen other important strongholds. To the eastward she gained Nancy in Lorraine and Fribourg across the Rhine. But Vauban was not satisfied with the western end of the frontier; he felt that the recent peace had disrupted it and left it open toward the Lowlands. In November 1678, three months after Nim-wegen, he wrote the first of a series of important general statements on the organization of the northern frontier from the channel to the Meuse.[34]

Vauban opens by discussing the purposes of a fortified frontier: it should close to the enemy all the points of entry into the kingdom and at the same time facilitate an attack upon enemy territory. Vauban never thought that fortresses were important solely for defense; he was careful to stress their importance as bases for offensive operations against the enemy. The fortified places should be situated so as to command the means of communication within one's own terri-tory and to provide access to enemy soil by controlling important roads or bridge-heads. They should be large enough to hold not only the supplies necessary for their defense, but the stores required to support and sustain an offensive based upon them. These ideas, enunciated tersely in this memoir, were later elaborated and systematized by one of Vauban's eighteenth-century disciples, the engineer and adventurer Maigret, whom Voltaire mentions in his *Charles XII*, and whose *Treatise on Preserving the Security of States by Means of Fortresses* became the standard work dealing with the strategic significance of fortifications. This book, all too little known, was used by the famous French school of military engineer-ing, the Ecole de Mézières. In this work Maigret writes that "the best kind of fortresses are those that forbid access to one's country while at the same time giving an opportunity to attack the enemy in his own territory."[35] He lists the characteristics that give value and importance to fortresses: control of key routes into the kingdom, such as a mountain gorge or pass; control of the bridgeheads on great rivers, a condition eminently fulfilled by Strasbourg, for example; con-trol of important communication lines within the state, as for example Luxem-bourg, which secured the emperor's communications with the Lowlands.

There were still other factors which could make a fort important. It might be a base of supplies for offensive action, or a refuge for the people of the sur-rounding countryside; perhaps it could dominate trade and commerce, exacting tolls from the foreigner; or it might be a fortified seaport with a good and safe harbor; a great frontier city with wealth, more than able to contribute the cost of fortification and sustaining the garrison; or a city capable of serving the king as a place to store his treasure against internal or external enemies.[36] The value of a fortress depends in large part, of course, upon the nature of its local situa-tion. Art or science may make up for certain defects in the terrain but they can do little with respect to the matter of communication. Thus certain fortresses are advantageously situated because the defenders have the communications leading to it well under their control, whereas the enemy, in consequence, will have diffi-culty in bringing up the supplies necessary for a sustained siege.[37]

These criteria make it possible to select certain fortresses in preference to others, but there still remains the question of their relation one to the other, of liaison. Vauban, in the memoir of 1678, concluded that the frontier would be adequately fortified if the strongholds were limited to two lines, each composed of about thirteen places, stretched across the northern frontier in imitation of infantry battle order.[38] This first line could be further strengthened and unified by the use of a water line stretching from the sea to the Scheldt. Canals or canalized streams or rivers would link one fort with another, and the canals themselves would be protected at regular intervals by redoubts. This scheme was not original with Vauban; in fact it was in operation over part of the frontier even as he wrote. He was under no illusions as to the strength of the water lines, for he saw that their chief purpose was to ward off the harassing raids by which small enemy detachments plagued the countryside. Should an enemy decide to attack the lines with an army, then the lines must be defended with an army.[39]

Such a project would of course necessitate new construction, but Vauban was careful to point out that it would also mean the elimination of numerous ancient strongholds, and he accordingly urged the razing of all fortresses remote from the frontier and not included in the two lines. This would be not only a saving for the treasury but, he urged, also a saving in manpower: with the elimination of their garrisons, ten fewer strongholds would mean about 30,000 soldiers free for duty elsewhere.

This famous memoir of 1678 also embodied a consideration of possible future conquests, and these indicate that, so far as the northern and eastern frontiers were concerned, Vauban was willing to pave the way for something more ambitious than a mere local rectification of a line. In the event of a future war, he said, certain enemy fortresses should be immediately seized. Dixmude, Courtrai, and Charlemont would open up the lowlands, while to the east, Strasbourg and Luxembourg were the supremely important cities to acquire. Not only did these fortresses have the most admirable features of size, wealth, and situation—in these matters they were the best in Europe—but they were the keys to France's expansion to its natural boundaries. Vauban would not have been Frenchman and patriot had he not accepted the familiar and tempting principle that France's natural frontier to the north and east was the Rhine. We know that he held this view and we can suspect that it was already clearly formulated in his mind early in his career. Later it certainly was. Just before the peace of Ryswick, when he was terrified for fear France was about to lose both Strasbourg and Luxembourg, he wrote: "If we do not take them again we shall lose forever the chance of having the Rhine for our boundary."[40]

It is not easy to say with certainty whether this memoir of 1678 represents Vauban's mature and final view on the matter of permanent fortification; his later memoirs leave much to be desired as examples of strategic thinking about the role of fortresses. Except for a memoir on the fortification of Paris, in which he discusses at length the strategic importance of a nation's capital, most of the later studies are lacking in genuine strategic interest. They are concerned chiefly with detailed recommendations as to which fortresses should be condemned and which enlarged or rebuilt.

Despite these handicaps it is not hard to detect a series of changes in Vauban's opinions, due partly to a gradual evolution of his ideas, but chiefly to the changed conditions under which he was obliged to work in the later years of the reign. Increasing financial stringency and a growing drain on the manpower supply encouraged Vauban to stress the razing of fortifications as much if not more than new construction.[41] This led him to urge the destruction of many of the places that had been listed in his second line of defense in the memoir of 1678. At the same time the armies of Louis XIV were being thrown more and more on the defensive and Vauban adapted himself increasingly to defensive thinking. He followed the trend, that was becoming evident at the close of the century, toward still greater reliance upon a continuous waterline along the northern frontier. But he was aware of the peculiar weakness of this sort of defense. In 1696 he wrote a memoir in which he urged the creation of *camps retranchés*, fortified encampments to supplement the fortresses and to strengthen the waterline. The purpose of these encampments was either to guard the water-line in the interval between the fortresses or to strengthen the forts themselves by producing a veritable external defense. With a small army—smaller than the ordinary field army—camped beyond the outworks of a fortress and protected by elaborate earthworks, it was possible either to interfere with any besieging forces unwise enough to tackle the fortress directly or to impose upon them a wider perimeter to be invested.

Taken together these two factors—first, the stress upon the continuous line supplemented by the fortified encampments; and second, the willingness to sacrifice the second line of forts he had favored in 1678—do not offer support to Lazard's assertion that Vauban was a pioneer advocate of the "fortified zone" which later strategy adopted. Quite the contrary, Vauban's thinking seems to have evolved in the direction of favoring a thinner and thinner line. He simplified that disorganized parody on a fortified zone which he had inherited from his predecessors. At first he reduced it to a double line of fortifications, a palpable imitation of the familiar infantry line, and then proceeded to simplify this still further into a single cordon, based on strong points linked by a continuous water-line and supported by troops. Perhaps it is not too farfetched to see in this a sign that the great engineer, near the close of his career, was led gradually to lay more emphasis upon armies and less upon fortification. He seems almost to have come close to the idea of Guibert that the true defense of a country is its army, not its fortifications; that the fortified points are merely the bastions of that greater fortress of which the army forms a living and flexible curtain.

Notes

1. In this and the following section I have relied heavily upon the works of Boutaric, Camille Rousset, and Susane listed in the bibliography. Louis André's *Michel Le Tellier* proved the most valuable single work concerned with army reform in the seventeenth century.
2. In this section I have relied upon my own unpublished doctoral dissertation: "Science and War in the Old Régime" (Harvard, 1941).

3. F. Artz, *Les débuts de l'éducation technique en France, 1500–1700* (Paris: F. Alcan, 1938), 38–43.
4. J. J. Fahie, "The Scientific Works of Galileo," in *Studies in the History and Method of Science*, ed. Charles Singer, 2 vols. (Oxford, 1921), II, 217.
5. Lieutenant Colonel Augoyat, *Aperçu historique sur les fortifications*, 3 vols. (Paris, 1860–64), I, 13–21.
6. The notion has been stressed by Hoffman Nickerson, *The Armed Horde, 1793–1939* (New York: G. P. Putnam's, 1940), 34–40.
7. An eighteenth-century writer on the education of the nobility suggests that the most important military authors a student should study are Rohan, Santa Cruz, Feuquières, Montecuccoli, and Vauban. See Chevalier de Brucourt, *Essai sur l'éducation de la noblesse, nouvelle édition corrigée et augmentée* (Paris, 1748), II, 262–63.
8. The works published in his lifetime were two: a work on administrative problems, called the *Directeur général des fortifications* (The Hague, 1685, reprinted in Paris, 1725), and his *Dixme Royale* (The Hague [?], 1707). A number of spurious works, however, had appeared before his death, purporting to expound his methods of fortification. His three treatises best known to the eighteenth century were printed for the first time in a slovenly combined edition titled *Traité de l'attaque et de la défense des places suivi d'un traité des mines* (The Hague, 1737). This was reprinted in 1742 and again in 1771. *The Traité de la défense des places* was published separately by Jombert in Paris in 1769. No carefully prepared editions were published until 1795. See the bibliography.
9. Vauban, *Lettres intimes inédites adressées au Marquis de Puyzieulx (1699–1705), Introduction et notes de Hyrvoix de Landosle* (Paris, 1924), 16–17.
10. Voltaire, *Le siècle de Louis XIV*, chap. 21.
11. A. Wolf, *History of Science, Technology and Philosophy in the Eighteenth Century* (New York: Macmillan Co., 1939), 531–32; Bernard Forest de Bélidor, *La science des ingénieurs*, 6 bks. (Paris, 1739), bk. 1, pp. 67–79.
12. His right to pioneer status in meteorology rests upon a memoir on rainfall which he submitted to the Académie des Sciences. See Bélidor, *La science des ingénieurs*, bk. 4, pp. 87–88.
13. Georges Michel, *Histoire de Vauban* (Paris, 1879), 447–51.
14. P. Lazard, *Vauban, 1633–1707* (Paris: F. Alcan, 1934), 445–500.
15. H. Chotard, "Louis XIV, Louvois, Vauban et les fortifications du nord de la France, d'après les lettres inédites de Louvois adressées à M. de Chazerat, Gentilhomme d'Auvergne," *Annales du Comité Flamand de France* 18 (1889–1890), 16–20.
16. Bélidor, *La science des ingénieurs*, bk. 4, p. 73.
17. Lazard, *Vauban*, 501–24; La Roncière, *Histoire de la marine française*, 6 vols. (Paris: Plon-Nourrit, 1909–1932), VI (1932), 164–69.
18. For a description of early methods, see Gaston Zeller, *L'organization défensive des frontières du nord et de l'est au XVIIᵉ siècle* (Paris, 1928), 54–55.
19. Cited by Eugène Asse in Didot-Hoefer, *Nouvelle Biographie Générale*, 46 vols. (Paris, 1855–1866), XLV, col. 1010.
20. *Ibid.*, but compare A. Allent, *Histoire du Corps Impériale du Génie* (Paris, 1805), I (only one volume published), 209–10.
21. Lazard, *Vauban*, 377–94.
22. Lazard, *Vauban*, 389–90; A. de Zastrow, *Histoire de la fortification permanente*, translated from the German by Ed. de la Barre-Duparcq, 3d ed. (1856), II, 62–208.
23. Bélidor, *La science des ingénieurs*, bk. 3, pp. 29–34, 35–43, 90–96.
24. Letter to Louvois, October 7, 1687, cited by Zeller, *L'organization défensive*, 144.
25. Chotard, "Les fortifications," 30–35; Zeller, *L'organization défensive*, 96–117; Lazard, *Vauban*, 49–50, 202–4.
26. Chotard, "Les fortifications," 36.
27. Zeller, *L'organization défensive*, 118.
28. Zeller, *L'organization défensive*, 103–4; Th. Lavallée, *Les frontières de France* (Paris, 1864), 83–85.
29. Hennebert, cited by Chotard, "Les fortifications," 42.
30. Zeller, *L'organization défensive*, 2.
31. *Ibid.*, 123.

32. Lazard, *Vauban*, 408–21.
33. Lazard, *Vauban*, 155; Colonel A. de Rochas, *Vauban, sa famille, et ses écrits*, 2 vols. (Paris, 1910), II, 89.
34. Lazard, *Vauban*, 409–14; Zeller, *L'organization défensive*, 96–98. This important memoir is printed *in extenso* in Rochas, I, 189 f.
35. *Traité de la sûreté et conservation des états, par le moyen des fortresses. Par M. Maigret, Ingénieur en Chef, Chevalier de l'ordre Royal et Militaire de Saint Louis* (Paris, 1725), 149.
36. Maigret, *Traité*, 129–48.
37. *Ibid.*, 152 f. and 221–22.
38. The first line: Dunkirk, Bergues, Furnes, Fort de la Kenoque, Ypres, Menin, Lille, Tournai, Fort de Mortagne, Condé, Valenciennes, Le Quesnoy, Maubeuge, Philippeville, and Dinant. The second line: Gravelines, Saint-Omer, Aire, Béthune, Arras, Douai, Bouchain, Cambrai, Landrecies, Avesnes, Marienbourg, Rocroi, and Charleville.
39. Lazard, *Vauban*, 282–84. Augoyat, *Aperçu historique*, I, 229.
40. Lavallée, *Les frontières de France*, 83–85.
41. Zeller, *L'organization défensive*, 98–107.

Bibliography

The plate illustrating siegecraft is from *Traité de l'attaque et de la défense des places*, by M. le Maréchal de Vauban, 2 vols. (The Hague: Pierre de Hondt, 1742), reproduced by permission of the New York Public Library. The one of Belfort, taken from Nicolas de Fer, *Introduction à la fortification dédiée à Monseigneur le Duc de Bourgogne* (Paris 1693), appears by courtesy of the Harvard Library.

Useful for the early period are: F. R. Taylor, *The Art of War in Italy, 1484–1529* (Cambridge, 1921), and Sir Charles Oman, *The History of the Art of War in the Sixteenth Century* (New York: Dutton, 1937).

The best general survey of the development of French military institutions down to the time of Louis XIV is the old work of Edgard Boutaric, *Institutions militaires de la France avant les armées permanentes* (Paris, 1863). For the reforms of Le Tellier and the organization of the French army in the seventeenth century the indispensable study is Louis André, *Michel le Tellier et l'organization de l'armée monarchique* (Montpellier, 1906); and for the period of Louis XIV's personal reign an important work is Camille Rousset, *Histoire de Louvois et de son administration politique et militaire*, 4 vols. (1862–1864). Among works which describe the general characteristics of the French royal army may be mentioned Albert Duruy, *L'armée royale en 1789* (Paris, 1888) and Albert Babeau's two volumes, *La vie militaire sous l'ancien régime* (Paris, 1889–90).

For the histories of the separate arms one is still obliged to consult General Susane's *Histoire de l'ancienne infanterie française* (Paris, 1849–53), his *Histoire de la cavalerie française* (Paris, 1874), and his parallel study of the French artillery arm. The standard work on the French engineers is of about the same vintage. It is Lieutenant Colonel Augoyat's *Aperçu historique sur les fortifications, les ingénieurs, et sur le corps du génie*, 3 vols. (Paris, 1860–64). There is valuable information in E. Legrand-Girade, "Etude historique sur le corps du génie," *Revue du génie militaire* (1897–98), and in C. Lecomte's *Les ingénieurs militaires en France sous le règne de Louis XIV* (1904).

Nothing is more conspicuously lacking in the field of military studies than a well-illustrated history of the arts of fortification and siegecraft. The best studies all date from the period of the Second Empire: J. Tripier, *La fortification déduite de son histoire* (1866); Cosseron de Villenoisy, *Essai historique sur la fortification* (1869); Prévost du Vernois, *De la fortification depuis Vauban* (1861). But the only work generally available in American libraries is A. de Zastrow, *Histoire de la fortification permanente*, 3d ed., translated from the German by Ed. de la Barre-Duparcq (1856).

A vivid way to grasp in its entirety the scope of the fortification system of France under Louis XIV is to consult a contemporary *atlas des places fortes* such as the beautifully illustrated work of Nicolas de Fer, *Introduction à la fortification, dédiée à Monseigneur le Duc de Bourgogne* (Paris, 1705). The best statement of the prevailing view as to the role of fortresses in national defense is given by the little-known work which was used as the official text during the eighteenth century in the Ecole de Mézières: *Traité de la Sûreté et Conservation des Etats, par le moyen des Fortresses, par M. Maigret, Ingénieur en Chef* (1725). The New York Public Library has a copy of the 1705 De Fer, and the Maigret can be consulted in the Library of Congress.

A recent work on Vauban in English is Sir Reginald Blomfield's *Sébastien le Prestre de Vauban, 1633–1707* (London: Methuen, 1938); though mention should be made of E. M. Lloyd's *Vauban, Montalembert, Carnot, Engineer Studies* (London, 1887). Blomfield's book relies heavily upon the best French work available: Colonel du génie P. Lazard's *Vauban* (Paris: F. Alcan, 1934), which is a brilliant study covering all aspects of Vauban's career. Both works are admirably illustrated, Blomfield's with his own drawings of Vauban forts in their present state of preservation, and Lazard's with sketches by Vauban and with photographs of models from the *Musée des plans reliefs* of the French army's geographical service. George Michel's *Histoire de Vauban* (Paris, 1879) remains a useful connected narrative, while Daniel Halévy's *Vauban* (Paris, 1923) is a readable but uncritical work available in English translation. Fontenelle's *Eloge du Maréchal de Vauban* and Voltaire's *Siècle de Louis XIV* are classics that may be consulted with profit.

There is important material in H. Chotard, "Louis XIV, Louvois, Vauban et les fortifications du Nord de la France, d'après les lettres inédites de Louvois adressées à M. de Chazerat, Gentilhomme d'Auvergne," *Annales du Comité Flamand de France* 18 (1889–90). An important critical study on Vauban's strategic contributions is the work of Gaston Zeller, *L'organization défensive des frontières du Nord et de l'Est au XVIIᵉ siècle* (Paris, 1928).

There is regrettably no standard edition of Vauban's collected works. The best editions of the two main treatises, the *Traité des sièges et de l'attaque des places* are those published by Latour-Foissac in 1795 and by Augoyat and Valazé in 1829. Both editions are rare and many libraries have only the imperfect editions of the eighteenth century, of which there are several. Of the mass of Vauban manuscripts a large part remains unpublished, and even those that were published in the eighteenth or nineteenth centuries are not found in most libraries. These are Vauban's autobiographical fragment, the *Abrégé des services du Maréchal de Vauban*, written in 1703, published by Augoyat in 1839; his earliest published work, the *Directeur général des fortifications*, published in The Hague in 1685, reprinted in Paris, 1725; the *Mémoire pour servir d'instruction dans la conduite des sièges*, written in 1669, printed Leyden, 1740; his *De l'importance dont Paris est à la France, et le soin que l'on doit prendre de sa conservation* (Paris, 1825); *Mémoires militaires de Vauban . . . Précédés d'un avant-propos par M. Favé* (Paris, 1847–54); *Mémoires inédites de Vauban sur Landau, Luxembourg et divers, sujets*, ed. by Augoyat (Paris, 1841); *Mémoire du Maréchal de Vauban sur les fortifications de Cherbourg, 1686* (Paris, 1851). The first four volumes of the *Oisivetés* were published by Augoyat from 1842 to 1845.

For personalia the best assemblage of materials is the work of Colonel de Rochas, *Vauban, sa famille, et ses écrits*, 2 vols. (Paris, 1910). Unfortunately this work is extremely rare. Much of Vauban's correspondence, including letters reprinted in Rochas, has been published in the *Revue de génie militaire* in the numbers from 1897 to 1901. Vauban's *Lettres intimes inédites adressées au Marquis de Puyzieulx* have been printed with introduction and notes by Hyrvoix de Landosle (Paris, 1924).

[A different emphasis, and a more up-to-date bibliography, are to be found in my article "Vauban" in the *Dictionary of Scientific Biography*, vol. 13 (1976), 590–95.]

29

Guy de La Brosse:
Botanist, Chemist, and Libertine

✿

The founder and first director (*Intendant*) of the Royal Botanical Garden in Paris, Guy de La Brosse, was born about 1586, during the reign of Henry III, probably in Paris, not in Rouen as is often claimed. He died—of epicurean over-indulgence, if his archenemy, Guy Patin, can be trusted—at his house in the Jardin des Plantes during the night of August 30–31, 1641.[1]

The La Brosse name was by no means uncommon, and it is not easy to sort out and identify his ancestors and relatives. But the La Brosse mentioned by the poet, historian, and chemist Jacques Gohory (*d.* 1576) as a learned "mathé-maticien du Roy," possessed of a fine botanical garden, may have been Guy's grandfather.[2] About the father we are on firmer ground: Isaïe de Vireneau, sieur de La Brosse, is described by his son as a respected physician and a fine medical botanist ("très bon simpliste").[3] Isaïe, who died about 1610, was long survived by Guy's stepmother, Judith de la Rivoire. These given names have suggested that the family was originally Protestant,[4] although Guy was at least a nominal Catholic; he built a chapel in the Jardin des Plantes, where Mass was said on feast days and where he was eventually buried.

In his youth, Guy may have been a soldier;[5] in any case, his major book testifies to extensive travels in France. Yet by 1614 he had settled in Paris, had begun the study of chemistry, and was botanizing on Mont Valérien. Although we are ignorant about his medical training,[6] we know that by 1619 he was physician to Henry II de Bourbon, Prince de Condé, and that in 1626 he had become one of the physicians in ordinary to Louis XIII. Like many of the doctors of the royal household, a number of whom were products of the medical university of Montpellier, Guy was highly critical of the Paris medical faculty: its conservatism, its worship of Galen, its addiction to venesection, its relative neglect of botany and anatomy, and its distrust of the newly emerging, and highly contro-versial, field of medical chemistry.[7] By 1616 Guy had begun his efforts to secure the establishment in Paris of a royal botanical garden, not merely for the study of medicinal herbs, but where chemistry would be taught as a handmaiden to medicine.

From *Dictionary of Scientific Biography*, vol. 7 (1973), 536–41. Copyright © 1973 American Council of Learned Societies.

Several small botanical gardens had been established in Paris by private persons—mostly physicians and apothecaries—during the sixteenth century. In Guy's youth the only Parisian garden of importance was the modest one of Jean Robin.[8] Guy hoped for something larger and more elaborate, and when he made his first overtures about 1616 to Louis XIII, through the good offices of Jean Hérouard, chief body physician of the king,[9] his models were the botanical garden of Montpellier, established by Henry IV but recently fallen into ruin, and those of Padua and Leiden. What Guy envisaged was a teaching and research institution, designed to raise medical standards and advance the art. Besides a collection of living plants, Guy planned a herbarium of dried specimens and a *droguier*, or laboratory, where students could learn distillation and the preparation of herbal remedies. A royal edict of January 6, 1626, authorized the establishment, in one of the suburbs of Paris, of such a royal garden of medicinal plants, and designated Jean Hérouard as superintendent. Six months later, when the edict was registered by the Parlement, La Brosse received his appointment as *intendant*.

In the next two years, to assure support and financing for the project, Guy published a series of pleas to government officials, among them Richelieu, and brought out his *Advis defensif*, defending his plan and severely criticizing the Paris medical faculty. Most of these pamphlets were reprinted in 1628 in his major book, *De la Nature, Vertu et utilité des Plantes*.

For several years there was no sign of progress; in the meantime Hérouard had died and Charles Bouvard succeeded him as superintendent of the proposed garden. At last, on February 21, 1633, there was purchased in the king's name, for the sum of 67,000 livres, a house and grounds in the Faubourg Saint-Victor. A year later Guy was able to show the king a plan of the new garden, where 1,500 species of plants were already growing. The act which detailed the organization and staffing of the Jardin des Plantes was a further royal edict of May 15, 1635. This specified that Guy, aided by a *sous-démonstrateur*, was to teach the "exterior" of plants—that is, their identification and taxonomic characteristics. Also authorized was the appointment of three demonstrators, to teach the "interior" of plants, in other words their pharmaceutical properties.

Meanwhile, to supervise the work, Guy moved into the building that was much later to serve as the zoological galleries. The ground was cleared and leveled, and garden plots and parterres were laid out. Many plants were provided by Vespasien Robin, the heir to his father's garden as well as to his title of *arboriste du Roi*, whom Guy appointed in 1635 as *sous-démonstrateur*.[10] Through active correspondence with botanists abroad, Guy obtained seeds and plants from foreign lands, notably from the East Indies and America. In 1636, when he published his *Description du Jardin Royal des Plantes medecinales* with two plans of the garden, he was able to list some 1,800 species and varieties under cultivation. Four years later, in 1640, came the formal opening of the institution, marked by the publication of a pamphlet of thirty-eight pages describing the foundation of the garden, comparing it with those of Padua, Pisa, Leiden, and Montpellier, printing some introductory remarks about the study of botany, and, finally, laying down regulations for the students. The following year, the year of his death, Guy published a second catalogue of the plants growing in the garden, with a handsome perspective plan drawn and finely engraved by Abraham Bosse. Its appearance could hardly have changed much when the young John Evelyn recorded in his travel diary his visit to Guy's establishment in February 1644:

> The 8th I tooke Coach and went to see the famous Garden Royale, which is an Enclosure wall'd in, consisting of all sorts of varietys of grounds, for the planting & culture of Medical simples. It is certainly for all advantages very well chosen, having within it both hills, meadows, growne Wood, & Upland, both artificial and naturall; nor is the furniture inferiour, being very richly stord with exotic plants: has a fayre fountaine in the middle of the Parterre, a very noble house, Chapel, Laboratory, Orangerie & other accommodations for the Praesident, who is allwayes one of the Kings chiefe Physitians.[11]

We know something of Guy's friendships and his ties with the intellectual and free-thinking circles of his day. Descartes knew about him, and mentions in his letters Guy's refutation of the *Géostatique* of Jean Beaugrand.[12] Guy was at least an occasional visitor to the cell of the famous scientist and Minorite friar Father Mersenne.[13] He was a familiar, too, of other learned circles like the "Cabinet" of the brothers Dupuy, and the "Tétrade" of Elie Diodati, Gabriel Naudé, Pierre Gassendi, and La Mothe le Vayer, who, while keeping up a discreet front as an early member of the French Academy, set forth his free-thinking views under the pseudonym of Orasius Tubero. Guy also formed part of the pleasure-loving group around François Lullier, financier and *maître des comptes*, and was perhaps closest of all to the libertine poet Théophile de Viau and his intimate and pupil Jacques Vallée, sieur des Barreaux, for Guy was Théophile's personal physician as well as friend. It was from Guy that the poet in his last illness received the narcotic pill that ended his life.[14]

Guy's earliest book was a short monograph on the causes of the plague, the *Traicté de la peste* (1623). In the following year (1624) appeared his *Traicté contre la mesdisance*, a work in which he defended various persons—among them, and perhaps notably, Théophile de Viau—who had been unjustly persecuted for their opinions.

In his work on the plague, Guy already showed his affinity with the Paracelsian doctors and his rejection of traditional medical theory. There are, he remarked, two different opinions about the efficient cause of the plague: (1) that it depends upon the active and passive qualities of the elements, a view held by all those who attempt to explain nature by the manifest qualities of things; and (2) that the cause is hidden, proceeding from agencies beyond the reach of our senses. Men in the first category follow Galen in making putrefaction the "principal and unique cause" of the plague. But Guy, opposing a philosophy "that knows the motions and changes of nature only through books," clearly preferred the second alternative, urging that the cause of the plague "is a venomous and contagious substance," which in turn is the cause of putrefaction. As Allen Debus has pointed out, Paracelsian doctors sought the causes of disease less in internal imbalances of fluids than in external factors.[15]

The major concern, however, of Guy de La Brosse was with medical botany, a field to which he was doubtless introduced by his father. Guy's *De la nature des plantes* is not only a defense of his project for creating a center for the study of medical botany, but also a theoretical work about plants in general. In it he raises questions that would be meaningful today—about the generation, growth, and nutrition of plants—as well as asking whether plants have souls, a subject to which he devotes considerable space, discussing also the influence of the stars, yet criticizing the doctrine of signatures. His belief in the essential unity of plant and animal life led him to catalogue their similarities: growth is observed in both; motion is not peculiar to animals, for indeed some animals are motionless or sessile; both plants and animals suffer disease; both animals and plants hibernate and plants even sleep. Plants, he would have us believe, seem to have more vital force than animals, yet they are readily fatigued by the process of nutrition and by tempestuous weather. Extending this analogy further, he was convinced that plants, like animals, must differ in sex (the vernal rise of sap, he argued, "testifies to their amorous desires"); and he urged that an effort be made to examine plants closely to discover distinctive sexual features.

As to whether plants, like animals, display feeling and sensation, La Brosse disagreed with Aristotle and returned to, and cited, the earlier views of Empedocles and Anaxagoras. Although sense organs, he remarked, have not been observed in plants, several species—notably *Mimosa pudica*, the sensitive plant, which he claimed to have been the first to introduce into France—markedly display the quality of sensibility.[16] What the soul of the plant is, Guy did not pretend to know, although he could note its operations. The individualizing agent that he adduced as the cause of specific differences, what others would call the plant's "form," he called the "Artisan" or "esprit artiste."

Perhaps Guy's most interesting suggestion concerned plant nutrition. Plants, like animals, derive their nutriment not only from solid food (*viande*) drawn from the earth, and from aqueous liquid (*breuvage*) from water, but also from air. Air is as necessary to nourish and sustain plants as it is for the life of animals. Deprived of air, plants die; to be sure, they have no lungs, but in this they resemble insects, which nevertheless need air to live. It is not necessary to have lungs to draw in air; it suffices to be supplied with pores. If plants need earth, it is for the nitrous and saline juices it contains ("*la terre sans sels est inutile à la génération*"); manure is nothing but the salt of the urine of animals. Water by itself is not a nutrient (*pace* Van Helmont), but serves chiefly as the vehicle for the salts and the manna. Guy suggested an experiment to show that "pure" earth and distilled water cannot sustain the growth of plants: rich earth is leached with warm water and put into a large glass vessel; if seeds are then planted and watered with distilled water, they may sprout, but they will not grow. Similarly, it is for the *esprit*—the dew and the manna contained in it—that plants need air to live. That plants seek air, Guy attempted to prove by pointing to a shrub which, growing close to a wall or otherwise sheltered, sends its branches towards the open air. This was, of course, a misreading of the phenomenon of phototropism; such plants were seeking sunlight rather than air. Nevertheless, a century before Stephen Hales, although on wholly inadequate evidence, La Brosse argued for the nutritive role of air in plants.[17]

Guy's interest in chemistry was keen; the third of the five books into which *De la nature des plantes* is divided is devoted to the subject; Guy himself described it as "un traicté général de la Chimie." For Guy, chemistry was an important adjunct both to medicine and botany. It is through chemistry, rather than by ordinary dissection, that one learns the causes of the virtues of plants. Fire if guided by an experienced hand produces marvels, for it has the ability to disclose those things that are hidden from the senses.[18]

The fundamental assumption of the chemist is that every substance can be reduced to those entities out of which it is formed; only when we have reduced bodies to their principles and elements can we truly understand them. All natural compound bodies, Guy tells us, can be reduced into five simple bodies of different natures: into three principles—salt, sulphur, and mercury, the *tria prima* of the Paracelsians—and into two elements—water and earth.[19] Neither air nor fire (as in the old doctrine of the four elements) should be thought of as an element or principle.

Guy devotes an entire chapter to explaining why the chemist refuses to include air as one of the elements. This may seem odd, he admits, since air serves as an excellent and necessary food for man, for other animals, and indeed for plants, "which cannot live without respiring it." But the chemist would reply that when compound substances (*mixtes*) are dissected by fire, no air appears. Air, moreover, is not an element or simple body, but ought better to be called *chaos*, because of the great number of substances that it contains, and of which it is composed; atoms of earth, the vapor of water, and the three principles subtilized make up "that mixture of fine, subtile, and diaphanous substances that we respire." Air is the "magazine" of all the sensible substances which evaporate and are subtilized. Chemistry concerns itself only with sensible phenomena; it is from the senses that the chemist learns, Guy claims, that all compound bodies "contain and are made of salt, sulphur and mercury," and that water and earth occur in the chemical dissection of all substances. Water and earth, however, are not to be considered principles, for without the capacity to produce the seeds which account for the specific forms and virtues of things, they are to be thought of as universal matrices, wombs, or "generous receptacles" found in all bodies, "not as contained in them, but as containing them." Chemical change, in sum, is the result of the action of two agents: "the Form, or as we call it, the Artisan," and fire, "the universal instrument" or the "Great Artist," which in turn acts in some mysterious way upon the three seminal principles.

Even if Guy had not mentioned Paracelsus and his disciples in the *De la nature des plantes*, the influence of the Swiss doctor would be quite apparent. In his use of the word *chaos*—he spelled it "cahos"—to describe aerial matter, and in his references to "dew" and "manna," Guy clearly echoed the Paracelsians. His "artisans," although they foreshadow the "plastic natures" imagined later in the century by Ralph Cudworth and John Ray, are the Paracelsian "Archei" under another name. Guy admired Paracelsus, not only as the enemy of medical biblio-latry and the first to dramatize an opposition to Aristotle and Galen, but because —as a revolutionary in the study of nature—he stressed experience and experi-ment, and because of his advocacy of chemistry as a key to nature and, through a knowledge of nature, to medical reform. The chemical doctrines of Paracelsus appealed strongly to Guy, but even in chemistry he cherished his independence and refused to follow blindly either the man "to whom first place is given in this excellent art" or Severinus, the best of the followers of Paracelsus. Rather than accept what they wrote, Guy insisted, "I have rather chosen to delve into the bowels of Nature," to test their assertions. "Using my hands," he continued, "I found that many of them wrote falsely; that even Paracelsus, at least if all the books bearing his name are his, was not always trustworthy . . . and that all the others did the same or even worse."[20]

Guy de La Brosse, it should be evident, shared a number of the attitudes and preconceptions we associate with the new learning of the seventeenth century: a distrust of authority, especially that of Aristotle; a preference, if one must choose, for the views of pre-Socratic philosophers; and a belief in the capacity of the natural sciences, guided by a critical use of human reason and a respect for ex-perience, to move steadily forward. Pintard sees in one of Guy's basic doctrines— his trust in human reason—an anticipation of Descartes.[21] Like Descartes, Guy believed error to result not from some innate weakness of the human intellect but from the defective way the mind is used. If man can overcome in his think-ing the influence of prejudice and the tyranny of "opinion," he may discover truth, "cette fille du temps." Like Descartes, and thirteen years before him, Guy announced his faith in that faculty of the human mind which, unhampered, allows men to distinguish truth from falsehood. This faculty—Descartes was to call it "le bon sens"—is found in all men and all climates; it works for the Parisian as well as for "the Indian, the Moor, the Chinese, the Jew, the Christian, the Mohammedan, even the Deist and the Atheist."

But the basis for the judgment of reason can only be experience, the real "maistresse des choses," and the only true foundation of the sciences. Here in his reiterated empiricism and eclecticism he diverges from Descartes and more closely resembles his friend Gassendi and Francis Bacon, whose writings he had in his library.

Guy's attitude toward medicine and science is neatly summed up in the handsome frontispiece of *De la nature des plantes*.[22] Four symmetrically arranged shields contain the portraits of Hippocrates, Dioscorides, Theophrastus, and Paracelsus, each accompanied by an appropriate motto. For Theophrastus, it is "Medicine is useless without plants." for Paracelsus, it is "Each thing has its heaven and its stars." Indeed the mottoes of Hippocrates and Dioscorides pretty well sum up Guy's empiricist position: For Hippocrates, it is "From effects to causes"; and for Dioscorides, it is "From experience to knowledge." Good doctrine for a man who could write, "It is difficult to have conceptions of things which have not entered the understanding through the senses." In this and in other matters, Guy may have echoed the dictum of Aristotle, yet Galen and Aristotle are conspicuously absent from his frontispiece. And the reason is evident: centered at the top of the page is a radiant sun, and below it the legend, "Truth, not authority." At the bottom is Guy's own device, "De bien en mieux," which well epitomizes his faith in scientific progress.

Guy de La Brosse was in a number of respects a confused child of his time, echoing its aspirations and its intellectual discontents. His book—an odd mixture of the antiquated, the perverse, and the novel—cannot be said to have exerted a marked influence on scientific thought. The book was rarely cited.[23] Indeed Guy himself was largely forgotten, in a truly physical sense, for some two and a half centuries. In 1797, when the chapel he had built adjoining the main building of what had become the Muséum d'Histoire Naturelle was demolished to enlarge the zoological galleries (to accommodate, we must suppose, the collections of Cuvier and Lamarck), workers came upon the crypt containing the coffin of La Brosse, easily identified by a crude inscription written on the wall by his niece. The coffin was unceremoniously stored in a convenient basement; and if there were plans for a suitable reburial, they were long deferred. It was not until 1893, nearly a century later, that Guy was reinterred with seemly honors.[24]

Notes

1. *Lettres de Gui Patin*, ed. J. H. Revillé-Parise, 3 vols. (Paris, 1846), I, 87–82. See also E. T. Hamy, "Quelques notes sur la mort et la succession de Guy de la Brosse," *Bulletin du Muséum d'Histoire Naturelle* 3 (1897), 152–54, and "La famille de Guy de la Brosse," *ibid.* 6 (1900), 1–3. Patin's enmity was as an impassioned defender of the Paris Faculty of Medicine. [The most recent study of Guy is the Cornell doctoral dissertation of my student Rio Howard, "Guy de La Brosse, The Founder of the Jardin des Plantes in Paris" (Ithaca, N.Y. 1974).]

2. In his *Instruction sur l'herbe petum* (Paris, 1572), Gohory mentions a certain La Brosse, "mathématicien du Roy" and his "beau jardin garny d'une infinité de simples rares et de fleurs esquises" from which he obtained tobacco leaf for his experiments. See Hamy, "Un précurseur de Guy de la Brosse. Jacques Gohory et le Lycium philosophal de Saint-Marceau-les-Paris (1571–1576)," *Nouvelles Archives du Muséum*, 4th ser., 1 (1899), 4–5.

3. Guy de La Brosse, *De la nature des plantes* (Paris, 1628), 767.

4. See Hamy, "La famille," 2. Cornelis de Waard calls Guy a Calvinist: see *Correspondance du P. Marin Mersenne*, ed. and annotated by Cornelis de Waard, 11 vols. (Paris, 1932–1970), V, 195. [The marriage certificate of Guy's parents, discovered by Miss Howard in the Minutier Central in Paris, proves that Guy's mother was not Judith de la Rivoire, as Hamy believed, but Isaïe's second wife, one Nicolle Barbin. Judith, Isaïe's third wife, was the mother of Guy's half-sister (and sole heir) Louise.]

5. Albrecht von Haller, without supporting evidence, describes Guy as "ex milite

botanicus et medicus," in *Bibliotheca botanica*, I (1771), 440. See also René Pintard, *Le libertinage érudit*, 2 vols. in 1 (Paris: Bonin, 1943), II, 605.

6. Hilarion de la Coste, in Tamisey de Larroque, ed. *Lettres écrites de Paris à Peiresc* (Paris, 1892), 59, mentions, as a visitor to Père Mersenne, a physician named La Brosse whom he describes as a "docteur de la Faculté de Montpellier." But there is no trace of Guy in the records of that medical university. Enemies called Guy an empiric, and doubted that he had ever received a medical degree.

7. See his "Advis défensif," in *De la nature des plantes*, 754–99.

8. See M. Bouvet, "Les anciens jardins botaniques médicaux de Paris," *Revue d' histoire de la pharmacie* (December 1947), 221–28. The garden of Jean Robin (1550–1629) first occupied, as Bouvet tells us (p. 226), "the western point of the Cité, where the Place Dauphine is located today." Late sixteenth-century plans of Paris show such a garden on that spot, but it must have moved to another location after the building (c. 1607) of the Place Dauphine. Where it was located after that time is hard to determine.

9. For this Montpellier doctor whose name also appears as Héroard or Erouard, see Hamy, "Jean Héroard, premier surintendant du Jardin Royal des Plantes médecinales (1626–28)," *Bulletin du Muséum d'Histoire Naturelle* 2 (1896), 171–76. For Guy's letter to Hérouard, see Louis Denise, *Bibliographie Historique et Iconographique du Jardin des Plantes* (Paris, 1903), no. 13.

10. The Robins introduced into Europe the first acacia tree (*Robinia*), which Vespasien planted in Guy's garden in 1636, and which still survives. For the younger Robin see Hamy, "Vespasien Robin, arboriste du Roy . . . ," *Nouvelles Archives du Muséum d'Histoire Naturelle* 8 (1896), 1–24.

11. John Evelyn, *Diary*, ed. E. S. de Beer, 6 vols. (Oxford, 1955), II, 102. By "Praesident" Evelyn means the *Intendant*.

12. For the Beaugrand episode see *Oeuvres de Descartes*, 13 vols. (Paris, 1908–1957), I (*Correspondance*), and *Correspondance du Mersenne*, V. See also Adrien Baillet, *Vie de Descartes* (Paris, 1691), bk. 4, chap. 12. For Beaugrand see Robert Lenoble, *Mersenne ou la naissance du mécanisme* (Paris: J. Vrin, 1943), 472; and Lynn Thorndike, *A History of Magic and Experimental Science*, 8 vols. (New York: Macmillan Co. and Columbia University Press, 1923–58), VII, 437–38.

13. Hilarion de la Coste's list of visitors to Mersenne is printed *in extenso* in *Correspondance du Mersenne*, I, pp. xxx–xlii.

14. For Guy's *libertin* associations see Pintard, *Le libertinage érudit*, 193–208.

15. Allen G. Debus, "The Medico-chemical World of the Paracelsians," to appear in essays in honor of Joseph Needham. [Subsequently published in *Changing Perspectives in the History of Science: Essays in Honour of Joseph Needham*, ed., Mikuláš Teich and Robert Young (London: Heinemann, 1973).]

16. Mersenne wrote Descartes in 1638 about "l'herbe sensitive" he had seen "chez Mr. de la Brosse." *Correspondance du Mersenne*, VIII (1963), 56–57. For the discovery of *Mimosa pudica* and others of the genus see Charles Webster, "The Recognition of Plant Sensitivity by English Botanists in the Seventeenth Century," in *Isis* 57 (1966), 5–23.

17. Guy's confidence in the role of air in plant nutrition surely has its origin in Paracelsian speculations. For this background, consult Allen G. Debus, "The Paracelsian Aerial Niter," in *Isis* 55 (1964), 43–61.

18. The lower level of the chief building of the Jardin des Plantes was to be the laboratory "pour les distillations"; see *De la nature des plantes* ("Epistre au Roy"), 699. Distillation in early medical chemistry is described by Robert Multhauf, "Significance of Distillation in Renaissance Medical Chemistry," *Bulletin of the History of Medicine* 30 (1956), 329–46.

19. The five-element theory, which dominated the speculations of seventeenth-century chemists, was early set forth by Joseph Duchesne, or Quercetanus, with whose work Guy de La Brosse was familiar. See R. Hooykaas, "Die Elementenlehre der Iatrochemiker," in *Janus* 41 (1937), 1–28, and Allan G. Debus, *The English Paracelsians* (New York: F. Watts, 1966), 90.

20. *De la nature des plantes*, "Argument du troisiesme livre" (inserted between pp. 288 and 289), fol. 2 *v.*

21. Pintard, *Le libertinage érudit*, 196.
22. This frontispiece was designed by the artist and engraver Michel l'Asne (or Lasne). See Denise, *Bibliographie*, no. 39.
23. It was nevertheless referred to by William Harvey's disciple George Ent, in his *Apologia pro circulatione sanguinis* (London, 1641).
24. "Translation et inhumation des restes de Guy de La Brosse et de Victor Jacquement faites au Muséum d'histoire naturelle, le 29 November 1893," *Nouvelles Archives du Muséum d'Histoire Naturelle*, 3d ser., 6 (1894), pp. iii–xvi. On this occasion the principal discourse was delivered by the director of the Muséum, Henri Milne-Edwards.

Bibliography

I. Original Works

The published works of Guy de La Brosse are the following:

Traicté de la peste (Paris, 1623); *Traicté contre la mesdisance* (Paris, 1624); and his most important, *De la nature, vertu et utilité des Plantes* (Paris, 1628). Several of Guy's previously published but undated pamphlets concerned with the proposed Jardin des Plantes Médecinales are reprinted in *De la nature des plantes*. These are *A Monseigneur le tres reverend et le tres illustre cardinal, Monseigneur le cardinal de Richelieu*; the letters *Au Roy, A Monseigneur le garde des Sceaux, A Monseigneur le superintendant des Finances de France*; the *Advis defensif du Jardin Royal des Plantes medecinales à Paris*; and the *Mémoire des plantes usagères et de leurs parties que l'on doit trouver à toutes occurrences soit récentes ou sèches, selon la saison, au Jardin Royal des plantes médecinales*.

Also printed is the royal edict of January 1626 authorizing the establishment of the garden. But the earliest of Guy's pamphlets concerning the garden, his letter *A Monsieur Erouard, premier médecin du Roy* (n.p., n.d., but written c. 1616), was not among those reprinted.

The following pamphlets are posterior to 1628 but published before the opening of the garden: *A Monsieur Bouvard, conseiller du Roy en ses conseils et son premier médecin* (n.p., n.d.); *Advis pour le Jardin royal des plantes medecinales que le Roy veut establir à Paris. Présenté à Nosseigneurs du Parlement par Guy de La Brosse, médecin ordinaire du Roy et intendant dudit jardin* (Paris, 1631); *Pour parfaictement accomplir le dessein de la construction du Jardin Royal, pour la culture des Plantes medecinales* (n.p., n.d.); *A Monseigneur le Chancelier* (n.p., n.d.). After the garden came into being, Guy published his *Description du Jardin Royal des Plantes medecinales estably par le Roy Louis le Juste à Paris; contenant le catalogue des plantes qui y sont de présent cultivées* (Paris, 1636) with an over-all plan of the garden (by Scalberge) and a plan of the four great flower beds.

With a single exception, Guy's later publications all dealt with the development of the Jardin des Plantes. The exception is his *Eclaircissement d'une partie des paralogismes ou fautes contre les loix du raisonnement et de la démonstration, que Monsieur de Beaugrand a commis en sa pretendue Demonstration de la première partie de la quatriesme proposition de son livre intitulé Geostatique. Adressé au mesme Monsieur de Beaugrand* (Paris, 1637).

Perhaps the two most important of his publications concerning the new garden are the following: *L'ouverture du Jardin royal de Paris pour la demonstration des plantes medecinales, par Guy de La Brosse, conseiller et medecin ordinaire du Roy, intendant du Jardin et demonstrateur de ses plantes, suivant les ordres de M. Bouvard, surintendant* (Paris, 1640), which summarizes the history of the garden, compares it with those of Padua, Pisa, Leiden, and Montpellier, refers to the acclimatizing of the *Mimosa pudica*, and prints the regulations for the students; and his *Catalogue des plantes cultivées à présent au Jardin royal des plantes medecinales estably par Louis le Juste, à Paris. Ensemble le plan de ce jardin en perspective orisontale. Par Guy de La Brosse, médecin ordinaire du roy et intendant dudit jardin* (Paris, 1641).

II. Secondary Literature

There is no book-length biography of Guy de La Brosse, and he has been largely neglected by modern historians of botany and almost totally so by historians of chemistry. There are short sketches (not always reliable) in N. F. J. Eloy, *Dictionnaire historique de la médecine*, 4 vols. (Mons, 1778), I, 456–57; E. Gurlt, A. Wernich, and August Hirsch, *Biographisches Lexicon der hervorragenden Arzte*, 2d ed. by W. Haberling et al., 5 vols. (1929–34), I, 715; Albrecht von Haller, *Bibliotheca botanica*, 2 vols. (Zurich, 1771–72), I, 440–41; Curt Sprengel, *Historia rei herbariae*, 2 vols. (Amsterdam, 1807–8), II, 111–12.

The common error that makes Rouen the birthplace of Guy is repeated by F. Hoefer in the *Nouvelle biographie générale*; by Théodore Lebreton, *Biographie normande*, 3 vols. (Rouen, 1857–61), II, 316; and by Jules Roger, *Les médecins normands* (Paris, 1890), 36–39.

A series of articles by E. T. Hamy, professor of anthropology at the Muséum National d'Histoire Naturelle, has clarified a number of points about Guy's life. See especially his "La famille de Guy de la Brosse," in *Bulletin du Muséum d'Historie Naturelle* 6 (1900), 13–16, and his "Quelques notes sur la mort et la succession de Guy de la Brosse," *ibid.* 3 (1897), 152–54.

The only study of Guy's botanical theories is by Agnes Arber, "The Botanical Philosophy of Guy de la Brosse," in *Isis* 1 (1913), 359–69. See also her *Herbals: Their Origin and Evolution*, new ed., rev. (Cambridge, 1938), 144–45, 250, 255. Miss Arber remarks that Guy was deeply influenced by Aristotelian thought, although he "inveighed against the authority of the classics."

For Guy's associations with the *libertins* see René Pintard, *La Mothe de Vayer, Gassendi, Guy Patin* (Paris, n.d.), 23, 79, 128; and his *Le libertinage érudit*, 2 vols. in 1, continuously paginated (Paris: Bonin, 1943), 195–200, 437–41, 605–6.

For Guy's comments on the Paracelsians, and his interest in chemistry, see Henry Guerlac, "Guy de La Brosse and the French Paracelsians," in Allen G. Debus, ed., *Science, Medicine and Society in the Renaissance: Essays to Honor Walter Pagel*, 2 vols. (New York, 1972), I, 177–99.

Essential for any study of the garden founded by Guy de La Brosse is Louis Denise, *Bibliographie Historique & Iconographique du Jardin des Plantes* (Paris, 1903), where the early pamphlets of Guy are listed and briefly described. For a short seventeenth-century description of the newly established garden, see Claude de Varennes, *Le voyage de France* (Paris, 1639 and later eds.). An early historical study of the garden, from its origins to the death of Buffon (1788), is by the famous botanist Antoine-Laurent Jussieu, whose "Notices historiques sur le Muséum d'histoire naturelle" appeared in the *Annales du Muséum* from 1802 to 1808; the first of these articles, covering the establishment of the Jardin and its development to 1643, was published in *Annales* 1 (1802), 1–14.

Other accounts are by Gotthelf Fischer von Waldheim, *Das Nationalmuseum der Naturgeschichte zu Paris*, 2 vols. (Frankfurt am Main, 1802–3), I, 21–42; and J. P. F. Deleuze, *Histoire et description du Muséum royal d'histoire naturelle*, 2 vols. (Paris, 1823).

For special aspects of the early history of the Jardin, see E. T. Hamy, "Recherches sur les origines de l'enseignement de l'anatomie humaine et de l'anthropologie au Jardin des Plantes," in *Nouvelles Archives du Muséum d'Historie Naturelle*, 3d ser., 7 (1895), 1–29; "Vespasien Robin, arboriste du Roy, premier sous-démonstrateur de botanique au Jardin royal des plantes (1635–1662)," *ibid.* 8 (1896), 1–24; and "Jean Héroard, premier superintendant du Jardin royal des plantes médecinales (1626–1628)," in *Bulletin du Muséum d'Historie Naturelle* 2 (1896), 171–76. Worth consulting is Jean-Paul Contant, *L'enseignement de la chimie au Jardin royal des plantes de Paris* (Cahors: A. Coueslant, 1952).

For early botanical gardens, see M. Bouvet, "Les anciens jardins botaniques médicaux de Paris," in *Revue d'histoire de la pharmacie* (December 1947), 221–28. E. T. Hamy has corrected a persistent error that the garden of Jacques Gohory was located on the site of the labyrinth of the Jardin des Plantes, and that the garden of the man who may have been Guy's grandfather was close by. See Hamy, "Un précurseur de Guy de la

Brosse. Jacques Gohory et le Lycium philosophal de Saint-Marceau-les-Paris (1571–1576)," *Nouvelles Archives du Muséum*, 4th ser., 1 (1899), 1–26.

The errors which originated with Gobet's *Anciens minéralogistes du royaume de France* (Paris, 1779) have been repeated by F. Hoefer, in *Histoire de la chimie*, 2d ed., 2 vols. (Paris, 1866–69), II, 102–3, and in his article on Gohory in *Nouvelle biographie générale*. Hoefer, in turn, is relied upon by J. R. Partington, *A History of Chemistry*, 4 vols. (London: Macmillan & Co., 1961–70), II, 162–63.

30

Three Eighteenth-Century Social Philosophers:
Scientific Influences on Their Thought

⚙

We have been asked to discuss "the manner in which the rise of science in the seventeenth century affected the culture and world view of the eighteenth century." I should like to raise the question of how great a "common understanding of science," and of its meaning for man, the eighteenth century possessed, by taking up briefly three well-known social philosophers of the Enlightenment: Montesquieu, Voltaire, and the Baron d'Holbach. I propose to examine whether an interest in science was in each case a chief influence upon, or a principal source of, their social philosophy; and to ask whether science meant the same thing to each of them, and to what extent Newton, whose name and accomplishments are so frequently invoked by the *philosophes* and their historians, provided a common inspiration.

⚙

Of Montesquieu, the first of my triad, I shall speak less fully than he deserves, for he was surely one of the most original, if not perhaps the most influential, minds of his century. Montesquieu's scientific interests were the avocation of his early manhood, when as a young magistrate of Bordeaux in the 1720s he was active in the affairs of the town's Academy. From the scientific juvenilia of this period; from the later entries in his workbook, the *Pensées*, and from his masterpiece, the *Esprit des lois*, a rather clear picture emerges of the scientific warp of his thought and of the motives that engendered this passing enthusiasm.[1]

Like his contemporaries, the young Montesquieu was moved by pride and wonder at the scientific accomplishments of the seventeenth century, which had cleared away so many superstitious beliefs and offered such a precise and simple picture of the world around us. But to Montesquieu, science was less significant as a remedy for the disease of credulity and superstition, less valuable because of

From *Daedalus* 87 (1958), 6–24. © 1958 by the American Academy of Arts and Sciences. Reprinted in Gerald Holton, ed., *Science and the Modern Mind* (Boston: Beacon Press, 1958). Copyright, 1958, by the Beacon Press.

the useful applications which might flow from it, than as a dramatic proof of the power of the human mind. For the man of humanistic training like himself, he saw a task of special urgency: to civilize and domesticate the recent discoveries of science, and fit them into place in the great heritage of humane learning, by removing that artificial barrier which technical jargon, poor writing, a dry and forbidding language, had erected. But for the bolder purpose that began to shape itself in his mind, he might well have sought to emulate Fontenelle, whom he so much admired, and have anticipated Voltaire as a humane expositor of the physical sciences. Instead he took a more fruitful path, that of attempting to demonstrate in a work of great erudition and creative power that science and humane learning could be combined in a search for the *Spirit of the Laws*. For nothing is more characteristic of Montesquieu than his conviction that all spheres of man's knowledge are compatible and harmonious; that "all the sciences are good and support each other."[2]

If the facts and conclusions of science, as Montesquieu understood them, contributed less to the *Spirit of the Laws* than he obviously hoped, the general spirit of scientific inquiry was, for perhaps the first time in history, invoked in the study of man in society. Cool objectivity, detached relativism, painstaking accumulation of fact and observation in support of his generalizations—these seem quite clearly to reflect the example of the students of nature. "Observations," Montesquieu once wrote, "are the history of physical science; systems are its fables."[3] If natural science had been remade by adhering to such principles as these—and he might well have drawn them in part from Thomas Sprat's *History of the Royal Society*, a work he admired—could not a science of man emerge from the application to the study of man of these tenets of the New Experimental Philosophy?[4]

It is when we turn to the substance of his scientific borrowings that we are disappointed. Montesquieu remained to his death an unwavering disciple of Descartes, though by the time the *Spirit of the Laws* appeared in 1748 Newtonian views had triumphed in France. Montesquieu's physical philosophy—I do not say his physics, for an understanding of the mathematical nature of physics was beyond his capabilities—was Cartesian, strictly mechanistic and deterministic. In natural history, physiology, and medical theory—the scientific fields that attracted him most—he accepted without question the teachings of the iatrophysical school which had emerged in the wake of Descartes's speculations. Employed in the pages where Montesquieu discusses his theory of the role of climate, this already obsolescent physiology contributes nothing but confusion.

Much has been made of Montesquieu's sweeping definition of laws, "Laws, in the widest sense, are the necessary relations which derive from the nature of things," and of the fact that Montesquieu is the first modern thinker to apply deterministic principles, not only to physical nature, but to man in his social and political life. For this—as much as for his more empirical side—the founders of sociology, Comte and Durkheim, mark him as a precursor.[5]

Yet it is Cartesian dualism, the dichotomy between mind (or soul) and body, between *l'homme physique* and *l'homme moral*, which is the key to Montesquieu's view of natural law as it regulates the conduct of men. While all beings, living or inert, are subject to the inexorable operation of physical law—the laws of the physical world—man is also under a law peculiar to himself: the natural moral law, the law of God, which is perceived by human reason and regulates man's behavior as a rational being. The diverse positive laws of different peoples are the special applications of the general moral law; or, better still, of the moral law as modified by time and place, by historical tradition and local custom, by the wisdom and moral frailty of law givers, and by the circumstantial operation of physical laws upon the minds and bodies of men.

In the sections of his book which are devoted to his famous theory of the influence of climate and topography, Montesquieu tries to show how physical laws interact with and modify the interpretations and applications of moral law.[6] But this scientific ingredient really plays a very small part in Montesquieu's total plan. It is the *moral* law of nature which really concerns him, and this (as we well know) lay ready at hand, expounded in the writings of Stoic philosophers, Roman jurists, and Christian theologians.

Montesquieu found the intellectual and moral forces, illustrated by examples of the traditional sort and familiar to anyone steeped in the classics, most useful in explaining the variations of positive law in different times and places. Psychological and social causes—*les causes morales*—are far more important, he feels he must warn his reader, than physical and environmental causes, for it is these that mainly determine the character of a people and the laws and institutions under which they live. Just as in an earlier work Montesquieu found the central cause for the rise and fall of the Roman Republic in the changing character of the Roman people, so in his description of different forms of government he discovers for each a psychological determinant or principle: for the Monarchy, "honor"; for the Republic, "virtue"; and for Despotism, "fear." It is these so-called principles which determine the pattern of laws a people shall have and it is their decay that threatens the stability of a society.

I believe Montesquieu's scientific veneer contributed less to the *Esprit des lois* than some have supposed. It was from the ready resources of the humanistic scholar—from omnivorous reading in the literature of travel, from his own perceptive observation of men and manners, and above all from his devotion to the Greek and Roman classics—that he drew his wealth of illustration and most of his inspiration. It was from the old, prescientific natural-law tradition and from the bookish riches of "moral philosophy"—from Saint Thomas Aquinas, from Plutarch, Seneca and his beloved Cicero, and of course from Aristotle and the historians of Greece and Rome—that he drew his conceptual framework and his basic outlook.[7] Steeped in the classics from his school days (as were all Frenchmen of this supposedly scientific century), he lived at the Château de La Brède the life of a Roman gentleman, close to his fields and his vines, and inseparable from his magnificent library. It is too much to suppose that the scientific interests of his youth—in which he maintained a desultory, if receding, curiosity all his life—could outweigh his deeper allegiance to the humanities. Montesquieu did not need Pope to tell him that the proper study of mankind is man.

❁

It comes as something of a surprise that Voltaire—incredibly versatile man of letters, tireless pamphleteer, and sibylline court jester—should have taken as one of his strongest claims to fame the fact that, as much as any man, he had introduced the French public to the discoveries, the philosophy, and the method of Sir Isaac Newton.

The trip to England in 1726–29 first brought to Voltaire's attention the two English thinkers who were to influence so profoundly his life and thought: Newton and Locke. But his full conversion to the physical doctrines of Newton and his rejection of the prevailing Cartesianism took place only after his return to France and was due in large measure to the influence of that gifted blue-stocking and competent Newtonian, the Marquise du Châtelet.[8] Under the eye of the *immortelle Emilie*, and indeed with considerable assistance from the Minerva of France, Voltaire published in 1738 his *Elements of Newton's Philosophy*. Sharing Montesquieu's conviction that scientific discoveries should be humanized and spread abroad, yet disdainful of Fontenelle's cloying sentimentality, Voltaire set out to "remove the thorns" from Newton's writings, "but without loading them with flowers which do not suit them."[9]

The *Elements of Newton's Philosophy* is not precisely the sort of book we might expect. It is not just a simple exposition of Newton's discoveries: of his *System of the World*, of the doctrine of attraction, and of the experiments on light and color. There is, at least in the later revisions, considerable space given to the religious implications of Newtonian thought and the support it accorded to Voltaire's deistical convictions. The chapters on light are not confined to Newton's own discoveries, but include some historical information and a good deal of later material. Most striking is the interest Voltaire displays in the physiology of vision. A chapter on the human eye as an optical instrument is followed by chapters treating the psychology of visual perception, a subject which fascinated Voltaire. Here he draws upon Locke and upon Bishop Berkeley's *New Theory of Vision*.[10] He recounts the famous case of Dr. Cheselden's successful operation on the young man born blind, and elaborates at unnecessary length Newton's casual comparison of the colors of the spectrum with the tones of the diatonic scale, a subject of timely interest because of Father Castel's invention of his color organ.

Voltaire's devotion to the sensationalist philosophy of John Locke apparently finds an echo in these passages, and Voltaire seems concerned to emphasize, here as elsewhere, the essential harmony of the doctrines of these two great English thinkers.

To Voltaire, Newton was supremely important for having demonstrated the effectiveness of a new method of mathematico-experimental discovery in science, the famous method of analysis and composition. To dissect nature—even, as Hermann Weyl pointed out, the perceptively or intuitively simple, like a beam of white light or the path of a planet—into constituent elements, and then to confirm this dissection by successfully recombining the elements to restore the original phenomenon: this was the method that Newton not only practiced but here and there elucidated in the *Opticks* and the *Principia*.

Locke (so it seemed to Voltaire) had applied so far as he could the same method. He had laid open to man the anatomy of his mental processes "just as a skilled anatomist explains the workings of the human body."[11] And the conclusions Locke reached—that all knowledge depends upon sense experience; that accordingly most synthetic or a priori thinking is groundless; that much verbal discourse is without meaning—all this supplied Voltaire with a sharp and devastating implement with which to cut down the luxuriant overgrowth of theological assertion and philosophical system.

That Locke had set bounds to the pretensions of the human intellect was for Voltaire his outstanding accomplishment. But it was easy to pursue Locke's clues and end in skepticism, as Hume was to demonstrate. It was possible also to derive from Locke—as did the Abbé Buffier[12] whom Voltaire read approvingly—the conclusion that all the basic truths to which the human mind habitually gives consent are mere probabilities, excepting only the knowledge of our own existence. But if in philosophy we may perhaps enjoy the indolent luxury of suspended judgment, life requires action, and men—from all time—have acted in accord with probability, which comes nearest to Truth. Voltaire was obviously attracted to this early form of the "philosophy of common sense"; and I believe he found in Newton's accomplishments a remarkable exemplification of it.

Locke did not draw these conclusions except in one significant instance. With respect to the moral sciences, he was confident that man could find a pathway to demonstrative knowledge; but as to the sciences of nature he was wholly pessimistic:

> I am apt to doubt that, how far soever human industry may advance useful and experimental philosophy in physical things, *scientifical* will still be out of our reach. . . . *Certainty* and *demonstration* are things we must not, in these matters, pretend to.[13]

Elsewhere, he is still more emphatic, asserting that natural philosophy can never become a science, that is, never be completely certain, like a demonstration in geometry:

> Experiments and historical observations we may have, from which we may draw advantages of ease and health, and thereby increase our stock of conveniences for this life; but beyond this I fear our talents reach not, nor are our faculties, as I guess, able to advance.[14]

The situation with Newton was similar, but also in a manner different. Those eighteenth-century devotees of Newtonian thought who believed that because Newton used a mathematical method he held his conclusions to be eternally valid and perfectly true, like the demonstrations of geometry, had not read their Newton. But Voltaire had, and he well knew that the knowledge gained through the method of analysis did not have the cogency of a geometrical proof. And he was familiar with those passages where Newton admits that the propositions arrived at by induction are "very nearly true," and that "although the arguing from Experiments and Observations by Induction be no Demonstration of general conclusions; yet it is the best way of arguing which the Nature of Things admits of...."[15]

To Voltaire it must have seemed that Newton—so serenely confident, so indifferent to this scandalous state of affairs—was a useful corrective to Locke and the implied threat of a sterilizing skepticism. Newton's dramatic demonstration of the power of the new science—its success in ordering our knowledge of the physical world; its striking ability to predict the unforeseen—was accompanied by his frank avowal that the entire edifice rested upon probabilities. Here was a superlative illustration of Buffier's pragmatic, common-sense theory; and it convinced Voltaire that philosophical doubt need not paralyze action but in fact was the road to successful action.

I cannot agree with those who have suggested that Voltaire's brief excursion into the physical sciences had little influence on his subsequent thought and writing and that he returned without regret to literary pursuits. Some have thought *Micromégas*, his little science-fiction tale, to be his swan song, recording his disillusionment with the sciences.[16] Instead, it seems to me merely another assertion of his basic outlook: another assault on the makers of systems, whose futile answers he contrasts—citing the achievements of Huygens, Maupertuis, Leeuwenhoek, Swammerdam, and Réaumur—with the success of the tiny humans in scientific observation and measurement.

Voltaire's later work shows in a number of places the impress of his early contact with science; nor did he ever abandon his scientific interests altogether.[17] Allusions to the superiority of the analytical method are frequent in his writings, always with appropriate references to Newton and Locke. And numerous are the reminders that absolute truth is a chimera, that our knowledge can only be probable knowledge, which is all that man in his middling state can aspire to. The true method of thought, in philosophy and human affairs alike, is that which Newton introduced, with such success, in natural science.[18] As Cassirer has summarized it so well, analysis was to Voltaire "the staff which a benevolent nature has placed in the blind man's hands. Equipped with this instrument he can feel his way forward among appearances, discovering their sequences and arrangement; and this is all he needs for his intellectual orientation to life and knowledge."[19]

There was a matter of deep import to Voltaire, which commanded his chief attention for nearly a decade late in life, and on which he brought to bear the ideas we have been discussing. This was his campaign on behalf of those men—Calas, Sirven, and the others—who were the victims of religious intolerance and judicial oppression. In the 1760s, aroused to fighting pitch by the juridical murder of Calas, Voltaire turned his attention, with the same single-mindedness he had devoted earlier to science, to the study of French criminal law, with particular reference to the strange rules of evidence employed in capital trials. His grasp of the technical side of the problem has earned him the admiration of historians of French law, Esmein for example.[20]

The logical force of the arguments Voltaire directed against some of the bizarre legal practices of his time stems directly from his views about the certitude of our knowledge. Behind the formalized rules of evidence employed by the judges, strictly codified during the previous century, he discerned the same intellectual arrogance, the same confidence in the power of the human mind to attain absolute truth, that he pilloried in theology and philosophy.

"There is no year," he wrote in this period, "when some provincial judges do not condemn to a frightful death some innocent father of a family; and that peacefully, gaily even, as one slits the throat of a turkey in the farm-yard."[21] The judges, he remarked, do not suffer from doubts or misgivings and think that guilt can be proved like a theorem in geometry. But can we attain in human affairs such a certitude as will allow "seven men to enjoy legally the amusement of putting an eighth man to death in public?"[22] Voltaire is convinced that we cannot. "I am certain," he says in the *Philosophical Dictionary*, "I have friends; my fortune is secure; . . . my lover will be faithful; these are phrases which any man with some experience in life strikes from his lexicon."

The history of the human mind throws a revealing light on this problem of certitude. Before Copernicus, everyone was *certain* that the earth was at rest and that the sun rose and set. Not long ago witchcraft, divination, possession by devils, were things deemed most certain in the eyes of all men. Yet today this certitude has, to say the least, somewhat diminished. Indeed there is no certitude so long as it is physically or morally possible that things might be otherwise.

Yet many judicial cases, Voltaire continues in the same vein, have been settled as certainties when on further examination they have turned out to be errors. "When the judges condemned Langlade, Lebrun, Calas, Sirven . . . and so many others, all later shown to be innocent, they were certain, or should have been, that all these unfortunate men were guilty; and yet they were wrong. . . . If such is the misfortune of humanity that one must be content with extreme probabilities, one should at least take into consideration the age, social position, the bearing of the accused, the motive he might have had to commit the crime."[23]

Against the quasi-mechanical system of proof used in the criminal courts—where two half-proofs, as they were called, or four quarter-proofs, added together were deemed to constitute full proof—Voltaire had nothing but scorn. The whole system appeared archaic, fallacious, and cruel, especially so in the case of capital offenses. When life is at stake, he insisted, even the greatest probability ought not to be thought sufficient for conviction.

It is perhaps Voltaire's least-known contribution that he sought to introduce this principle of uncertainty into the rules of criminal evidence; and that in pamphlet after pamphlet during this legal debate he reiterated his conviction that in the realm of human affairs, probabilities and not certainties are the most we can expect. His mature opinion is to be found temperately and gravely stated in a little-read but important piece, his *Essay on Probabilities Applied to the Law* (1772). Here, after insisting upon his main point, he writes:

> One must take a stand, but one should not take it at random. It is therefore necessary to our weak and blind human nature, always subject to error, to study probability with as much care as we learn arithmetic and geometry. To judges, this study is peculiarly important. . . . A judge passes his life in weighing probabilities against each other, in calculating them, in evaluating their force.[24]

I have not introduced this topic to suggest that Voltaire is a forgotten pioneer in probability theory, for I am not sure how much he knew of this very important current in eighteenth-century science. I doubt whether he bothered his head with the mortality tables of Antoine Deparcieux, the first in France to continue the pioneer work of Graunt, Petty, and Edmond Halley, though he cites him approvingly on one occasion. And he certainly would have made little headway with the work of James Bernoulli or Abraham Demoivre on the mathematics of probability. Yet it is perhaps no accident that a devoted disciple of Voltaire, the mathematician Condorcet, attempted to apply probability theory to the sort of problems that interested Voltaire—the decisions of courts and legislatures—and urged that statistics and probability might someday form the basis of a reliable science of human affairs.

I have tried to illustrate here, by some concrete examples, what Voltaire may have got from his study of Newton. Paradoxical as it must seem, Voltaire appears—for all his technical limitations and his mathematical ineptitude—to have grasped the implications of Newtonian thought and method better than any other French man of letters of the eighteenth century. From Newtonians and Lockeians of the strict observance—chiefly from Voltaire and Condillac—the much-abused *ideologists* of the Revolutionary period took their start. Yet this tradition, culminating in the *Idéologues*, was buried for a time under an avalanche of materialist systems of moral and social philosophy; this line, which extends from Voltaire and Condillac through Turgot and Condorcet, seems to me to embody the deepest and most persistent elements of the Enlightenment's faith in science. At least, so the eighteenth-century scientists appeared to believe; for it was to this current of the Age of Reason that men like Lavoisier and Laplace, among other scientific figures who wished to play their role as renovators of society, appear to have lent their allegiance.

⚙

It is only fair and proper to introduce an out-and-out materialist, if only in contrast to those more moderate men I have already discussed. My candidate is the

Baron d'Holbach, whose powerful *Système de la nature* is frequently referred to and, quite evidently, not often read with attention.

The *Système de la nature* is not a work of science or natural philosophy, or even primarily a work of social theory, but a thundering attack on all supernatural beliefs. Having devoted a decade of his life to assaulting Christianity in a series of books and pamphlets, most of them anonymous, d'Holbach was now in 1770 prepared to take the next step and give to the world his refutation of all deists and theists, Voltaire and Newton included.[25] Somewhat incidental to his main purpose d'Holbach pretended to advance a rigorously secular view of man and society, derived—as he sincerely hoped—from the axioms of materialism and of science. From a set of assertions about the natural world—confident, combative, and uncompromising—d'Holbach draws out his view of man and his formula for man's secular redemption. Man is unhappy because he is out of tune with Nature; Nature—*le grand tout*, as d'Holbach calls it—consists only of material bodies in motion; and man is not only a part of nature, as Montesquieu would have agreed, but is a purely physical being. Just as belief in a realm of spirits, or even in a single spiritual creator, is illusory, so also is the habitual distinction between *l'homme physique* and *l'homme moral*. *L'homme physique* is man acting according to those laws of cause and effect which our senses can observe; *l'homme moral* is man acting in response to physical causes which our ignorance conceals from us.

In the early chapters d'Holbach sets forth his view of physical nature from which all this is supposedly derived:

> It is to physics and to experience that man must have recourse in all his investigations: he must consult them in matters of religion, ethics, legislation, political government, the sciences and the arts, even in his pleasures and sufferings. Nature acts by simple, uniform and invariable laws which experience enables us to know.[26]

In view of these words, we should expect d'Holbach to share with Voltaire some knowledge of contemporary physics, or at least of the popularized Newtonian physics of the eighteenth century. Yet the "immortal Newton" is mentioned seldom and then with scarcely veiled impatience. And the physics we are offered would scarcely pass as Newtonian. Newton, to be sure, is praised for having done away with the chimerical causes which had been invoked to explain the motion of the planets, as for example the vortices of Descartes; but d'Holbach is unhappy with Newton and his followers for regarding the cause of gravitation as unexplained or inexplicable. D'Holbach suggests that gravitation may be merely a special case of that propensity toward motion which is a property of all matter, and which depends upon "the inner and outer configuration of the bodies," a view of motion which is as far from Newton as it is from Descartes. Elsewhere, with bland inconsistency, he cites the followers of Newton as identifying attraction and repulsion with sympathies and antipathies, with affinities and *rapports*. But, once again, this is not the language of eighteenth-century physics, any more than is d'Holbach's definition of inertia as "gravitation on self" or his use of Newton's Third Law as an example of *nisus*, the urge of bodies at rest to move.

Perhaps d'Holbach's physical theory owes a good deal to his earlier experience as a chemist, more indeed than to his knowledge, or lack of knowledge, of physics. Chemistry had in truth been his first love; he was the author of many of the chemical and mineralogical articles that appeared in Diderot's *Encyclopédie*; and like Diderot he was a disciple and close friend of Rouelle, the outstanding chemist of the day. More obviously d'Holbach's views about matter and motion and his interpretation of Newton derive from John Toland's *Letters to Serena* (1704), which he published in French translation in 1768.

But there is another current still more evident in d'Holbach's thought: his debt to classical antiquity. This was clearly perceived by the prosecuting attorney who read the indictment that led to the condemnation and public burning of the *Système de la nature*. The author of this nefarious and sacrilegious work is not accused of having misused the results of contemporary science, but of having revived and extended the materialistic system of Epicurus and Lucretius.

That d'Holbach was thoroughly steeped in the Greek and Latin classics is evident from the most cursory inspection of the *Système de la nature*. It is, in fact, upon classical authorities rather than scientific ones that his paganism and materialism are explicitly based; and it is just in the pretentiously scientific preliminaries of his book that the classical references are most evident. Plutarch, Lucretius, Cicero—especially the *De divinatione* and the *De natura deorum*— and above all Seneca, provide the heaviest field pieces. Democritus and Aristoxenes are cited to show that in antiquity the soul was thought of as material. Not merely Lucretius but also Manilius and Pythagoras (the latter courtesy of Ovid) are authorities for the eternity of matter. Lucretius is quoted, as one would expect, on the evils that flow from the fear of death; while Seneca provides him with arguments in defense of suicide and with support for the principle that man is inherently neither good nor evil, as well as with useful quotations on fatalism.

The ancients are even made to share in the glory of modern scientific discoveries and doctrines, much in the spirit of Dutens's recent history.[27] Citing Diogenes Laërtius as his source, d'Holbach asserts that the Newtonian "system of attraction" is very old, having been anticipated by Empedocles in his doctrine of Love and Strife. Again he rather labors the point that Aristotle, long before the "profound Locke," insisted that nothing enters the mind except through the senses. Indeed, he seems to take a special pleasure in claiming that the two most popular philosophic doctrines of his day were of ancient origin.

As much as by any other influence, including that of the Greek materialists, d'Holbach's thought is shaped by the teachings of the Roman followers of the Stoa: Cicero, Seneca, and Marcus Aurelius.[28] It is from these men, rather than primarily from Newton or Descartes, that he draws his deep conviction of an inexorable order of physical nature, an order to which man, as a part of nature, must submit:

> Let man cease to look outside of the world in which he lives for beings who will give him a happiness that nature refuses him: let him study nature, learn its laws, contemplate its energy and the immutable way in which it acts. Let him apply his discoveries to his own happiness . . . and submit in silence to those laws from which nothing can shield him. Let him agree to remain in ignorance of those causes which for him are enveloped in an impenetrable veil; let him suffer without complaint the decrees of a universal force which cannot turn back or ever depart from the rules which its essence has imposed on it.[29]

If d'Holbach makes his bow to the science of his day, it is both awkward and perfunctory. He genuflects appropriately before the busts of Locke and Newton, mumbles the suitable incantations, and passes on to more important matters, and to the search for more convincing authority. It is as if he contented himself with showing that contemporary findings of science appeared to confirm the insight of the ancients that there is an inexorable order of material nature.

If the method and spirit of Newtonian science meant anything to d'Holbach, it is not apparent, not even in his manner of exposition. While Montesquieu's method is comparative and historical, and whereas other writers, like the physiocrats, affected an abstract, pseudomathematical method of presentation, such as Tocqueville and Taine deplored, d'Holbach's method can best be described as rhetorical, dialectical, and hortatory. The style moves in whirlpools and eddies: it is a series of loosely linked affirmations, challenges, questions, and indictments. It is the work of a prosecuting attorney familiar with the forms and devices of Roman forensic. It begins not really with axioms, but with an accusation; it ends not with a *quod erat demonstrandum*, but with a summation of the case and a final, impassioned charge to the jury.

❁

Regrettably, I shall end with neither, but only with some tentative suggestions. If my account of these three social thinkers of the eighteenth century has any validity, it may perhaps suggest that historians of the Age of Reason have been rather too sanguine in treating the thinkers of this complex age as if they did indeed possess a "common understanding of science." I am not too concerned to stress the point that of these three men only one, Voltaire, seems really to have got the drift of Newton. Newton's name was after all only a symbol or catchword of the age, though one of great evocative power. It can do little harm, when treating the eighteenth century in general terms, to use the Newtonian symbol, as did the men of that century, to stand for the complex framework of science itself.

But if we have been wrong in making science so exclusively the intellectual force which shaped the eighteenth-century mind (and so in consequence much of our modern thought), our interpretations will have to be radically altered. Perhaps we can no longer lay solely at the door of science all the social nostrums, all the flights of fancy, all the naïve confidence in the power of reason so characteristic of that century. Science may have to share the credit and the responsibility with certain less spectacular forces, and one of these I have hinted at. I referred several times to the deep-rooted classical formation of the French mind of the eighteenth century; and I would have liked to develop more fully, especially in my discussion of d'Holbach, the implications of the "new paganism": that peculiar movement which has its roots in the open or clandestine writings of the earlier *libertins* and which is so marked a characteristic of the last years of the century.[30] The vision of an age before the advent of Christianity—a golden age of free thought and free inquiry when men were men and not fallen angels— certainly served to express the secular aspirations of the Enlightenment fully as well as the vision of a world of the future, guided and improved by the light of science.

Notes

1. The old study of Montesquieu's early scientific writings, Désiré André's *Sur les écrits scientifiques de Montesquieu* (Paris, 1880), is still worth consulting. But see Sergio Cotta, *Montesquieu e la scienza della società* (Turin, 1953), chap. 2. For the Academy, see Pierre Barrière, *L'Académie de Bordeaux, centre de culture internationale au XVIIIᵉ siècle, 1712–1792* (Bordeaux: Editions Bière, 1951).

2. *Oeuvres complètes de Montesquieu*, ed. André Masson, 3 vols. (Paris: Nagel, 1950–55), II, 228.

3. *Pensées et fragments inédits de Montesquieu*, publiés par le baron Gaston de Montesquieu, 2 vols. (Bordeaux, 1899), I, 461.

4. *Oeuvres complètes de Montesquieu*, II, 847. "Livres bien écrits en Anglois: le Dr. Bangor, Tillotson, Praats [sic], *Histoire de la Société royale*." Sprat's work had been translated into French as early as 1670.

5. For Comte, see the *Cours de philosophie positive*, as cited by Cotta, *Montesquieu*, 10 and 76. For Durkheim, see his *Montesquieu et Rousseau, précurseurs de la sociologie. Note introductive de Georges Davy* (Paris: M. Rivière, 1953), with a translation of Durkheim's Latin thesis of 1892. Roger Caillois, in his useful preface to the *Pléiade* edition of the *Oeuvres complètes* of Montesquieu (Paris, 1947), speaks of the "révolution sociologique" as already heralded by Montesquieu's *Lettres persanes* (Rouen, 1721).

6. For the scientific sources and inspiration of Montesquieu's doctrine of the role of climate and his theory of the action of air on the fiber of the body, see Joseph Dedieu, *Montesquieu et la tradition politique anglaise en France; les sources anglaises de l'Esprit des Lois* (Paris, 1909), chap. 7; and the present writer's chapter, "Humanism in Science," in Julian Harris, ed., *The Humanities: An Appraisal* (Madison, Wis.: University of Wisconsin Press, 1950), 104–15.

7. A detailed account of the classical influence upon Montesquieu is given by L. M. Levin, *The Political Doctrine of Montesquieu's Esprit des Lois: Its Classical Background* (New York: Columbia University Press, 1936). For Montesquieu's education at the Collège de Juilly see Pierre Barrière, *Un Grand Provincial: Charles-Louis de Secondat, Baron de La Brède et de Montesquieu* (Bordeaux: Editions Delmas, 1946), 12–20.

8. Ira O. Wade, *Voltaire and Madame du Châtelet: An Essay on the Intellectual Activity at Cirey* (Princeton: Princeton University Press, 1941); and the section entitled "Some Aspects of Newtonian Study at Cirey," in Wade's *Studies on Voltaire* (Princeton: Princeton University Press, 1947). See also Margaret S. Libby, *The Attitude of Voltaire to Magic and the Sciences* (New York: Columbia University Press, 1935), a largely factual study of Voltaire's scientific ideas, and the older work of E. Saigey, *Les sciences au XVIII⁰ siècle: la physique de Voltaire* (Paris, 1873).

9. "Conseils à un journaliste," in *Oeuvres*, ed. Louis Moland, 52 vols. (Paris, 1885), XXII, 242. In several places—as in the opening lines of the dedication to the Marquise du Châtelet in the first edition of *Elémens de la philosophie de Neuton* (1738), and in his *Micromégas*—Voltaire made transparent fun of the style and method of Fontenelle.

10. Voltaire's interest in Berkeley is described by Libby, *Attitude of Voltaire*, 105–6; and see T. E. Jessop, *A Bibliography of George Berkeley* (Oxford, 1934), 8–9, 65. Voltaire mentions having had several conversations with Berkeley during his stay in England.

11. Voltaire, *Lettres philosophiques; édition critique, avec une introduction et un commentaire par Gustave Lanson*, 2 vols. (Paris, 1909), I, 168.

12. See *Oeuvres de Voltaire*, XIV, 49. The Jesuit Father Claude Buffier (1661–1737), Locke's first disciple in France, deserves more attention than he has received. Author of numerous works, his most important book is his *Traité des premières vérités et de la source de nos jugements, où l'on examine le sentiment des philosophes de ce temps sur les premières notions des choses* (Paris, 1724).

13. *An Essay Concerning Human Understanding*, ed., Alexander Campbell Frazer, 2 vols. (Oxford, 1894), II, 217–18.

14. *Ibid.*, 350.

15. *Mathematical Principles of Natural Philosophy*, Andrew Motte's translation, ed. Florian Cajori (Berkeley, Calif.: University of California Press, 1934), 400; *Opticks*, 3d ed. rev. (London, 1721), 380.

16. Ira O. Wade, *Voltaire's Micromégas: A Study in the Fusion of Science, Myth, and Art* (Princeton: Princeton University Press, 1950), 170–8.

17. At Ferney in the 1760s he carried out some experiments, inspired by the findings of Spallanzani, on regeneration in snails. In 1768 he published his *Singularités de la Nature*, a short collection of pieces in which he attacked what he believed to be the irresponsible speculations of naturalists: the animal nature of coral; Buffon's theory of organic molecules; J. Turberville Needham's defense of spontaneous generation; the organic origin of fossils; the historical geology of Burnet, Woodward, Whiston, de Maillet, and Buffon. "Dans la physique, comme dans toutes les affaires du monde, commençons par douter. . . . C'est la raison éclairée et soumise qui sait qu'un être chétif ne peut pénétrer l'infini. Un fétu suffit pour nous démontrer nôtre impuissance. Il nous est donné de mesurer, calculer, peser, et faire des expériences." *Singularités* (ed. 1768), 3–5.

18. See his letter to s'Gravesande (June 1, 1741): "*Vanitas vanitatum, et metaphysica vanitas!* Nous sommes faits pour compter, mesurer, peser: voilà ce qu'a fait Newton; voilà ce que vous faites avec Musschenbroeck." Voltaire, *Oeuvres*, XXXVI, 65; cited, with other variations on this theme, by Wade, *Voltaire's Micromégas*, 164, n. 68.

19. Ernst Cassirer, *The Philosophy of the Enlightenment*, trans. Fritz C. A. Koelln and James P. Pettegrove (Princeton: Princeton University Press, 1951), 12.

20. A. Esmein, *Histoire de la procédure criminelle en France* (Paris, 1882), 363–70.

21. *Dictionnaire Philosophique*, "Lois criminelles," *Oeuvres*, XIX, 626.

22. *Ibid.*, "Crimes," *Oeuvres*, XVIII, 275–76.

23. *Ibid.*, "Certain, Certitude," *Oeuvres*, XVIII, 117–21.

24. *Oeuvres*, XXVIII, 497.

25. Pierre Naville, *Paul Thiry d'Holbach et la philosophie scientifique au XVIII⁰ siècle* (Paris: Gallimard, 1943), 411–17.

26. *Système de la nature, ou des lois du monde physique & du monde moral, nouvelle édition, augmentée par l'Auteur*, 2 vols. (London, 1774), I, 5.

27. Louis Dutens, *Recherches sur l'origine des découvertes attribuées aux modernes,*

> *où l'on démontre que nos plus célèbres philosophes ont puisé la plupart de leurs connoissances dans les ouvrages des anciens* (Paris, 1766).

28. This enthusiasm for the Stoic philosophers was shared by others of d'Holbach's circle. Diderot wrote a *Vie de Sénèque* (1779); his disciple Naigeon published a French translation (by a certain de Lagrange) of the works of Seneca. Diderot, also like d'Holbach a student of chemistry, was in general agreement with d'Holbach's theories of matter. See the *Principes philosophiques sur la matière et le mouvement*, which was evidently inspired by d'Holbach's *Système de la nature*, as we gather from Naigeon's introductory remarks. *Oeuvres complètes de Diderot*, ed. J. Assézat and Maurice Tourneux, 20 vols. (Paris, 1875–77), II, 64–70.

29. *Système de la nature*, I, 2.

30. See, for example, Louis Bertrand, *La fin du classicisme et le retour à l'antique* (Paris, 1897), especially chap. 1, "La renaissance de l'idée païenne." The need for further studies of this sort has been emphasized by Henri Peyre, *L'influence des littératures antiques sur la littérature française moderne—Etat des travaux* (New Haven: Yale University Press, 1941).

31

Some Aspects of Science During the French Revolution

❀

The late eighteenth century, including the period of the French Revolution, is a richly rewarding field of study for anyone concerned with the influence of science upon society, or with the impact of social change upon the work and thought of scientists. Never before, and rarely since, has science enjoyed such unalloyed esteem as it did in the Europe, especially the France, of the Age of Enlightenment, when it had for its advocates and propagandists the outstanding men of letters and social theorists from Montesquieu and Voltaire to Condorcet and Volney. Inspired by the writings of Descartes and Newton, these men drew confident arguments from the realm of physical law in their campaign to bring a similar rule of reason, law, and harmony into the inherited social institutions of their day. From science and its recent history, moreover, they took their most compelling examples of intellectual progress, finding support therein for their gilded vision of indefinite human perfectibility.

This favorable climate of opinion helps explain the mysterious concatenation of events discussed in this essay: the fact that the greatest period of French scientific leadership coincided almost precisely with the Age of Revolution; and that the time of Mirabeau, Danton, Robespierre, and Bonaparte was also, I need hardly point out, that of Lagrange, Laplace, Monge, Condorcet, and many other illustrious names in mathematics, physics, and astronomy; of A. L. de Jussieu, Lamarck, Cuvier, and Geoffroy Saint-Hilaire in botany, zoology, and paleontology; of Bichat in physiology and anatomy; and of Lavoisier, Berthollet, and the other French founders of modern chemistry.

What is difficult to comprehend—especially in view of the persistent tradition that the spirit of the Revolution was detrimental, if not actually antagonistic, to science—is that this scientific flowering was not fatally arrested or totally destroyed by the distractions of the Revolution, by the bloodbath of the Terror, by the mounting wave of emigration, or by the endless wars of the Republic, the Consulate, and the Empire. Yet this was clearly not the case. The scientific gen-

From *Scientific Monthly* 80 (1955), 93–101. Copyright, 1955, by the American Association of Science.

Paper read at a symposium on the validation of scientific theories, held in Boston in December 1953. Also appeared in Philipp Frank, ed., *The Validation of Scientific Theories* (Boston: Beacon Press). © 1954, 1955, 1956 by the American Association for the Advancement of Science.

eration of the Napoleonic period and the Restoration—that of Ampère, Arago, Poisson, Magendie, Gay-Lussac, Sadi Carnot, Cauchy, Fresnel, and the rest—is as rich if not richer in talent than the generations that came before. Yet we look in vain for truly comparable achievements in the art, the music, or the *belles lettres* of this revolutionary period; and we are forced to the conclusion that the national energy and the great social ferment that overthrew the Old Regime, spreading a new democratic gospel across Europe, found its greatest cultural expression in scientific accomplishment.

Science and its practitioners played a notable role in the intellectual preparation for the Revolution as well as in the events of the Revolution itself. What this may have amounted to I can only summarize, in full realization of the complexity of the problem and the monographic work that remains to be done. My main purpose is to examine in preliminary fashion what happened to scientific progress, to scientific institutions, and to scientists themselves during the great Revolution, and to offer a general picture, tentative at best, that may help awaken the interest of other scholars in these problems and reveal, perhaps through my own errors, the gaps in our knowledge.

❖

In the decade before the Revolution, European science felt the loss of such outstanding figures as Euler, Linnaeus, Daniel Bernoulli, and the great northern chemists Scheele and Bergman. In France, d'Alembert died in 1783 and Buffon in 1788. Despite these losses, this was throughout Europe a time of extraordinary productivity and promise in the world of science, with France and England unquestionably in the lead, but with Switzerland, Italy, the German states, and Scandinavia boasting many proud names. As if to put the seal on France's acknowledged leadership, the eminent mathematician Lagrange, a native of Turin who for twenty years had been the beacon light of the Prussian Academy, left Berlin in 1787 after the death of Frederick the Great and took up his residence in Paris. It was here that he published in the following year his great *Mécanique analytique* under the auspices of the Royal Academy of Sciences. In 1789, the year of Revolution, there appeared three of the acknowledged classics of French science: A. L. de Jussieu's *Genera plantarum*; Philippe Pinel's *Nosographie philosophique*; and Lavoisier's epoch-making *Traité élémentaire de chimie*. Each work in its own domain—botany, medicine, and chemistry—was both a fulfillment and a new departure.

The swiftly moving events of the Revolution's first phase—the convening of the Estates General, the disorders in Paris and the provinces, the abolition of inherited privilege, the creation of a constitutional monarchy—found the scientists neither aloof nor unprepared. Having taken an active part in the liberal movement of the previous decades, they welcomed the first phase of the Revolution with enthusiasm; indeed men like Bailly, Condorcet, and Lavoisier had played their modest part in bringing it about. Politics already infringed upon science, disturbing the tranquility of the laboratory and penetrating the fastnesses of the Academy of Sciences. On July 4, 1789, the Academicians took the unprecedented step of expressing to their fellow member, the astronomer Bailly, their satisfaction at the manner in which he had performed his duties as president of the National Assembly; and later in the month the members of the Academy went in a body to Bailly's residence in Chaillot to pay their respects to him. Yet on the day following the storming of the Bastille the Academy held its regular meeting with twenty-three members present; technical papers were presented, and there is no echo in the *Procès-verbaux* of the storm raging without. Throughout the remainder of 1789 and well into 1790, fundamental scientific questions continued to dominate the meetings of the Academy.[1] Laplace read a series of important papers on celestial mechanics; Coulomb presented his sixth memoir on electrostatic experiments with the torsion balance; Lavoisier and Seguin described their classic experiments on respiration and heat regulation in man and other animals, the last work on pure science carried out by the senior partner.[2] There was great interest in current English work in observational astronomy; money was even set aside to build a large reflecting telescope, modeled upon the huge instrument with which the great English astronomer William Herschel was busy charting the nebulae and observing the rings and satellites of Saturn.[3]

But concern about the effect of the revolutionary tensions and of the obsession with political events is reflected in the correspondence of the scientists. In August 1789 the chemist Berthollet wrote to James Watt:

> While you are occupied tranquilly with science and the useful arts which owe you such great obligations, we have been obliged to lose sight of them. The ferocity of the great nobles, the insurrection of the citizens, the fury of the people, the scourge of famine have absorbed all our attention; yet one must return to peaceful occupations, and one can begin to enjoy the pleasures of study. I am taking up my experiments once more.[4]

And the mathematician Gaspard Monge, toward the end of the same year, commented to the same correspondent:

> Our Revolution occupies every mind, each in his own fashion; and science is the loser. May God bring it to a swift conclusion, for we shall lose the habit of work and the love of science.[5]

Lavoisier wrote in like vein to the elderly Scottish chemist Joseph Black, lamenting the interruption of scientific activity and expressing the hope that calm and prosperity would soon allow the scientists to return to their laboratories.[6] While from Chaptal, the founder of French industrial chemistry, we have the following cautious appraisal of the opportunities and dangers that lay ahead:

> The revolution which is taking place is a beautiful thing, but I wish it had arrived twenty years ago. It is annoying to find oneself under a house that is being torn down, but that is precisely our position. . . . In this general con-

fusion, in this torrent of passions, the intelligent man studies the role he should play: but it seems just as dangerous to remain outside of the excitement as to participate.[7]

As these letters indicate, the scientists were being inexorably drawn into the revolutionary turmoil. In 1788 Lavoisier had prepared a long memorandum on the proper constitution of the Estates General; early in 1789 he took a major part in drawing up the *cahier* of grievances to instruct the representatives of the nobility of Blois; and in May he was chosen alternate deputy to the Estates General. The astronomer Bailly was one of the political leaders of the first Revolutionary assembly; and shortly afterward was chosen the first mayor of Paris. The mathematician Condorcet, Perpetual Secretary of the Academy of Sciences, plunged at once into the journalistic and political activity that led to his election—along with such other scientists as Tenon, Lacépède, Fourcroy, and Guyton de Morveau—to one or another of the succeeding assemblies. Other scientists were active in the Paris *sections*, or Revolutionary districts, and served in battalions of the National Guard.

By the time of the acceptance of the constitution in September 1791, the moderates—and they included all but a handful of the younger scientists—hoped, as we have just seen, that the violent and disruptive phase of the Revolution was at an end and that the time had come to plan constructively for the future. To this end there was founded in April of 1790 a short-lived but influential association called the Society of 1789, which aspired to be the intellectual guide and official planning agency of the new society and its elected assembly. Besides well-known liberals and reformers of the pre-Revolutionary period—Brissot, Dupont de Nemours, Mirabeau, Talleyrand—it included influential members of the Academy of Sciences: Condorcet and Lavoisier; the mathematician Gaspard Monge; the biologists Lacépède and Lamarck; and others.[8] Briefly, from June through September 1790, this society published a journal, edited by Condorcet, which is our only direct evidence of what transpired in their meetings and for the philosophy that pervaded them. The society's avowed purpose was to aid and promote

> . . . all discoveries useful to the progress of *l'art social*, to encourage those being made in these sciences themselves and to gather together suggestions relative to public institutions that may be formed for public welfare and for education.[9]

A perusal of this short-lived journal shows that, besides treating in rather high-flown and abstract language such basic social problems as the rights of women and the proper foreign policy for a free nation, it devoted space to the discussion of the national economy and the importance of a scientific technology.[10] Lavoisier read a famous paper on the *assignats* and the inflationary dangers of a paper currency. The chemist Hassenfratz wrote on the importance of promoting the useful arts, contributed a long article on the mineral resources of France and the possibilities of developing them, and described in another number recent advances in chemical industry: LeBlanc's famous soda process and Berthollet's use of hypochlorites for bleaching.

In the discussions of this society are to be found, I believe, the germs of many of the constructive revolutionary accomplishments: the various efforts undertaken to stimulate productive industry and invention; possibly also the great reform in weights and measures and the creation of the metric system, although this had earlier antecedents; but above all, although direct evidence is lacking in the *Journal*, the plans for new educational and scientific institutions, such as those later elaborated by Talleyrand and Condorcet.

The philosophy of science, or rather of social science, that guided these men is worthy of a moment's attention. It centers upon Condorcet's conception of a unified social science, an *art social*, based upon a collaboration and unification of the sciences according to a common spirit and, where possible, a common methodology. The vision of what Comte was to call *sociology* is clearly discernible in the manifesto of the society drawn up by Condorcet:

> There should exist for all societies a science of maintaining and extending their happiness: this is what has been called *l'art social*. This science, to which all others are contributors, has not been treated as a whole. The science of agriculture, the science of economics, the science of government . . . are only portions of this greater science. These separate sciences will not reach their complete development until they have been made into a well-organized whole. . . . And this result will be obtained sooner if all the workers are led to follow a constant and uniform method of work.[11]

If one had asked of Condorcet how such a unification of the sciences could be brought about, how the experimental and mathematical spirit of the natural sciences could be transferred to the sciences of man and society, he would not have agreed with the early system-builders of the eighteenth century, or with John Stuart Mill, that it is enough to build aprioristic deductive systems in imitation of the great scheme of classical mechanics. Condorcet placed his faith in what he called social mathematics, embodying the twin disciplines of social statistics and mathematical probability, subjects to which he had contributed, as had Laplace and Lavoisier, in the years before the Revolution. Just at this time he was preparing a popular exposition of social mathematics in his *Elémens du calcul des probabilités*, for he saw in it a useful instrument of social improvement and reform. In his preface he explains why he feels that social mathematics was at this moment indispensable; and he continues:

> When a Revolution has ended, this method of treating the social sciences takes a new direction and acquires a greater degree of utility. In fact, to repair promptly the dislocations inseparable from every great movement, to restore general prosperity, one needs stronger methods [than mere argument], means calculated with greater precision, supported by unattackable proofs in order to ensure the adoption of needed reforms in the face of selfish interests and base faith.[12]

We hardly need to remind ourselves that when Condorcet wrote the Revolution had not ended but was moving with torrential rapidity toward greater confusions and dangers, in which science and scientists alike suffered. Later men like Quetelet, Cournot, and Auguste Comte in the nineteenth century were to pick up, each in his own fashion, the prophetic program that Condorcet was forced to abandon.

✺

Even before the outbreak of war in 1792, the demands of a succession of revolutionary governments upon the Academy of Sciences and its members left little time for normal activities. By all odds the most time-consuming and exacting responsibility—overshadowing such requests as that the Academy examine and test silver vessels taken from *ci-devant* churches and recommend the proper method of converting the secularized church bells of bronze and bell metal to military use—was the great project for the standardization and rationalization of the system of weights and measures. By a decree of May 8, 1790, the National Assembly charged the Academy of Sciences with determining the best scheme, based upon some universal standard found in nature, for adoption by all nations. Early in 1791 the Academy recommended a decimal system of units, derived from a unit of length, the meter, to be established by geodetic measurements. After a favorable report by Talleyrand to the Assembly, the Academy was assigned the task of making the basic measurements and preparing a reliable set of primary standards. This involved a long and tedious series of operations, still not completed when the Academy was abolished.

Despite this drain on its personnel and energy, the Academy continued regular sessions until the summer of 1793. Even in the final six months of its existence, regular meetings were still being held, although the exigencies of national defense and the mobilization of science for war—one of the earliest such phenomena in history—sometimes reduced the participants to a mere handful. The Academy even continued its practice of announcing the subjects of annual prize contests.[13] A subject proposed for the year 1793 is of special interest. The prize was to be awarded for the best theoretical analysis of the operation of steam engines, with a discussion of methods for their improvement,[14] surely one of the most important technical problems of the time. No prize was actually awarded, although it was again announced in 1793 for the year 1795, no memoirs having been received, and the problem was finally attacked for the first time a generation later by Sadi Carnot, the son of the man who in these years was organizing the victory of the Republican armies.

During this crucial period the collapse and destruction of many of the venerable scientific institutions had an equally damaging effect upon the progress of science. The *Imprimerie Nationale*, now flooded with job printing for the government, could no longer serve, as it had throughout the eighteenth century, for the publication of scientific books. Important serial publications, among them the *Journal des savants*, the *Mémoires* of the Academy of Sciences, and even the newly founded *Annales de chimie*, were suspended for lack of funds or contributors. But the most serious blow was the suppression, in August 1793, of the venerable Academy of Sciences.

The detailed story of the Academy's fall has yet to be written, and I shall not attempt it here. Its ultimate fate, and that of the other royal academies, was heatedly debated from 1790 to 1793. While the monarchy lasted, eloquent voices were raised to preserve it virtually unchanged, but vitriolic attacks had already begun, both within and without the assemblies, demanding its immediate abolition as an aristocratic remnant of the past, and as a "school of servility and falsehood." Effective pamphleteers, chief among them J. P. Marat, attacked the Academy and its members unmercifully in the public press. As far as I have been able to judge the plan most widely favored—for example, by men like Talleyrand and Condorcet—was to effect a peaceable transformation of academies, including the Academy of Sciences, into learned bodies more acceptable to the new climate of opinion; until this could be effected it was hoped to continue the Academy virtually unchanged. These tactics were very nearly successful, as they proved to be in the case of the Jardin du Roi, which emerged enlarged and strengthened as the Muséum d'Histoire Naturelle, a research center of great importance. Disagreement within the Academy of Sciences brought delay, and delay was fatal. While defending the academies before the Legislative Assembly in 1791 the Abbé Grégoire made known that the academies were, of their own accord, reforming their statutes to put them in harmony with the new era and erasing traces of their monarchical past. There is evidence that a draft of new statutes was actually prepared by the Academy of Sciences, but it seems to have been without effect.[15]

On August 8, 1793, Grégoire read to the Convention a report on behalf of the Committee of Public Instruction in which he proposed the suppression of the academies, in order to reorganize these bodies, as he put it, in the light of human wisdom and progress. The Academy of Sciences alone, by virtue of its special utility, was to escape suppression. But the Convention was in no mood to brook exceptions; after a vituperative speech by the painter Louis David, the Academy of Sciences was extinguished with the others. The members were even denied the privilege of constituting themselves a Free Scientific Society and of using their accustomed meeting place in the Louvre. The doors were closed and sealed; soon after, these echoing chambers were invaded by an army of tailors, busy stitching uniforms for the Revolutionary armies, while workers removed the last vestiges of monarchical symbolism from the walls.

Several years were to elapse before the constructive plans of the Convention replaced or successfully remodeled the older scientific institutions. Under the Directory, harried as it was by inflation and war, there nevertheless were miraculously established those scientific institutions which were to be the boast and pride of France during the succeeding century: the Institut de France, the Ecole Polytechnique, the Conservatoire des Arts et Métiers, the brilliant but ephemeral

Ecole Normale, the Muséum d'Histoire Naturelle. The result of prolonged planning and debate, going back at least to the discussions in the Society of 1789 and in the early Revolutionary assemblies, the final formulation of these plans must be credited to the Convention, the most ruthless and determined of the Revolutionary governments. This fact was conveniently forgotten by the writers and propagandists of Napoleonic days, who left the impression that Bonaparte, almost singlehandedly, had saved French science, which the Jacobins had sought to destroy.

Soon after the collapse of the Academy, organized scientific work came to a virtual standstill. Under the Jacobins' iron rule, the Republican conservatives—the Girondist opposition—were herded to public execution in the Place de la Révolution. The astronomer Bailly joined Philippe Egalité, Mme. Roland, and lesser enemies of Robespierre in the tumbrils of the guillotine. The members of the General Farm, the tax-collecting corporation of the Old Régime, were arrested in the fall of 1793, the chemist Lavoisier among them. Tried before the Revolutionary Tribunal on May 8, 1794, and convicted on a specious charge of conspiring with the enemies of the Republic, all were executed on the same day. The scientific community stood in appalled confusion. Many, like Laplace, found hiding places in the country (it was in such circumstances, for example, that he completed his popular Système du monde). Trapped on the outskirts of Paris, Condorcet is said to have taken his own life with poison foresightedly obtained from his friend Cabanis, the physician and philosopher.[16] Yet it is astonishing to record that no scientist of note joined the flood of emigrés, which reached its peak in these years. Against none of the scientists, moreover, can a charge of counterrevolutionary activity be seriously maintained. On the other hand, few of Lavoisier's erstwhile co-workers—not Fourcroy, Guyton-Morveau, Monge, or Berthollet, all of whom were serving the Convention—raised effective voices in his defense, nor had they openly protested the arrest of Bailly. Political passion, fear, and perhaps personal resentments may explain, but cannot condone, this conduct.

After 9 Thermidor (July 27, 1794), when Robespierre fell, sanity returned and the scientists could survey the wreckage. The vandalism of the scientific institutions, the execution of Bailly and Lavoisier, were at once held up as among the most abominable of Robespierre's crimes. Condorcet was accorded his apotheosis as patron saint of a new learned publication, the Décade philosophique, founded in 1794, organ of the so-called "Idéologues." In Millin's Magasin encyclopédique there appeared, the year following the execution of the Farmers General, the first biographical sketch of Lavoisier, a factual but moving account by his long-time friend, the astronomer Lalande.[17] The same year a somewhat nauseating memorial service was held for Lavoisier at the Lycée des Arts, the main feature of which was a ponderous eulogy by Fourcroy, the erstwhile Jacobin, who sought thereby to obliterate his failure to aid Lavoisier.[18]

❧

It is of some interest that even during these trying years, and before the creation of the new scientific institutions, a thin but persistent thread of scientific activity is clearly evident. Private initiative took over where the public institutions gave way or were destroyed. The old Lycée de Paris, a center for public lectures founded in 1780, took on considerable importance until it was shunned for harboring men suspected of antirevolutionary proclivities. Its more scientific and utilitarian competitor, the Lycée des Arts, founded in 1792, flourished through the darkest days of the Terror. It stressed the application of science to the useful arts, and among its outstanding lecturers were Fourcroy, Berthollet, Daubenton, and Jussieu.[19] Millin's *Magasin encyclopédique*, begun in 1792 but not firmly established until 1795, gives a picture of its activities and was in some respects its organ.

Of more importance for fundamental scientific work were two new societies whose rebirth, in one case, and prosperity, in the other, were due to the conditions I have described, The first of these, the Societé d'Histoire Naturelle, had been founded, only to disband, in 1788. It was revived during the Terror, and an English commentator wrote of it as follows in 1793:

> . . . the disadvantages to which it was exposed, in common with the non-privileged societies, under the old government, and the jealousy of some of the protected literary bodies [i.e., the Academies] soon caused its dissolution.
>
> After the revolution, its founders, however, reunited, extended their plan, and instituted the present Society of Natural History, which was joined by all the naturalists of the capital. . . . The object of their labours is Natural History in general, but especially that of France, and in particular of the environs of Paris. . . . New researches are to be made by means of periodical excursions taken by the Society, either in the country, at the proper seasons of the year, or to gardens, museums, etc. . . .[20]

This is rather too peaceful and bucolic a picture for this period of general harassment, but it is certain that the society became genuinely active after Thermidor. It was frequented by the professors of the Muséum d'Histoire Naturelle, which had early made its peace with the Revolution and where substantial scientific work was being accomplished by men like Lacépède, Lamarck, and Cuvier. Significant papers by these men and others were published in the society's *Journal*, which appeared briefly in 1792, and in its *Mémoires*, first published in 1799.[21]

A second and distinctly more important scientific society owed its inception, like the Natural History Society, to the fact that it was no longer necessary to obtain royal approval (and, in addition, at least the passive acquiescence of the Academy of Sciences) for a society holding scientific meetings and issuing a regular publication. This was the Société Philomatique, which played a very useful role in the scientific life of this troubled period and has continued to this day. Beginning in 1788 as an informal discussion group of six almost unknown physicians and scientific amateurs, it was joined, in September 1789, by the young chemist Vauquelin and a few others. These men constituted themselves a regularly organized scientific society, with dues, correspondents, and the project of publishing a monthly *Bulletin* or journal. Its membership increased slowly between 1790 and 1792, but as yet it attracted no important scientists. But suddenly, after the suppression of the Academy of Sciences, distinguished names were added to the roster. In 1793 the Société Philomatique was joined by Berthollet, Fourcroy, Monge, Lamarck, and Lavoisier, that is to say, by acknowledged luminaries of the *ci-devant* Academy of Sciences. Between the time the Academy was abolished in August 1793 and the autumn of 1796, when the Institut de France was formally established, the Société Philomatique was the principal haven of the dispossessed scientists. It maintained close ties with the Société d'Histoire Naturelle, and with the scientists at the Muséum. After Thermidor, its president referred to it as the only society officially recognized as having offered during the period of terrorism "un point fixe de réunion aux sciences et aux arts."[22]

In 1791 its *Bulletin* was launched as a monthly summary of scientific progress that was circulated in manuscript among the members. In 1792 it was printed in a few copies but consisted only of short abstracts. Its first real issue as a learned journal is that of April 1797. The printer, of whom we must say a few words, was the economist and publicist, Dupont de Nemours, erstwhile member of the Society of 1789 and close friend of many scientists.

Private initiative once again filled the gap left by the loss to the Academy of Sciences of the facilities of the Imprimerie Nationale, which during the eighteenth century had printed the official publications of the Academy, and many individual works of science bearing the seal of its approval. Private printers like the Jomberts and the Didots had made something of a specialty of scientific printing during the eighteenth century, and they were by no means inactive during the Revolution. But the man who should be notable for coming to the aid of the scientists in this capacity during the period of crisis is Pierre Samuel Dupont de Nemours. In June 1791, as he was leaving the Constituent Assembly, Dupont issued a prospectus informing the public that he was opening a well-equipped publishing house where, he said, he proposed to do "good and inexpensive work for those who are chiefly interested in the contents of a book."[23] This venture of Dupont's is well known, but it is usually assumed that he printed only political tracts and his *Correspondance patriotique*. Nor is it widely known that the Lavoisiers, husband and wife, were among his sponsors.[24] From them Dupont borrowed the sum of 710,000 francs for the purchase of his printing house. It is therefore not surprising to find that Dupont had a share in publishing works of his earlier associates in the Society of 1789. In 1791 he helped bring out Talleyrand's famous report on public education and printed Lavoisier's *Etat des finances de France*. In 1793 he published the *Réflexions sur l'instruction publique*, which Lavoisier drafted in the name of the Bureau de Consultation des Arts et Métiers.

A number of scientific books also appeared over Dupont's imprint: the first edition of Fourcroy's *Philosophie chimique* (1792) and a second edition in

1795; a treatise by Antoine Portal on tuberculosis (1792); two editions of Daubenton's *Tableau méthodique des minéraux*; a *Flora* of the Pyrenees by Picot de la Peyrouse (1795), and other works. Dupont was the official printer for the Academy of Sciences from 1791 until its dissolution (he is so listed in the *Almanach Royal*), and in this capacity did such job printing as the prize announcement of the Academy mentioned in the preceding section. In 1794 Dupont published the belated volume of the *Mémoires* of the Academy for 1789 and, in 1797, the volume for 1790. He was also, as we have learned, publisher of the *Bulletin* of the Société Philomatique.[25]

❂

Some general remarks about the character of scientific work in this period seem appropriate here. If we cite at random, as I did at the beginning of this article, the great names that illuminated these decades, we convey the impression of distinguished and virtually uninterrupted scientific progress. A closer examination does not confirm this. The first years of the Revolution, perhaps to 1792, were still quite productive, because the momentum of the previous years was not immediately arrested. The really creative period of the men of the Revolutionary generation—men who, like Lagrange, Lavoisier, Monge, Berthollet, and Laplace, were in their forties or early fifties in 1789—falls in the years before the Revolution. Monge presented in his lectures at the Ecole Normale his great invention of descriptive geometry, but this had been worked out long before when he taught at the Ecole de Mézières. The monumental *Mécanique céleste* of Laplace did not begin to appear until 1799, but the work seems to have been well advanced by 1790 and, but for the Revolution, might have been completed much sooner. The same rule seems to hold for Legendre's *Essay on the Theory of Numbers* (1794) and his work on elliptic functions (1798), both of which grew out of earlier work.

Those scientists of the older generation who survived the turmoil of the Revolution—and they were the great majority—made their greatest contributions in this period as inspiring teachers of the Napoleonic generation. Indeed the production of brilliant pedagogic works is a marked feature of the period from 1794 to 1800. Monge's lectures are in this category, and so were the later books of Lagrange, based on his teaching at the Ecole Normale and the Ecole Polytechnique. Laplace's famous *Essai philosophique sur les probabilités* grew out of lectures delivered at the Ecole Normale in 1795. An outstandingly successful example is Legendre's *Eléments de géométrie*, a skillful reworking of Euclid. And on a lower plane were the immensely popular mathematical textbooks of Sylvestre Lacroix, widely used at one time in this country. Laplace's readable *Exposition du système du monde*, like his general discussion of probability just mentioned, was clearly a manifestation of a desire to present serious scientific speculation to a wide audience, yet in a spirit markedly different from the glib popularizations of the eighteenth century.

A similar phenomenon is observed in chemistry, where the great textbooks of Chaptal and Fourcroy, and the latter's *Dictionary of Chemistry* in the *Encyclopédie méthodique*, sought to present in intelligent order the facts of the new chemistry. The pioneer works on applied chemistry by Chaptal and Berthollet emphasize another aspect of this new orientation.

The naturalists of the Muséum, chief among them Lamarck, form something of a special case. Sustained, even pampered, by the Revolutionaries of the Left, the workers at the Muséum were less adversely affected by events. It was the revolutionists who called Lamarck, a man of fifty and a botanist, to the newly created chair of invertebrate zoology; and here during the subsequent years he did his best and most famous work, developing his theory of biological evolution and collecting the materials for his pioneer descriptive treatise of invertebrate zoology (1815–22).

That science lost much that the Revolutionary generation intended to accomplish is suggested by what we know of the work of leading Academicians on the eve of the Terror. Had the great reflecting telescope been built, it might have turned French astronomy into channels of observational astronomy in which, in the nineteenth century, other countries, including America, surpassed her. But the money set aside for this purpose was presented to the Convention as a *don patriotique* in a frantic effort of ingratiation.

From what we have recently learned of the plans and projects of Lavoisier at the time he was lost to science, we see that his creative energies had in no sense flagged. He was forty-six when the Revolution broke out, and if he could have matured his scientific plans uninterruptedly the results might have been incalculable. At the time of his imprisonment he was preparing a fundamental revision of his *Traité élémentaire de chimie* and an edition of his collected works; and only recently it has been pointed out that in addition he had outlined a great work of chemical theory, or, as he said, of *philosophie chimique*.[26] More important still, he considered his last work on respiration and body heat control as a starting point for a research program in what we would now call medical biochemistry. A passage in Lalande's sketch of Lavoisier, which has been generally overlooked, makes this point with pardonable exaggeration:

> By these curious and difficult experiments [on body chemistry] he had already acquired insight into the causes of different diseases and on ways of supplementing nature in their cure, and he was preparing to attack the revered and ancient colossus of medical prejudice and error. Nothing was more important than this work of Lavoisier; and one can say that if the sciences have experienced an irreparable loss, all humanity should join us in lamenting this privation.[27]

Except for Bichat and some of the younger naturalists I have mentioned, the men who were between the ages of eighteen and thirty in 1789 belonged to a lost generation: men old enough to have their earliest productive years blighted by the storm, too old to benefit by the great schools and the illustrious masters that prepared the Napoleonic generation. Yet the brilliance and the diversity of talent that blossomed in the first decades of the nineteenth century testifies to the fundamental vitality of science in Revolutionary France. Clearly there were shifts of emphasis toward a broader democratic base of scientific instruction, toward a greater preoccupation of scientists with problems of industrial application and questions of social utility. Those men of science, a small though illustrious minority, who lost their lives during the Terror, were men who had, to a large extent, given up science for politics. It was as politicians, financiers, and public officials that they were executed, not as men of science. Although the Revolution, as any such painful crisis must, produced profound dislocations, it yielded also enduring benefits in industrial progress and new scientific institutions. At no point can we simply affirm, as did the men who wished to blacken still further the men of the Revolution, that the Revolution felt it had no need of men of science.

Notes

1. *Procès-verbaux de l'Académie Royale des Sciences,* vol. 108 (1789); vol. 109 (1790–93), in Archives de l'Académie des Sciences.
2. *Oeuvres de Lavoisier,* 6 vols. (Paris, 1862–93), II, 688–714.
3. J. Lalande, "Histoire de l'astronomie, pour 1792," in *Connaissance des Tems pour l'année sextile VII° de la République* (Paris, 1797), 236.
4. J. P Muirhead, *Mechanical Inventions of James Watt,* 3 vols. (London, 1844), II, 228.
5. Letter of December 10, 1789, in Muirhead, *Mechanical Inventions of James Watt,* II, 237.
6. Letter of July 5, 1790, cited by E. Grimaux, *Lavoisier,* 2d ed. (Paris, 1896), 201–2.
7. J. Pigeire, *La vie et l'oeuvre de Chaptal* (Paris: Editions Spès, 1932), 124.
8. A. Challamel, *Les clubs contre-révolutionnaires* (Paris, 1885), 390–443. See also Condorcet, "A Monsieur * * *, sur la société de 1789," where we learn that an informal nucleus of this society had been in existence since October 1789. *Oeuvres de Condorcet,* ed. A. Condorcet O'Connor and F. Arago, 12 vols. (Paris, 1847–49), X, 69.
9. "Règlemens de la Société de 1789 et liste des membres," in Challamel, *Les clubs contre-révolutionnaires,* 393.
10. The Andrew D. White Collection of books and pamphlets on the French Revolution at Cornell University has a bound volume of the 15 numbers of the *Journal de la Société de 1789,* from June 5 to September 15 1790, together with the prospectus of the *Journal.*
11. Challamel, *Les clubs contre-révolutionnaires,* 393. See also M. J. Laboulle, "La mathématique sociale: Condorcet et ses prédécesseurs," *Revue d'histoire littéraire de la France* 46 (1939), 33.
12. Condorcet, *Elémens du calcul des probabilités, et son application aux jeux de hazard, à la lotterie, et aux jugemens des hommes, par Feu M. de Condorcet, Avec un discours sur les avantages des mathématiques sociales et une notice sur M. de Condorcet* (Paris, An XIII, 1805).
13. A "prix national d'utilité" had been awarded in 1791 to the English astronomer William Herschel and in 1792 to Paul Mascagni (1752–1815) for his magnificent illustrated work on the lymphatics. See Lalande, "Histoire de l'astronomie, pour 1792," 249.

14. This prize of 1080 livres, the Prix Montyon, for the best memoir tending to simplify the processes of some mechanical art had been awarded in 1792 to a M. Girard, an engineer of Poitiers, for his study of the best method of constructing locks for canals and harbors. See *Prix Proposé par l'Académie Royale des Sciences, Pour l'année 1793*, 2 pp. (l'Imprimerie de Dupont, Imprimeur de l'Académie des Sciences, 1792). Collection de Chazelles, Bibliothèque de Clermont-Ferrand.

15. When this paper was delivered, the existence of these new statutes was a mere surmise. After a search in the records of the Académie des Sciences, in the summer of 1954, two working copies were discovered. These regulations were debated for nearly six months and were finally approved by the Academy on September 13, 1790; they were never officially promulgated or printed. [For a careful, documented treatment of the *Académie des Sciences* during the years of the Revolution, see Roger Hahn, *The Anatomy of a Scientific Institution: The Paris Academy of Sciences, 1666–1803* (Berkeley: University of California Press, 1971).]

16. It is more likely that Condorcet died of a circulatory disorder brought about by fatigue, exposure, and hunger. See J. S. Schapiro, *Condorcet* (New York: Harcourt, Brace, and Co., 1934), 106–7.

17. *Magasin encyclopédique* 5 (1795), 174.

18. *Notice sur la vie et les travaux de Lavoisier, précédée d'un discours sur les funérailles, et suivie d'une ode sur l'immortalité de l'âme* (Paris, An IV, 1795).

19. Ch. Dejob, *De l'établissement connu sous le nom de Lycée et d'Athénée et de quelques établissements analogues* (Paris, 1889).

20. "A discourse on the origin and progress of natural history in France," *Memoirs of Science & the Arts, etc.* (London, 1793), vol. 1, pt. 2, p. 448. On the early days of the society, see C. G. Krafft, *Notice sur Aubin-Louis Millin* (Paris, 1818), 8–10.

21. *Mémoires de la Société d'Histoire Naturelle de Paris* (Paris, Prairial An VII, 1799), especially pp. iii–ix. The activities of this society can be followed in the *Magasin encyclopédique* from 1795 onward.

22. Marcellin Berthelot, "Origines et histoire de la Société philomatique," *Mémoires publiées par la Société philomatique à l'occasion du centenaire de sa fondation, 1788–1888* (Paris, 1888), pp. i–xv. For the constituent articles and the list of early members, see the *Rapports généraux des travaux de la Société philomatique de Paris* (Paris, n.d., 1800?).

23. See his prospectus entitled *Imprimerie de Dupont Député de Nemours à l'Assemblée Nationale*, dated June 8, 1791 (Bib. Nat. Vp 21199). An English version of this document is published in B. G. du Pont, *Life of Eleuthère Irenée du Pont from Contemporary Correspondence*, 11 vols. (Newark, Del., 1923), I, 141–45.

24. See B. G. du Pont, *Eleuthère Irenée du Pont*, I, 185, n. 1.

25. Dupont's undated *Notice sur l'institution de la Société philomatique*, an 8-page pamphlet, probably dates from the period of the society's rapid expansion, 1792 or 1793. A copy of this was found among the Lavoisier papers (dossier 162) in the Archives de l'Académie des Sciences.

26. M. Daumas, "L'élaboration du Traité de chimie de Lavoisier," *Archives Internationales d'Histoire des Sciences* 29 (1950), 570–90.

27. *Magasin encyclopédique* 5 (1795), 183.

32
The Newtonianism of Dortous de Mairan

⚙

Pierre Brunet's well-known study, *L'Introduction des théories de Newton en France au XVIIIe siècle* (1931), treats in almost excessive detail the defense by the Cartesians of the Paris Academy of Sciences of their vortex theories of planetary motion and their debates with the earliest French disciples of Newton. Brunet, who carries the story down to 1738, recalling Fontenelle's praise of Newton's *Opticks* in his *Eloge* of 1727, remarks that those who attacked the notion of attraction and the concept of empty space, taken to be the hallmarks of Newtonianism, had no trouble accepting Newton's explanations about other matters, notably the nature of light and the origin of colors.[1] Although he does not say so explicitly, Brunet can be read as suggesting that Newton's election as an *associé étranger* of the Academy in 1699 was due to the popularity of his early work on light and color.[2] This is almost certainly incorrect, for in that year it was surely as a gifted mathematician that Newton was so honored. On the Continent Newton's early optical experiments were still generally ignored or distrusted.

To be sure, although the *Opticks* did not appear until 1704 (and in a Latin translation two years later), Newton's classic paper of 1672, which set forth his new theory about light and color, was made known to European scientists, chiefly through the persistence of Henry Oldenburg, soon after its appearance in the *Philosophical Transactions* of the Royal Society of London. But these optical discoveries were received with skepticism; the earliest attempts to confirm them—attempts by the English Jesuits of Liège, for example—were inconclusive. What appeared to be the definitive refutation was not long in coming. About the year 1679 Edme Mariotte, judged to be France's leading experimenter, set out to repeat Newton's experiments. Some of Mariotte's results, so he tells us, accorded well enough with the *hypothèse nouvelle* of the learned M. Newton, but others appeared to refute it. Chief among the latter was the famous *experimentum crucis* (as Newton called his early experiment with the two prisms), which Mariotte believed he had performed after Newton's fashion, but which did not succeed as Newton said it should; and so, Mariotte concluded, "the ingenious hypothesis of M. Newton should not be accepted."[3] Mariotte's great reputation

---

This essay appears in less expanded form in a Festschrift entitled *Essays on the Age of Enlightenment in Honor of Ira O. Wade*, ed. Jean Macary (Geneva and Paris: Librairie Droz, 1977).

479

lent authority to his conclusion; and we have ample testimony that his failure to repeat Newton's most famous experiment long delayed the acceptance in France of the revolutionary theory of light and color.[4] In sum, more than forty years elapsed after Newton first announced his results before his doctrine of the heterogeneity of white light and the origin of colors gained acceptance in France.

۞

Among the earliest to adopt Newton's theory of color was one whom Brunet lists among the leading Cartesians and the firm opponents of the ideas of attraction and the void: Jean-Baptiste Dortous de Mairan (1678–1771). Although not a scientist of the first, or perhaps even the second, rank, Mairan was by no means a negligible figure in the world of eighteenth-century science. The author of a widely known work of experimental and speculative physics, his *Dissertation sur la glace*, also of a major compendium of information on the aurora borealis[5] and of numerous other works (including an excursion into the debate over the *forces vives*[6]), Mairan entered the Royal Academy of Sciences in 1718, became a *pensionnaire géomètre* a year later, served briefly as Fontenelle's successor as perpetual secretary (January 7, 1741, to August 23, 1743), and was chosen *Directeur* of the Academy—that is, its chief scientific officer—on five different occasions.

Some years ago, in the course of some desultory book-buying in Paris, I purchased a copy of the 1749 edition of Mairan's *Dissertation sur la glace*, not a particularly rare book. On glancing through it, I noted a number of laudatory references to Newton and his discoveries. This occasioned mild surprise, since I seemed to recall that Brunet had devoted considerable space to Mairan as an outstanding antagonist of Newton. Especially prominent in the *Dissertation* were references to Newton's *Opticks*: to refraction, the transparency of bodies, and the physical explanation of color.[7] Now the presence of such passages in a book published in 1749 has, on the face of it, little that is surprising: for nearly a generation, Newton's ideas on light and color had been common coin, even among the so-called Cartesians. But what of the earlier editions of Mairan's book, especially the first edition, published in Bordeaux more than thirty years earlier? Although it seemed possible that the references to Newton had been inserted in later editions, forming perhaps a record of Mairan's capitulation to Newtonianism, the chance that some of these passages could be found in the edition of 1716 raised interesting possibilities. With not very high expectations I sought out a copy of this first edition in the Bibliothèque Nationale and was amply rewarded.

In this slim work of 1716 one finds the sentence: "For, according to the opinion of the illustrious M. Newton, the parts of nearly all bodies are naturally transparent, and their opacity comes only from the many reflections of these parts." A note refers the reader to the Latin *Optice* of 1706, page 210.[8] Further on appears the following remark: "Now we know the extreme dependence of colors upon the force and the different rates of vibration of the subtile matter, or of the rays of the sun. The excellent works that have appeared on the subject in recent years, no longer permit of any doubt."[9] The *Optice* is again cited, also the discussion of light and color in Malebranche's *Recherche de la vérité* in the edition of 1712. The reference to Malebranche is extremely significant, for Mairan's words echo the undulatory theory of Malebranche, rather than anything that could be found in Newton's *Optice*.

And finally we have this sentence: "This phenomenon is altogether consistent with the general theory of refractions explained in the books I have cited above."[10] Here again a note refers the reader to Newton and Malebranche. This modest success, these vague clues, indicated the wisdom of looking into other of Mairan's early writings, of exploring his relations with Malebranche, and indeed of learning what one could of Mairan's early years as a scientist.

✻

Jean-Jacques d'Ortous, Ecuyer, Sieur de Mairan (as Fouchy calls him at the start of his *éloge*), was born in Béziers in Languedoc in 1678 to parents of the provincial gentry.[11] Losing his father when he was four years old, he was educated at home by his mother until her death, when, at the age of nineteen and apparently on his own initiative, he betook himself to Toulouse to continue his studies. Soon thereafter he made his way to Paris, where he remained for four years, came to know those whom Fouchy resoundingly calls "les habiles gens qui ornoient alors cette Capitale," and acquired a taste for mathematics and physics. Although Fouchy does not identify these "habiles gens," he probably had in mind members of the circle around the philosopher Malebranche, for it was they, as we shall soon see, who were mainly responsible for kindling Mairan's scientific interest.

In 1702 Mairan returned to Béziers, where he enjoyed the protection of the Bishop, applied himself to his studies, and scrupulously avoided the pleasure-seeking life of the young nobles, with whom—as Fouchy put it—he had natural connections. Only towards the end of 1713 do we have documentary trace of him; in September of that year he entered into a spirited philosophic correspondence with the venerable Father Malebranche in Paris.[12] Mairan's first letter, dated from Béziers, September 17, 1713, opens with the following long but informative sentence:

> That young man who was attending the Academy of Longpray, and whom M. de Romainval, your relative, sometimes brought to your house; to whom you had the kindness to explain the book of M. de l'Hôpital, and to give several other lessons in mathematics and physics, is the same one who today has the honor of writing this letter to you.[13]

Mairan goes on to explain that in the past year or two he has put aside mathematics and physics (he was not to do so for long) in favor of the study of religion, in which he has been chiefly guided by the book of Malebranche, and by the writings of Descartes, Pascal, and Labadie. Of late he has been reading Spinoza, who has impressed him greatly by the cogency of his arguments but who has led him inexorably to the most dangerous conclusions. Would Father Malebranche be kind enough to point out the hidden fallacies in Spinoza's persuasive dialectic?

Before the end of the month Malebranche replied in very general, somewhat evasive, and to Mairan quite unsatisfactory, terms. A second importunate letter from Mairan brought a more detailed answer. The correspondence, clearly not greatly to the taste of the aged philosopher, continued until the following September. To certain of the letters—there were four inquiries from Mairan and four replies from Malebranche—we shall return in a moment.

Not long after this correspondence had run its course, the young provincial *savant* took the first step that led eventually to his recognition by the world of science: the submission of his earliest scientific communication. Some years before, the Académie de Bordeaux had transformed itself from a musical society into a learned group under the patronage of the Duc de la Force.[14] In 1712, it was granted royal letters patent and began to attract to its membership such young men of learning and talent as Montesquieu. To obtain wider recognition for the society, the Duc de la Force proposed that the Academy should establish an annual prize, to be awarded for the best memoir submitted on some aspect of physics.[15] The subject selected for the competition of 1715 was the cause of the variations of the barometer. Mairan emerged from obscurity to compete for and win this prize with his *Dissertation sur les variations du baromètre*. In the short piece, Mairan's maiden effort, Newton was cited; there appears this sentence: "The need for shortening the pendulum as one approaches the Equator made certain celebrated mathematicians suspect at first that the earth was a globe flattened towards the poles."[16] A note identifies the celebrated mathematicians and cites "Mr. Hugens dis. de la pesanteur. Mr. Newton, Principia Philos. natur."

The following year (1716) Mairan again competed and became the Academy's laureate for the short work on ice and freezing from which we have already quoted, and where he displays some acquaintance with the *Principia* and the *Optice* of Isaac Newton, and with Malebranche's speculations about color.

To the consternation of the Bordelais academicians, their third prize of 1717, for the best essay on the cause of the light produced by phosphorescent substances, had again to be awarded to the tireless young man from Béziers. The academy felt obliged to apologize to the contestants because one man had taken their prize three times running, and henceforth to declare Mairan *hors concours* by the simple expedient of admitting him to membership.[17]

The fullest of Mairan's early references to Newton's view concerning light and color appears in this short essay on phosphorescence, where we find the following reasonably clear account of the Newtonian theory of dispersion:

> Every kind of light has its definite refractions—that is to say, each color—in passing obliquely from one medium into another, from air, for example, into crystal, is bent [*se rompt*] at a particular angle different from that of the other colors. That is what Mr. Newton, author of this discovery, calls the "different refrangibility of the colors of light." It was principally by this property that he

found out all the others; and the ingenious experiments which he used to convince himself of it, could by themselves immortalize a name less celebrated than his.[18]

Mairan then recounted in some detail Newton's classical experiments "in order to acquaint those who have not seen the *Optics* of Mr. Newton, with what I shall have to say on this matter."

The Newtonian influence on Mairan is not far to seek. It has been amply demonstrated by Pierre Duhem that Malebranche, Mairan's distinguished correspondent, was the first Frenchman to give close and sympathetic attention to Newton's optical investigations, although his interpretation of Newton's experiments was very peculiarly his own.[19] Newton's early papers on light and color played no part in converting Malebranche; he seems not to have known them, or at least not to have been persuaded by them. As Duhem pointed out, when Malebranche communicated his paper entitled "Réflexions sur la lumière et les couleurs" to the Paris Academy of Sciences in 1699, soon after his election to that body as an *académicien honoraire*, he had as yet no knowledge of, or had not accepted, Newton's experiments on the composition of white light or the properties of the elementary colors (*les couleurs simples*).[20] In his paper Malebranche elaborated a theory, about which he had previously dropped some hints in his *Recherche de la vérité*, that departed from the generally accepted views of Descartes. More particularly, he undertook to do precisely what a generation earlier Huygens had criticized Newton for not attempting: i.e., to supply a plausible hypothesis, a mechanical model, to account for the origin of the different colors, a matter which Huygens himself did not presume to tackle in his *Traité de la lumière* (1690), and concerning which he remarks in the preface to that work "no one up to now can boast of having succeeded."[21]

That Malebranche owed a considerable debt for his optical theories to Huygens as well as to Descartes and Mariotte is quite apparent; but his theory is uniquely his own and might be described as an early exercise in the undulatory theory *of color*. Light, to Malebranche as to Huygens, is nothing but a pulse or vibratory motion of the aether or subtile matter; and colors (or at least certain of them) are the result of the differing frequencies of these pulses, just as in music pitch is determined by the frequency of sound waves in a vibrating medium. Malebranche's subtile matter, like the "second matter" of Descartes, is particulate; but unlike the little hard bodies of the Cartesian aether, the particles imagined by Malebranche are compressed fluid masses, tiny vortices which, because they are fluid, can transmit vibrations simultaneously in all directions. To be sure, not all the hues of the spectrum are produced by this mechanism, only the primaries (*primitives*) red, yellow, and blue. From these, the other hues arise by mixture. The sensation of black is produced when no vibrations at all are transmitted to the retina; white, when the original vibrations set up by a luminous source are unaltered by reflection, refraction, or the conditions of transmission. This is manifestly what has been called a "modification" theory of color, since the primary hues are produced when white light is in some way altered in frequency by some medium encountered in its path. Yet the model has the interesting feature that it could be, and subsequently was (by Malebranche himself), rather drastically changed to accommodate Newton's experimental results and the notion of the prior existence in a beam of white light of an infinite range of properties capable of producing the infinite range of colors; all one had to do was to imagine a corresponding infinitude of possible frequencies.

Paul Mouy has ascertained that Malebranche discovered Newton's work on color between the year 1700, when there appeared the fifth edition of the *Recherche de la vérité*, and 1712 when the sixth edition was published.[22] As early as the third edition of his famous work, Malebranche had adopted the practice of adding appendices that he called elucidations (*éclaircissements*). In the fifth edition of 1700 he reprinted, virtually unaltered as a new "elucidation," his 1699 paper on light and color. But in the edition of 1712, the sixth, this "XVIᵉ éclaircissement" is markedly changed. The "primitive colors" of the earlier paper are now referred to, in Newton's terminology, as "simple" and "homogeneous," and white light ("the most composite of all") is described, according to his own adaptation of Newton, as "composed of an assemblage of different vibrations" of the subtile matter. Each color, Malebranche insisted, has its own characteristic refraction; to be convinced of this, one need only consult the experiments in the "excellent work of M. Newton." And to the original *éclaircissement* so emended Malebranche further added several pages of a "Proof of the Supposition I have made," in the course of which he again cited Newton's experiments.[23]

It can only have been the appearance of Newton's *Opticks* (in English in 1704 and in Latin in 1706) that explains Malebranche's conversion to Newton's doctrine, as indeed his own words seem to indicate. In any case, there is supporting evidence for this. In October 1707, J. Lelong, Leibniz's faithful correspondent in France, wrote an important letter in which he remarked to Leibniz that the Reverend Father Malebranche "has been for some time in the country. He has taken with him a work of Mr. Newton, printed in London, 1704, *De Quadratis Curvarum*, in which this author pushed the integral calculus farther than everything that has been printed up to the present."[24] A variant of this letter mentions the fact that his short mathematical treatise was appended to Newton's "treatise on colors"—that is to say, the *Opticks*. If Malebranche did not read English, and if what he took with him to the country was the original *Opticks* of 1704, he could only have concerned himself with the two appended mathematical tracts which were printed in Latin.[25] But he soon studied the main work on light and color as well, and since the Latin version had appeared the year before Lelong wrote, it is possible that Lelong erred and that it was this book that Malebranche took with him *en villégiature*. At all events, it was the Latin *Optice* that Malebranche subsequently cited in referring to Newton's theory of color.

Not long after, on December 13, 1707, Leibniz wrote to Lelong:

> I should be curious to learn what the Reverend Father de Malebranche may have observed about colors. The subject is important. There is an experiment

of M. Newton which M. Mariotte has disputed, and which should be examined above all. For M. Newton claims that one can separate the colored rays from one another in such a way that after this separation refraction does not make them change color any further.[26]

It seems to have been in the same year that Malebranche wrote to a friend:

> Although M. Newton is not a physicist, his book is very curious and very useful to those who have sound principles in physics. He is moreover an excellent mathematician. Everything I believe concerning the properties of light fits all his experiments (*s'ajuste à toutes ses expériences*).[27]

That members of Malebranche's scientific circle did more than simply read Newton's book, and may have undertaken to repeat and confirm the dispersion experiments, is implied in the following passage in a letter, dated April 9, 1708, from Leibniz to Lelong:

> I fear that M. de la Hire may have tried the experiments of M. Newton on colors with some prejudice, and may not have used all the care that could be given to them. For, since M. Newton has worked at them for so many years, and one cannot doubt his merit, it is not credible that he has recounted imaginary experiments. So I should wish that persons who have all the necessary leisure, and who are willing to apply themselves sufficiently (which one should not ask of persons of the age and merit of the Reverend Father Malebranche and M. de la Hire), might be entrusted with this inquiry: this is what I have written to M. l'Abbé Bignon.[28]

From this it appears that Philippe de La Hire (1640–1718), the well-known mathematician and the friend of Mariotte, tried, but without success, to repeat Newton's experiments. But whether Malebranche also attempted an experimental confirmation is doubtful. There is no evidence that he did; it is more likely that he simply took Newton's account at face value. As Duhem put it:

> The reading of the *Opticks* of Newton inspired him [Malebranche] with the greatest confidence in the admirable experiments that the book reported; he hastened to touch up his theory of colors, so that it would accord fully with the truths put forward by the great English physicist.[29]

It is not difficult to infer that it was Malebranche who first called young Dortous de Mairan's attention to the importance of Newton's achievements. Not improbably Mairan had learned something of the *Principia* and of the Newtonian calculus from Malebranche during his years in Paris, since Malebranche, and the men of his circle—Louis Carré, the Marquis de l'Hôpital, Charles Reyneau, and Pierre Varignon—were all familiar with the *Principia* (although none fully accepted the physical principles there set forth) and were active in learning and elaborating the calculus. Varignon, we know, was the first to express Newton's fundamental laws of physics in algebraic analytical form. Yet it was certainly after his return to Béziers, in consequence of his later correspondence with Malebranche, that Mairan learned about Newton's doctrine of light and color. The letters in which Malebranche dealt largely with metaphysics, but in which he could not remain aloof from scientific questions, would seem to have rekindled Mairan's interest in the sciences. Our chief evidence is a letter written by Malebranche to Mairan on August 26, 1714; here, in the course of discussing the question whether for the generation of plants and animals a substance must be divisible to infinity, he remarks that he has treated his subject in an "optics which I gave in the last edition of the *Recherche de la Vérité*."[30]

In his reply, Mairan writes that his curiosity has been aroused to see the optical treatise Malebranche mentioned and the new additions to be found in the last edition of Malebranche's famous book. The edition he owns, Mairan adds, is that of 1700 in three volumes (that is, the fifth edition where the Newtonian material is absent); accordingly, he is having the new edition of 1712 sent to him from Paris.[31] Doubtless Mairan, soon after reading the "XVI*e* *éclaircissement*" in Malebranche's book, took steps to acquire a copy of Newton's *Optice*; for it can hardly be a coincidence that within a year, and afterwards in each of his later prize essays for the Academy of Bordeaux, Mairan shows himself familiar not only with Malebranche's theory of color but also directly with the Latin version of Newton's *Opticks*.

It would seem, if we can believe Pierre Coste, the well-informed translator of Newton's *Opticks*, that Mairan did more than read Newton's book and set about to confirm the accuracy of Newton's principal experiments; for after describing the successful repetition of Newton's experiments in Paris in 1719 by Jean Truchet (Père Sébastien) under the auspices of the Cardinal de Polignac, and later by Nicolas Gauger, Coste writes: "*M. de Mairan* les avoit aussi verifiées à Beziers en 1716," and apparently repeated them in 1717 "avec le meme succez." If this is so, and if Coste's dating is exact, then Mairan was the earliest in France to confirm by experiment the unalterability in the color of rays carefully separated by refraction.[32]

This did not end Mairan's interest in Newton's theories of light and color. A few years after his promotion to the status of *pensionnaire géomètre* in the Royal Academy of Sciences in Paris, Mairan read a memoir entitled "Recherches physico-mathématiques sur la réflexion des corps," which Brunet has abstracted at length. Here Mairan invokes Newton's ideas in opposition to Descartes to explain why white light reflected from plane surfaces does not produce color.[33] And in his massive study of the aurora borealis he interprets the colors observed in the northern lights by appealing to the "admirable system" of Mr. Newton, as Mairan rather freely interprets it.[34]

At one point Brunet writes that although Mairan was beguiled (*séduit*) by the ideas of Newton, chiefly because of the persuasiveness of the optical experiments, he nevertheless remained firmly loyal to Cartesian principles in the majority of fundamental matters.[35] This is overstating the case. There are instances where he clearly cannot decide between the explanations of Descartes and of Newton. In an early paper Mairan accepts the principle, which he willingly attributed to Huygens and Newton, that a rotating sphere experiences a greater centrifugal tendency at the equator than at the poles, and tries to reconcile this principle—which led Newton to assert that the earth is flattened at the poles and distended at the equator—with the conflicting geodetic measurements of the French astronomers.[36]

Mairan had certainly studied and used Newton's *Principia*. From Newton's gravitational data he gives the relative weight of an identical mass on the surface of the sun and on the earth, noting that the numbers Newton uses to represent the centripetal forces of the earth and the sun are different in the three successive editions of the *Principia*.[37] This, Mairan explains, is because Newton used different values for the solar parallax. A memoir which Brunet cites from 1729 shows Mairan confronting the problem of the direction of the earth's rotation and of the revolutions of the planets, and concluding that neither Descartes nor Newton had solved the mystery. Yet we find Mairan accepting the inverse square law of gravitation (a supposition, he remarks, which "me paraît généralement reçue aujourd'hui") and attempting to combine it with the vortices (*tourbillons*) of Descartes.[38]

Perhaps the most impressive evidence to convince us that by 1733 Dortous de Mairan, despite his attachment to vortices, was not the unregenerate Cartesian we have been led to believe, that besides having consulted the *Principia* he had gone some distance towards accepting Newton's celestial physics, is a passage with which I should like to conclude this article. He is willing, Mairan says, to accept the Keplerian law of the periodic revolution of the planets about the sun, and the Newtonian principle that central forces operate according to the inverse square of the distance. These laws are well known and fit modern observations:

> We therefore admit these principles in conformity with what one finds about them in the *Mathematical Principles* of Newton as far as the facts are concerned, and without claiming to enter in any way into the discussion of causes, or obliging the reader to choose between the general systems on which one could make them depend; because they are, in my opinion, just so many truths or premises which belong henceforth to all of natural philosophy, not even excepting the one that seems to oppose most forcibly Newton's philosophy. The heavens better understood, the laws of motion better developed, gave to this great man an advantage over Descartes and the early Cartesians which cannot deprive them of the glory they have justly gained, and which should even less be damaging to their successors, or forbid them the use of knowledge that time has brought forth, on the pretext that this knowledge did not emanate from their school.[39]

This, I believe, should suggest to us that Brunet was far too eager to classify Dortous de Mairan as a Cartesian of the strict observance. Brunet's third chapter, dealing with what he calls the "grands cartésiens," deals at particular length with two men: Privat de Molières and Dortous de Mairan.[40] Yet neither man quite deserves this label; both could far more correctly be called Malebranchistes, for each was profoundly influenced by that most open-minded and scientifically curious of seventeenth-century philosophers. This is to some degree true of other so-called eighteenth-century Cartesians as well. Malebranche and his followers, I suspect, ought properly to be seen as paving the way for the militant French Newtonians like Clairaut, Maupertuis, and, of course, Voltaire and the Minerva of France, his divine Emilie. This may be a subject well worth exploring further.

Notes

1. Pierre Brunet, *L'Introduction des théories de Newton en France au XVIIIe siècle* (Paris: A. Blanchard, 1931), 7. Brunet often found Fontenelle's summaries in the annual *Histoire* of the Academy more convenient than laboring through the memoirs of Villemot, Saurin, and the others.
2. *Ibid.*, 8.
3. E. Mariotte, Quatrième Essay. *De la nature des couleurs* (Paris, 1681), 211. See also *Oeuvres de Mariotte*, 2 vols. (Leiden, 1717), I, 227–28. In the 1740 edition of Mariotte's *Oeuvres*, 2 vols. in 1, consecutively paginated, the *Traité des couleurs* is reprinted on pp. 196–320; the discussion of Newton's theory and Mariotte's experiments are found on pp. 226–28. A reference to Mariotte's theory of color read to the Academy in 1679 is by J. B. Du Hamel, "Tractatum suum de coloribus legit D. Mariotte, quem postea in publicum emisit." *Regiae scientiarum academiae historia*, 2d ed. (1701), 184. A longer account of Mariotte's work on color is given in *Histoire de l'Académie Royale des Sciences*, I, 1666–86 (Paris, 1733), 291–303, but Newton's theory is not mentioned.
4. Voltaire wrote: "On s'est d'abord révolté contre le fait [the unequal refrangibility of the different rays], & on l'a nié long-tems, parce que Mr. Mariote [*sic*] avoit manqué en France les expériences de Neuton." *Elémens de la philosophie de Neuton* (Amsterdam, 1738), 121. See also J. E. Montucla, *Histoire des mathématiques*, 2 vols. (Paris, 1758), II, 622; and Pierre Coste's preface to his French translation of Newton's *Opticks, Traité d'Optique* (Paris, 1722), fol. e2 recto and verso. See also Condorcet, *Eloges des académiciens de l'Académie Royale des Sciences morts depuis 1666, jusqu'en 1699* (Paris, 1773), 60; and Joseph Priestley, *The History and Present State of Discoveries relating to Vision, Light and Colours* (London, 1772), 350.
5. *Traité physique et historique de l'aurore boréale par Mr de Mairan. Suite des Mémoires d l'Académie Royale des Sciences, Année MDCCXXXI* (Paris, Imprimerie Royale, 1733). A second edition, "Revue, and augmentée de plusieurs Eclaircissemens," was published in 1754.
6. Carolyn Iltis, "The Decline of Cartesianism in Mechanics: The Leibnizian-Cartesian Debates," *Isis* 64 (1973), 356–73, esp. section 8, entitled "Mairan and the Elimination of Force."
7. *Dissertation sur la Glace, ou explication physique de la formation de la Glace, & de ses divers phénomènes. Par M. Dortous de Mairan, l'un des Quarante de l'Académie Françoise, de l'Académie Royale des Sciences*, &c. (Paris, 1749), pp. ix, xviii–xxi, 292.
8. *Dissertation sur la glace* (Bordeaux, 1716), 78. This first edition was also reprinted in Paris in 1717.
9. *Ibid.*, 79.
10. *Ibid.*, 83.
11. For biographical facts about Dortous de Mairan see the *éloge* by Grandjean de Fouchy in *Hist. Acad. Roy. Sci.*, 1771 (1774), 89; J. Duboul, "Dortous de Mairan, étude sur sa vie et sur ses travaux," *Mémoires de l'Académie de Bordeaux* (1863, 2e trimestre), 163–97; Didot-Hoefer, *Nouvelle Biographie Générale*, 32 (1860); and the short article by Sigalia C. Dostrovsky in *Dictionary of Scientific Biography*, 9 (1974), 33–34, with a useful bibliography citing recent secondary sources.
12. *Méditations métaphysiques et Correspondance de N. Malebranche Prêtre de l'Oratoire, avec J. J. Dortous de Mairan* (Paris, 1841). See also *Malebranche—Correspondance*

avec J. J. Dortous de Mairan, ed. Joseph Moreau. Bibliothèque des Textes Philosophiques (Paris: J. Vrin, 1947). I shall cite the earlier work as *Méditations et Correspondance*.

13. *Méditations et Correspondance*, 93; *Malebranche—Correspondance*, 101. According to Moreau (n. 12 above) the Academy of the sieur de Longpray was an establishment where horsemanship, fencing, and dancing were taught. "M. de l'Hôpital" was the author of the famous *Analyse des infiniment petits* (Paris, 1696), a pioneer textbook of the Leibnizian calculus. For Malebranche's scientific activity and the group of mathematically minded men gathered around him, see André Robinet, *Malebranche de l'Académie des Sciences* (Paris: J. Vrin, 1970).

14. For the Academy of Bordeaux see P. Barrière, *L'Académie de Bordeaux, centre de culture internationale au XVIII*e *siècle (1712-1792)* (Bordeaux: Editions Bière, 1951).

15. According to Barrière, *L'Académie de Bordeaux*, 120, the Academy of Bordeaux was the earliest such body in France to offer a scientific prize. The award consisted of a gold medal.

16. *Dissertation sur les variations du baromètre* (Bordeaux, 1715), 38. The catalogue of the Bibliothèque Nationale lists another edition (Béziers, 1715).

17. Barrière, *L'Académie de Bordeaux*, 121. Mairan had the title of "académicien associé." *Ibid.*, 43.

18. *Dissertation sur la cause de la lumière des phosphores et des noctiluques* (Bordeaux, 1717), 48. Brunet, *L'Introduction des théories de Newton*, 84, n. 2, refers to this early "exposé de la théorie de Newton sur la lumière," but incorrectly gives the date as 1715.

19. Pierre Duhem, "L'Optique de Malebranche," *Revue de Métaphysique et de Morale* 23 (1916), 37-91. See also the important paper of P. Mouy, "Malebranche et Newton," *Revue de Métaphysique et de Morale* 45 (1938), 411-35.

20. Duhem, "L'Optique de Malebranche," 84.

21. Christiaan Huygens, *Traité de la Lumière* (Leiden, 1690), fol. *3 v.

22. Mouy, "Malebranche et Newton," 419. See also Duhem, "L'Optique de Malebranche," 85, who misnumbered the successive editions.

23. *Recherche de la Vérité*, ed. Francisque Bouillier, 2 vols. (Paris, 1880), II, 480-81 and 459-530.

24. *Oeuvres de Malebranche*, collected under direction of André Robinet, 20 vols. (Paris: J. Vrin, 1958-1970), XIX (*Correspondance, Actes et Documents 1690-1715*) (1961), 768-69. The variant is cited by Mouy, "Malebranche et Newton," 491, from L'Abbé E. A. Blampignon, *Etude sur Malebranche—Correspondance inédite* (Paris, 1862), 138-39. The correct title of Newton's mathematical treatise is *Tractatus de Quadratura Curvarum*.

25. Mouy, "Malebranche et Newton," 420, writes "Il semble bien que Malebranche ne lût pas l'anglais, et, en tous cas, lorsqu'il citera l'*Optique* de Newton, ce sera d'après l'édition latine de 1706, non d'après l'édition anglaise de 1704." Mouy had reached the same conclusion in his *Développement de la physique cartésienne, 1646-1712* (Paris, J. Vrin, 1934), 308, n. 1.

    Robinet is convinced that Malebranche used Newton's Latin *Optice* of 1706 (not the first English edition of 1704) for the revision of his optical theories. See *Malebranche de l'Académie des Sciences*, 299-323, especially 300-301 and n. 1. Malebranche seems to have studied the *Optice* in the summer of 1707, which he spent at the country estate of his disciple Edmond de Montmort. It is likely, however, that he heard E.-F. Geoffroy summarize the contents of the English *Opticks* during sessions of the Académie des Sciences of 1706 and 1707, for he faithfully attended the meetings.

26. *Correspondance, Actes et Documents*, 769.

27. *Ibid.*, 771-72. This fragmentary document, not dated by Blampignon, *Etude sur Malebranche—Correspondance inédite*, 25, is here assigned the year 1707, which would seem to be correct.

28. *Correspondance, Actes et Documents*, 784. The Abbé Bignon, mentioned by Leibniz, was the president of the Academy of Sciences, chiefly responsible for its reorganization in 1699. An oratorian, like Malebranche, he died in 1743 and his *éloge* was written by Mairan.

29. Duhem, "L'Optique de Malebranche," 90.
30. *Méditations et Correspondance*, 145; *Malebranche—Correspondance*, 140.
31. *Méditations et Correspondance*, P.S. of letter 7 by Mairan, pp. 167–68; Moreau, *Malebranche—Correspondance*, 167.
32. *Traité d'Optique*, trans. Pierre Coste, Préface, fol. e³.
33. Brunet, *L'Introduction des théories de Newton*, 113–16.
34. *Traité physique et historique de l'aurore boréale* (see n. 5 above). The Newtonian passages are identical in both editions. Mairan's theory of the nature of light is an amalgam of the ideas of Newton and of Malebranche.
35. Brunet, *L'Introduction des théories de Newton*, 118–19.
36. Dortous de Mairan, "Recherches géométriques sur la diminution des degrès terrestres en allant de l'équateur vers les pôles, etc." *Mémoires de l'Académie des Sciences* (1720), II, 33–95, cited by Brunet, *L'Introduction des théories de Newton*, 86–89, who mentions only Huygens. But see Mairan's reference to this early paper in his *Traité de l'aurore boréale* (1733), 97.
37. *Traité de l'aurore boréale* (1733), 88 n., 137; 2d ed. (1754), 96 n., 149.
38. "Nouvelles conjectures sur la cause du mouvement diurne de la terre sur son axe d'occident en orient," *Mémoires de l'Académie des Sciences*, 1729 (1731), 41–68. Summarized by Brunet, *L'Introduction des théories de Newton*, 165–76. See esp. p. 170.
39. *Traité physique et historique de l'aurore boréale* (1733), 88; 2d ed. (1754), 96.
40. The chapter is entitled "L'effort des grands cartésiens (1728–1732)"; see *L'Introduction des théories de Newton*, 153–202. See especially for Privat and Mairan, pp. 153–77.

33
Science and French National Strength

✺

It is not the purpose of this paper to speculate concerning the ways in which scientific progress might contribute to a program of French industrial or military recovery. That is a problem for the economist or the military specialist. It may be taken for granted that a flourishing state of science, as well as energy in developing its useful applications, is essential to the survival and prosperity of a modern nation and at the same time a sure index of that nation's cultural vitality. I wish simply to call attention to some notable features of the history of science in France and to some of the institutional and traditional factors which are relevant to the present situation. I shall conclude with a discussion of the recent policies and concrete proposals which the French nation has adopted since World War II to reinvigorate her scientific life.

✺

The man on the street and the man of science commonly give quite different meanings to the term "science." It is through its palpable, useful applications that the average person gains his only familiarity with its mysteries. Yet the majority of scientists view science not as the milch-cow of the modern world, but as a noble intellectual good, as John Tyndall once emphasized to an audience of materialistic Americans, to be "cultivated for its own sake, for the pure love of truth, rather than for the applause or profit that it brings."[1]

I emphasize this definition of science because, though it is widely expressed as an ideal, the French scientist has made it very specially his own; science to him is part of culture. It is in France that the cult of pure science, of ivory-tower science, has been most eloquently expounded; in fact, it is from the words of Frenchmen that Tyndall documented his plea that science should be viewed in this fashion. From Cuvier, for example, he translated with Victorian stiffness a statement which perhaps puts the doctrine in extreme form:

From *Modern France: Problems of the Third and Fourth Republics* (Princeton: Princeton University Press, 1951). Copyright, 1951, by Princeton University Press.

The reader should keep in mind the date at which this paper was first published. There have been many changes since then in the structure of French science.

These grand practical innovations are the mere applications of truths of a higher order, not sought with a practical intent, but which were pursued for their own sake, and solely through an ardour for knowledge. Those who applied them could not have discovered them; those who discovered them had no inclination to pursue them to a practical end. Engaged in the high regions whither their thoughts had carried them they hardly perceived these practical issues, though born of their own deeds. These rising workshops, these peopled colonies, those ships which furrow the seas—this abundance, this luxury, this tumult—all this comes from discoveries in science, and it all remains strange to them. At the point where science merges into practice they abandon it; it concerns them no more.[2]

The advancement of science is admittedly a complex international activity; it is a rare event in the history of science which can be attributed to one man or one nation. Yet to deny national variations altogether is quite unrealistic; for there are subtle cultural tendencies which set apart the scientific achievements of every nation. The French animus in favor of pure science is just a tendency; so also is the American scientific genius which manifests itself best in engineering application, in carrying through to profitable completion the ideas of others, or in conducting research in which massive equipment, heavy financing, organized activity, and engineering skills all enter in. By comparison, England and France have a record of producing eminently important results in pure science with very modest equipment. If, further, we contrast France with England, France's achievement appears to be more theoretical and less empirical, more mathematical and less experimental, than England's.

Very possibly France's greatest contribution has not been in natural science at all, but in mathematics. It could be argued that in no other scientific field can France produce an aggregate of names as impressive as those of Viète, Descartes, Pascal, Lagrange, Laplace, Cauchy, Hermite, Henri Poincaré, Lebesgue, and d'Ocagne. But even if this is an exaggeration, the typical French achievement in the sciences seems nonetheless to be rationalistic, a synthesis or great theoretical insight: such for example as Laplace's great masterpiece, the *Mécanique céleste*; Carnot's pioneer speculations into the nature of heat and energy; Jussieu's natural system of classifying plants; Cuvier's great unifying studies in comparative anatomy and paleontology; Lavoisier's reform of chemistry. Despite the weighty instances of Claude Bernard and Pasteur, it is the Anglo-Saxons, not the French, who seem to have produced the mainly experimental men, the Franklins, the Faradays, and the Joules.

If we view matters at close range, all the countries of Europe at every period have shared in the progress of every field of science; but, taking a longer view, we discover that different nations, in different times, have assumed the chief burden of advancing it. From about 1500 to the middle of the sixteenth century the leadership in science, as in so much else, belonged to Italy; for a brief period —from the Restoration to the death of Queen Anne—England predominated, and held a position in science she did not regain until after the middle of the nineteenth century. The ascendancy of France in science lagged behind her greatest period of literary and artistic domination; the Age of Louis XIV marked, it is true, the beginning of an official approval of science in France, but Colbert's Academy owed its greatest renown chiefly to foreigners, like Huygens, Cassini, and Olaus Roemer. The period of French scientific supremacy extends from about 1750 through the 1830s and 1840s, only temporarily interrupted by the great crisis of the Revolution. During this long hegemony, French leadership extended into every field of natural science, and Paris was sought out by students of chemistry, mathematics, natural history, and medicine from other European countries and from America. After the mid-century mark, and especially after 1870, there is an appearance of decline, which is partly a matter of comparative position and the swift rise of German and British scientific achievement. After the First World War a decline set in which is indisputably real, related to the exhaustion of French resources in manpower and morale, and in part to the inadequacy of her scientific institutions and to what I shall call an intellectual devaluation of science. These last two factors we must consider briefly.

※

All aspects of intellectual and artistic life in France—fine arts, music, higher education, and scientific research—are and have been to a large extent government-sponsored, and of course overwhelmingly concentrated in Paris, the one true center of all that, to a Frenchman, is worthwhile. Whatever the prevailing fashion in government the pattern has remained the same. In these conditions of work French science differed greatly from that of nineteenth-century Germany, which was government-sponsored but widely decentralized, and from that of Great Britain where private institutions have predominated—up to and including the Royal Society itself—and where some degree of decentralization has always existed.

Most of the scientific and educational institutions of France are creations of the Revolutionary Convention, or recreations of institutions of the Old Régime. Leaving aside the Academy of Sciences, which is not a center of research, the most famous and still perhaps the most important centers of scientific work are the faculties of the University, the great special schools like Polytechnique and the Ecole Normale Supérieure, and venerable centers like the illustrious Collège de France and the Muséum d'Histoire Naturelle, which is not just a museum but a training ground and many-sided research center as well. All of these institutions except Polytechnique, which is a military establishment, are under the eye and the budgetary control of the Ministère de l'Instruction Publique.

Let us look first at the laboratory facilities which these institutions provided. The first important chemical laboratory was that associated with the Ecole Polytechnique—the famous "X" (as the Polytechnique has always been called) from which radiated French influence in mathematics and the exact sciences. This laboratory dated back to the time of Berthollet, Fourcroy, Gay-Lussac, and the other great chemists of the Napoleonic period; but it exerted its wide influence upon the spread of chemical studies in France through the great investigator and teacher, J.-B. Dumas (1800–1884). During the Second Empire, Dumas's pupils founded laboratories in all the principal educational centers just mentioned. The most famous laboratory was perhaps that of organic chemistry at the Collège de France, where Marcellin Berthelot worked for nearly half a century.[3]

Laboratories of physics, modest and designed primarily to allow the professor to prepare his lecture demonstrations, existed in Polytechnique, the Ecole Normale, the Faculté des Sciences, the Collège de France, and the Muséum d'Histoire Naturelle. At the Muséum the Becquerel family enjoyed a monopoly of the chair of physics through four generations. Here on a modest building a plaque commemorates the achievement of Henri Becquerel (1852–1909), the third in line, who ushered in the Atomic Age in 1896 by discovering the radioactivity of uranium salts.

Until the end of the Second Empire, according to the testimony of Louis Pasteur, not a penny of the budget of the Ministère de l'Instruction Publique was earmarked for the support of research laboratories, and only an administrative fiction and official tolerance allowed the investigators to use some of the funds intended for teaching for their private research. Pasteur further pointed out that numerous laboratories and their equipment—notably those of Dumas, the physicists Foucault and Fizeau, and the agricultural chemist Boussingault—were the private property of the scientists themselves.[4]

In many respects the most discussed laboratories in nineteenth-century Paris were those of the Collège de France, in which were carried out by Magendie, Claude Bernard, Brown-Séquard, and Marey some of the noblest and most revolutionary experiments in the history of biological science. This brings us to an important point: the extraordinary inadequacy of scientific facilities during the nineteenth century. There were exceptions—for example, Frémy's chemical laboratory at the Muséum which was something of a showpiece, for he had one laboratory for his own use, one for his preparators, and one for his students. But in general there is a story of indifference and material starvation that repeats itself not only in obscure corners but even in the august Collège de France. The greatest figures in French science struggled against this handicap, and repeatedly demanded the sort of facilities that already existed in German and even English universities. Until 1840, the great physiologist, Magendie, had as a laboratory only a very small room which Claude Bernard, his preparator, later described as a closet that could scarcely hold both men. When in 1854 Claude Bernard became a professor at the Faculté des Sciences he was unable to secure a laboratory or provision for an assistant. A year later, when he succeeded Magendie at the Collège de France, he fell heir to what Paul Bert described as "a dark, damp tannery" and which led Bernard himself to characterize laboratories as the "tombs of scientists." In 1864 after an audience of Bernard with the Emperor, the Minister of Public Instruction, Victor Duruy, was told to give the great physiologist anything he wanted. When he mentioned a well-equipped laboratory, Duruy replied that his Imperial Majesty had in mind something more personal.[5] In 1867 Claude Bernard took his campaign to the public in his famous report on the progress of general physiology in France and in an article in the *Revue des Deux Mondes*.[6] Moreover, he contrasted the obstacles his own new science had encountered in France—official neglect and the hostility of the established naturalists and anatomists—with the flourishing condition of physiological studies in Germany.

In the meantime Pasteur lent his voice and great influence. He had suffered the same sort of neglect. His first quarters at the Ecole Normale had been inconvenient and primitive in the extreme—attic rooms that were freezing in winter and unbearably hot in summer. When he finally obtained possession of a small wing, he was obliged to install his drying oven under the stairs and could reach it only on hands and knees. In 1867 he wrote directly to Louis Napoleon asking for an adequate laboratory with facilities for experiments as dangerous to health as his were likely to be. Napoleon was well disposed; so also was Duruy, but the necessary credits were nonetheless eventually refused.[7] This was a hard blow, since millions were being spent at this time (1861–1874) on the erection of the Opéra with lavish adherence to Charles Garnier's plans to make it the most sumptuous in Europe. Pasteur brought matters to a head by a brief but impassioned article published in February 1868, in which he eloquently described the "tombs of the scientists" and the material handicaps under which French science was laboring.[8] The article created a sensation and led Napoleon to call a conference on March 16, 1868, at which French science was represented by Claude Bernard, Milne-Edwards, Sainte-Claire-Deville, and Pasteur. The whole question of France's need for research laboratories was discussed, and as a consequence work was begun on a new laboratory for Pasteur at the Ecole Normale and for Claude Bernard at the Muséum. A still more important consequence of the agitation was the creation of that unique institution known as the Ecole Pratique des Hautes Etudes to be discussed below.[9]

The tragedy of neglect was repeated in the well-known case of the Curies. Pierre Curie spent nearly all of his scientific life at the Ecole de Physique et de Chimie Industrielle de la Ville de Paris in the dreary buildings of the Collège Rollin, first as director of laboratory work, then as professor, with no funds and no personal laboratory, not even a room reserved entirely to himself. As much as he dared, he used the space and sums available for the teaching laboratory. His important experiments on magnetism were conducted mainly in an outside corridor running between a stairway and a laboratory. As *chef de laboratoire* for twelve years, he received roughly the salary of a factory hand of those times, about 300 francs a month.[10] The inadequate and unhealthy conditions under which Marie Curie began her work on radium and the notorious *hangar* of the rue Lhomond in which the two workers undertook the large-scale extraction from pitchblende are familiar to all. When Pierre Curie was given a chair of physics at the Sorbonne in 1900 he had no laboratory, but continued to work at the Ecole de Physique. Two years before his death, he accepted his professorate at the Faculté des Sciences only upon condition that he be given a satisfactory laboratory.[11]

The most serious effect of this early stringency was that lack of space and of funds made it difficult for French scientists to bring advanced students into their laboratories in the German fashion. A sort of apprentice system, in which the professor was usually forced to confine himself to a single able student as preparator, was the common practice. Two important institutional innovations, though wholly different in conception, were inspired by an attempt to improve laboratory facilities and increase the opportunities for advanced study in pure science. The first of these was the Ecole Pratique des Hautes Etudes, established by a decree of July 31, 1868; the second was the Institut Pasteur, which opened its doors twenty years later.

The Ecole Pratique, largely the achievement of Victor Duruy, grew out of the famous Napoleonic conference with the scientists. It was not a new physical creation, but consisted of machinery for giving grants-in-aid to existing research centers so that more students of high caliber could work under the immediate direction of the ablest professors in their laboratories and seminars.[12] In the physical sciences the major Paris laboratories just mentioned received important grants, but the only significant institutions outside of Paris to be assisted before World War I were marine biological stations that depended upon the Sorbonne or the Faculté des Sciences, like the famous Station Maritime de Roscoff.[13] The exaggerated centralization of resources for research which had been, in Duruy's mind, one of the merits of the French system, was never seriously threatened.

The Institut Pasteur is the only wholly private scientific research institution of great importance in France; it was one of the models for our own Rockefeller Institute. After Pasteur's dramatic victory over rabies, an international subscription was launched to provide what Pasteur had long desired, a research center where he could be complete master and work according to his own methods.[14] Contributions of all sorts poured in, including gifts from the Tsar of Russia, the Emperor of Brazil, and the Sultan of Turkey. The French Chamber voted the then respectable sum of 200,000 francs. In little time, the Institut became self-supporting through the sale of sera and other by-products of its research activities. Even before World War I the Institut Pasteur was able to share equally with the University in endowing Mme. Curie's Institut de Radium, which was thus supported partly by public and partly by private funds.[15]

❖

It is curious that no country in the past has written more in praise of science or has set a higher intellectual value upon scientific accomplishment than France, and yet none has been more unimaginative and parsimonious in providing scientists with the facilities and resources they require. Since the time of Condorcet, a succession of writers has emphasized the social importance of science, and the central role it was destined to play in the progress of the modern world. This thread of scientific rationalism inherited from the eighteenth century, and nurtured by the scientific successes of the nineteenth, has influenced and conditioned the thinking even of those Frenchmen who inveigh against it. It has become an important intellectual tradition to be reckoned with.

Yet, more than elsewhere, the subject of science and its importance is a controversial one in France. For brief periods England and America were divided into warring camps by the heated, but usually superficial, debates over Darwinian evolution; until recently, however, in the Anglo-Saxon world science has enjoyed a position *au-dessus de la mêlée*. But in France the profound division between Right and Left, the great fissure that on every subject long divided France into two, has defined the proper attitude for the partisans on each important issue, and has not allowed science to escape. A crucial debate on the meaning and ethical significance of science has continued in France until our own time.

For the Third Republic, in the course of which so many men of scientific training at one time or other held office—men like Paul Bert, Marcellin Berthelot, Freycinet, Scheurer-Kestner, and Painlevé—for the anticlerical Third Republic, the cult of reason and science provided, as we all know, the central *mystique*, and served as a useful political blunderbuss. In his assault on the outdated medievalism of the Right, in his attacks on the clerical enemy, the good Radical Socialist appealed to the light of reason and to faith in scientific determinism. In an ascending order of subtlety and sophistication, Berthelot, Taine, and Renan made up his triumvirate of sages. In many respects the most enduring and influential statement of this doctrine is Renan's *L'Avenir de la science*. Though a rambling, ill-written work of his impassioned youth, produced in 1849, its real position in intellectual history is that of a document in the scientific disputes of the Third Republic. Renan fished it from a dusty carton in 1890 and tossed it into the boiling controversy set off by the publication of Paul Bourget's *Le Disciple*.

Renan's *L'Avenir* was an epoch-making book, sensitive and penetrating and, in a clumsy kind of way, deeply moving. Because he did not fully accept the prevailing materialism or the naïve determinism of his own time, and did not present the case of science with quite the customary egotism and overconfidence, the work has survived its century and even today is a center of controversy. The faith in what science for all its philosophical limitations, its fumblings and half-knowledge, can mean for mankind, has never been more eloquently stated. Its first achievements seemed to Renan negative, critical; science has performed a useful sanitary operation by destroying the world of childlike dreams and alluring falsehoods; but the fulfillment of its great mission lies far in the future: "La science seule fera désormais les symboles; la science seule peut résoudre à l'homme les éternels problèmes dont sa nature exige impérieusement la solution."[16]

Science, says Renan, must penetrate the educational system. Since it is the work of many hands, scientific research should be coordinated and supplied with resources beyond anything yet proposed. The state must patronize and support science as lavishly as it does the arts; but it must do so with the most complete neutrality, since liberty is the condition of scientific progress. From the government must come the great scientific workshops, the observatories, laboratories, and libraries which cannot be supplied by individual initiative.[17]

If Renan's *L'Avenir*, published belatedly in 1890, was the strongest statement on behalf of science, the outstanding attack on the prevailing cult of science was Paul Bourget's psychological novel, *Le Disciple*, which had appeared the year before. Until the appearance of *Le Disciple*, attacks on science had been largely confined to sporadic sorties from the ultramontanist camp. Clerical authoritarians from De Maistre and Bonald to Louis Veuillot had each claimed their pound of flesh.[18] But Paul Bourget's abandonment of the prevailing cult of scientific determinism, like Brunetière's conversion, signified a deeper and more serious cleavage in opinion than had been evident before. Bourget had been a disciple of Taine and Claude Bernard, and a member of the reigning school of naturalistic writers led by Emile Zola, which included—besides the exponent of the *roman expérimental* himself—Flaubert, Alphonse Daudet, Hector Malot, the early Huysmans, and Maupassant. Not only scientific naturalism but also the reigning psychological determinism of Taine and Ribot, attracted Bourget. Yet it was against this, and what he believed to be its moral implications, that he at last rebelled. The tragedy of Robert Greslou, the central figure in his novel, is a consequence of the moral irresponsibility of science. Disciple of the mild, harmless philosopher Adrien Sixte, from whom he learned a rigid uncompromising psychological determinism like that of Taine, Robert Greslou becomes the tutor in a noble family, and puts these theories into practice in the scientific seduction of the sister of his young pupil. Greslou's experiments result in the suicide of his victim, his trial for murder, and his own death at the avenging hand of the girl's older brother. By means of this central figure, whom he once described as a Julien Sorel inspired by Renan rather than by Napoleon, Bourget hoped to show that scientific and philosophic ideas are not neutral but can—when they take possession of unstable minds and weak characters—become implements of disaster.[19]

In the *Revue des Deux Mondes*, Ferdinand Brunetière by his article "A Propos du *Disciple*" launched his campaign against the cult of science and urged the moral necessity of putting limits to man's speculative audacity, a proposal that brought heated replies from Anatole France, Ribot, Janet, and others.[20] In 1895 François de Curel made the conflict between the scientific and the religious views of the world the theme of his problem play *La Nouvelle Idole*. The force of this famous work is that it puts into action precisely the deepest conflicting elements in the French tradition. In the breast of the perplexed Dr. Donnat is fought the battle between scientific devotion and religious faith, between reason and emotion. Significantly the dilemma remains unresolved, but before the curtain falls Louise Donnat has eloquently expressed in her troubled words ideas destined to appeal with increasing force to the next literary generation. As far as its influence on literature is concerned, Brunetière's heralded "bankruptcy of science" is an accomplished fact. The fashion is changing; the new generation prefers the literature of moral responsibility, the *"culte de moi"* and the neo-traditionalism of Barrès, the romantic neo-Catholicism of Péguy: the movement summed up by Julien Benda as the *trahison des clercs*. The new generation turns away from science; Henri Bergson, whose philosophy sets such sharp limits to the legitimate pretensions of science, is that generation's prophet and savior.

✿

Some of the Catholic scientists were troubled by the conflict between their traditional faith and the cult of science. They found themselves on untenable ground between a Catholic clergy and laity generally ignorant and suspicious of science and scientific associates who were for the most part free thinkers. Louis Pasteur, in his famous *Discours de réception* delivered at the Académie Française when he succeeded the positivist scholar Littré, dutifully praised his predecessor, but stated the case of the Catholic savant when he attacked positivism and the widespread belief that science could solve all problems, moral and spiritual.[21] Catholic scientists, wishing to remain both Catholic and scientists, drew closer together. In 1875 a group of them founded the Société scientifique de Bruxelles, an international association of scientists who wished to affirm their inflexible Catholicism.[22] Their instinct that the loudly asserted incompatibility of science and faith was partly the result of a superficial philosophy of science and ignorance of its true history was to some extent borne out by events. It was precisely at this period that science was forced to abandon the simple mechanistic picture of the universe which had so long been held, and the criticism of Duhem, Mach, and Poincaré was fundamentally altering men's notion of scientific fact and scientific law. At the same time the work of Pierre Duhem and Paul Tannery was revealing how inadequate was our knowledge of the origins of modern science. Duhem's great discoveries in the medieval background of Galileo's thought were stimulated, if not motivated, by a desire to show that the official history of science of the positivists was an insubstantial myth, and that reputable science had existed in the Catholic Middle Ages.[23]

The Dreyfus affair divided the scientists as it did all professions, all groups, and even many families; but in the main the scientists tended, as one would expect, to be found among the Dreyfusards. I have in my possession one of those early printed petitions circulated in the Latin Quarter on behalf of Dreyfus. A number of leading scientific names appear on it. Emile Declaux, biographer of Pasteur and director of the Pasteur Institute, and Edouard Grimaux, the professor at Polytechnique who wrote the standard life of Lavoisier, were among the most active on behalf of Dreyfus. Many younger scientists were profoundly influenced by the *affaire*. Jean Perrin and Paul Langevin, both physicists under thirty in 1899, were members of the Socialist circle which gathered around Lucien Herr and Charles Andler at the Socialist Party bookstore opened by Péguy. Here they met Léon Blum and other young Socialists, occasionally glimpsing Jaurès himself; and here the doctrines of Renan and Berthelot experienced an important revival.[24] In the period after World War I, Perrin and Langevin took the lead in what their opponents called a *scientisme politique*, carrying on a campaign for ideas not greatly different from those of Renan, but with Socialist overtones: for a moral and intellectual liberation through science; for a progressive elimination of the inherited evils of society with emphasis upon what Langevin called the "human value of science."

These doctrines came into sharp conflict with the views of the Catholic scientists. In the 1920s and 1930s the ideological battle over science continued unabated. The attack on modernism gave rise to some debates of great interest and importance for the philosophy of science, but the net result was nonetheless an atmosphere not at all favorable to the popularity of science.

Generally, the position of the Catholic scientists was moderate and enlightened, compared at least with the attacks of Brunetière and his forerunners, but it involved both a devaluation and a sharp circumscription of science. It was an encircling, not a frontal movement, intended to confine science and its method to the laboratory, surrounding it with a *cordon sanitaire* to prevent it from spreading into the sensitive areas of life. In the first place, the prestige of science must be strictly limited to the natural sciences; it should not be allowed to inflate the confidence of the moral and social studies. Sociology is particularly dangerous, hence the scorching attack leveled at Durkheim by the rector of the Faculty of Philosophy of Louvain.[25] In the second place, science cannot do the work of philosophy; here Bergson's intuitionism, and the rise of scientific phenomenalism—the denial that science can deal with ultimate reality and the notion that it deals only with the relations between phenomena—provided a powerful, if often double-edged, weapon. Lastly, the great development of modern physics, with its seeming abandonment of strict determinism, was a godsend to those who hoped to deflate and devalue science by pointing to the inadequacy of the "eternal laws" of which the nineteenth century made so much. The epithet *scientisme* was coined to fit the doctrine of those who wished, by contrast, to inflate the importance of science—specifically, men like Perrin and Langevin.[26]

These views which I have summarized are well expressed in Louis de Launay's *L'Eglise et la science* (1936), which appeared just as *scientisme* was being adopted as an official policy by the Front Populaire. How welcome these views were to controversialists of the Right is evident from the long and effusive front-page review which Léon Daudet devoted to it in the notorious *Candide*. A more recent presentation of this view is the symposium entitled, significantly, *L'Avenir de la science*,[27] which appeared in a series of eminently Catholic books during the time of the Vichy government. It is an outright attack—as the title implies—on the tradition of Renan and on the presumption of science, or rather of the *scientistes*. It is introduced by an able and moderate essay on the state of physics by Prince Louis de Broglie, of whose scientific accomplishments we shall soon speak. His discussion of the insubstantial foundation of modern physics sets the stage for the violent attacks that follow: of Father Sertillanges on the *scientisme* of Renan and his kind; and of Raymond Charmet on modernism and the whole French rationalist tradition. M. Charmet finds the cause of the fall of France to be the same as the fall of Greece—for, as everyone knows, France is the

Greece of modern Europe. In both cases, the decline is attributable to excessive intellectuality and loquacity and what he sums up as the "*culte obscur de la raison claire*," whose most dangerous expositors were Descartes and Renan.

This attack from the Right upon the scientific humanism preached by the men of the Left is not avowedly antiscientific. It does not wish to destroy science but to tame it and disqualify it for any very profound intellectual mission in the modern world. By an odd coincidence, the Catholic philosophers of science appear rather narrowly materialistic, for they would reduce science to its useful applications. "La science pratique," writes one of them, "est peut-être la vraie science." It is plain to see that this negativist mood is not such as to lure the best young intellects into science; it may well have been a very real factor in the decline of French scientific achievement in our own time, by encouraging an indifference and contempt for science among young men whose background happens to be Catholic and conservative. A decline has undoubtedly taken place, but I do not want to exaggerate it. Let us at this point search for signs of real vitality in the scientific picture.

❖

To illustrate my point that French achievements in science, while no longer leading the world, still play an essential part in world scientific progress, I shall take my examples from pure science, from the history of modern atomic physics, where we will find a great French name at each of the three stages that were traversed in penetrating into the mysteries of atomic structure. These three stages are: (1) the proof of the existence of atoms and molecules; (2) the exploration of the electronic outer shell of atoms; (3) the penetration into the atomic nucleus. By a convenient historical accident, this will introduce us to the three principal personalities who are responsible for the great postwar scientific effort in France: Jean Perrin, Louis de Broglie, and Frédéric Joliot.

As late as 1895 there were still many who considered the hypothesis of atoms and molecules merely a convenient fiction. In this year Jean Perrin, a young preparator at the Ecole Normale, entered upon a research career that was largely devoted to marshaling unassailable proof of the existence of atoms and molecules. From his studies of Brownian movement—the perpetual and irregular dance of microscopic particles suspended in a fluid—and of the properties of thin soap films, Perrin was able about 1910 to determine the sizes and numbers of atoms and molecules, and supply independent quantitative confirmation of the molecular nature of fluids. Perrin's general treatise *Les Atomes* (1921) brought together the most important proofs and to all intents and purposes ended a long-standing argument.

Perrin took little part in the succession of discoveries which showed that the atoms were not the hard, unyielding particles of early speculation but complex microcosms with charged particles, the electrons, rotating about a central core, work we associate with the name of J. J. Thomson, Niels Bohr, Max Planck, and others. This brings us to our second historical stage, for in 1925 Louis de Broglie, now France's leading theoretical physicist, hit upon the idea that a moving electron, previously thought of only as a particle, might behave like a bundle of waves. The proof of de Broglie's daring hypothesis, that electrons should have the wavelike properties of light, was not long delayed, and led to that branch of modern physics known as wave mechanics.

Thus we find one French scientist contributing to understanding the properties of the atom as a whole, while a French physicist of the next generation was the first to expose, at least in theory, the mysterious properties of the electron, half particle and half wave. French physicists also have had an important share—and why not, since Becquerel and the Curies were the first to reveal the inner storehouse of atomic energy?—in the fateful discovery of atomic fission, which resulted from stage three, the study of the inner core or nucleus of the atom.

The sport of atom-smashing dates from the work of Ernest Rutherford in 1919, who showed that the heavy, positively charged particles emitted by radioactive substances could be used to bombard the compact cores of other atoms. Nuclear chips seemed to be knocked off, producing actual transmutations on a minute scale. During the 1930s it was found possible to use other atomic projectiles, the protons, and accelerate them to great velocities in devices like the cyclotron.

In 1932 the problem was fundamentally advanced by the discovery in England of a hitherto unsuspected atomic particle, the neutron, which had the same mass as the proton, but no charge. The fact that the neutron was uncharged, and hence would not be repelled by the positively charged atomic nuclei, at once led Enrico Fermi to reason that neutrons would be very effective in inducing nuclear transformation in heavy elements, where the repelling charges are very great. This proved to be the signpost pointing down the road to nuclear fission. It is precisely at this important moment that French science enters the nuclear picture in the person of Frédéric Joliot and his wife, Irène Curie.

Frédéric Joliot was born in 1900 and thus was eight years younger than the Prince de Broglie, the father of wave mechanics, and a whole generation younger than his teachers, Jean Perrin and Paul Langevin. Just after World War I Joliot began work on radioactivity in the Institut de Radium. In 1926 he married his co-worker, Irène Curie, three years his senior. Together they shared in the important discovery, for which they received the Nobel Prize of 1935, that substances would be made artificially radioactive in the course of nuclear disintegration.

When in 1936 the chair of physics at the Collège de France fell vacant, Joliot was chosen to fill it. A Laboratory of Nuclear Physics was built for him with great care and at considerable expense. Its most important piece of equipment was Europe's first cyclotron. At the time it was ready for use, early in 1939, the peculiar behavior of uranium was the principal subject of discussion. In Rome, Fermi had bombarded small quantities of uranium—the heaviest element then known—with neutrons, and believed he had produced newer and stranger, perhaps heavier, elements. In Germany the experiments had been repeated and extended by careful chemical analysis of the products. Starting from these results, Frédéric Joliot and his co-workers—independently of O. R. Frisch and Lise Meitner, whose work has been widely publicized—proved that what Fermi and the German workers had actually done was to bring about the rupture of the uranium atom into two approximately equal parts with the release of large amounts of energy. Still more important, in January 1939 two of Joliot's collaborators, Halban and Kowarski, demonstrated that when a uranium atom is split by neutron bombardment, other neutrons are released, which under proper conditions could be used to split still other uranium atoms. The French workers were thus among the earliest to discover that a nuclear chain reaction, leading to the release of significant quantities of atomic energy, was theoretically possible.[28] The French government patented the use of fission neutrons, which gave them the master patent in the field of atomic energy. The Laboratory at the Collège de France prepared at once to explore this fateful possibility, but the German invasion brought matters to a standstill, and when French resistance collapsed, Joliot dispatched Halban and Kowarski to England, where during the war they made useful contributions to the Allied atomic-energy program.

Let us now turn to the great scientific institutions these men were instrumental in creating.

※

Of greatest significance for the future of science in France is the extraordinary boldness and enthusiasm with which French scientists and their associates in the government have set about providing a great central organization, the sort envisaged by Renan. This may go far to offset the present difficulties under which French scientists are obliged to work—the crowded laboratories, the obsolete equipment, the disastrous effects of inflation—and in the long run it may eradicate some of the evils which have so long afflicted French science. So little is as yet known about this organization—one of the focal points, I believe, of French postwar activity—that I shall devote the remainder of this paper to a factual account of its origins, organization, and purposes.

In conception, it is typically French—a plan of sweeping lines and bold proportions—possible only in a country where even the scientists of the Right are to some extent convinced of the necessities of national planning in the present emergency, and where a tradition of government responsibility for science is fully accepted. It breaks sharply with the *petit bourgeois*, close-fisted spirit with which science had so often been treated in France. In the United States we saw the far more timid plan for our National Science Foundation fail of adoption, after passing the Senate three times. By comparison the French project, which was in full swing before the American measure was finally enacted, is bold indeed.

French attempts to modernize science date from the conclusion of World War I. Many scientists in France saw clearly at the end of that ordeal that countries faithful to habitual and conservative ways of doing things could not contend for long on equal terms with a country like Germany, where science was applied, with a thoroughness unknown elsewhere, to the enhancement of every aspect of national and industrial strength. Even in England and America this comparison gave rise to an active movement to stimulate industrial research. In France the movement had to contend with a traditional distaste for applied research, with the conservatism of French industry, and with the antiscientific tendencies of the postwar generation. In a symposium called *L'Avenir de la France* published in the hour of victory (1918), one writer called upon French scientists to leave their *"tour d'ivoire"* and help extend the benefits of science to industry.[29] In 1925 Henri Le Chatelier, well known as one who had made extensive scientific contributions to industry, published his classic *Science et industrie*, a fine theoretical analysis of the role of science in industry with vivid and ludicrous examples of the indifference of French industrialists and of their encyclopedic ignorance of the uses of science.

In the meantime, the French government took the first timid steps in the right direction by creating at Bellevue, near Meudon, a center where industrialists could receive scientific aid. The laboratories were installed in what had once been the luxurious Palace Hotel, which had briefly served as a wartime hospital, and before that as Isidora Duncan's Temple of the Dance of the Future. The laboratories were placed under a government body that was supposed to coordinate scientific and industrial research, investigate problems at the request of the government or of *sociétés industrielles*, and carry on developmental projects of public interest. The functions resembled those of the National Bureau of Standards or the British National Physical Laboratory.[30]

A similar movement, emphasizing pure or fundamental research even more than service to industry, took shape in the minds of the Socialists, who in this respect were the heirs of the old Radical Socialist party. The chief spokesman for the constructive regeneration of French science through government aid was the universally beloved and respected physicist Jean Perrin, the ardent Socialist and lifelong friend of Léon Blum.[31] Perrin, and his still more Leftist colleague in physics, Paul Langevin, were instrumental in persuading men like Herriot, Painlevé, and Blum in 1936 that the Front Populaire should make the encouragement of science an important plank in its platform. The Blum government was induced to merge two previous offices for the assistance of pure science into a single Caisse Nationale.[32] At the same time it was resolved to create a corps of government scientific workers with ranks corresponding to those of the Academic hierarchy, and to breathe life into the laboratories at Bellevue.[33]

The scientific aspect of the Blum program received considerable publicity, partly from the establishment of the post of Undersecretary of State for Scientific Research, which at Perrin's instigation was given to Mme. Irène Joliot-Curie; and partly from Perrin's creation, in connection with the famous Exposition of 1937, of the Palais de la Découverte, one of the finest science museums.

Early in 1939 the physiologist Henri Laugier succeeded Perrin (who in turn had followed Mme. Joliot-Curie) at the head of the scientific program. Laugier had much to do with drafting the legislation which in October 1939, soon after the outbreak of the war, finally established a single, comprehensive research agency under the Ministry of Education, called the Centre National de la Recherche Scientifique.[34]

The debacle of 1940 and the German occupation at first demoralized and disrupted the scientific life of France. Laugier and Perrin both escaped to America, where the former became an active leader in the Free French movement and taught physiology in Canada, while Perrin became vice-president of the Ecole Libre des Hautes Etudes in New York, dying in this country in 1942 at the age of seventy-two. In France the CNRS continued a shadowy existence. A Vichy decree of March 1941 gave the organization a new charter which merged the sections of pure and applied research and channeled the efforts of the scientists into a modest attempt to serve industry.[35] Rather amazingly, the Bellevue laboratories thrived, despite the shortage of raw materials and the restrictions imposed by the Germans. Its facilities grew rapidly during the Occupation; in 1942 six new laboratories were equipped and opened; and in 1943, seven more.[36]

The scientists who remained in France took an active part in the underground; few of their number collaborated with the enemy. The aging Langevin, an outspoken radical of the extreme Left, closely associated with the Communists, whom he later joined, became the official scientific martyr of the Resistance. In 1940, on returning to his laboratory at the Collège de France, he was arrested by the Germans and thrown into the prison of La Santé. His arrest provoked a great demonstration of students and a public protest by his colleague Frédéric Joliot, who temporarily closed his laboratory. The arrest of Langevin—soon released and allowed to live under strict surveillance at Troyes—greatly stimulated the formation of the secret Front National Universitaire in which the scientists were very active.[37]

Joliot's Laboratory of Nuclear Physics was visited by German scientists who had instructions to remove the cyclotron and other useful apparatus to German laboratories. This was somehow avoided, but Joliot was obliged to permit German scientists to work there. This fact, which gave rise to rumors that Joliot was collaborating with the enemy, actually allowed him to carry out his underground activities behind an impenetrable screen. His laboratory became an important arsenal for the production of "Molotov cocktails" and other types of homemade explosives.[38]

Much has been made, and I believe can legitimately be made, of the influence of the Resistance experience upon the social awareness and the political education of the French scientists. Politicians and working class leaders, for their part, had learned to understand and cooperate with the scientists; the doctrines of Perrin and Langevin became important in their planning. Even those scientists who were not Socialists or who did not follow Langevin and Joliot into the Communist Party, came to share their views in these matters. As it emerged from the Resistance, the scientific program was a revival of the Front Populaire spirit. The

Soviet Union had not yet fully revealed its determination to control its intellectuals and its scientists in all their activities; the Lysenko affair was only brewing; and the Left, even with the Communists, appeared to be the strongest bulwark of scientific progress.

Joliot emerged, at the war's end, the uncontested leader of the French scientists. His scientific prestige unassailable; a hero of the Resistance, a leading Communist intellectual, president of the largest Resistance group, he was entrusted by the Provisional Government with the task of mobilizing science to aid in French recovery. He took charge of the CNRS with a free hand to refashion it into a suitable instrument for the reinvigoration of French science. He brought with him men of common purpose who had served with him in the underground and who shared his conviction that a bold program of Socialist planning should be substituted for the Vichy policy of serving the industrialists. The postwar scientific program in France, while by no means the work of Socialists alone, was colored an ineluctable pink. The changes effected extralegally by Joliot and his aides were given legal sanction by a decree of November 2, 1945. In the meantime the atomic bombs had been loosed on Hiroshima and Nagasaki, and Joliot was swiftly withdrawn to direct the Commissariat of Atomic Energy.[39]

The CNRS has as its principal function the support and encouragement of pure and applied research. This is carried out in a variety of ways which I shall enumerate. But it also has a supervisory and planning function, for the charter of 1945 gives it authority "to undertake the coordination of research carried out by the government, private industry, and private individuals, by establishing liaison among the groups involved," and to organize inquiries into research being pursued in private and public laboratories.

So far, grants to individuals and laboratories and the support of a staff of professional researchers and technicians constitute the principal method of encouragement. In 1943–1944, at the close of the war in Europe, the Centre supported some six hundred researchers. By 1946–1947 this number had more than doubled.[40] Those who hold nonacademic research appointments from the Centre constitute a permanent professional staff of investigators—*directeurs, maîtres,* and *chargés de recherches*—who are paid the salaries of university personnel of roughly equivalent rank. Unlike the technicians who assist them, and for whom there is also a rigid hierarchy, the investigators are not civil servants.[41]

A great many of the Centre's scientific personnel carry on their research in the laboratories of the other government-supported institutions. At present the laboratories and institutions directly run by the Centre are few but important: an Institut d'Astrophysique; the Observatoire de Haute Provence; the Grenoble Laboratory of Metallurgy; and, perhaps most important of all, the mathematical and statistical laboratory of the Institut Henri Poincaré, to which scientists can go for special mathematical assistance and the use of modern computing devices.[42] The Bellevue laboratories have remained under the control of CNRS and have steadily expanded their resources for industrial research. They have not been afraid to undertake long-range projects of potential value to the national economy. For example, there is an important project to explore the industrial uses of direct solar energy.[43]

The CNRS envisages a great expansion of the laboratory facilities under its own direct control. It hopes to establish at Strasbourg, for example, a laboratory of nuclear studies applied to biological and medical problems. The most ambitious program, however, is for the group of laboratories to be erected at Gif-sur-Yvette. Some 130 acres of wooded and open estate about twenty miles from Paris have been acquired by lease, through Joliot's activities, from a banker friend interested in physics. Plans are already under way for a great genetics laboratory, where the fidelity of some Communist scientists to the party line may be publicly tested. Beyond this there are plans for laboratories of pure and applied entomology, geological laboratories, and laboratories for nuclear physics, electron optics, microchemistry, and microanalysis.[44]

Perhaps the most extraordinary aspect of the organization is that the social sciences are given a prominent place: historical, philological, and sociological studies receive virtually the same treatment as the natural sciences.[45] As Renan observed, these are precisely the kinds of investigation which need government support, for many of them have no conceivable utility except to add to the sum of human knowledge. The CNRS directly supports a Center of Sociological Studies and an Historical Institute.

In one very important respect, first things have been done first. Though the Centre has been unable to remedy the wretched economic plight of many of the country's investigators, it has tried to make up the deficit in properly trained personnel. For example, it has set up a training program at the University of Paris to give researchers certain accessory tools of their trade. Lectures are given on applied mathematics, scientific English, the use of electronic equipment, and the theory of measurement. Researchers of the CNRS, technicians of CNRS, and other scientific workers are admitted in that order of priority.[46] The Centre has furthermore made great efforts to promote scientific contracts with other nations, by sending scientific missions abroad, and by making it possible—this last with the assistance of American funds—for foreign scientists and scholars to come to France to take part in small work groups and colloquia.[47]

Second only to the inferior economic status of research workers, the difficulties and expense of scientific publication have been the most criticized and lamented aspect of the French situation. At first the CNRS confined itself to indirect support of scientific publication by occasional subsidies to publishers for important books and monographs. Increasingly, it has now gone into the publishing business itself. It issues a progress report, the *Bulletin du CNRS*, reporting the projecs of its staff, and a *Journal des recherches*. But its most famous achievement is the great abstract journal, the *Bulletin analytique*, which reports on over four thousand periodicals and lists about 100,000 titles of articles and books annually.[48]

Perhaps a paragraph should be devoted to a brief description of the very interesting organization of the CNRS. The Centre is placed under the Minister of Education, and is administered by a director and assistant directors, appointed by the President of the Republic on the recommendation of the Minister. The director works through two administrative bodies, one concerned solely with high policy and finance, the Administrative Council; another, the National Committee of Scientific Research, which is the scientific policy-making body of CNRS.

It is this National Committee which is the most impressive feature of this organization. One-third of its members are named by the Minister of Education on the recommendation of the director; but the other two-thirds are chosen by an electoral body composed of all the scientists working in government institutions, the University, the great schools, the Collège de France, and the CNRS itself; that is to say, an important fraction of the scientists in France.[49]

❄

The success of this program, which in the long run can have such a favorable influence on French economic prosperity and national prestige, depends in the immediate future upon the economic recovery of the country, and the sums which during this critical period can be spared for the CNRS from other more urgent tasks. It depends also on the continued support of scientists of all political allegiances, and on the solution of the very delicate problem of control. The democratic scheme of organization which on paper gives so much weight to scientific opinion, and the creation of a buffer organization of scientists between the politicians and the laboratory, must operate, as it is intended it shall, to prevent a restriction of liberty or the perversion of scientific purpose to the narrow and utilitarian needs of the state.

At present [1950] the picture is distinctly favorable. Though the CNRS is so much an achievement of the parties of the Left, and the voice of the French Communist Party is admittedly heard in its councils, there seems to have been no overt attempt to bias its activities, and up to now government officials have left the scientific planning entirely to the qualified scientific leaders. It is widely taken to be a nonpartisan project desperately needed in the national interest, conceived in long-range terms, and no one seems to expect from it short-run economic benefits. It seems to have widespread support, and is criticized only for doing too little. If political or party control should threaten the basic freedoms of the scientist, then indeed the broad support might melt away.

The CNRS has not been openly attacked as a product of the *scientisme* of the Left. The Prince de Broglie, *Secrétaire Perpétuel* of the Academy of Sciences, in his public utterances has been more than generous to it. Whatever his private misgivings, he has spoken up strongly in its favor, first in an interview of 1945, again in the earnest preface he contributed in 1948—side by side with one by Léon Blum—to the posthumous essays of Jean Perrin.[50] In his tribute to a great man and a great scientist of the older generation, de Broglie gives Perrin credit for having defended French science and for helping to inspire the Centre National, which he speaks of as having already pursued a fruitful career, and which he hopes will contribute powerfully in the future to a great revival of French science.

On the other hand, however, the problem of Communist Party members in high scientific councils has come to a head in France with the case of Frédéric Joliot. After his pronouncement on April 5, 1950, at the Twelfth Congress of the French Communist Party that "a truly progressive scientist will never donate a particle of his scientific knowledge to the purpose of making war against the Soviet Union," Joliot was summarily dismissed "with regret" by the government of Georges Bidault. Through a spokesman, M. Teitgen, M. Bidault explained on April 28 that Joliot's public statements and his unreserved acceptance of the pro-Russian resolution of the Communist Party made it impossible to maintain him in his post as High Commissioner. This has not affected his post at the Collège de France nor, as yet, his connection with the Centre National. Nevertheless the question remains whether, and to what extent, the Centre will be affected by such an open recognition of the split that exists. Joliot's scientific supporters on the Commission—Kowarski and François Perrin among others—were prompt to protest his dismissal. The conflict between political loyalties and scientific freedom remains unresolved and is part of the larger question of the obligations of the citizen to the truth, as well as to the state, in an ideologically divided world.

It remains to be seen what the future has in store. Whether the individualistic tradition of French science—in fact, of all science—can survive the new era of collective activity and government control, or whether it can survive without it, are serious questions. One thing is certain. There must be a new generation of scientists ready to make use of the new opportunities. And unless the educational tradition of France is drastically overhauled, and a general atmosphere is created that is favorable to science, the men will not be there to do the job. At the moment France badly needs some small portion of the eighteenth century's great faith in the value of scientific knowledge. There are signs that—unfashionable though this is at present—such a revaluation of science may be taking place.

Notes

1. John Tyndall, *Six Lectures on Light* (London, 1873), 212.
2. *Ibid.*, 221–22.
3. A complete list of French scientific laboratories—both teaching and research—as they existed at the close of the nineteenth century is given in Paul Melon, *L'Enseignement supérieur et l'enseignement technique en France* (Paris, 1893). The chemical laboratories at the Ecole des Mines and the Ecole Supérieure de Pharmacie are discussed at length in (Edmond) Fremy, Carnot, Jungfleisch et Terreil, *Les Laboratoires de chimie*, 2 vols. (Paris, 1881).
4. Louis Pasteur, "Le Budget de la science," *Revue des cours scientifiques de la France et de l'étranger*, 5ᵉ année, no. 9 (February 1, 1868), 138.

5. J. M. D. Olmsted, *Claude Bernard, Physiologist* (New York: Harper & Bros., 1938), 63 and 75, and *passim.*

6. Claude Bernard, *Rapport sur les progrès et la marche de la physiologie générale en France* (Paris, 1867). See especially the conclusion, pp. 143–49; and his article "Le problème de la physiologie générale," *Revue des Deux Mondes,* 72 (1867), 874–92.

7. Olmsted, *Claude Bernard,* 88–89.

8. Pasteur, "Le Budget de la science," 137–39. Pasteur's title calls attention to the fact that while there existed a *"budget des cultes,"* there was none for science. Compare Ernest Renan's remarks in *L'Avenir de la science* (Paris, 1890), chap. 14.

9. Olmsted, *Claude Bernard,* 89. Important documents concerning the burst of activity authorized by the decrees of July 1868—prepared with the approval of the leading French scientists—are to be found in Victor Duruy, *Notes et souvenirs,* 2 vols. (Paris, 1891), I, chap. 12; and in *L'Administration de l'instruction publique de 1863 à 1869,* Ministère de S. Exc. M. Duruy (Paris, n.d.), 592–603 and 644 ff.

10. Marie Curie discusses her own and Pierre's hardships at length in *Pierre Curie,* trans. Charlotte and Vernon Kellogg (New York, 1923). See particularly p. 95.

11. *Ibid.,* 99–100, 109–10; Eve Curie, *Madame Curie,* trans. Vincent Sheean (New York: Literary Guild, 1937), 236–39.

12. For the Ecole Pratique, see Duruy, *L'Administration de l'instruction publique,* 305–9; Ernest Lavisse, *Un Ministre—Victor Duruy* (Paris, 1895), 80–83. See also "Rapport de S. Exc. M. le Ministre à S. M. l'Empereur, précédant les deux décrets du 31 juillet 1868, relatifs aux laboratoires d'enseignement et de recherches et à la création d'une école pratique des hautes études," in *L'Administration de l'instruction publique,* 644–48.

13. Melon, *L'enseignement supérieur,* 22–29.

14. René Vallery-Radot, *The Life of Pasteur,* trans. Mrs. R. L. Devonshire (Garden City, N.Y., n.d.), 442; Louis Lumet, *Pasteur* (Paris, 1923), 129–30. For a full description of the Institute, see Lumet, 137 ff. On the early history of the Pasteur Institute, see Dr. Roux's address, "Le XXVᵉ Anniversaire de l'Institut Pasteur," *Revue scientifique,* 52ᵉ année, no. 7 (February 1914), 193–204.

15. See Eve Curie, *Madame Curie,* 285–86.

16. *Oeuvres complètes de Ernest Renan,* édition définitive établie par Henriette Psichari, 10 vols. (Paris: Calmann-Lévy, 1947–61), III, 814.

17. *Oeuvres complètes de Renan,* III, 928–33.

18. See, for example, Veuillot's *Odeurs de Paris,* bk. 5.

19. Albert Feuillerat, *Paul Bourget* (Paris: Plon, 1937), 136. Feuillerat has shown that Bourget was strongly influenced by an episode that had taken place ten years earlier, when a dairymaid had been brutally assassinated and dismembered by a psychopathic medical student who a short time before, in a public lecture, had appealed to the Darwinian struggle of existence to justify the murder of the weak by the strong.

20. A brief account of the quarrel over *Le Disciple* is to be found in Feuillerat, *Paul Bourget,* 135–48. Brunetière's important antiscientific essays are collected in his *Nouvelles questions de critique* (Paris, 1898) and in his *La Science et la religion,* new ed., rev. (Paris, 1913).

21. Louis Pasteur, *Discours de réception à l'académie française* (Paris, 1882). Reprinted in *Oeuvres de Pasteur,* ed. Pasteur Vallery-Radot, 7 vols. (Paris: Masson et cie., 1939), VII, 326–39.

22. The members included such distinguished scientific figures as the mathematicians Hermite, Humbert, Camille Jourdan, and d'Ocagne; the sociologist LePlay; and Louis Pasteur. On the Society, see Maurice d'Ocagne, *Hommes et choses de science—propos familiers,* 2d ser. (Paris: Voibert, 1932), 163 and 182; *Annales de la Société scientifique de Bruxelles, Table analytique . . . Précédée de l'histoire documentaire de la Société scientifique* (Louvain, 1904).

23. It is of some significance in this connection that both Paul Tannery and Pierre Duhem, two leading founders of the history-of-science movement, were members of the *Société scientifique de Bruxelles.*

24. See the preface by Léon Blum in Jean Perrin, *La Science et l'espérance* (Paris: Presses Universitaires de France, 1948).

25. M. Deploige, *Le Conflit de la morale et de la sociologie* (Paris, 1911).
26. The expression was given currency by the work of a Catholic physician, Jean Fiolle, *Science et scientisme*, 2 vols. (Paris, 1936).
27. Louis de Broglie, André Therive, et al., *L'Avenir de la science* (Paris: Plon, 1941).
28. For the general reader, the best nontechnical account of the discovery of nuclear fission is in Selig Hecht, *Explaining the Atom* (New York: Viking Press, 1947). The scientific reader should consult L. A. Turner, "Nuclear Fission," *Reviews of Modern Physics*, vol. 12, no. 1. This now famous review article summarizes the many papers published in the year that followed the experiments of Hahn and Strassmann.
29. Alphand, Belot, et al., *L'Avenir de la France, réformes nécessaires* (Paris, 1918).
30. L. Quevron, "Les Laboratoires de Bellevue," *Journal des recherches du Centre National de la Recherche Scientifique, Laboratoires de Bellevue*, special issue (1946). The chapter on France in J. G. Crowther, *Science in Liberated Europe* (London: Pilot Press, 1949), also gives some background material on Bellevue.
31. For Perrin's views on the place of science in society, see Blum in Perrin, *La Science et l'espérance*, pp. xxxi–xxxiv.
32. The *Caisse nationale de la recherche scientifique* and the two *Caisses* that preceded it are discussed in J. Delsarte, "De l'organisation de la recherche scientifique," *Revue scientifique* (January 15, 1939), 1–2, and Emmanuel Dubois, "Sur l'administration de la recherche scientifique en France," *Revue scientifique* (February 1939), 68–69.
33. The *Service central de la recherche scentifique* is discussed in the Delsarte and Dubois articles, and in J. D. Bernal, *The Social Function of Science* (London: G. Routledge & Sons, 1939), Appendix VI: "The Organization of Science in France."
34. The best article on the CNRS and its problems is L. Kopelmanas, "Le Centre national de la recherche scientifique et les problèmes actuels de la recherche française, *Revue socialiste* (June–July 1949), 120–34. Crowther, *Science in Liberated Europe*, contains information on the background and early history. For the text of the decree of October 19, 1939, establishing the CNRS, see the *Journal officiel* (October 24, 1939), 12594–95, or *Revue scientifique* (November–December 1939), 704–5. There is a chart showing the relationship of the Centre to the general organization of governmental research in France as of January 1940 in the article "Visit of French Scientists," *Notes and Records of the Royal Society* 3 (1940–41), 11–21.
35. See the "Exposé des motifs" preceding the *ordonnance* of November 2, 1945, *Journal officiel* (November 3, 1945), 7193.
36. Quevron, "Les Laboratoires de Bellevue," 44.
37. René Maublanc, "French Teachers in the Resistance Movement," *Science and Society* (Winter 1947); and Crowther, *Science in Liberated Europe*, 51–52.
38. Samuel A. Goudsmit, *Alsos* (New York: H. Schuman, 1947), 34–36.
39. The Commissariat was established in October 1945. *Journal officiel* (October 31, 1945), 7066 and 7079. On January 3, 1946, Joliot was named High Commissioner and Pierre Auger, Irène Joliot-Curie, and François Perrin, son of Jean, were named the other members. *Journal officiel* (January 4, 1946), 146.
40. Ministère de l'Education Nationale, Centre National de la Recherche Scientifique, *Séance plénière du comité national de la recherche scientifique* (Paris, June 2, 1948), 19.
41. The status of research personnel was determined by the decree of August 12, 1945. *Journal officiel* (August 21, 1945), 5201.
42. The *laboratoire de calcul* and *laboratoire de statistique* of the Institut Henri Poincaré had a full-page letter in the January–May 1941 issue of *Revue scientifique* inviting people to use its facilities for a small hourly fee.
43. Felix Trombe, Marx Foex, and Charlotte Henry La Blanchetais, "Utilisation de l'énergie solaire," *Journal des recherches*, nos. 4 and 5 (1949), 61–89.
44. Crowther, *Science in Liberated Europe*, 23, 30–31.
45. For a list of projects supported by the CNRS in what it calls the *"sciences humaines,"* in its own and outside research centers, see the *Bulletin du CNRS*, series B, no. 1 (1949). By the decree of June 11, 1949, these sciences were given 144 members on the *Comité nationale* to 228 for the other sciences. *Journal officiel* (June 14, 1949), 5866–68.

46. *Bulletin du CNRS*, series A, no. 1 (1949), 135–37. The training program was discussed at some length in the "Exposé des motifs" preceding the *ordonnance* of November 2, 1945.

47. See Emile F. Terroine, "Les Colloques internationaux du CNRS," *Revue scientifique* (November 1, 1946), 496–97. *Comptes-rendus* of the first three colloquia held under the program can be found in *Revue scientifique* for November 1, 1946 (pp. 497–500), December 1–15, 1946 (pp. 621–23), and May 15, 1947 (pp. 559–63), respectively.

48. The figures are from a broadside of the *Service des publications* of the CNRS. The *Bulletin analytique* appears in three parts: one for the physicochemical sciences, one for the biological sciences, and one for the philosophical sciences. The first two appear monthly, the third quarterly. For a full discussion of the role of the CNRS in the publication of books and periodicals, see Kopelmanas, "Le Centre national de la recherche scientifique," 120–34.

49. The latest decision available on the composition and functioning of the various parts of the administration of the CNRS was the decree of June 11, 1949.

50. See Jean Dumont, "Interview with M. Louis de Broglie," *Essais et études universitaires* 1 (1945), 84–98, and de Broglie's preface to Perrin, *La Science et l'espérance.*

Bibliography of Other Books and Papers
by
Henry Guerlac

✿

"Combined Action of Ethyl Urethane and Sodium Thiocyanate on the Living Cell," *Proc. Soc. Experimental Biology and Medicine* 30 (1932), 265–68.

"George Lincoln Burr," *Isis* 35 (1944), 147–52.

"The Radio Background of Radar," *Journal of the Franklin Institute* 250 (1950), 285–308.

"The Radio War Against the U-Boat," in collaboration with Marie Boas, *Military Affairs* 14 (1950), 99–111.

"Rapport de M. H. Guerlac" [sur l'histoire des sciences], *IX⁰ Congrès International des Sciences Historiques*. Paris, 28 août–3 septembre, 1950. *I. Rapports*. Paris: Armand Colin, 1950, pp. 182–211.

"Rapport de M. H. Guerlac" [comments and discussion], *IX⁰ Congrès International des Sciences Historiques, II. Actes*. Paris: Armand Colin, 1951, pp. 83–100.

Science in Western Civilization: A Syllabus. The Ronald Press, 1952.

Selected Readings in the History of Science. Vol. I: Ithaca, 1950. Vol. II: Ithaca, 1953. (Multilith.)

"A Proposed Revision of the *Isis* Critical Bibliography," *Isis* 44 (1953), 226–28.

"Foundation of Chemistry," *Cornell University Library Associates, Occasional Papers*. Ithaca, 1956, pp. 8–9.

"Joseph Black and Fixed Air: a Bicentenary Retrospective with Some New or Little Known Materials," *Isis* 48 (1957), 124–50, 433–56.

"Science for the Student in Humanities," *The Cornell Engineer* 23, no. 6 (March 1958), 27–28, 66.

"Award of the George Sarton Medal to Lynn Thorndike," *Isis* 49 (1958), 107–8.

"History of Science for Engineering Students at Cornell," in Marshall Clagett, ed., *Critical Problems in the History of Science*. Madison, Wis., 1959, pp. 235–40.

"Commentary on the Papers of Charles Coulston Gillispie and L. Pearce Williams," and "Commentary on the Papers of Cyril Stanley Smith and Marie Boas," *ibid.*, pp. 317–20, 515–18.

"Address of Retiring President," *Isis* 52 (1961), 3–6.

"Quantification in Chemistry," *Isis* 52 (1961), 194–214.

Lavoisier—The Crucial Year. Ithaca, Cornell University Press, 1961.

"Some Daltonian Doubts," *Isis* 52 (1961), 544–54.

"Newton in France—Two Minor Episodes," *Isis* 53 (1962), 219–22.

"Recent Studies in the History of Renaissance Science," *Renaissance News* 17 (1964), 1: 53–58.

"Sir Isaac and the Ingenious Mr. Hauksbee," in *Mélanges Alexandre Koyré*, 2 vols. (Paris, Hermann, 1964), vol. II, pp. 228–53.

"The Word 'Spectrum': A Lexicographic Note with a Query," *Isis* 56 (1965), 206–7.

"Copernicus and Aristotles Cosmos," *Journal of the History of Ideas* 29 (1968), 109–13.

"The Correspondence of Henry Oldenburg," *British Journal for the History of Science* 4 (1968), 166–72.

"Bentley, Newton and Providence," *Journal of the History of Ideas* 30 (1969), 307–18. (With Margaret Candee Jacob.)

"Lavoisier's Draft Memoir of July 1772," *Isis* 60 (1969), 380–82.

"Hauksbee, Francis," *Dictionary of Scientific Biography*, vol. VI, 1972, pp. 169–75.

"Hauksbee, Francis, the Younger," *Dictionary of Scientific Biography*, vol. VI, 1972, pp. 175–76.

"Guy de La Brosse and the French Paracelsians," in Allen G. Debus, ed., *Science, Medicine and Society in the Renaissance. Essays to Honour Walter Pagel*, 2 vols., New York, 1972; I, 177–99.

"Lavoisier, Antoine Laurent," *Dictionary of Scientific Biography*, vol. VIII, 1973, pp. 66–91.

Antoine-Laurent Lavoisier, Chemist and Revolutionary. New York: Charles Scribner's Sons, 1975.

"Sage, Balthazar-Georges," *Dictionary of Scientific Biography*, vol. XII, 1975, pp. 63–69.

"Vauban, Sébastien Le Prestre de," *Dictionary of Scientific Biography*, vol. XIII, 1976, pp. 590–95.

"Chemistry as a Branch of Physics: Laplace's Collaboration with Lavoisier, *Historical Studies in the Physical Sciences*, ed. Russell McCormmach, 7 (1976), 193–276.

Index

❀

Abat, Father, 117 n. 3

Abelard, Peter, 199

Académie de Bordeaux, 13, 482, 486

Académie des Sciences (Paris), 10, 61, 73, 76, 77, 131, 133, 228, 307, 314, 316, 318, 334–37, 341, 342, 344, 347, 348, 351–52, 362, 380–87, 399, 401–2, 422, 424, 480, 483, 486, 493, 508; Foreign Associates, 186, 278, 356, 479; founding of, 419; during French Revolution, 466, 467, 468, 470–71, 475; *Mémoires*, 79 n. 12, 278, 314, 401, 402, 471, 475; 1793 abolition of, 471, 474

Académie Française, 499

Academy of Sciences of Bologna, 278

Academy of Upsala, 356

Accademia del Cimento: on capillarity, 109; *Saggi* of, 279

Acid spirits, 17th-century terminology, 248

Actio in distantia, absurdity of, 96

Active principles, 83, 95, 99–100, 120

Adam, Charles, 60

Aerial nitre, xiv, 179, 183, 230, 245–47, 248, 250–55, 260–61; and meteorological phenomena, 262–70; not identical with oxygen, 319–20

Aeschylus, 44

Aether theories, 83, 84–85, 96, 97–99, 112, 117, 125, 158–62, 168 n. 63, 169 n. 67, 218, 224, 483; mechanical aether of Descartes, 120, 127, 483; optical aether of Newton, 120–28, 130 n. 25, 158–60, 162, 167 n. 48

Affinities, doctrine of. *See* Elective affinities, doctrine of

Agatharcus, 44

Age of Reason, 16, 31, 70–71, 72, 458, 461–62

Agricola, Georgius, 46, 263, 266, 273 n. 33, 355

Air: Boyle's description, 179; "dephlogisticated," 400; early investigations and views, of, 179–80, 246–55, 269, 276–78, 376–77; elasticity, 178–79, 180–81, 182–183, 185, 239 n. 26, 261, 280, 297, 376–77; "factitious," 179; "food of life," 247, 249, 250, 252, 254–55; Hales's experiments and views, 170, 178–79,

180–83, 185, 275, 276–79, 280, 336, 376, 379; Hooke's theory, 179, 248–49, 255; iatrophysical theory of action on "fibers," 13, 14–16; "inflammable," 304, 400, 402; Lavoisier's studies of its chemical role, 336, 361, 375–80, 384, 386–88, 396; Mayow's theory, 179, 245, 247, 248–49, 250, 255, 260–61 (*see also* Aerial nitre); views of, as "element," 179, 183–84, 336, 444; "vitiated," 182, 185. *See also* "Fixed air"

Airs, Waters, and Places (Hippocrates), 14

Akenside, Mark, 152, 156

Albert of Saxony, 61

Albertus Magnus, 230

Albineus, Nathan, 251

Alchemy, 58, 251–52, 254, 262; Islamic, 55

Alembert, Jean le Round d', 76–77, 131, 133, 137, 140–41, 193, 331 n. 1, 359, 466; positivistic view of, 142, 143, 221

Alexander of Aphrodisias, 197

Alexander the Great, 44

Algarotti, Francesco, 71, 76, 152

Algebra, 45, 55, 197

Algebra (Wallis), 150

Alkaline substances, Black's experiments with, 286, 289, 290–93, 299, 300

Allent, A., 428

Almagest (Ptolemy), 8

Almanach Royal, 475

Alston, Charles, 285, 289–90, 291

America: Newtonianism in, 76, 80 n. 23, 143; scientific achievement and conditions, 492, 497, 503, 504

American Philosophical Society, 305, 321

Amontons, Guillaume, 226

Ampère, André Marie, 466

Analogy, principle of, 136, 141, 195

Analysis (resolution), 196–205, 208, 456; Cartesian, 202–5, 222–23; in mathematics, 136–37, 194–95, 196–97, 205; in Newton's method, 52, 75, 132, 133, 135–37, 162–63, 193–95, 205–7, 211, 212, 214–15, 223

Anatomy, 55, 171–72, 440, 465, 492; plant, 180

Anatomy of Plants (Grew), 150

Anaxagoras, 443